New Trends in Technologies: Devices, Computer, Communication and Industrial Systems

edited by

Prof. Meng Joo Er

New Trends in Technologies: Devices, Computer, Communication and Industrial Systems
Edited by Prof. Meng Joo Er

Published by InTech

Janeza Trdine 9, 51000 Rijeka, Croatia

Publishing Process Manager Jelena Marusic

Technical Editor Teodora Smiljanic

Cover Designer InTech Design Team

Additional hard copies can be obtained from orders@intechopen.com

New Trends in Technologies: Devices, Computer, Communication and Industrial Systems,
Edited by Prof. Meng Joo Er
p. cm.
ISBN 978-953-307-212-8

For EU product safety concerns:
IN TECH d.o.o., Prolaz Marije Krucifikse Kozulić 3, 51000 Rijeka, Croatia, info@intechopen.com or visit our website at intechopen.com.

Contents

Preface

From the metallurgists who ended the Stone Age to the shipbuilders who united the world's peoples through travel and trade, the past witnessed many marvels of engineering prowess. As civilization grew, it was nourished and enhanced with the help of increasingly sophisticated tools for agriculture, technologies for producing textiles, and inventions transforming human interaction and communication. Inventions such as the mechanical clock and the printing press irrevocably changed civilization.

In the modern era, the Industrial Revolution brought engineering's influence to every niche of life, as machines supplemented and replaced human labor for countless tasks, improved systems for sanitation enhanced health, and the steam engine facilitated mining, powered trains and ships, and provided energy for factories.

In the century just ended, engineering recorded its grandest accomplishments. The widespread development and distribution of electricity and clean water, automobiles and airplanes, radio and television, spacecraft and lasers, antibiotics and medical imaging, and computers and the Internet are just some of the highlights from a century in which engineering revolutionized and improved virtually every aspect of human life.

In this book entitled "New Trends in Technologies: Devices, Computers, Communications and Industrial Systems', the authors provide a glimpse of the new trends of technologies pertaining to devices, computers, communications and industrial systems. The book comprised of 22 very interesting and excellent articles covers core topics ranging from "Overview of Multifucntional Materials" to "New Technologies and Devices to Increase Structures' Safety to Dynamic Actions in Devices", "Advances in CAD/CAM Technologies" to "Operating System Kernel Coprocessor for Embedded Applications" in Computers and "Modern Internet-based Production Technology" to "BEPU Approach in Licensing Framework including 3D NK Applications" in Communication. It also covers topics pertaining to industrial systems ranging from "A Review of Work on Oscillatory Problems in Francis Turbines" to "The Role of Gaming Theory and Open Source Philosophy in the Creation of Affordable Virtual Heritage Solutions". The book will serve a unique purpose through these multi-disciplinary topics to share different but interesting views on each of these exciting topics which form the backbone of Engineering Challenges for the 21St Century.

I would like to thank all the authors for their excellent contributions in the different areas of their expertise. It is their domain knowledge, enthusiastic collaboration and strong support which made the creation of this book possible. I sincerely hope that readers of all disciplines will find this book valuable.

Editor

Professor Meng Joo Er
School of Electrical and Electronic Engineering
Nanyang Technological University

1

Overview of Multi Functional Materials

Parul Gupta[1] and R.K.Srivastava[2]
[1]*Department of Mechanical Engineering, Moradabad Institute of Technology (Uttar Pradesh Technical University), Moradabad, Uttar Pradesh,*
[2]*Department of Mechanical Engineering, Motilal Nehru National Institute of Technology (Deemed University), Allahabad, Uttar Pradesh, India*

1. Introduction

Until relatively recent times, most periods of technological development have been linked to changes in the use of materials (eg the stone, bronze and iron ages). In more recent years the driving force for technological change in many respects has shifted towards information technology. This is amply illustrated by the way the humble microprocessor has built intelligence into everyday domestic appliances. However, it is important to note that the IT age has not left engineered materials untouched, and that the fusion between designer materials and the power of information storage and processing has led to a new family of engineered materials and structures.

The development of materials science has lead to the introduction of a new kind of material- "*smart*" material. Smart materials not only have the traditional structural material functions. But also have functions such as actuation, sensibility and microprocessing capability. in the materials family , we find three major materials : metals .ceramic and polymer and among those polymers being the youngest members find wide application in comparison to metals and ceramics In every sectors of our society . the recent progress in smart materials has taken the new momentum in the materials science and technology .the board spectrum of applications of smart materials cover the biomedical , environmental , communication defence , space , nanotechnolgical fields of research with tremendous progress . The present report describes the application of smart materials in terms of opportunity and future challenges.

Smart materials referred to materials that will undergo controlled transformations through physical interactions. Smart Materials are materials that respond to environmental stimuli, such as temperature, moisture, pH, or electric and magnetic fields. For example- photochromic materials that change colour in response to light; shape memory alloys and polymers which change/recover their shape in response to heat and electro- and magneto- rheological fluids that change viscosity in response to electric or magnetic stimuli. Smart Materials can be used directly to make smart systems or structures or embedded in structures whose inherent properties can be changed to meet high value-added performance needs.The different types of smart materials include piezoelectric, shape-memory alloys, electro-active conductive polymers, electrochromic materials, biomaterials,

1.1 Needs of smart materials

To achieve a specific objective for a particular function or application, a new material or alloy has to satisfy specific qualifications related to the following properties:-

a. Technical properties, including mechanical characteristics such as plastic flow, fatigue and yield strength and behavioural characteristics such as damage tolerance and electrical, heat and fire resistance.
b. Technological properties, encompassing manufacturing, forming, welding abilities, thermal processing, waste level, workability, automation and repair capacities.
c. Economic criteria, related to raw material and production costs, supply expenses and availability.
d. Environmental characteristics, including features such as toxicity and pollution.
e. Sustainable development criteria, implying reuse and recycling capacities. If the functions of sensing and actuation are added to the list, then the new material/alloy is considered a smart material.

2. Classification of smart materials

Types of materials

a. Piezoelectric Materials
b. Shape memory alloy (SMA)
c. Electro-active conductive polymers
d. Biomaterials,
e. Electro-rheological fluids

Piezoelectric materials, shape-memory alloys, electrostrictive materials, magnetostrictive materials, electro-rheological fluids, etc. are currently available smart materials.

2.1 Piezoelectric materials

Piezoelectric materials were first discovered by Jacques and Pierre Curie in the 19th century. The literal meaning of piezoelectric is "pressure electricity", and this meaning aptly describes these materials, as a deformation caused by pressure on the substance will create an electric current within the material. The mechanical pressure is therefore converted to voltage. Natural piezoelectric materials are crystalline materials that exhibit the piezoelectric effect. Often they are strong physically and chemically inert. The piezoelectric effect can also be found in synthetic polycrystalline ceramics which can be designed to have other specific properties that make individual ceramics useful in many different applications. Piezoelectric materials are often linked to pyroelectric effects. Often a material can undergo both effects - one converting mechanical stress energy into electrical energy and the other converting from heat energy. A piezoelectric material does not, however, always show the pyroelectric effect as well as the piezo.

Piezoelectric materials have two unique properties which are interrelated. When a piezoelectric material is deformed, it gives off a small but measurable electrical discharge. Alternately, when an electrical current is passed through a piezoelectric material it experiences a significant increase in size (up to a 4% change in volume). Piezoelectric materials are most widely used as sensors in different environments. They are often used to measure fluid compositions, fluid density, fluid viscosity, or the force of an impact. An example of a piezoelectric material in everyday life is the airbag sensor in your car. The material senses the force of an impact on the car and sends and electric charge deploying the airbag.

2.2 How do it works

Piezoelectric materials have two crystalline configurations. One structure is organized, while the other is not. Organization of the structure has to do with polarization of the molecules that make up the material. Hence, a non-polarized material has a non-organized structure, while the polarized material is organized. To polarize the material, voltage or electricity must be conducted through it. As a result of this electrical force, the molecules of the material reorient themselves, thus changing the shape of the material; this is called electrostriction. The picture below shows this process at a microscopic level. Change in shape can produce mechanical force, as well as changes in physical characteristics (like density, shown below).

Non-polarized material:

Electricity is produced with
Input of electricity

Polarized material:

shape change is produced with
input of shape change.

Fig. 1.

Similarly, if mechanical force is exerted on the material to change its shape, an electrical field is produced; this is called piezoelectric effect. Electrostriction and piezoelectric effect are opposite phenomena. In the graphic below, a thin piezoelectric material within a plastic sheath is being bent, and electricity is being generated and passed through the red wires at the PZT is the most popular piezoelectric material in use.

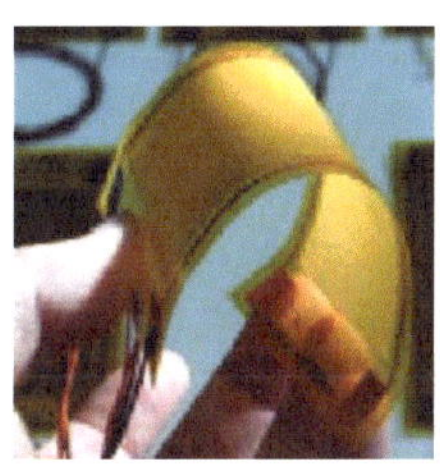

Fig. 2.

Its physical properties can be optimized for certain applications by controlling the chemistry and processing of this material. Therefore, it can have a variety of compositions, geometries, and applications. Limitations in its use are related to high excitation voltages needed, mechanical durability, and stability in coupling the material to the control system and/or structure.
Some piezoelectric materials are:

- Quartz
- Barium titanate
- Cadmium sulfide
- Lead zirconium titanate (PZT)
- Piezoelectric polymers (PVDF, PVC)

Piezoelectric effects. Electric charge is created when the material is mechanical stressed.

2.3 The piezoelectric effect

A piezoelectric substance is one that produces an electric charge when a mechanical stress is applied (the substance is squeezed or stretched). Conversely, a mechanical deformation (the substance shrinks or expands) is produced when an electric field is applied. This effect is formed in crystals that have no center of symmetry. To explain this, we have to look at the individual molecules that make up the crystal. Each molecule has a polarization, one end is more negatively charged and the other end is positively charged, and is called a dipole. This is a result of the atoms that make up the molecule and the way the molecules are shaped. The polar axis is an imaginary line that runs through the center of both charges on the molecule. In a monocrystal the polar axes of all of the dipoles lie in one direction. The crystal is said to be symmetrical because if you were to cut the crystal at any point, the resultant polar axes of the two pieces would lie in the same direction as the original. In a polycrystal, there are different regions within the material that have a different polar axis. It is asymmetrical because there is no point at which the crystal could be cut that would leave the two remaining pieces with the same resultant polar axis. Figure 3 illustrates this concept.

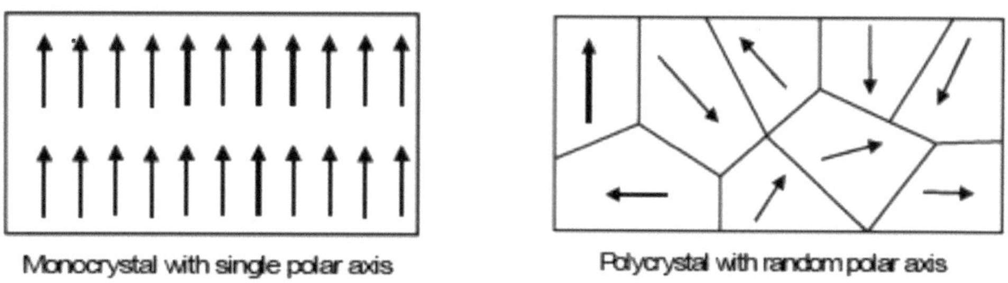

Fig. 3. Mono vs. Crystals

In order to produce the piezoelectric effect, the polycrystal is heated under the application of a strong electric field. The heat allows the molecules to move more freely and the electric field forces all of the dipoles in the crystal to line up and face in nearly the same direction (Figure 4).

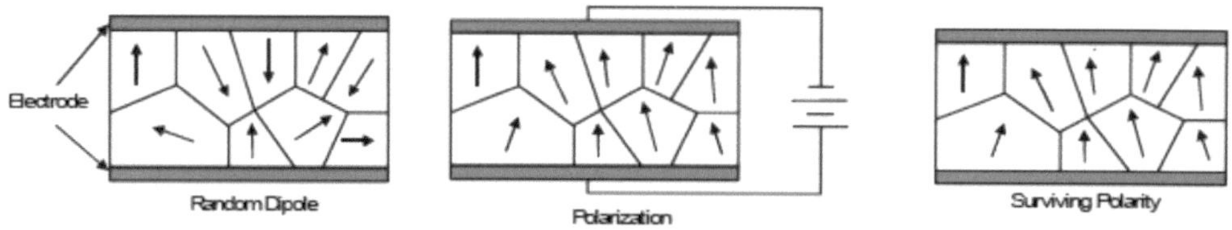

Fig. 4. Polarization of Ceramic Material to Generate Piezoelectric Effect

The piezoelectric effect can now be observed in the crystal. Figure 5 illustrates the piezoelectric effect. Figure 5a shows the piezoelectric material without a stress or charge. If the material is compressed, then a voltage of the same polarity as the poling voltage will appear between the electrodes (b). If stretched, a voltage of opposite polarity will appear (c). Conversely, if a voltage is applied the material will deform. A voltage with the opposite polarity as the poling voltage will cause the material to expand, (d), and a voltage with the same polarity will cause the material to compress (e). If an AC signal is applied then the material will vibrate at the same frequency as the signal (f).

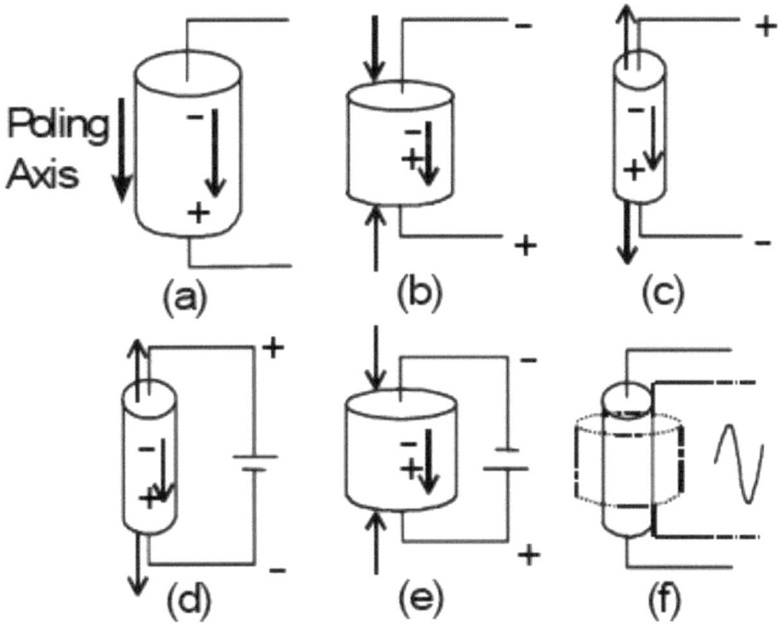

Fig. 5. Example of Piezoelectric Effect

Using the Piezoelectric Effect -The piezoelectric crystal bends in different ways at different frequencies. This bending is called the vibration mode. The crystal can be made into various shapes to achieve different vibration modes. To realize small, cost effective, and high performance products, several modes have been developed to operate over several frequency ranges. These modes allow us to make products working in the low kHz range up to the MHz range. Figure 6 shows the vibration modes and the frequencies over which they can work. An important group of piezoelectric materials are ceramics. Murata utilizes these various vibration modes and ceramics to make many useful products, such as ceramic resonators, ceramic bandpass filters, ceramic discriminators, ceramic traps, SAW filters, and buzzers.

2.4 Application of piezoelectric materials

Piezoelectric materials may be used passively as sensors, or actively as actuators. The piezoelectric sensors that will be used on this bridge include the PZT (lead-zirconate-titanate), a ceramic sensor, and PVDF (polyvinylidene fluoride), a polymeric sensor. The PZT's are extremely sensitive and very accurate, but due to their brittle nature, PZT's are restricted to being point sensors. Likewise, PVDF's are also very sensitive and accurate. However, PVDF's are not as brittle and may be integrated to perform distributed measurements. Therefore, PZT's will be located primarily in critical areas, whereas the PVDF's will be located along side the strain gauges.

2.5 Shape memory alloy (SMA)

Shape memory alloys were discovered in 1932, however it was not until 1961 when the most common form was discovered - nickel-titanium (NiTi)Several NiTi products are can be seen in the picture to the right. A shape memory alloy (SMA) is a material that has the ability to

return to its previous shape after being deformed (bent) by applying heat or in some cases just releasing the stress. Shape Memory Alloys (SMA's) are novel materials which have the ability to return to a predetermined shape when heated. When an SMA is cold, or below its transformation temperature, it has a very low yield strength and can be deformed quite easily into any new shape--which it will retain. However, when the material is heated above its transformation temperature it undergoes a change in crystal structure which causes it to return to its original shape. If the SMA encounters any resistance during this transformation, it can generate extremely large forces. This phenomenon provides a unique mechanism for remote actuation.
The most common shape memory material is an alloy of nickel and titanium called Nitinol. This particular alloy has very good electrical and mechanical properties, long fatigue life, and high corrosion resistance. As an actuator, it is capable of up to 5% strain recovery and 50,000 psi restoration stress with many cycles. By example, a Nitinol wire 0.020 inches in diameter can lift as much as 16 pounds. Nitinol also has the resistance properties which enable it to be actuated electrically by joule heating. When an electric current is passed directly through the wire, it can generate enough heat to cause the phase transformation. In most cases, the transition temperature of the SMA is chosen such that room temperature is well below the transformation point of the material. Only with the intentional addition of heat can the SMA exhibit actuation. In essence, Nitinol is anactuator, sensor, and heater all in one material.Shape memory alloys, however, are not for all applications. One must take into account the forces, displacements, temperature conditions, and cycle rates required of a particular actuator. The advantages of Nitinol become more pronounced as the size of the application decreases. Large mechanisms may find solenoids, motors, and electromagnets more appropriate. But in applications where such actuators can not be used, shape memory alloys provide an excellent alternative. There are few actuating mechanisms which produce more useful work per unit volume than Nitinol. Nitinol is available in the form of wire, rod and bar stock, and thin film. Examples of SMA products developed by TiNi Alloy Company include silicon micro-machined gas flow microvalves, non-explosive release devices, tactile feedback device (skin stimulators), and aerospace latching mechanisms. If you are considering an application for shape memory alloys, TiNi Alloy Company can assist you in the design, prototyping, and manufacture of actuators and devices.

Physical Properties of Nitinol

- Density: 6.45gms/cc
- Melting Temperature: 1240-1310° C
- Resistivity (hi-temp state): 82 uohm-cm
- Resistivity (lo-temp state): 76 uohm-cm
- Thermal Conductivity: 0.1 W/cm-° C
- Heat Capacity: 0.077 cal/gm-° C
- Latent Heat: 5.78 cal/gm; 24.2 J/gm
- Magnetic Susceptibility (hi-temp): 3.8 uemu/gm
- Magnetic Susceptibility (lo-temp): 2.5 uemu/gm

Mechanical Properties of Nitinol

- Ultimate Tensile Strength: 754 - 960 MPa or 110 - 140 ksi
- Typical Elongation to Fracture: 15.5 percent
- Typical Yield Strength (hi-temp): 560 MPa, 80 ksi

- Typical Yield Strength (lo-temp): 100 MPa, 15 ksi
- Approximate Elastic Modulus (hi-tem): 75 GPa, 11 Mpsi
- Approximate Elastic Modulus (lo-temp): 28 GPa, 4 Mpsi
- Approximate Poisson's Ratio: 0.3

Actuation

- Energy Conversion Efficiency: 5%
- Work Output: ~1 Joule/gram
- Available Transformation Temperatures: -100 to +100° C

Stress-Strain Characteristics of Nitinol at Various Temperatures

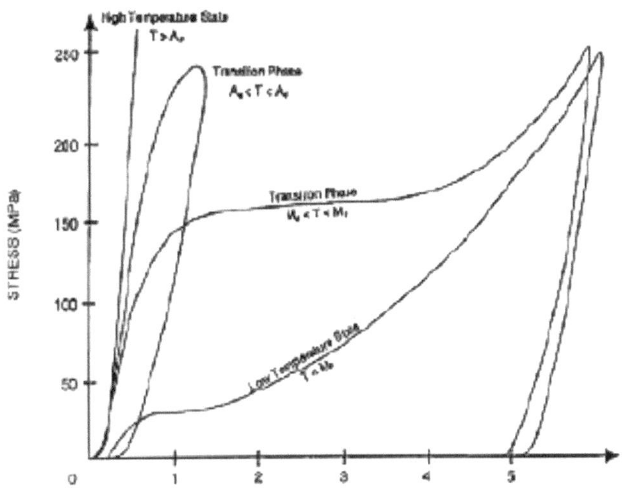

Fig. 6.

2.6 How do shape memory alloys works

In order to understand the way in which the shape memory effect occurs it is useful to understand the crystal structure of a SMA. All shape memory alloys exhibit two very distinct crystal structures. Which phase is present depends on the temperature and the amount of stress being applied to the SMA. These phases as known as marten site which exists at lower temperatures and austenite for higher temperatures. The properties of an SMA depends on which the amount of each crystal phase present.

The two unique properties described above are made possible through a solid state phase change that is a molecular rearrangement, which occurs in the shape memory alloy. Typically when one thinks of a phase change a solid to liquid or liquid to gas change is the first idea that comes to mind. A solid state phase change is similar in that a molecular rearrangement is occurring, but the molecules remain closely packed so that the substance remains a solid. In most shape memory alloys, a temperature change of only about 10°C is necessary to initiate this phase change. The two phases, which occur in shape memory alloys are Martensite, Austenite. Martensite is the relatively soft and easily deformed phase of shape memory alloys, which exists at lower temperatures. The molecular structure in this phase is twinned which the configuration is shown in the middle of Figure 4. Upon deformation this phase takes on the second form shown in Figure 4, on the right. Austenite, the stronger phase of shape memory alloys, occurs at higher temperatures. The shape of the Austenite structure is cubic, the structure shown on the left side of Figure 4. The un-deformed Martensite phase is the same size and shape as the cubic Austenite phase on a macroscopic scale, so that no change in size or shape is visible in shape memory alloys until the Martensite is deformed.

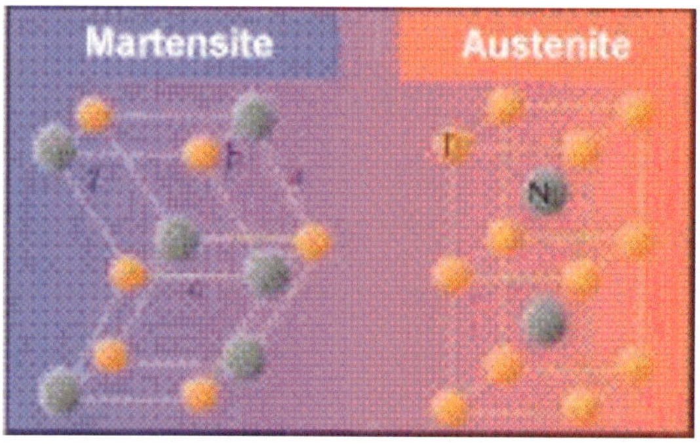

Fig. 7. The Martensite and Austenite phases

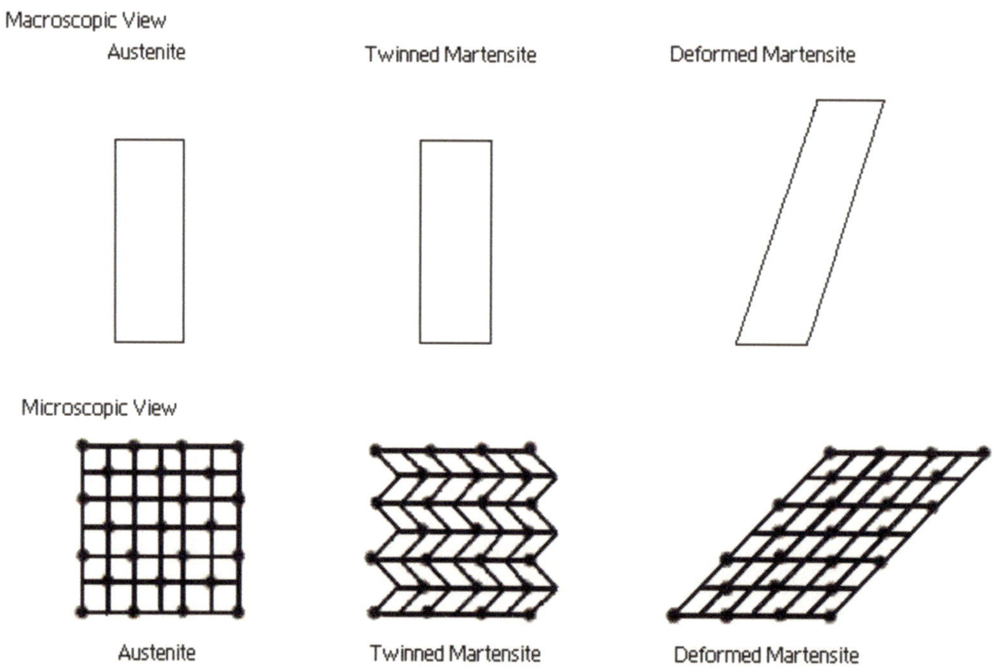

Fig. 8. Microscopic and Macroscopic Views of the Two Phases of Shape Memory Alloys

The temperatures at which each of these phases begin and finish forming are represented by the following variables: Ms, Mf, As, Af. The amount of loading placed on a piece of shape memory alloy increases the values of these four variables as shown in Figure 9. The initial values of these four variables are also dramatically affected by the composition of the wire (i.e. what amounts of each element are present).

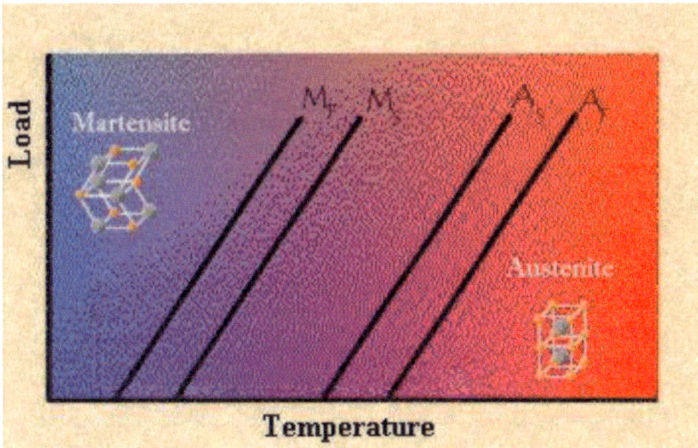

Fig. 9. The Dependency of Phase Change Temperature on Loading

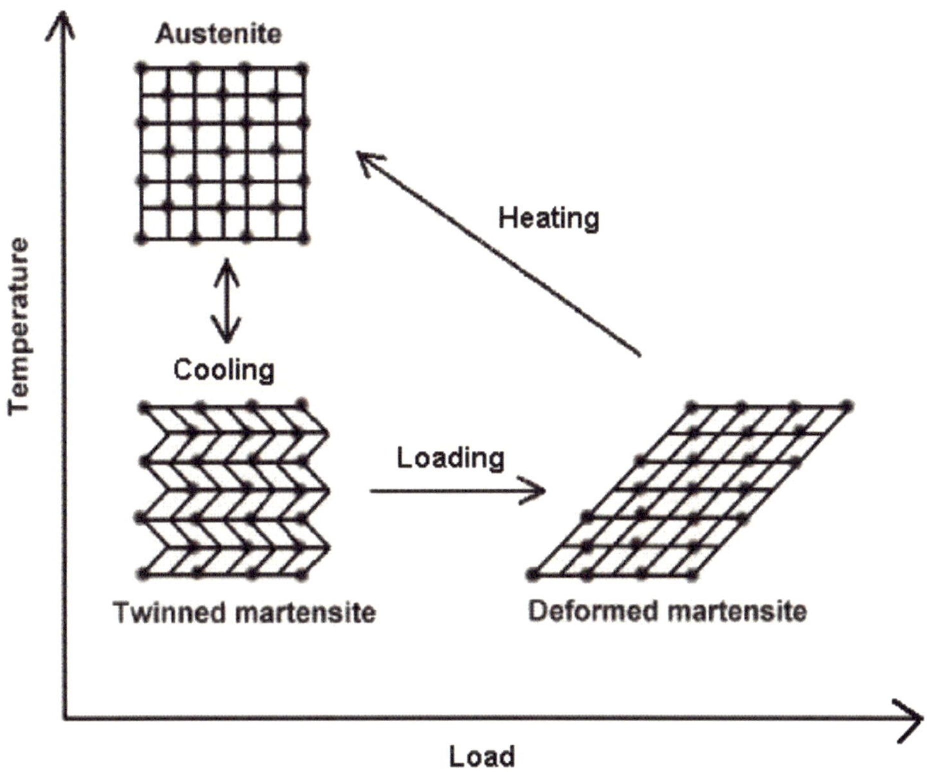

Fig. 10. Microscopic Diagram of the Shape Memory Effect

2.7 Shape memory effect

The shape memory effect is observed when the temperature of a piece of shape memory alloy is cooled to below the temperature Mf. At this stage the alloy is completely composed of Martensite which can be easily deformed. After distorting the SMA the original shape can be recovered simply by heating the wire above the temperature Af. The heat transferred to the

wire is the power driving the molecular rearrangement of the alloy, similar to heat melting ice into water, but the alloy remains solid. The deformed Martensite is now transformed to the cubic Austenite phase, which is configured in the original shape of the wire.
The Shape memory effect is currently being implemented in:

- Coffepots
- The space shuttle
- Thermostats
- Vascular Stents
- Hydraulic Fittings (for Airplanes

Shape Memory Effect (SME) is a unique property of certain alloys exhibiting martensitic transformation. Even though the alloy is deformed in the low temperature phase, it recovers its original shape upon heating to a critical temperature known as the Reverse Transformation Temperature (RTT). They are most commonly Nitinol, or nickel and titanium combined. Less popular but still possessing the shape memory effect are gold cadmium, silver cadmium, copper-aluminum-nickel, copper tin, copper zinc, and copper-zinc-aluminum. The same alloys have another unique property called superelasticity (SE) at a higher temperature. It is a large (0-18%) nonlinear recoverable strain upon loading (stretching) and unloading (unstretchiung).
The basic one-way shape memory effect-The actual mechanism of the shape memory effect can be described simply as a reversible, thermoelastic, phase transformation between a parent austenitic phase and a martensitic phase. The phase transformation occurs when a material in the austenitic phase is cooled below the martensite start temperature (Ms), where the two phases coexist. The material must then accommodate the two phases without changing shape through a mechanism called twinning. This is where mirror-image lattices form adjacent to the parent lattices. The phase transformation is completed upon reaching the martensite finish temperature (Mf). The material can then be plastically deformed into another shape. During this deformation the twinned martensite is converted to a deformed martensite. The material remains deformed until it is heated to the austenite start temperature (As), and at this point the martensite begins to transform back into austenite. Heating above the austenite finish temperature (Af) allows the material to regain its original shape. (The extent to which the shape is regained usually depends on the type of SMA, amount of deformation, and the material's thermomechanical history.) When cooled again the material does not automatically revert to the deformed shape. This is called the oneway shape memory effect. The entire shape memory process can be repeated.

2.8 Advantages and disadvantages of shape memory alloys

Some of the main advantages of shape memory alloys include:

- Bio-compatibility
- Diverse Fields of Application
- Good Mechanical Properties (strong, corrosion resistant)

There are still some difficulties with shape memory alloys that must be overcome before they can live up to their full potential. These alloys are still relatively expensive to manufacture and machine compared to other materials such as steel and aluminum. Most SMA's have poor fatigue properties; this means that while under the same loading conditions (i.e. twisting, bending, compressing) a steel component may survive for more than one hundred times more cycles than an SMA element.

2.9 Applications for shape memory alloys

Bioengineering:

Bones: Broken bones can be mended with shape memory alloys. The alloy plate has a memory transfer temperature that is close to body temperature, and is attached to both ends of the broken bone. From body heat, the plate wants to contract and retain its original shape, therefore exerting a compression force on the broken bone at the place of fracture. After the bone has healed, the plate continues exerting the compressive force, and aids in strengthening during rehabilitation. Memory metals also apply to hip replacements, considering the high level of super-elasticity. The photo above shows a hip replacement.
Reinforcement for Arteries and Veins: For clogged blood vessels, an alloy tube is crushed and inserted into the clogged veins. The memory metal has a memory transfer temperature close to body heat, so the memory metal expands to open the clogged arteries.

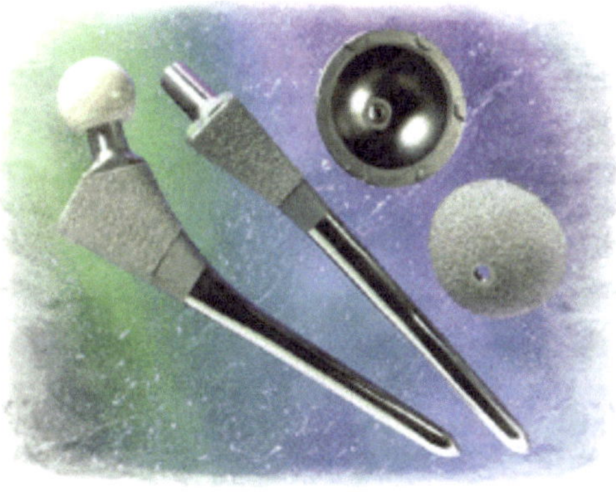

Fig. 11.

Dental wires: used for braces and dental arch wires, memory alloys maintain their shape since they are at a constant temperature, and because of the super elasticity of the memory metal, the wires retain their original shape after stress has been applied and removed.
Fire security and Protection systems: Lines that carry highly flammable and toxic fluids and gases must have a great amount of control to prevent catastrophic events. Systems can be programmed with memory metals to immediately shut down in the presence of increased heat. This can greatly decrease devastating problems in industries that involve petrochemicals, semiconductors, pharmaceuticals, and large oil and gas boilers.

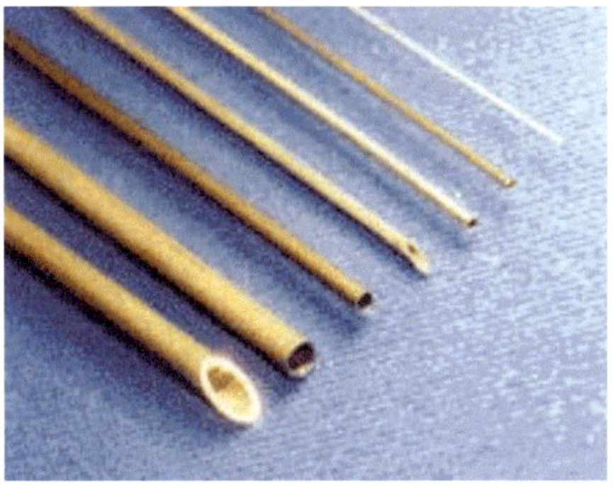

Fig. 12.

Tubes, Wires, and Ribbons: For many applications that deal with a heated fluid flowing through tubes, or wire and ribbon applications where it is crucial for the alloys to maintain their shape in the midst of a heated environment, memory metals are ideal.

3. Applications of smart materials

- Piezoelectric materials - These ceramics or polymers are character-ized by a swift, linear shape change in response to an electric field. The electricity makes the material expand or contract almost instantly. The materials have potential uses in actuators that control chatter in precision machine tools, improved robotic parts that move faster and with greater accuracy, smaller microelectronic circuits in machines ranging from computers to photolithography printers, and health-monitoring fibers for bridges, buildings, and wood utility poles.
- Electrostrictive and magnetostrictive materials - This refers to the material quality of changing size in response to either an electric or magnetic field, and conversely, producing a voltage when stretched. These materials show promise in applications ranging from pumps and valves, to aerospace wind tunnel and shock tube instrumentation and landing gear hydraulics, to biomechanics force measurement for ortho-pedic gait and posturography, sports, ergonomics, neurology, cardiology, and rehabilitation.
- Rheological materials - Smart materials encompass not only solids but also fluids, electrorheological and magnetorheological fluids that can change state instantly through the application of an electric or magnetic charge. These fluids show promise in shock absorbers, dampers for vehicle seats and exercise equipment, and optical finishing.
- Thermoresponsive materials - Shape memory alloys, the dominant smart material, change shape in response to heat or cold. They are most commonly Nitinol, or nickel and titanium combined. Less popular but still possessing the shape memory effect are gold cadmium, silver cad-mium, copper-aluminum-nickel, copper tin, copper zinc, and copper zinc aluminum. They are useful in couplers, thermostats, automobile, plane and helicopter parts.
- pH-sensitive materials - The most interesting of these are indicators that change colors as a function of pH, and show promise in paints that change color when the metal beneath begins to corrode.
- Electrochromic materials - Electrochromism is defined as the ability of a material to change its optical properties when a voltage is applied across it. These materials are used as antistatic layers, electrochrome layers in LCDs (liquid crystal displays), and cathodes in lithium batteries.
- Fullerenes - Spherically caged molecules with carbon atoms at the corner of a polyhedral structure consisting of pentagons and hexagons. In one application of fullerenes as a smart material, they are embedded into sol-gel matrices to enhance optical limiting properties.
- Smart gels - Engineered response gels that shrink or swell by a factor of 1000, and that can be programmed to absorb or release fluids in response to almost any chemical or physical stimulus. These gels are used in many applications in agriculture, food, drug delivery, prostheses, cosmetics, and chemical processing.

4. The future for smart materials-

The development of true smart materials at the atomic scale is still some way off, although the enabling technologies are under development. These require novel aspects of nanotechnology (technologies associated with materials and processes at the nanometre scale, 10^{-9}m) and the newly developing science of shape chemistry.
Worldwide, considerable effort is being deployed to develop smart materials and structures. The technological benefits of such systems have begun to be identified and, demonstrators are under construction for a wide range of applications from space and aerospace, to civil engineering and domestic products. In many of these applications, the cost benefit analyses of such systems have yet to be fully demonstrated.
The Office of Science and Technology's Foresight Programme has recognised these systems as a strategic technology for the future, having considerable potential for wealth creation through the development of hitherto unknown products, and performance enhancement of existing products in a broad range of industrial sectors.
The core of Yanagida's philosophy of ken materials is such a concept. This is 'techno-democracy' where the general public understand and 'own' the technology. Techno-democracy can come about only through education and exposure of the general public to these technologies. However, such general acceptance of smart materials and structures may in fact be more difficult than some of the technological hurdles associated with their development.

5. Conclusion

Smart materials are poised to emerged from the laboratory of medical, defence and industrial applications. Understanding and using these advanced materials in new product development efforts may make the difference between success and failure in today's intensely competitive markets.
It's the profile job of the technocrats and management personnel to find out the promising materials for specific applications-when the use of smart memory alloys is to be replaced by a smart polymer, the primary laboratories and companies who are developing these materials, to identify the key researchers and engineers in those fields .With smart materials research taking place in hundreds of public and private sector labs across the globe, to get them available at once is difficult - yet they are vital for the advancement of technology and to profit from new developments in this fast moving field.
The concept of engineering materials and structures which respond to their environment, including their human owners, is a somewhat alien concept. It is therefore not only important that the technological and financial implications of these materials and structures are addressed, but also issues associated with public understanding and acceptance.
There are many possibilities for such materials and structures in the man made world. Engineering structures could operate at the very limit of their performance envelopes and to their structural limits without fear of exceeding either. These structures could also give maintenance engineers a full report on performance history, as well as the location of defects, whilst having the ability to counteract unwanted or potentially dangerous conditions such as excessive vibration, and effect self repair. The Office of Science and Technology Foresight Programme has stated that `Smart materials will have an increasing range of applications (and) the underlying sciences in this area must be maintained at a

standard which helps achieve technological objectives', which means that smart materials and structures must solve engineering problems with hitherto unachievable efficiency, and provide an opportunity for new wealth creating products.

6. References

[1] J.S. Harrison, Z. Ounaies, 'Piezoelectric Polymers', NASA/CR-2001-211422, ICASE Report No. 2001-43 (2001).

[2] K. Ullakko, 'Magnetically Controlled Shape Memory Alloys: A New Class of Actuator Materials', Journal of Material Engineering and Performance, Vol. 5, 405-409 (1996).

[3] Akhras, G., "Advanced Composites for Smart Structures", *Proceedings, ICCM-12*, 12th International Conference on Composite Materials, Paris, July 5-9, 1999.

[4] Sun, G. and Sun, C.T., *Bending of Shape Memory Alloy Reinforced Composite Beam*, Journal of Materials Science, Vol-30, No.13, pp5750-5754.

[5] P.k.dutta, I.K.bhat, "smart materials opportunity & challenges", Proceedings, winter school on smart materials, MNNIT Allahabad, 29 Nov- 10 Dec, 2004.

[6] PZT Application manual.

[7] Humbeeck, Jan Van. "Non-medical applications of shape memory alloys." *Materials Science and Engineering* A273-275 (1999): 134-148.

[8] http://www.piezo.com/tech4history.html

[9] http://www.nanomotion.com

[10] H. Horikawa, in *Proc. SMST99* (Shape Memory and Superelastic Technologies), edited by W. Van Moorleghem, P. Besselink, and D. Aslandis (Shape Memory and Superelastic Technologies Europe, Antwerp, 1999) p. 256.

[11] http://www.mrs.org/publications/bulletin

[12] http:// www.ceramics.queensu.ca

[13] Active Materials Laboratory, MIT, USA. http://aml.seas.ucla.edu/

[14] Otsuka, K. & Ren, X. Recent developments in the research of shape memory alloys. Intermetallics 1999, 7, pp. 511–528.

[15] Manz, H. & Schmidt, K. On the application of smart materials in automotive structures, Euromat99, Wiley-VCH Verlag Gmbh, Weinheim, Germany, 2000. Vol. 13, pp. 539–544.

[16] Ball, Philip. *Made to Measure, New materials for the 21st Century*. Princeton, NJ: Princeton University Press, 1997

[17] http://www.freepatentsonline.com

[18] http://www.specialmetals.com

[19] http://www.fzs.tu-berlin.de/html/en/ausr_werkstoffe.html

[20] http://www.nobles.se/

[21] Oulu University - http://herkules.oulu.fi/

[22] S. Ramamurthy, M.V. Gandhi, and B.S. Thompson, *Smart Materials for Army Structures*, Quantum Consultants, Inc.,Michigan 1992; DTIC Doc. AD-A300 215.

[23] Carlson, J. D. *et al.*, 1994, "Magnetorheological Fluid Dampers", U.S. Patent 5,277,282. (1994) and "Magnetorheological Fluid Devices", U.S. Patent 5,284,330.

[24] "MRB-2107-3 Brake", Lord Corporation product bulletin, Cary, NC, http://www.mrfluid.com/ devices_brake_begin.htm 2001.

[25] Gupta P, Srivastava R.K., Gangwar M., "Opportunity and Challenges of Smart Materials: A Review", National seminar on APD, MNNIT allahabad, Feb.2006.

2

Polymer Materials and Its Properties

Jana Šišáková
Alexander Dubček University
Faculty of Industrial Technologies in Púchov
Slovakia

1. Introduction

The technological process of rubber blend preparation plays an important role in the product quality, and mainly in traffic security. Nevertheless, after more than two hundred years of tyre production, the technology of rubber blend mixing is still a current and intensively developed process. The principle of rubber blend preparation in mechanical mixing is implemented by screw mechanism and extrusion of blend manipulated in such way. The success of the whole technology depends on proper mechanical construction of the masticator's components for every mixture. Nowadays the polymer materials used in rubber or automotive industry are submitted to new research activities by the miscellaneous angle of vision. Of coarse the final purpose is to provide the best required properties. Tire as a composite imply a lot of materials which have quite disparate properties and are able to appear as unit - consequently they are up to required standard. For example in areas of safety characteristics as an (wet road holding, tire grip, stability), the achievement of high speeds, low power consumption (rolling resistance), low noise level, comfort, attractive design and the tendency of increase lifetime (abrasion resistance, ozone and heat ageing) of these products with a purpose of economical and protection the environment. The interest in polymer materials and its application rising, e.g. in Audi A6 there is represented mass percentage contribution of polymer materials nearly 14 % on the present.

2. Mixing aspects

There are common a lot of factors influence the final properties of article in manufacturing process. These baseline impacts occur in case of quality and kind of deliver chemicals, particle size, its concentration, particle morphology, its specific surface. Consequently these chemicals are mixing. The main base of rubber mixture is a caoutchouc, according as natural or synthetic. By caoutchouc relates the term of plasticizing - this represents the process of reduction the molecular length of caoutchouc and occur from the elastic state to the plastic state during this time. After the caoutchouc feeding into the masticator or calender (figure 1) occur to the heat generation and stress transmission on the polymer chains in consequence of frictional forces between the machine components and visco-elastic material.
Thus the caoutchouc plasticizing is mechanical - chemical process characterizing degrade of caoutchouc which leads to structure change and consequently change on mechanical and physical - chemical properties of caoutchouc.

Abnormal plasticizing is companied by bad mechanical properties of vulcanizates, decrease fatique and ageing resistance. Process of plasticizing follows large noise in consequence the shear stress of caoutchouc masticated between rotors and mixing chamber wall.

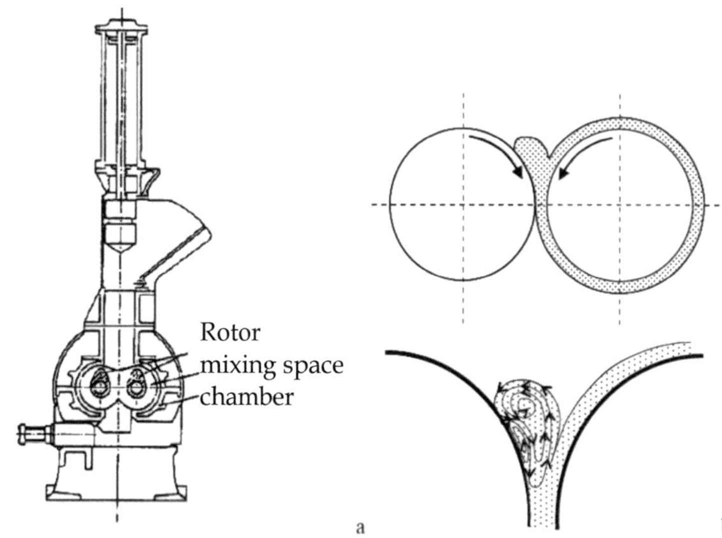

Fig. 1. Schema of masticator (a) and calender (b)

The efficiency of mechanical tearing of caoutchouc molecules decreases by temperature and consequently is process of plasticizing delayed. A rare behavior of temperature dependency of minimal efficiency is observed by temperature scope of 115 - 120 °C on figure 2. There is occurred decrease efficiency of mechanical tearing of molecules in case on the left side and rising effect of heat oxidizing degradation of polyisoprene chains in case of right side.
Maximal plasticizing efficiency is achieved by temperature of 55 °C or temperature over 140 °C by case of natural caoutchouc. Lower temperatures are used by calender plasticizing, higher temperatures are characterized for masticators.

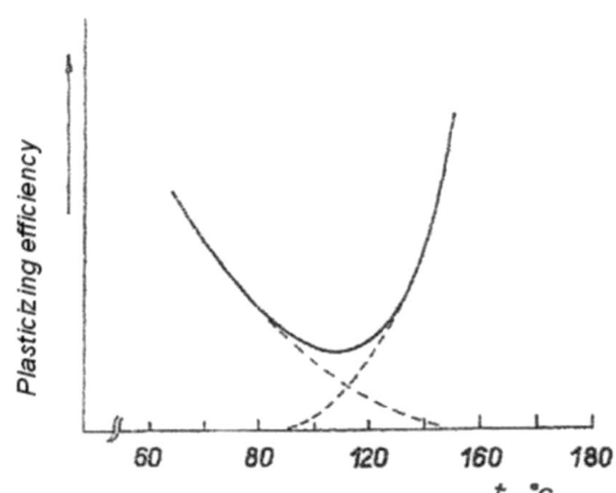

Fig. 2. Plasticizing efficiency temperature dependence

At higher temperatures occur to radical creation by bonds refracting which is necessary chemical stabilized. Long storage of this treated caoutchouc results in re-meshing and viscosity rising in consequence of un-stabilization. Chemical chain plasticizing and stabilization practically ends by carbon black feeding. There is occurred only physical chain reduction in consequence of re - mixing (reprocessing). Only natural caoutchouc is practically plasticizing. Synthetic caoutchouc and caoutchouc with constant viscosity isn't necessary to plasticize. Altogether the physical plasticizing is first mixing step of each mixture but this effect isn't as large as plasticizing of straight caoutchouc. The oils decrease the plasticizing effect in mixtures.

Temperature influence on plasticizing efficiency

The quality of caoutchouc plasticizing process in masticator affected following factors: dimensions of masticator, geometry of rotors, the dimension between rotor and chamber wall, state of machine filling, stress on the upper stop element, rotor speed, caoutchouc temperature, kind and concentration of plasticizing agent, time of plasticizing.
Mixture preparation and especially its mixing is one of important processes in rubber industry because following mixtures processing, products properties and producing economy largely dependence on mixtures quality.
Besides caoutchouc rubber mixture contains next approximately 10 chemicals (additions), table 1. Each chemical has specific role and therefore mixing purpose must provide the best homogeneity of mixture (most uniform distribution of chemicals in the whole caoutchouc mixture).

Compound of mixture	PHR *
Elastomers - natural and synthetic caoutchouc, respectively their ratio	100
Fillers - active and inactive	0 - 200
Softeners	0 - 40
Activators of vulcanizing	0 - 40
Stabilizers - antioxidants, antiozonants a protective wax	0 - 9
Vulcanizing agents	0,3 - 50
Accelerators of vulcanizing	0,3 - 4
Assistant rubber additives - plastic and adhesive agents	0 - 10

* PHR - parts per hundred rubbers (means elastomers - caoutchouc)

Table 1. Standard recipe for rubber mixture

After the caoutchouc plasticizing is coming the next step - blending - together the plasticizing there is a preliminary step of mixing where are the separate additions (caoutchouc and fillers) moulded by rotors or calender into a coherent mass. The blending action is described as a carbon black wetting by caoutchouc together an air voids expelling which was internally incorporated along with chemicals. Before blending process with powder fillers occur to a large deformation of caoutchouc follow by bond chain tearing (plasticizing). This conditioned material is able to flow in the chamber of masticator during all the mixing process round. The process of blending has two parallel mechanisms. Primarily occurs to caoutchouc formation follow by its surface rising with consecutive filler wetting by caoutchouc matrix. In the second mechanism is caoutchouc sharply deformed in consequence of bonds tearing, during which time occurred to transmission of tearing forces on adjacent chains. Bonds tearing allow rapid agglomerates coating by caoutchouc.

During the initializing stage of carbon black blending process in consequence of high shear stress created in masticator chamber thereafter of polymer pressure which coated the agglomerates are these smashed. Separate agglomerates are air filled by creation of weakly, partly composites. Caoutchouc is gradually embossed to agglomerates what causes air elimination.
The progressive carbon black wetting by polymer can be observed in case of machinery indirectly as an immediate loss of power the engine which specifies shear stress between mixture and rotor. Where this time period is the time of carbon black incorporation (so called the incorporation time - BIT).
On macroscopic scale the dispersion of filler into a polymer matrix shows the following stages:

- The filler smears into striations following the deformation pattern of the polymer
- Agglomerates up to 10 to 100 micrometer in size appear
- Agglomerates are continuously broken and aggregates with an average size of 100 nanometers till 0.5 micrometers appear
- Smaller aggregates and primary particles appear on the expense of larger aggregates and agglomerates

Figure 3 illustrates the transition from large agglomerates into smaller aggregates and primary particles, and gives an indication of their dimensions.

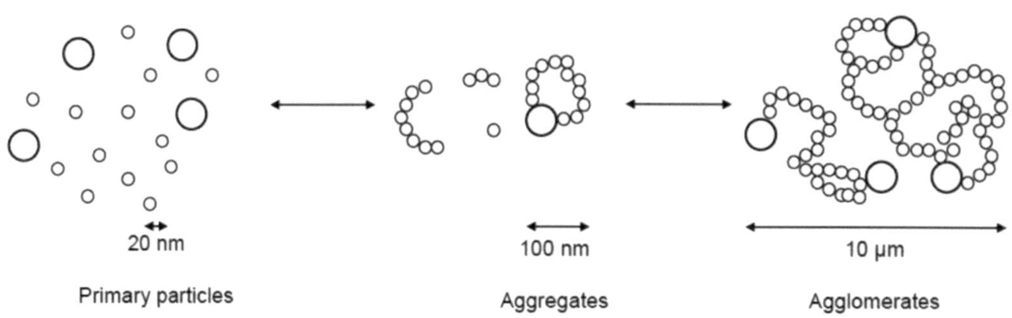

Fig. 3. The filler aggregation and dispersion

Based on these phenomena, Yoshida described mixing as a three step process: transposition, insertion and breaking of the disperse system. During transposition the system is subjected to stretching deformation by shearing forces, which increases the interface between the disperse phase and the matrix, and results in a gradual insertion of the disperse phase into the matrix. The particles of the dispersed phase are disrupted by shearing forces, and the size of agglomerates and aggregates is reduced. The degree to which the filler finally has to be dispersed depends on the quality requirements of the compound: the higher the degree of dispersion, the better the properties. But there is a lower limit to the aggregate size as the properties deteriorate with very small aggregate sizes and an increasing amount of primary particles.
A more refined model of dispersive mixing separates the process into four different steps:

- Incorporation
- Plasticization
- Dispersion
- Distribution

In the initial stages of mixing before incorporation starts, two processes take place: The first process involves large deformations and subsequent relaxation of rubber domains, during which filler aggregates are sandwiched between rubber layers. The second mechanism is based on the disintegration of the rubber into small particles, which are blended with the filler agglomerates and finally seal them inside.

Incorporation

The incorporation of the filler is subdivided into two phases: formation of a rubber shell around the filler particles followed by filling up the voids within the filler agglomerates, in other words between the filler aggregates. It includes a subdivision step: breaking of large agglomerates into smaller entities.

Mastication and plasticization

Mastication and plasticization take place during the whole mixing process and result in a change of the rheological properties of the compound, especially a viscosity decrease of the polymer matrix by degradation of the elastomer chains. Figure 4 gives a schematic view of the viscosity changes during mixing and the contributions of temperature increase and polymer breakdown to the viscosity decrease.

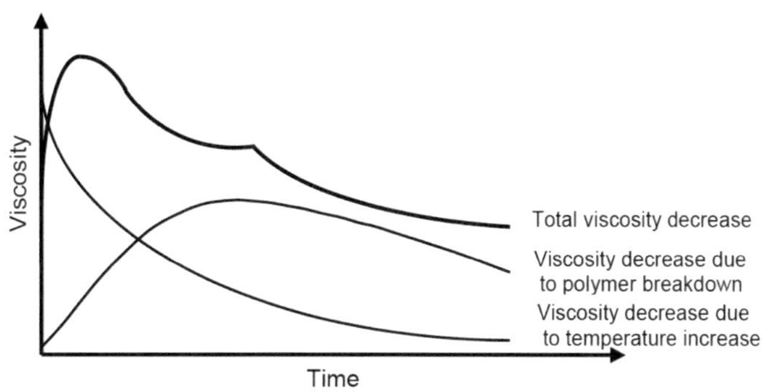

Fig. 4. Contributions of polymer breakdown and temperature increase to the viscosity decrease

Dispersion

At the end of the incorporation stage the majority of the filler is present as agglomerates. They act as large, rigid particles, whose effective volume is higher than that of the filler alone due to rubber trapped inside the filler voids and the rubber immobilized on the surface. The bound and occluded rubber increases the rate of dispersive mixing by increasing the effective radii of the filler particles: larger effective radii lead to higher stress during mixing. The filler agglomerates are successively broken to their ultimate size, mainly by shear stress. Parallel with the reduction of the agglomerate size the interface between the matrix and the filler is increased, and the filler particles are distributed homogenously throughout the rubber matrix. When the filler agglomerates decrease in size, the occluded rubber concentration is reduced. The viscosity of the compound decreases and finally reaches a plateau region. In general, the average particle size reaches a minimum value and a further energy input does not result in a reduction of the size of the filler aggregates any more, as the mixing and dispersion efficiency is decreasing with reduced viscosity of the compound.

Distribution

In the distributive mixing step particles are spread homogenously throughout the polymer matrix without changing their size and physical appearance. The thermodynamic driving force for this process is the entropy increase of the blend.
Figure 5 illustrates the different mixing stages with respect to filler subdivision, incorporation, dispersion and distribution.

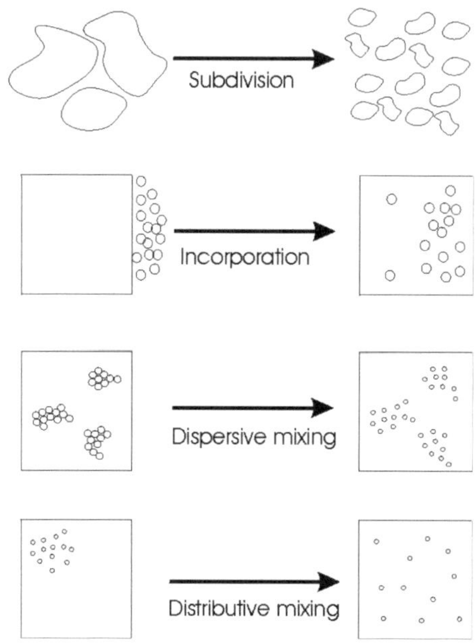

Fig. 5. Illustration of the different mixing stages for filler-polymer systems

An optimal dispersion is one that evenly distributes carbon black throughout the polymer down to the smallest carbon black unit, the aggregates. A poorer dispersion results in larger agglomerates, figure 6.

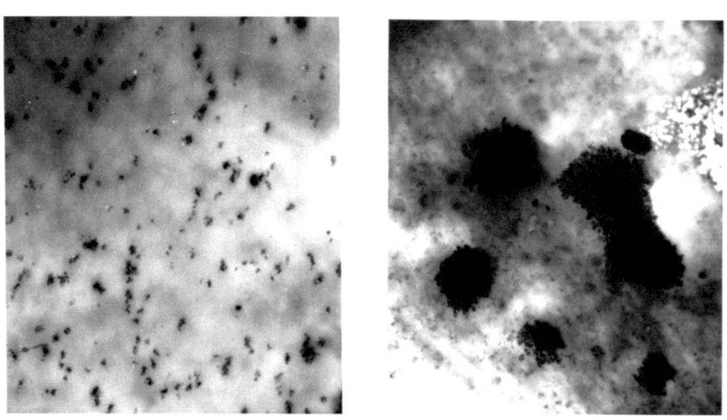

Fig. 6. Scheme of aggregates and agglomerates

The stage of dispersion characterized the process efficiency demands the higher energy consumption than blending process. Generally the dispersion stage is evaluated in according to methodic of CABOT Company. By using of this methodic is possibly to determine the quantify of un-dispersed carbon black and following particle size is possible to merge the mixture into six different dispersion qualities accordance to figure 7.

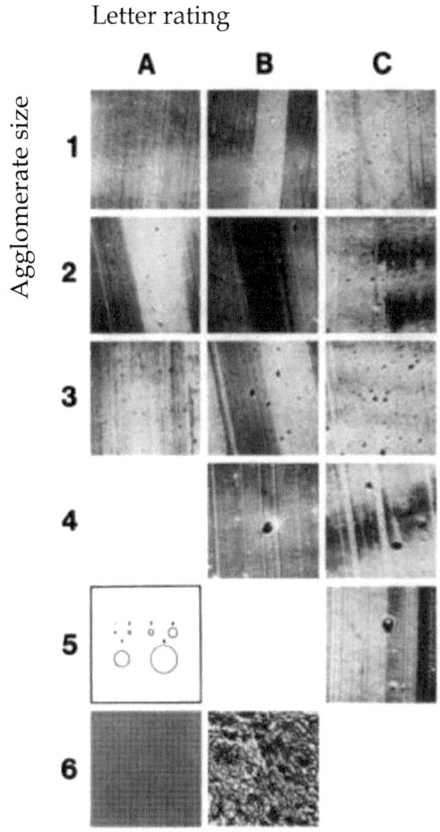

Fig. 7. Quality ranking of dispersion

The behavior of energy consumption the masticator is characterized by two marked maximums. The first maximum is created by caoutchouc feeding and ram down mixing. The second is characterized by fillers feeding (carbon black or white fillers). First peak relates by caoutchouc tearing and raising the active surface, the second marked the agglomerates creation of bond caoutchouc - filler. By sequential agglomerates smashing, the energy consumption decrease and the stage of filler dispersion accrue (mixing uniformity of primarily units). The mixture properties are changed by increased filler dispersion. We can allege that Mooney viscosity decrease by improved filler dispersion. Dispersion filler have different chemical contain, morphology (shape), density, color, hardness, specific surface and other physical - mechanical properties. The morphology of dispersion fillers can be regular (spherical, fibrous, platy) or irregularly, figure 8.

On behalf of observing purposes the mixing process were created various models studies which are able to ordered stage dispersion in rubber mixture e.g. through the particle flow in masticator. Similarly rotor design, slit dimension and chamber profile affected the mixing process and therefore these parameters are optimizes by producers.

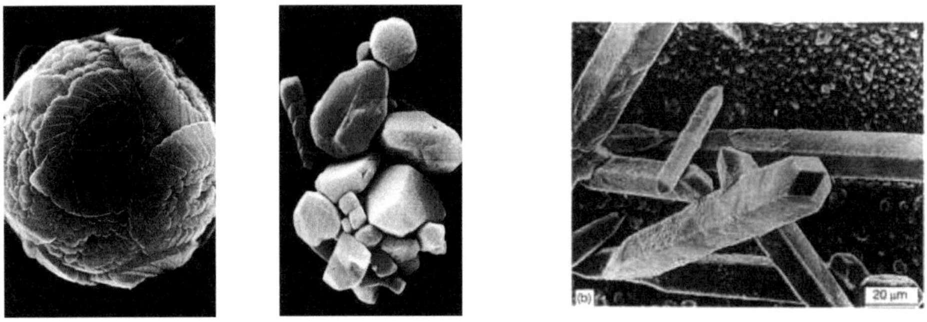

Spherical Anisotropy Platy

Fig. 8. The particle shape of dispersive fillers

Last step inducing the mixing process is the homogenization and viscosity reduction. For a good uniformity is necessary excellent filler distribution in the whole volume of mixture. This effect is possibly to achieve by becomingly processed mixing settlement together the time step of particular components feeding.

For the first degree preparation of mixture (without the vulcanizing agents) are apply two ways of mixing:

- Conventional,
- Upside - down.

By using of a conventional mixing occurs to filler incorporation into caoutchouc matrix at first. Mixing time is longer and follows to temperature increase during mixing. The final temperature can reach the value of 160°C. Owing to the achievement of high stage of filler homogenization and consecutive achievement of necessary processed properties are feeding oils in the final mixing phase.

In order to dispersive degree increase together homogeneity improvement are used so called processing additives (PA) in manufacturing process. At the same time these additives minimize the energy state of process. However is possibly to feed PA only in the first degree of mixing before fillers feeding. The processing additives also decrease viscosity of mixture, but the efficiency isn't the case of softeners. To achievement of a minimal tolerance in case of viscosity is important for the next processing operations - e.g. extrusion, calandering.

The UPSIDE-DOWN method of mixing uses the reverse practice and there is realize by high feeding of softeners and big particles of fillers. Primarily are feeding the fillers, then are mixed softeners and last chemicals added into masticator are caoutchouc.

However this method isn't very useful for the mixtures includes high-activated fillers (high structural carbon black, active SiO_2) or mixtures the high contains of soft mineral filler and oil together polymer with a high viscosity.

The second mixing stage is characterized by vulcanizing agents mixing off. Time of its adequate dispersion and homogeneity achievement considering physical properties is shorter than mixture preparation in the first stage. The maximal temperature for finishing of mixing process is much below (max. 120 °C) owing to the possibly reaction of sulphur with caoutchouc. The decrease of viscosity is not so high than the first mixing stage.

Besides the vulcanizing agents is convenient to adding the vulcanization inhibitors and retarders in the second stage of mixing process needs for consequently repeated processes linked to mixture warming - profile extruding, multi calandering, returnable wastes processing. The progressive methods enable mixing of the first and second stage in the same line equipment.

Figure 9 shows a generalized torque-time profile of a mixing cycle for a carbon black compound, which is characterized by two torque maxima. The first torque maximum corresponds to the point when all ingredients so far added to the compound are forced into the mixing chamber and the ram reaches its final position. Time and height of the maximum depend on the composition of the compound, the fill factor of the mixing chamber, the activity and structure of the filler as well as the bound rubber fraction. The region between the start of the mixing cycle and the first minimum is the region, in which mainly mastication of the polymer takes place and dispersion starts. In the next zone dispersion and distribution of the additives including the filler occurs, and a second torque maximum is observed. This torque maximum corresponds to the incorporation of the filler: the agglomerates are wetted by the polymer, entrapped air is removed and the compound reaches its final density. Filler dispersion and polymer breakdown result in a fast torque decrease after the second peak down to a dispersion degree above 90%, and the kinetics can be described by a first order law: The rate of agglomerate breakdown is proportional to the concentration of the agglomerates. This period of steep torque reduction is followed by a period of a slower torque decrease, during which polymer breakdown and dispersion occur to a very limited degree, shearing is reduced and homogenization starts.

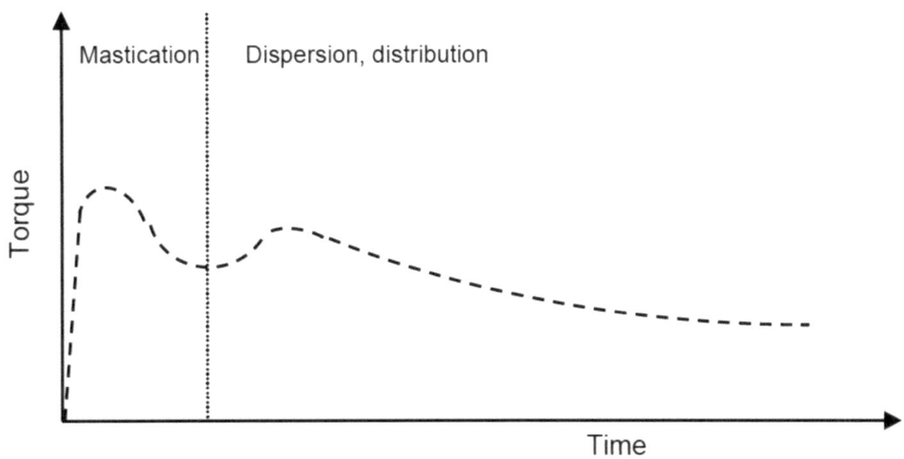

Fig. 9. Generalized torque-time profile of a mixing cycle

The carbon black construction of units and particles influences the final properties. Base particles have approximately sphere morphology, the furnace carbon black have a more complicated construction. Elementary sphere particles are coupled into big formations, called as primary structural aggregates which created chains or 3-D branched formations. The aggregation of elementary particles into big formations represents a "structure". Primary structure means the joining of basic particles into primary structural aggregates which are resistant to mechanically destruction. Primary structural aggregates may create bigger formations holding together by Van Der Waals forces, marked as secondary structure. This structure has low strength and easier mechanically destruction, especially by mixing of carbon black to caoutchouc matrix. The baseline particle contains the ligament of graphite layer. Accordingly the created aggregate and chain have compact, cohesive formation where are

occurred areas by concentric oriented plane layer; the outer layer have a high graphite contain than inner, there are oriented parallels to surface and have relatively few defects.
Interaction stage of caoutchouc and carbon black dependences on three factors:

- Size of interface between carbon black surface and caoutchouc, so called extensive factor. The interface between carbon black and caoutchouc equals to composition of specific surface carbon black and its concentration in mixture.
- Carbon black structure, so called geometric factor.
- Surface activity or intensive factor.

The size of particles and specific surface occurs; the softer are particles (i.e. the more is specific surface) the difficulty are mixed into a caoutchouc matrix; the viscosity of mixture increases, there accrue the energy consumption for processing, the mixtures are more heated and the safety processing time shorten. Particle size hasn't an important effect on the mixture precipitation but meaningful influence affected the mechanical properties of vulcanizates. By increasing of specific surface size (the decrease is particle size) are improving the properties detected by destructive examination: tensile strength, structure strength, wear resistance, hardness of vulcanizates increases. The modulus of vulcanizates hasn't directly dependency on particles size; the dynamical properties worsen by particle size decrease, the hysteresis increases.
Carbon black structure has more effect on prepared mixtures processing as a final vulcanizates properties. Otherwise the high structured carbon black are slower mixing into caoutchouc, but easier cause of good dispersion achievement and during mixing is created more joining caoutchouc. Viscosity of carbon black mixtures increase by growing structure and consequently the mixtures are more heated.
High structured carbon black increases the mixture anisotropy, mixtures are better extruding, they are less porous, the surface is smoother and profile edges are sharper. Similar behavior is characterized by calandering, mixtures filled high structured carbon black have smoother surface and smaller precipitation. Both causes induced due to structure influence and joining caoutchouc.
The structure of vulcanizates has a respectable effect on modulus; the influence is linked to surface activity. Strength and structural strength aren't considerably affecting by vulcanizates structure. A favorable effect is noted for wear resistance due to dispersion improvement. Hysteresis dependences on particles size and isn't markedly affected by structure.
The surface activity a factor influences besides vulcanizing process, composition of carbon black surface also electric conductivity of vulcanizates. By fewer oxygen compounds of carbon black, by softer particles and high structure the more conductive is filled vulcanizates.
When the convectional carbon black is treated high temperatures (approximately 2700 °C) there occur a big transferring of carbon atoms in the inert atmosphere. As a result is graphite carbon black with big crystalline planes on the surface. The surface energy of basic planes is low and small is also the reinforcement effect. For the smallest ordering and the highest surface energy was development so-called inverse carbon black. Compared to conventional carbon black have inverse carbon black more mutual traversing planes and edges which leads to surface energy increasing.
This special carbon black plays an important part by composition of treading mixtures with good adhesion and reduced rolling resistance.

Vulcanization

The next step influenced the quality of preparing mixture is vulcanizing (curing) process. This term is defined as a process when the temperature and vulcanizing system influence occur to the creation of chemical cross linked bonds among the caoutchouc chains. Throughout the vulcanization the linear chain structure is changed on 3-D during which times are markedly affected physical - mechanical properties and caoutchouc is converting from the plastic into the elastic state. The vulcanizates is characterized by its high reversible deformation by relatively low value of elasticity modulus due to creating cross linked structure. By cross linking mesh creation is limited the caoutchouc macromolecules mobility especially occurring the:

- Insolubility of cross linked polymer, just swelling vulcanizates.
- Polymer strength increase in the definite value by vulcanizing, by overrunning the optimal stage the strength decrease, but the modulus and hardness increase.
- Increasing vulcanizing degree improving the strain resistance and dynamic fatique resistance. There is mended structural strength, i.e. resistance related with additional tearing of aborted sample.
- Low vulcanizates sensitivity on temperature changes.
- Elasticity and stiffness vulcanizates retaining in the wide temperature region.

Vulcanization course and its basic characteristics (scorch time, optimal vulcanization time, difference between maximum and minimum torsion moment, cure rate index, eventually reverse rate etc.) is most often evaluated on the basis of so called vulcanization curves measured on different types of rheometers, figure 10. There could be free sulphur content in a case of chemical changes, which decreases in vulcanization running. In case of physical - mechanical values there are strength, tensile elasticity modulus, plastic deformation, and swelling.

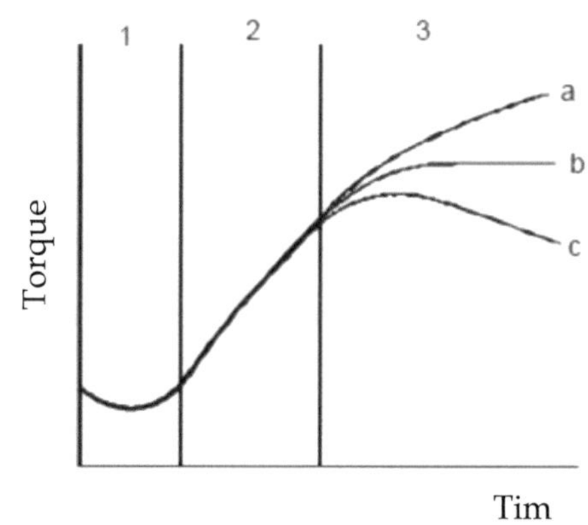

1- Scorch period, 2- fundamental meshing reaction, 3- structure changes created by meshing: a - marching, b - relaxation, c - reversion

Fig. 10. The schematically dependence of torque versus vulcanizing time

The induction period (scorch) characterized the time interval; there occur to very slowly reaction of vulcanizing agents with caoutchouc and other additives, during which time are created soluble caoutchouc intermediate products. This period length is important in term of vulcanizing system options for its safety mixture preparation and the next processing. Accordingly influences the processing economy and vulcanizates quality.

Fundamental meshing reaction is defined as a time interval of induction period and optimum of vulcanization. In this phase occurs to mesh creation and changes of physical - mechanical properties of caoutchouc mixtures by sequential transfer to vulcanizate.
The last area of vulcanization curve is characterized by other reactions than vulcanizing agent meshing and therefore practically occurs to no affection of the total meshing stage. This shape of vulcanizing curve dependences on used caoutchouc, vulcanizing system and vulcanizing temperature. The rate of meshing reaction is decreasing or aborting in this phase. These changes relate by the changes of cross links character. Relaxation is an area on vulcanizing curve where are no property changes in case of vulcanizate. The amount is presented by constant vulcanizate properties the heat stress for a term. Especially this property consequential from the vulcanizate thermal stability enables the achievement of optimal properties practically in all tire sections by automotive production. Reversion is the area of vulcanizing curve where occurs to the meshing density decrement and the vulcanizing properties become worsen.
Marching is called the region area of vulcanizing curve where the meshing moderate increases; this type is vulcanization with a "walking" modulus. By vulcanizing process is used an ability to heat transformation (from metallic parts of vulcanizing press and mould, figure 11) from vulcanizing medium (steam, hot water) into tire. Consequently occur these phases in process of compression moulding and vulcanizing:

- Caoutchouc mixtures flow - the mixture must perfect flowing into the parts of compression mould in this processing phase during which time the air is removing through the air channels. The efficiency of moulding and vulcanizing process also affected on the flow rate of caoutchouc mixtures and rate of conversion to the meshing phase. Ideal processing representation in a mould characterized the faster mixture flowing into the parts of mould and follow transition into a meshing phase. By the very fast starting of meshing process is possibly risk in technical praxis where the imperfect mould filling can menace the flowing phase, whereas the cross link creation increase the viscosity and eliminate the flowing rate. Because there is very important to choose the correct rate and time of steps so can occurred a worsen visualization and total quality decline of a tire.
- The meshing of caoutchouc mixtures.
- Vulcanization finishing by achievement of the vulcanizing optimum time.

Fig. 11. Segment moulds

Some of vulcanizates properties reach optimum values even before vulcanization optimum achievement, figure 12.

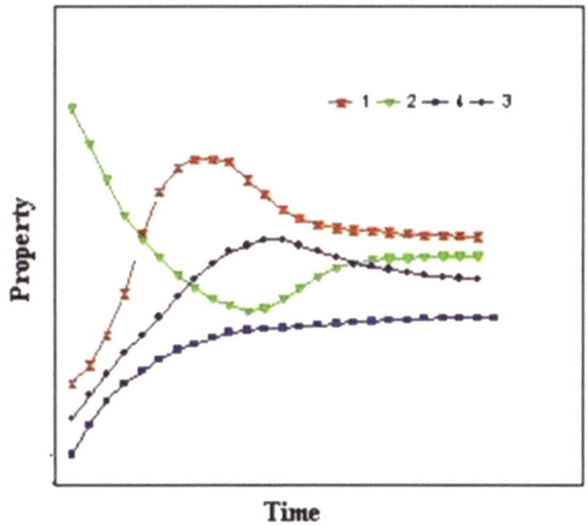

1 – Tensile strength, 2 – Elongation at break, 3 – Hardness, 4 - Elasticity

Fig. 12. Change of some properties of rubbers compounds during vulcanization.

Modulus and the tensile strength at breaking are at low elongation proportional to their crosslink density v, eventually reciprocal value of average molecular weight of rubber macromolecules segments between two cross-links Mc. Their relative connection is described by the relation:

$$\sigma = \frac{\rho}{M_c} RTA^*(\lambda - \lambda^{-2})$$

(σ is stress, λ is relative extension, R is gas constant, T is absolute temperature, A * is the cross sectional area of the test specimen in non-deformable status).
At higher crosslink density tensile strength does not proportionally increase with the crosslink density increasing but after achievement of optimal value decreases.
Elongation at break is with the increasing of crosslink density lowered at first and then is asymptotically approached to minimum value. Vulcanizates hardness is by the vulcanization time increased like their crosslink density. The highest structural strength has soft under-vulcanized vulcanizate. After vulcanization optimum achievement this characteristic together with the time decreased. Elasticity change is similar to change of modulus. It is proportional to crosslink density and to relative elongation in all three coordinates. After vulcanization optimum achievement or at high content of cross-links between rubbers macromolecules can be with the vulcanization time decreased.
Figure 13 – the mechanism of vulcanization with sulfur, accelerators and activators') shows the variation of modulus for natural rubber (NR) and synthetic styrene butadiene rubber (SBR) during vulcanization. In both cases there is an induction period (called "scorch time") and an optimum vulcanization time, corresponding to the maximum value of the modulus. If vulcanization goes on, the modulus value decreases for NR, phenomenon which was

called reversion, while for synthetic elastomers, the modulus stays constant or gradually increases, phenomenon called overcure.

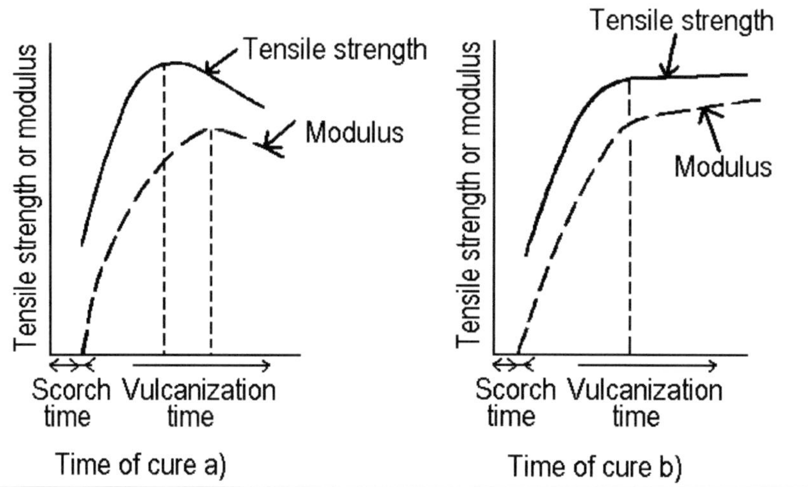

Fig. 13. Variation of tensile strength and modulus with vulcanization time: (a) NR; (b) SBR

The dependency of shear modulus G and temperature is shown in figure 14. Sulphur in position of vulcanizing agent is responsible and importantly factor for meshing of polymer chains, markedly affected also the temperature dependence of shear modulus. In case of un-vulcanized caoutchouc the shear modulus decreasing by temperature dependency, by rising of sulphur content over 0°C is constant or moderate increasing following sulphur content.

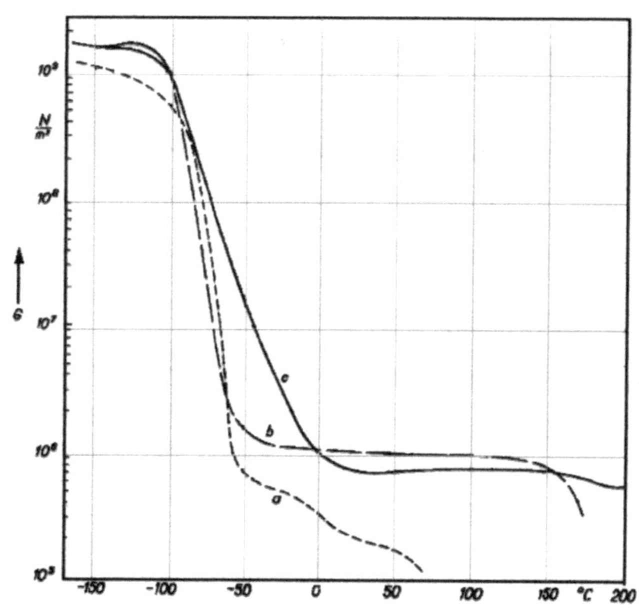

Fig. 14. Shear modulus of caoutchouc vulcanized by different stages in temperature dependence, a) un-vulcanized, b) 0,5% of sulphur, c) 5% of sulphur

The influence of more polar Acrylonitrile Butadiene Rubber (NBR) caoutchouc versus lesser-polar caoutchouc SBR is shown by real dependence the elastic part torque moment (S´) versus time (t), figure 15 or elastic part of tear modulus (G´), figure 16.
In both cases is the rising of elastic part torsion moment S´ following the adding the polar caoutchouc and the meshing become much faster.

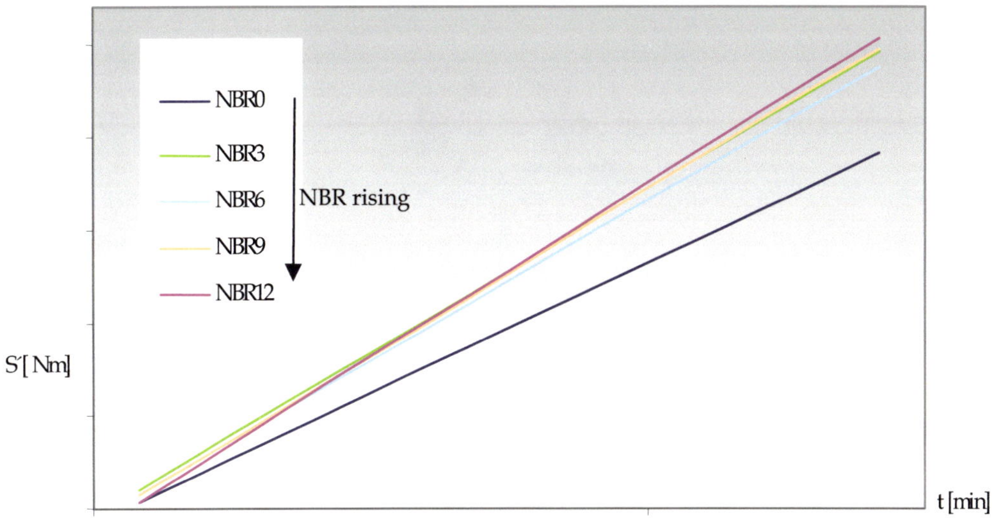

Fig. 15. The elastic part of torque moment versus time

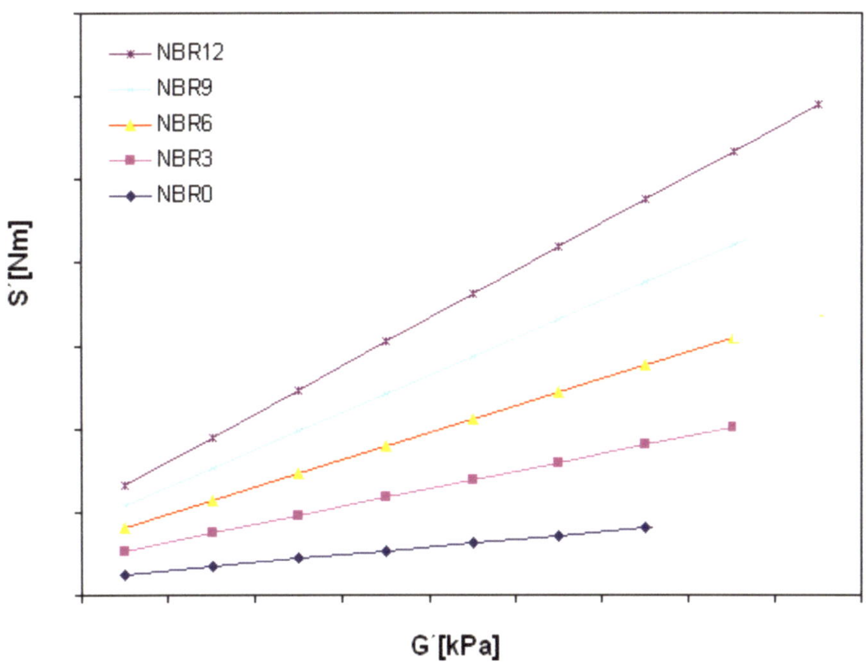

Fig. 16. The elastic part of torque moment versus real part of torsion modulus

A new progressive method in order to quality and economy mixing improving is the on-line rubber blend monitoring. By construction a following production processes of a measurement probe is possibly to perform this method, figure 17. The chamber wall acted as the second electrode.

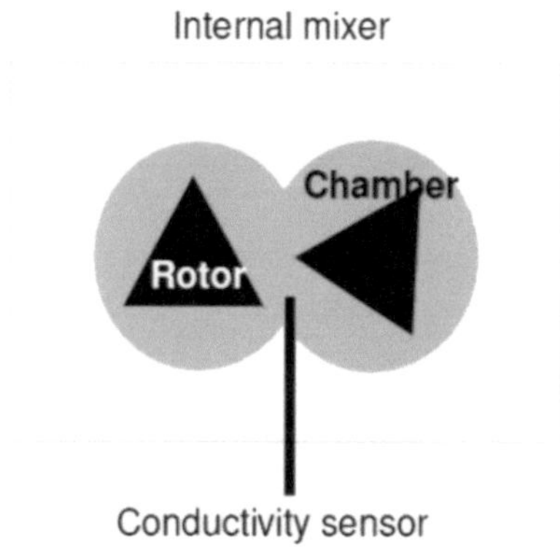

Fig. 17. The probe fixation in the mixer chamber

The sensor (probe) location is important factor. By the sensor location in the wall of chamber is occurred the problem with a periodically washing by prepared mixture (mainly when the mixing chamber isn't exactly filled). The "short circuit" following the carbon black and oil adding can occur in case of softening mixtures. At the same time the sensor doesn't provide the right information about the electrical conductivity of mixing blend. The preferable case of monitoring occurred if is the sensor located in the bottom part of mixing chamber where is provide the continuous contact between sensor and mixing blend (holds also for not exactly filled chamber).
Case of figure 18 gives the electrical parameters information's about the chemical coupling behavior into the mixing chamber; the capacity (C), impedance (Z) and resistance (R) measurements during mixing provide the relevant on-line characteristics.
As was mentioned, at firs are feeding caoutchouc into a mixing chamber, the observed running is without some properties changes. By adding of fillers and oil are the properties fluctuated, the impedance decrease, the capacity increase, following the high conductance fillers. By the next mixing are the electrical parameters changed following the probability of no segregated caoutchouc - creation of agglomerates or aggregates into the chamber by zigzag character. The end of mixing process is considering by the little changes of parameters, approximately in time of 640 s – there is a probability of a good mixing. Usually are the on-line mixing periods shorter in consequence of behavior observing than if it's using classical mixture preparation following the achievement of energy consumption and quality rising.
The quality aspect is important by the question of sample preparation – if the samples would be mixed or prepared by qualitative advance the "good" influence will be noted by its useful properties.
Next are introducing the measurements of electrical parameters and atomic force microscopy as the "echo testing" of the quality preparation.
The quality of prepared samples together a filler influence was evaluated by electrical measurements method. The sample without filler content together the next three samples

filled by different chemicals (standard - no filler, barium sulphate - $BaSO_4$ filler, carbon nanotubes - CNT fillers, metal powder filler) were used and specifically was evaluated the homogeneity by using the characteristics of real part of permitivity versus frequency investigated on three different points, as interpreted the figures 19-22.

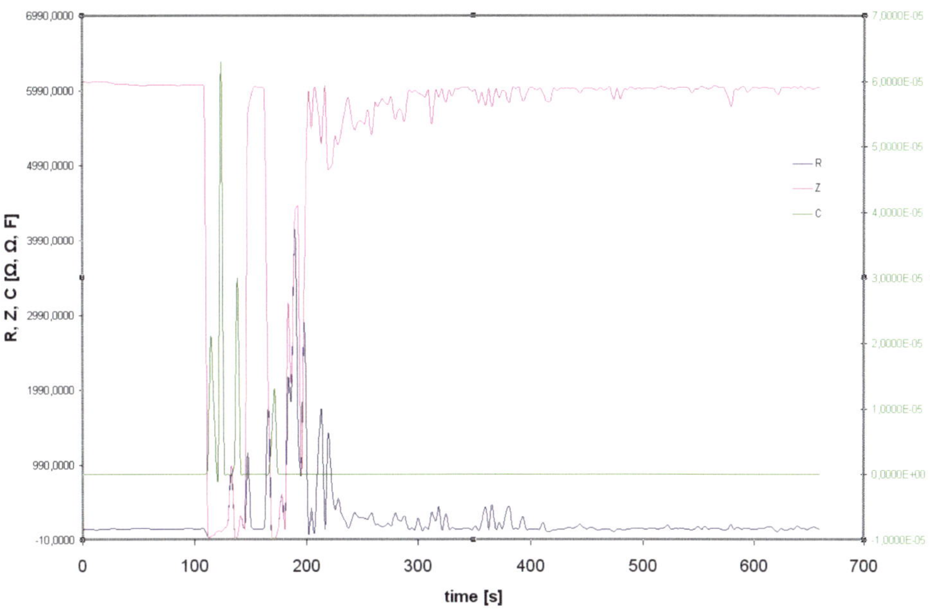

Fig. 18. On-line monitoring the chemicals mixing into the mixing chamber

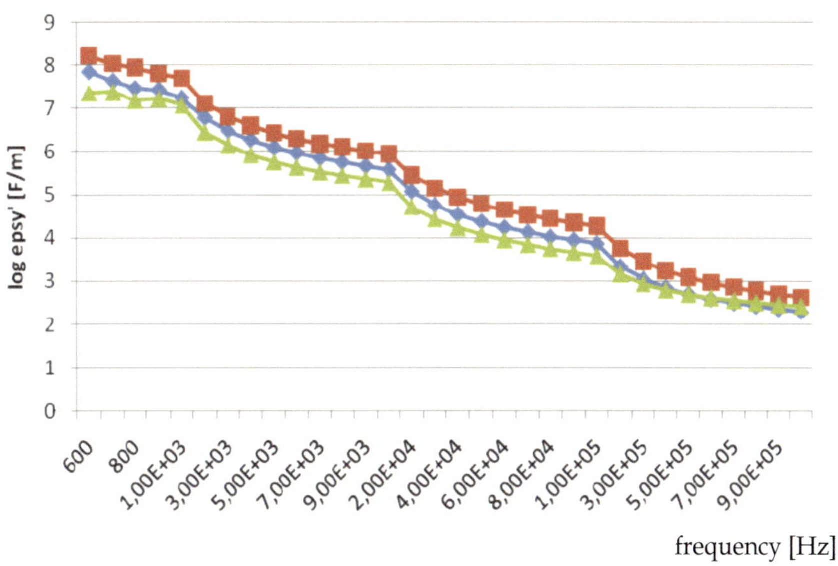

Fig. 19. Sample without filler

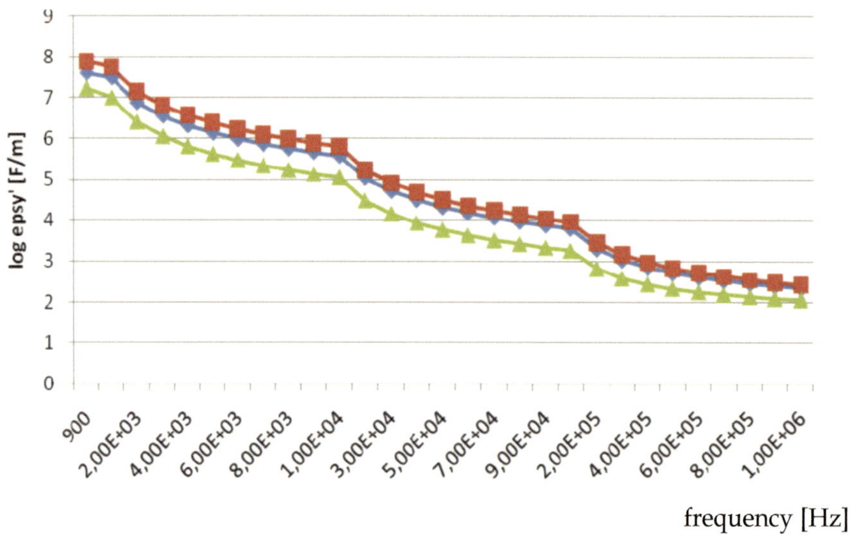

Fig. 20. Sample with $BaSO_4$ filler

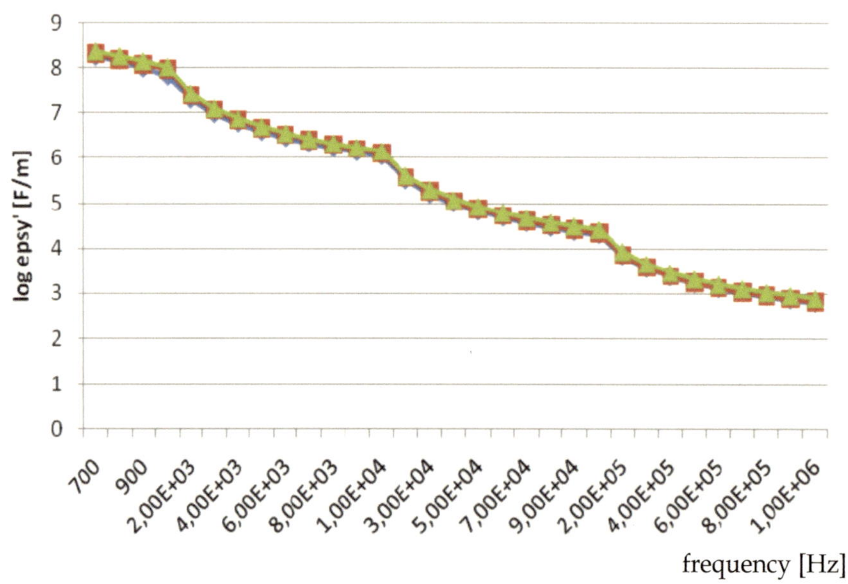

Fig. 21. Sample with CNT filler

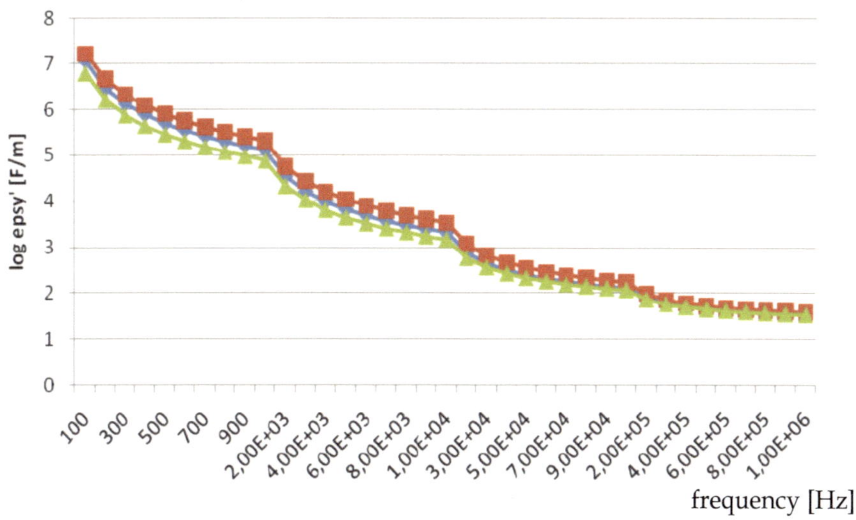

Fig. 22. Sample with metal powder filler

In connection with present results the best homogeneity has the sample with CNT and metal powder filler. There is visible the influence of mixing process for each sample compared the impedance and phase angle values during mixing time, figures 23 - 26 characterized by peaks of electrical values.

On the other hand, modern imagining methods, such as atom force microscopy (AFM), can visualize structure inhomogeneity caused by imperfection of the technological process. The next influence is visible by sample with carbon black filler.

The carbon black fines contents in this case consist of pellet fragments. In such fine materials the probability of forming bigger agglomerates increases again, because Van der Waals´ forces are influenced by the particle diameter, the distance between the particles and the number of contact points.

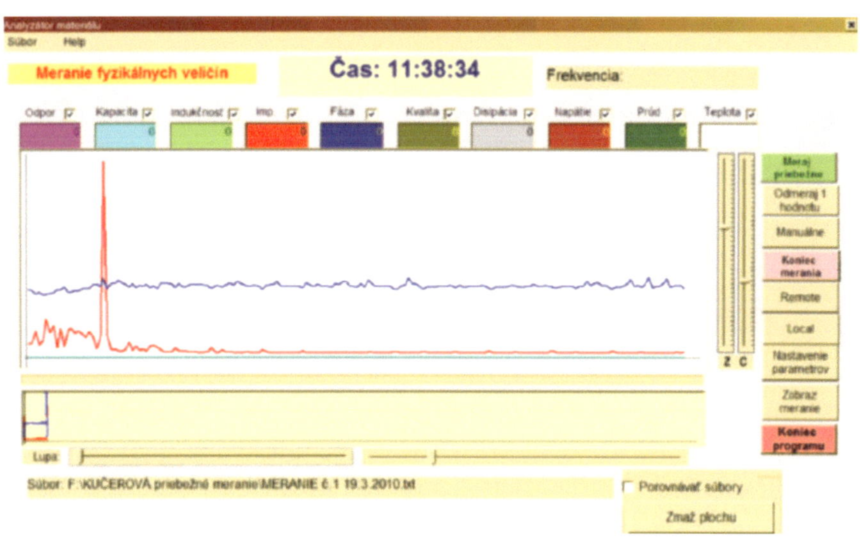

Fig. 23. The mixing behavior of caoutchouc

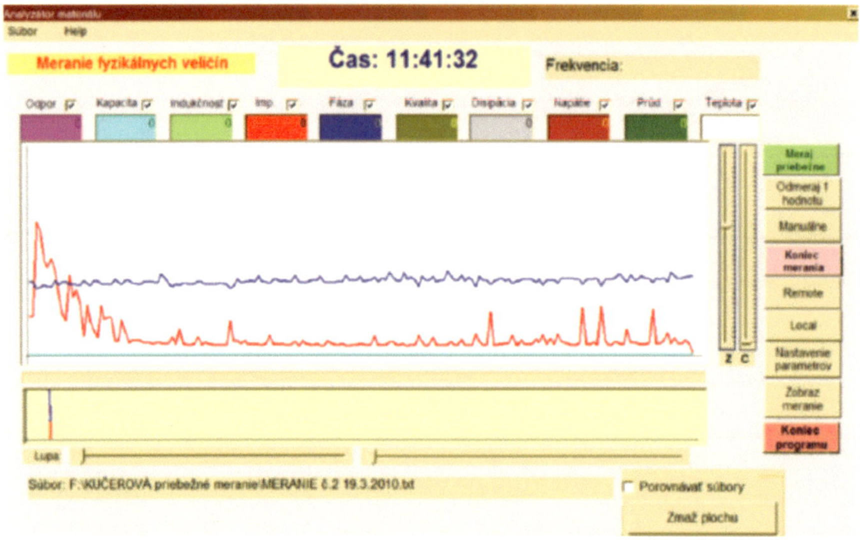

Fig. 24. The mixing behavior of caoutchouc and $BaSO_4$

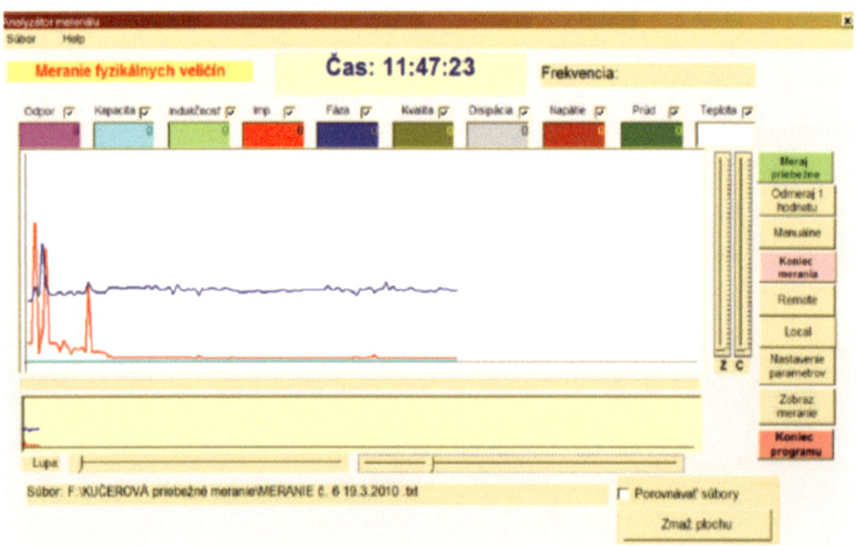

Fig. 25. The mixing behavior of caoutchouc and CNT

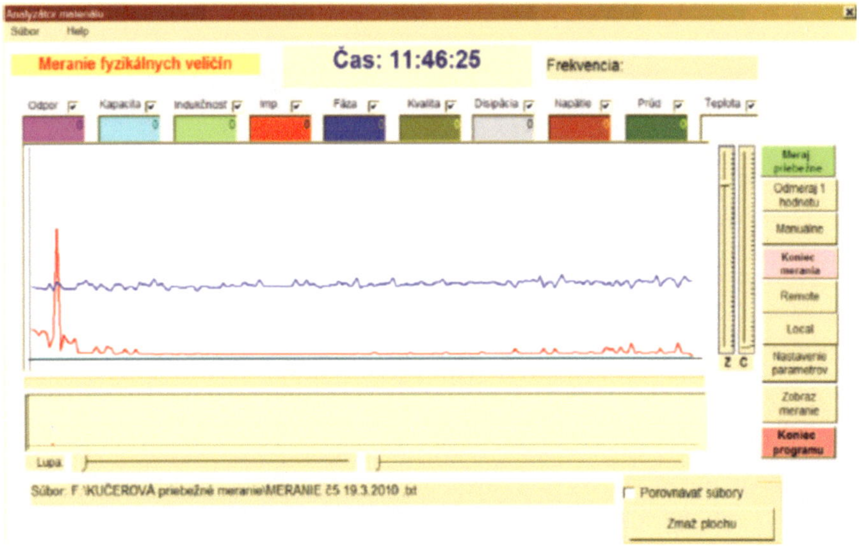

Fig. 26. The mixing behavior of caoutchouc and metal powder

In picture 27, we can see the AFM image of tested rubber sample, where three different areas are marked by arrows. The characteristic AFM curves for all three regions are in Fig. 28-30. The first rectangular breakdown (if we follow the right - left direction with the testing point), represents attraction of the tested material and point by Van der Waals forces. It is clearly seen that for all three tested regions, the values of this change are quite different, which supports the idea of sample inhomogeneity. This relative "macro" inhomogeneity (in the scale of micrometers) needs to be reflected also on electrical properties of the material, which we have presented above.

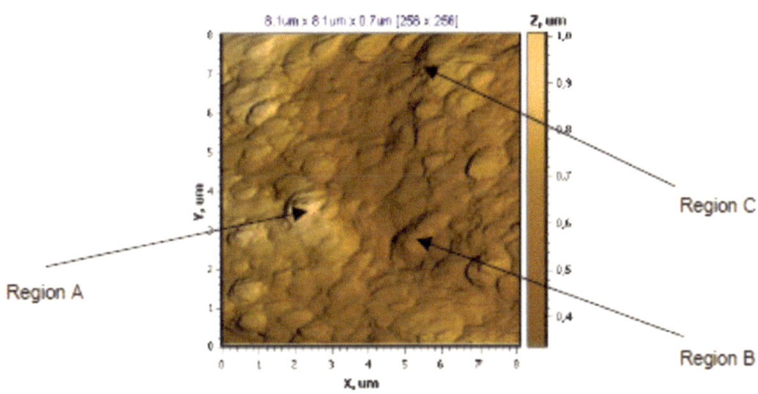

Fig. 27. AFM picture of sample with adding of carbon black fillers

On the other hand the activation energy calculated was found to be highly affected by both the type and concentration of the filler. The activation energies for the five different regions are calculated by the Arrhenious equation.

The large inhomogeneities in activation energies (E_a) of the samples with different filler composition are seen in table 2.

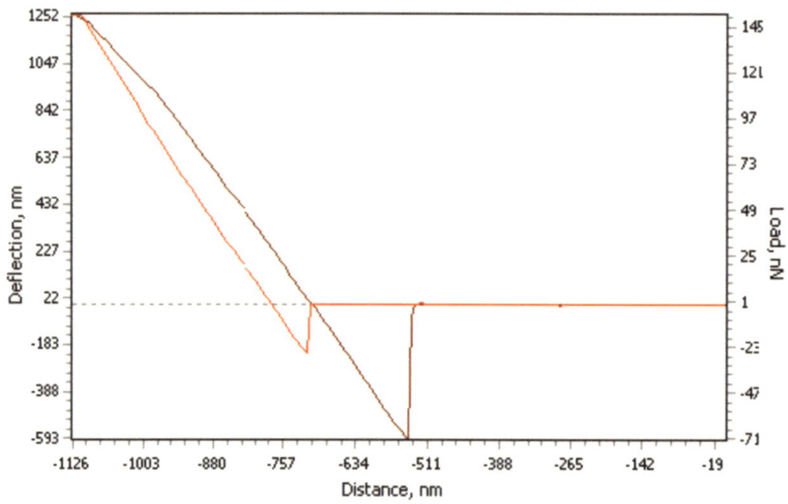

Fig. 28. Representation of Van der Waals forces for region A

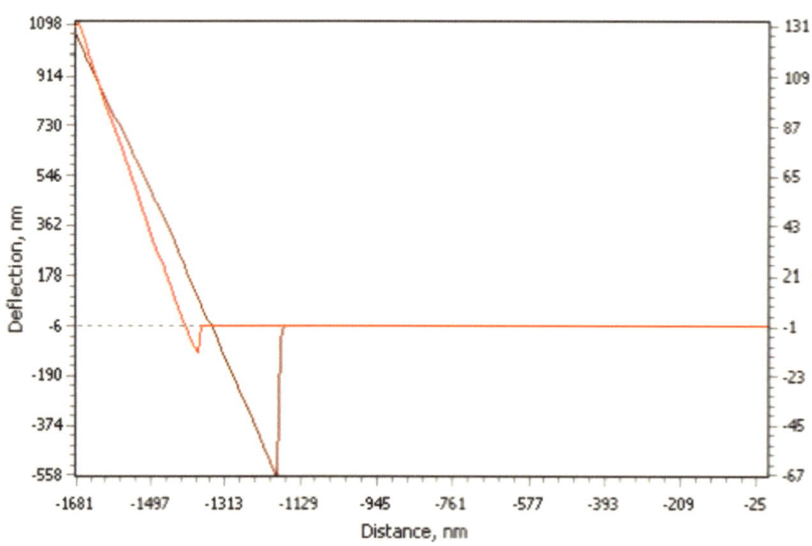

Fig. 29. Representation of Van der Waals forces for region B

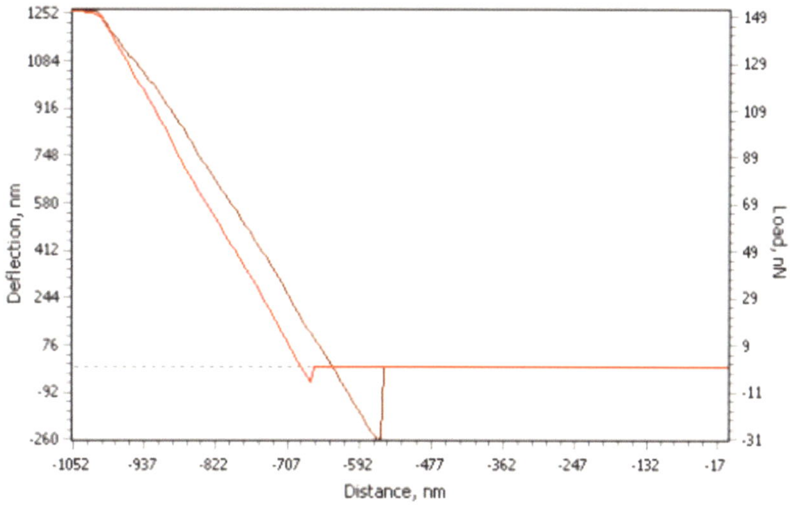

Fig. 30. Representation of Van der Waals forces for region C

Sample	Filler ratio: carbon black/SiO_2	E_a (eV)				
		Region 1	Region 2	Region 3	Region 4	Region 5
A1	6/30	3,01E+00	5,95E+00	-	-	-
A2	6/30	1,20E+00	4,46E+00	2,82E+00	-	-
A3	6/30	1,47E+00	1,95E+00	-	-	-
B1	20,5/13,5	1,89E+00	2,43E-01	-	-	-
B2	20,5/13,5	2,17E+00	5,15E+00	1,18E+00	-	-
B3	20,5/13,5	1,84E+00	4,52E+00	-	-	-
C1	35/0	8,62E-01	2,47E+00	1,13E+00	4,26E+00	2,25E-01
C2	35/0	1,44E+00	5,76E+00	3,34E+00	-	-
C3	35/0	5,58E+00	1,80E+00	2,94E+00	-	-

Table 2. Values of the activation energy calculated from the dc conductivity measurements.

The reason for such behavior is in different mechanism of electrical conductivity which is in narrow connection with chemical composition. The other effect which influences electrical properties is homogeneity of the samples. The mismatch of electrical parameters is caused by bad dispersion of silica and carbon black respectively. For example this is demonstrated by large distribution of Van der Waals forces presented in figures 28-30 for case of carbon black fillers influence.

From presented results it can be clearly seen that electrical measurements are very sensitive tools for evaluation of chemical composition and homogeneity of sample preparation which is a crucial problem in rubber technology.

As a conclusion we can say that the ideal mixing process can be achieved good homogeneity stage without the waste caoutchouc matrix straining; when is possibly the chain reduction and following degradation of vulcanizate properties.

By observing of electrical measurements behavior this method can supply to time specification when the mixture is perfectly homogeneous and can be removed from masticator. This exactly statement of rubber blend mixing can short the mixing time is following the energy consumption, productivity rising and the most important improving the vulcanizate properties.

At last, interesting dynamical mechanical dependency of standard mixture composition and CNT is described. The mixture composition is different by using of the triplex adding of softener owing to CNT filler mudding. The Payne effect of standard and CNT mixture is illustrated on figures 31-34 which include the influence before and after vulcanizing process as well as the loading or unloading state.

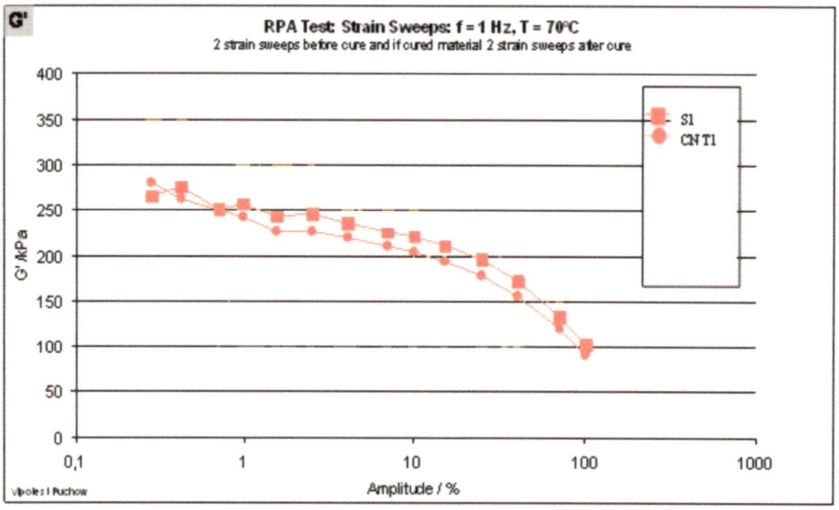

Fig. 31. Payne effect before vulcanizing in the unloading state

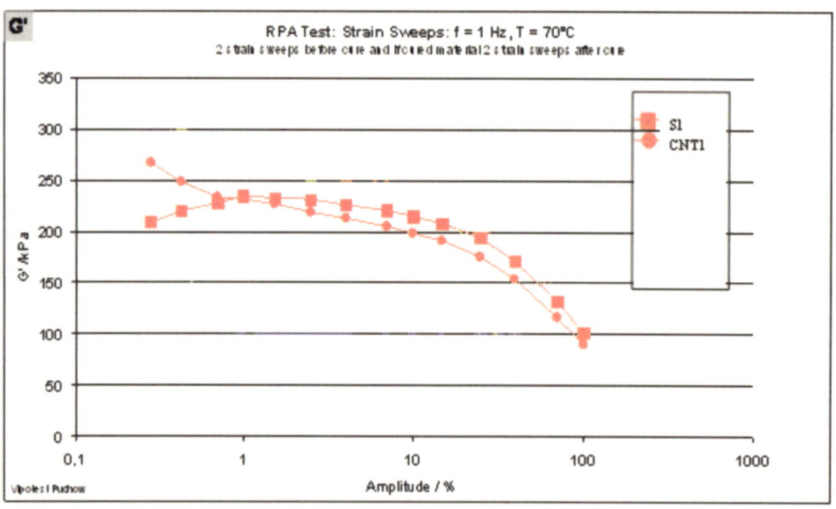

Fig. 32. Payne effect before in the loading state

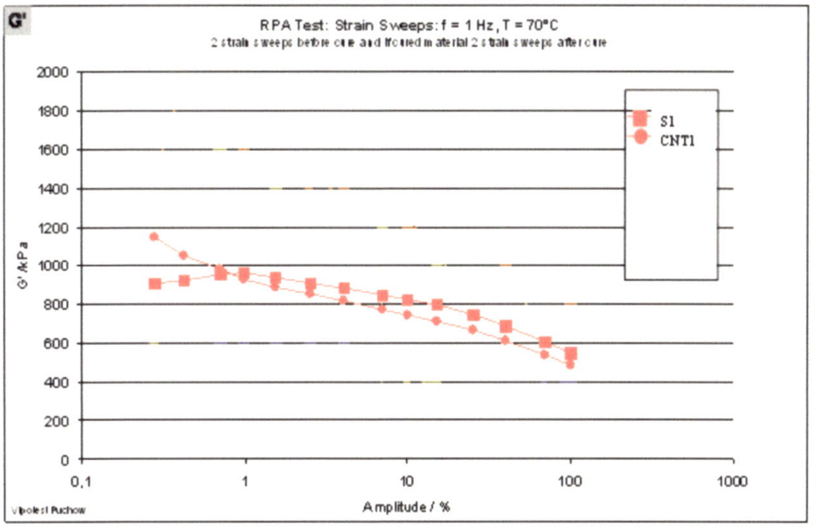

Fig. 33. Payne effect before vulcanizing in the unloading state

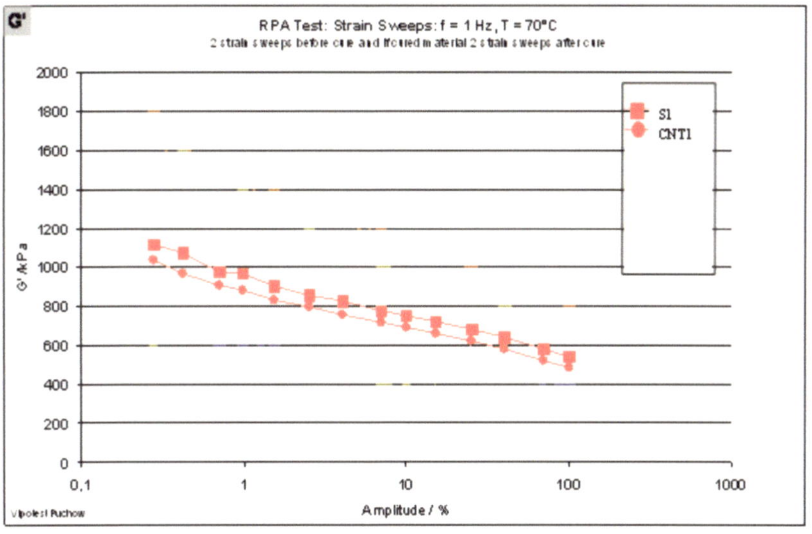

Fig. 34. Payne effect before in the loading state

Last but not least preparing mixture quality is influenced by the contamination of the entrance raw materials, which can negative affected the quality of product in the final phase. This problem relates by systematical preparation of mixing process - controlling of entrance raw materials, storage conditions consecutive manipulation and feeding into a masticator.

Polymer materials are requisite for industry at all in spite but relatively the young science. The first mention was finding by Columbus sailors the observed Americans aborigines plays with a flexible ball. This was made from dry liquid which flowing into a tree called Hevea (Latin) or "Cau-Uchu"(caoutchouc).

The primary caoutchouc was used for production of waterproof canvas and shoes, but there haven't properly behavior - the silky and adhesive behavior during the summer, the hardened and embrittlement manner during the winter time.

To Europe was a natural caoutchouc transported in 1736, the first processing was in 1791 for sails and mailbags. Later in 1839 Charles Goodyear approached a problem with improvement of quality of caoutchouc products by its impregnation in sulphur solutions.

Thomas Hancock found that the Goodyear's good does sulphur smell and then investigated the property changes by the caoutchouc warming in melting-down sulphur. Hancock and Goodyear found the vulcanization process, vulcanization means term of god Vulcan from a Greek mythology, process is characterized by heat influence and sulphur smelling.

The various processes of a good product and described problems by this production which we have presented above from the raw material to final product are summarized by the term of "Rubber goods process" in the figure 35.

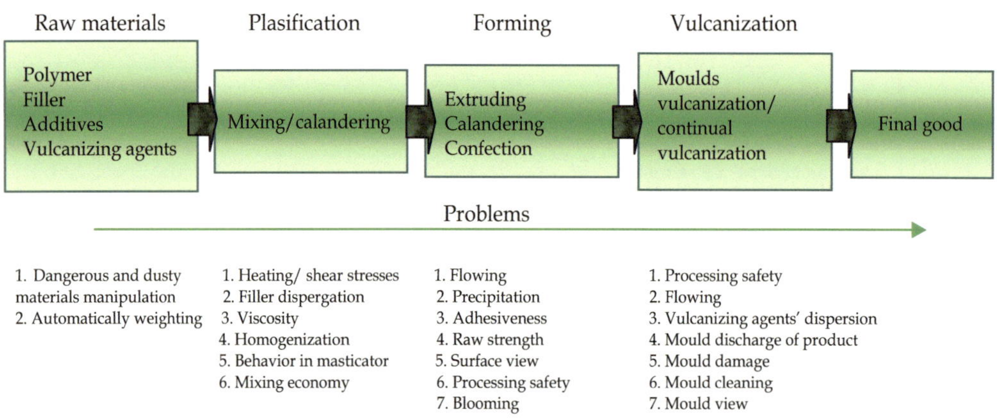

Fig. 35. The stages and processing problems occurred by rubber goods preparation

There is possibly to allege that the claims of rubber producers for a production rate together the quality of masticator are in connection into a whole process controlling increasing and are conditioned by achievement of the high quality of prepared mixture.

3. References

Čokinová A. (2003). *Diploma work,* Slovak Technical University of Bratislava, /in Slovak/.

Dierkes, W. (2006). *Economic mixing of silica-rubber compounds Interaction between the chemistry of the silica-silane reaction and the physics of mixing,* Holland, ISBN 90-365-2185-8.

Holzmüller., W.; Altenburg., K. (1966). *Fyzika polymérov,* Praha /in Czech/.

Hrenák T. (2009). *Bachelor work,* Alexander Dubček University, FPT Púchov, /in Slovak/.

Jahnátek, Ľ. et all. (2007). *Štruktúra a vlastnosti polymérov.* Materiálovotechnologická fakulta STU v Trnave, /in Slovak/.

Jakob B. (2002). Conductivity measurement for rubber compound qualification, *Polymeric Materials,* Halle (Saale), Germany 2002.

Jurčiová J. et all. (2008). The alternating current circuit parameters measurement as a tool for the monitoring of carbon black dispersion in a rubber blend, 7th Youth Symposium on Experimental Solid Mechanics -YSESM, Poland 2008.

Knapec P. (2009). *Diploma work,* Tomas Bata University in Zlín, /in Slovak/.

Koštial P. et all. (2006). The combined study of fillers dispersion in rubber blends, 5th Youth Symposium on Experimental Solid Mechanics - YSESM, Slovakia 2006.

Kučerová J. (2006). *Dissertation thesis,* Alexander Dubček University, FPT Púchov,/in Slovak/.

Kučerová J., Bakošová D., Mičúch M. (2006). Electric conductivity measurement of inhomogeneities of rubber blends, *18th Slovak Rubber Conference,* ISBN 80 - 969189-6-6, Púchov, May 2006.

Kučerová L. (2010). *Diploma work,* Alexander Dubček University, FPT Púchov,/in Slovak/.

Le H. H., Ilisch S. & Radusch H. - J. (2006). Correlation between morphology and properties of carbon black filled rubber blends, *18th Slovak Rubber Conference,* ISBN 80 - 969189-6-6, Púchov, May 2006.

Le H. H., Ilisch S., Jakob B., Radusch H. - J. (2004). Online characterization of the effect of mixing parameters on carbon black dispersion in rubber compounds using electrical conductivity, *Rubber chemistry and technology,* vol. 77, no 1, pp. 147-160.

Le H. H., Ilisch S., Prodanova I., RADUSCH H. - J. (2004). Russdispersion in Kautschuk - mischungen in Anwesenheit von Weichmachrölen. *Kautschuk Gummi Kunststoffe KGK,* 57. Jahrgang, Nr. 7-8.

Liptáková T. et all. (2009). *Plastics - technical materials,* Hypertext textbook, STU Bratislava.

Materials of Technical University of Tampere www.tut.fi/plastics/tyreschool

Materials of Technical University of Tampere www.tut.fi/plastics/vert

Nicolas P. Cheremisinoff. (1987). Polymer *mixing and extrusion technology.* Marcel dekker, inc./New york - Basel.

Prekop, Š. et all. (1998). *Rubber technology I,* University in Žilina, /in Slovak/.

Prekop, Š. et all (2003). *Rubber technology II,* GC TECH, Trenčín,/in Slovak/.

Presentation of ATEAM
http://www.ateam.zcu.cz/download/pryz.pdf, /in Czech/.

Presentation of CABOT Company
http://www.cabot-corp.com/wcm/download/en-us/sb/INTERWIRE%2020011.pdf
Research project FPT-2/2006 - 3rd part, /in Slovak/.
Šišáková J. et all. (2010). The high density PE and its modifications, XI International conference on machine modeling and simulation - MMS 2010, Poland 2010.

3

New Technologies and Devices to Increase Structures' Safety to Dynamic Actions

Viorel Serban, Madalina Zamfir, Marian Androne and George Ciocan
Subsidiary of Technology and Engineering for Nuclear Projects
Romania

1. Introduction

A dynamic action may transfer to a structure a quantity of energy equal or smaller than the excitation energy associated to a vibration cycle, function of the harmonization or de-harmonization of the structure eigen movement with the dynamic action kinetics. The transferred energy may build-up as kinetic and potential energy in the structure, which in point of its dynamic behavior can be over-harmonized, under-harmonized or in resonance with the excitation.

In order to see how to limit the energy built-up in a structure and how to reduce it, the dynamic response of the structure in the time and frequency ranges with an oscillating system with a single degree of freedom subjected to a harmonic dynamic action, should be analyzed (Fig. 1.1.).

The oscillating system of a 'm' mass, 'k' stiffness and 'c' damping is subjected to a transportation oscillating movement $u_s(t)$ of a harmonic type with period T_S.

Analyzing the diagram one it results that the reduction of the seismic response of a structure may be obtained by increasing the structure damping capacity. For example, the response of energy built-up in the structure on dynamic actions is reduced by about 4 times if the structure damping is increased from 5% to about 10% and of about 16 times if the structure damping is increased from 5% to 20%.

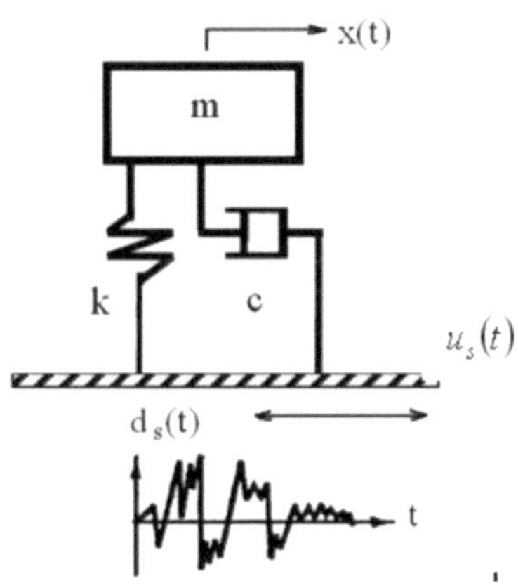

Fig. 1.1. Oscillating system with one degree of freedom driven by excitation $u_s(t)$

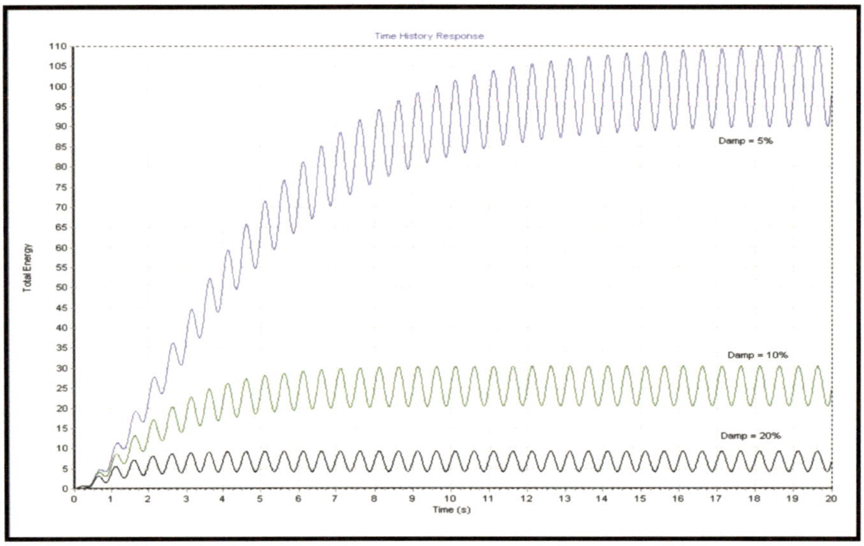

Fig. 1.2. The total energy built-up in the oscillating system in resonance regime with excitation for the relative damping β=5%, 10% and 20%.

In order to point-out the great differences that may occur in the behavior of some structures when such structures are affected by dynamic actions of the same intensity (amplitude) but with different spectral components, an analysis of the structure response in the frequency range is conducted to offer a better qualitative analysis of the amplification phenomena as to the in-time analysis. Such diagrams also allow a substantiation of the innovative solutions to strengthen the structures in order to withstand dynamic actions.

The variation with the eigen period of the kinetic energy, E_c, and potential energy, W_p amplitude specific to the oscillating system mass, m, for the amplitude of the source energy E_s (a harmonic component, as well) are given in Figs. 1.3. - 1.4., for β = 5% and β = 20%, respectively.

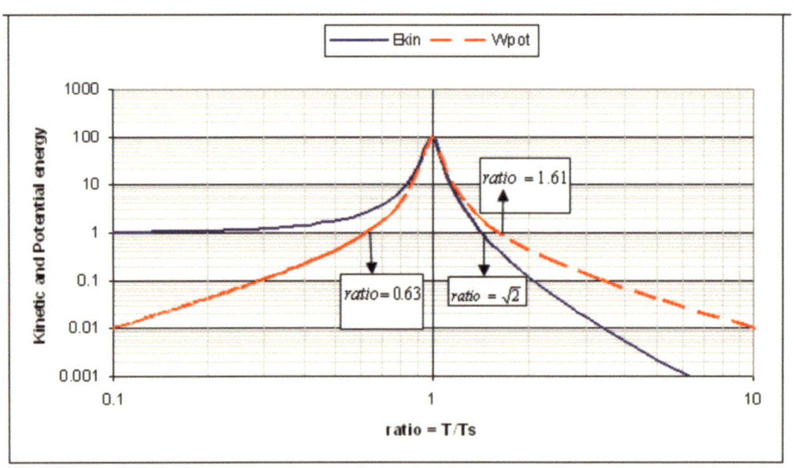

Fig. 1.3. The amplitude of the kinetic and potential energy of the oscillating system as to the excitation amplitude function of T/T_s for the relative damping β = 5%.

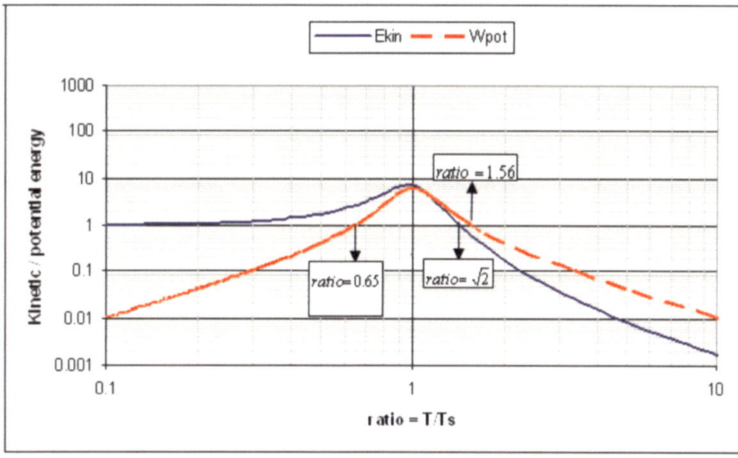

Fig. 1.4. The amplitude of the kinetic and potential energy of the oscillating system as to the excitation amplitude function of T/T_s for the relative damping $\beta = 20\%$.

Based on these diagrams one may determine the requirement that an oscillating system responds to dynamic loads of smaller or equal amplitudes than with static loads of the same intensity (without amplification). In point of energy, the kinetic and potential energy amplitude built-up in the oscillating system should be smaller or equal to the excitation energy amplitude. The kinetic energy is smaller than the source energy (irrespective of the oscillating system damping) if the eigen period of the oscillating system, T, is greater by 41% than the dominant repetition period of the excitation, T_s.

The requirement that the maximum potential energy of the oscillating system be smaller or equal to the maximum energy of the source is dependent on the oscillating system damping and such a thing can be obtained for the oscillating system vibration periods are greater than 1.41 T_s. For $\beta = 30\%$ the result is $T > 1.88T_s$. So such a requirement is satisfied if the eigen dominant vibration period of the structure T is 2 times greater than the dominant period of the dynamic action Ts. Considering the evaluations errors in the computation programs, the presence of several periodic components in the seismic response of structures as well as the errors in the input data or modeling, it is necessary to impose the design requirement $T > 3T_s$, a requirement that is satisfying the design standards related to the seismic isolation design.

Analyzing the diagrams it results that the most efficient solution to reduce the energy built up in a structure is to make a connection having large elasticity and small damping between the structure and foundation so that its eigen period should be greater by at least 3 times the repetition period in the dynamic action.

Figure 1.5 and 1.6 shows the variation of the dissipation power amplitude $|P_d|$ in the oscillating system and the variation of the total power amplitude $|P_e|$ transferred from the excitation to the oscillating system, related to the system mass unit and source power $|P_s|$ function of the ratio between the oscillation system period and the excitation period for a fraction of critical damping of 5% and 20%.

Analyzing the diagrams it results that the amplitude of the total power transferred from the excitation source to the oscillation system and its built up is actually equal with the excitation power amplitude for the oscillation system vibration periods smaller than 0.4 T_s. The amplitude of the power transferred and dissipated is much increased in the vicinity of resonance and the dissipated power (which usually is smaller than the transferred power) becomes actually equal with the power transferred from the excitation to the oscillating system on a resonance. For periods of the oscillating system greater than the excitation dominant period, the power transferred to the oscillating system is decreasing with the increase of the ratio between the two periods.

The dissipated power is decreasing faster in the vicinity of resonance and, for periods $T >> 2T_s$, the speed in decreasing the dissipated power is getting reduced. For high critical damping (e.g. $\beta = 30$ %) the power transferred from the excitation is obvious as dissipated power for $T > T_s$.

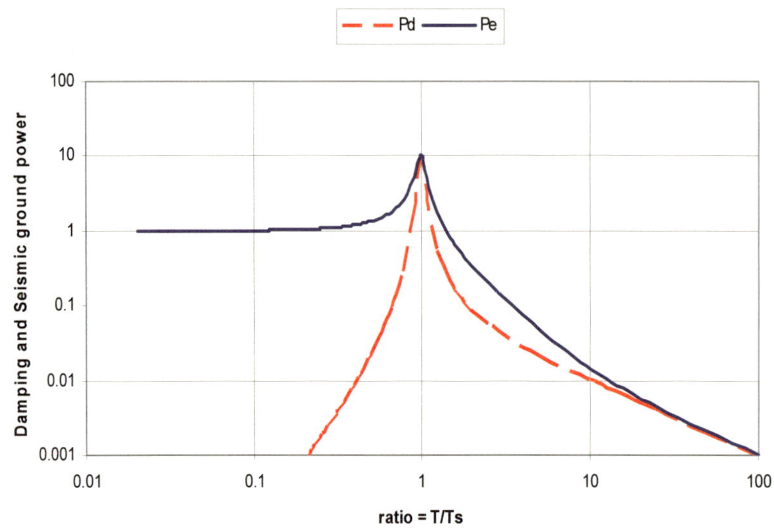

Fig. 1.5. Amplitude of power transferred from the excitation to the oscillating system and the power dissipated in the system, function of T/T_s and relative damping $\beta = 5\%$

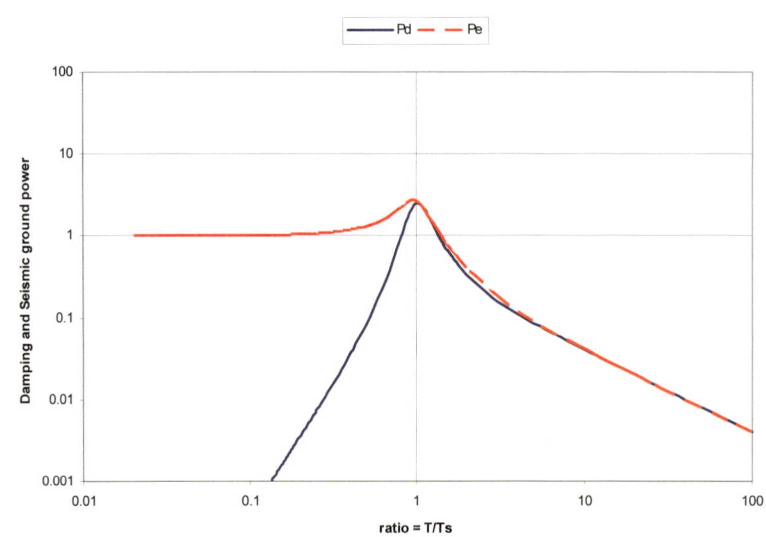

Fig. 1.6. Amplitude of power transferred from the excitation to the oscillating system and the power dissipated in the system, function of T/T_s and relative damping $\beta = 20\%$.

Analyzing the diagrams in Figs. 1.5.-1.6. it results that the increase of the structure damping capacity with flexible systems evidences positive effects in case of resonance and negative effects in case of the structure isolation because the power transferred to structure is accomplished by damping forces.
For that reason, the structure damping is limited to 30% of the critical damping in the design codes (Romanian Seismic Design Code 2006). For isolation system it is recommendable that the damping should be smallest possible. With this case, the damping shows a positive effect only for the relative displacements between the isolated supra-structure and foundation.

2. Reduction of structure dynamic response employing SERB-SITON method

Considering the large diversity of structures, the analysis of solutions to reduce dynamic response at different excitations is conducted separately for buildings, equipment and pipe-network. In function of the type of structure and/or the kinetic characteristics of the dynamic action, the reduction of the dynamic response may be obtained in several ways.
Bearing in mind that the most important dynamic action that may affect a building is the seismic action, the alternatives to accomplish a small seismic response are presented below.

2.1 Solutions to reduce the seismic response of buildings

2.1.1 Alternative 1

Increase of building damping capacity while also limiting the relative distortions in the linear range of behavior
The solution consists in the control, limitation and damping of level relative distortions by the installation of elastic devices with damping called "telescopic devices", in the structure and/or between the structure segments (Figs. 2.1. - 2.3.).
The telescopic devices are usually installed at the building lower levels in central or excentral braces or around the nodes that make up a symmetrical network of braced panels at each level of the building and which are continued vertically with possible reductions symmetrically arranged.
SERB-SITON telescopic devices are capable of overtaking forces ranging between 1000÷5000kN and to limit the level relative distortions to values usually ranging between ± 10 mm to ±20 mm or other values imposed by the building, (Serban, 2005).
The devices can be fabricated in a large variety of typo-dimensions, Figs. 2.4.-2.5.
The force-distortion characteristic of the devices is nonlinear type, with strengthening in order to limit the structure distortion and their damping may be accomplished for preset values ranging between 30% and 80% of the elastic energy associated to one cycle.
Force-deformation characteristics may actually be accomplished as per any desired shape, Figs. 2.6. - 2.7.
For building rehabilitation the columns, beams or nodes of the braced panels are strengthened by lining with metal profiles tightened to the reinforced concrete structure and the braces by means of SERB-SITON telescopic devices are arranged as per 1 of the 3 alternatives presented in Figs. 2.1.-2.3.
This alternative may be applied to the construction of new buildings or to the rehabilitation of buildings in a nuclear or classic unit without interrupting the operation.
In case of building strengthening, 5% and 10% of the useful surfaces on a building level is affected by the strengthening solution for a period of 30 and 45 days, the rest of the building being useful without restrictions.

By the application of this alternative the important advantages, compared with the classic strengthening solutions are: necessary materials: *1/10÷1/20;* resulted wasted: *1/10÷1/20;* strengthening duration: *1/2÷1/4;* surfaces of site temporary organization: *1/10÷1/50* of the surface required employing the classic strengthening solution; price: *0.7÷0.9.*
Buildings constructed or strengthened by use of SERB SITON method provide a behavior of the building structure in the elastic range during an earthquake.
The control, limitation and damping of the building seismic response is provided by the telescopic devices inserted in the building structure rather than the building damaged structure with plastic hinges as the case with classic solutions.

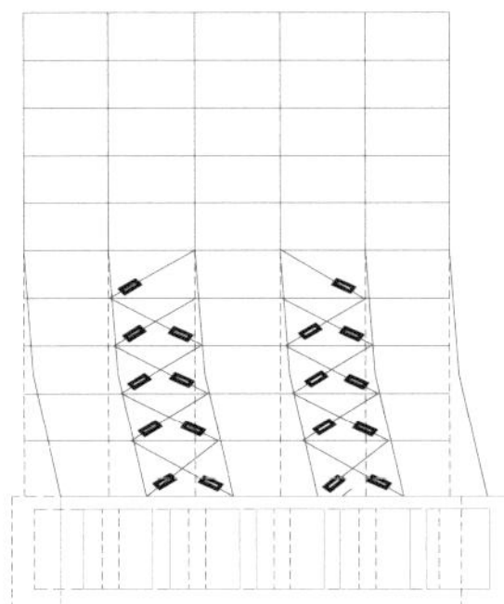

Fig. 2.1. Alternative 1 - central braces

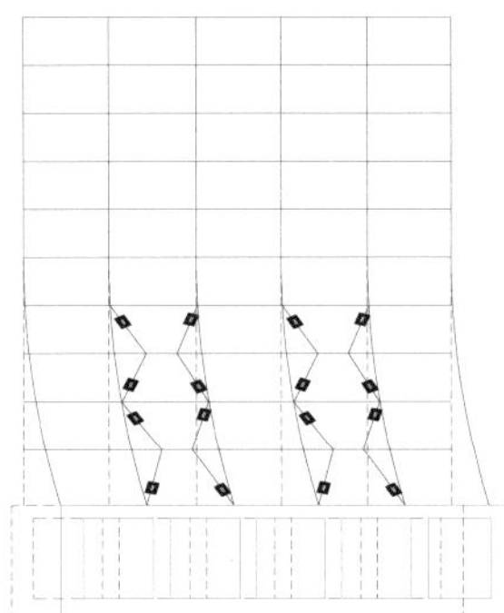

Fig. 2.2. Alternative - eccentric braces

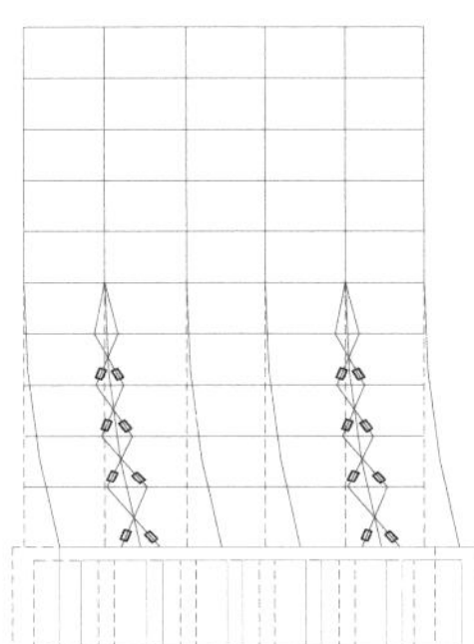

Fig. 2.3. Alternative 2 - strengthening around the nodes

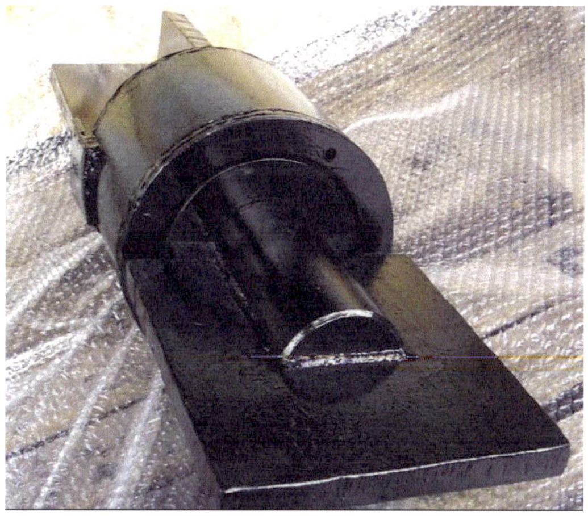

Fig. 2.4. SERB SITON V1 telescopic devices

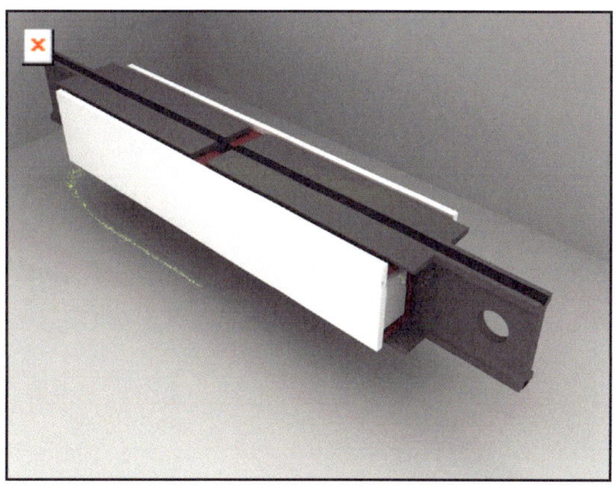

Fig. 2.5. SERB SITON V2 telescopic devices

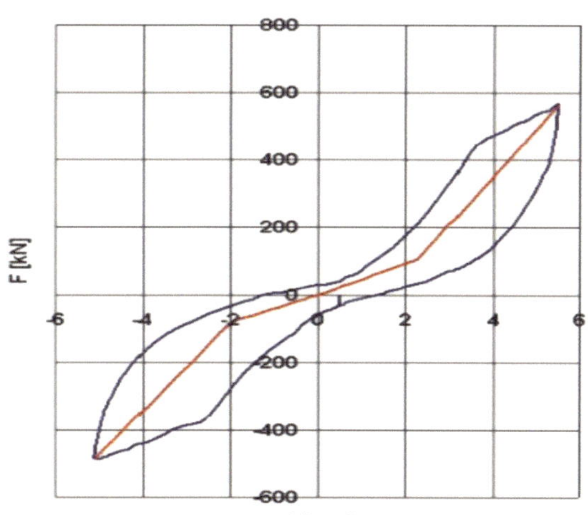

Fig. 2.6. Force-distortion diagram for V1

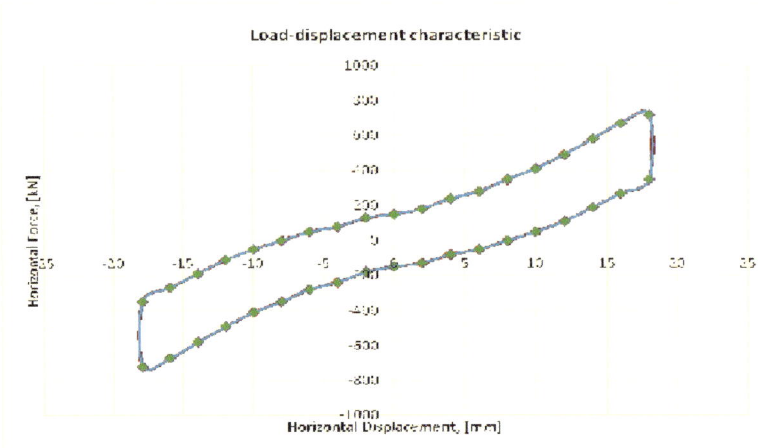

Fig. 2.7. SERB-SITON V2 isolation device

Fig. 2.8. NAVROM GALATI - ward B - strengthened, extended and rehabilitated.

2.1.2 Alternative 2 - building seismic isolation

The most efficient solution to reduce seismic loads on buildings is "to cut off the transfer of the seismic action from the ground to the building by isolating the building. With this case, the loads on the structures may be reduced tens of times function of the dynamic characteristic of the isolation systems, compared with the kinetic characteristics of the dynamic action.

For the isolation system to be efficient it is necessary that the system should satisfy the following requirements:

- the eigen vibration period T_i of the supra-structure - isolation device assembly should be about three times greater than the eigen vibration period T_r of the supra-structure embedded at the isolation level and respectively three times greater than the corner period T_c in the ground response spectrum;

- the isolation devices should provide the overtaking of permanent loads with no distortion on vertical, over which tensile and compression dynamic loads overlap also providing a sufficient displacement on horizontal plane;

SERB SITON isolation system is a non-linear system which can provide very large vibration periods, with large capacity of reverting to initial position and rather small relative damping lest the seismic action should be transferred from the ground to the building via the damping forces.

Limitation of site displacements to reset values as well as the revert to the initial balanced position are provided by non-linear elements with the stiffness increasing with the displacement increase, making part of the isolation device.

SERB-SITON type isolation system is made by mechanical devices which do not include safety components that may be negatively affected by ageing and sometimes radiations, humidity or temperature. They are capable to overtake permanent loads, up to 5000KN overwhich tensile and compression dynamic loads of ±1500 KN are overloaded and which are also allowing translations in horizontal plane, usually up to ±300 mm. Figures 2.9. - 2.11 show 3 alternatives of isolation devices developed by SITON.

SERB-SITON LP 800 x 800 capsulated isolation device with sliding on plane surfaces is made up of 2 identical boxes connected between them by a central body with a possibility of relative sliding between themselves, Fig. 2.9.

Sliding between the central body and the box is provided by 5 sliding surfaces between teflon plates confined in steel rings and stainless steel plates which operate in parallel and in series. On vertical direction, the device is stiff and allows compression loads up to 3000 kN over which tensile and compression dynamic loads are overlapped and which, for size reasons, have been limited to 1000 kN.

SERB SITON RP 1000 X 1000 rolling or sliding capsulated device is made of 2 identical peripheral boxes which are coupled to a central box. These boxes are connected between them by axial ball bearings or metal profiles Teflon coated which provide the overtanking of a permanent load up to 5000 kN over which tensile and compression loads of ± 1500 kN may overlap. Displacement on any direction in horizontal plane is ± 300 mm, Fig. 2.10.

Isolation device of swinging (pendular) pillar type with controlled elasticity and damping top and bearing. Swinging pillars, usually made of reinforced concrete, have a top and a bearing made of mechanical devices with controlled elasticity and damping for all the degrees of freedom. The translation of the isolated supra-structure in the horizontal plane is accomplished by the controlled and dumped rotation of the mechanical devices on top and bearing, Fig. 2.11. Each device is provided with 4 non-linear elastic stops with strengthening in order to elastically limit the swinging vibrations to preset values. As an additional safety measure for the swinging pillars, the isolation system may be provided with a brace system with tilted telescopic devices capable to provide additional control and safety.

Strengthening of a building by isolation employing swinging pillars shows the advantage that it can be applied by substituting the existing pillars with swinging pillars step by step for a level of frame structure buildings.

Stability of the building is given by the stiffness of the hinges on top and pillar bearing as well as by the isolated supra-structure weight due to the fact that rotation on top and bearing is accomplished around some rigid elements with preset geometry arranged in elastic blade packages which can provide rotation with a slight rise of supra-structure. Reverting to undistorted position is provided both by the elastic system bearings and top and the supra-structure weight which operate as a stability element through the stability moment.

Fig. 2.9. SERB - SITON LP 800 x 800 capsulated isolation device with sliding on plane surfaces

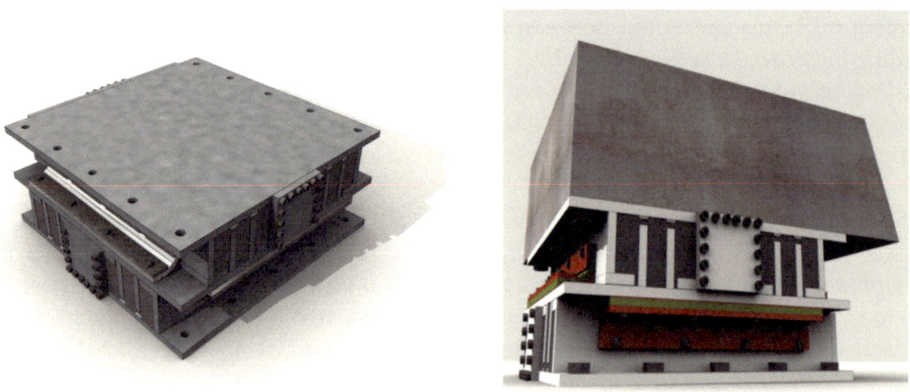

Fig. 2.10. SERB SITON RP 1000 x 1000 rolling or sliding capsulated device

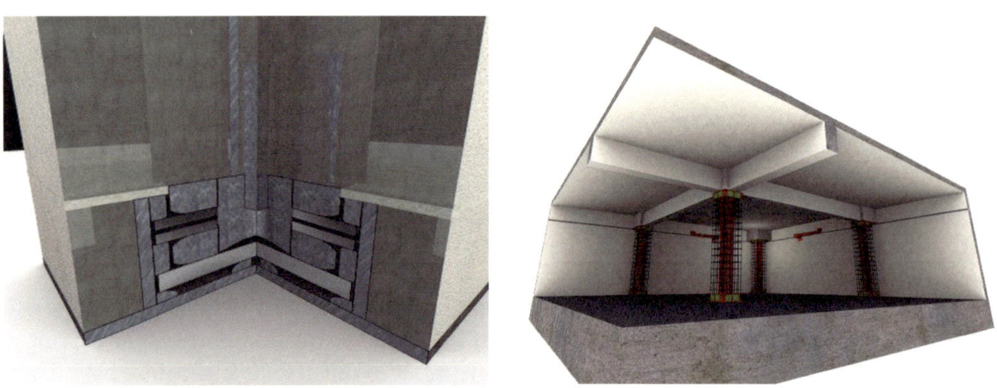

Fig. 2.11. Isolation device of swinging (pendular) pillar type with controlled elasticity and damping top and bearing

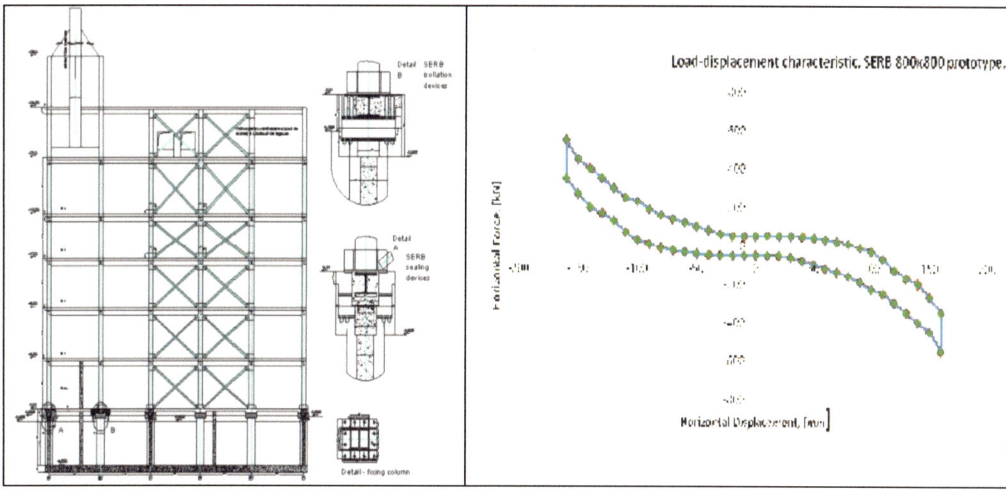

Fig. 2.12. Case study and horizontal hysteresis diagram of the insulator with slides

In order to evaluate the insulating system's efficiency, a case study was conducted for the NPP Cernavoda detritiation building. Fig. 2.12 presents the analyzed insulating system and the horizontal hysteresis diagram of the insulating system.

An analysis was conducted for a site specific design accelerating diagram and for a sinusoidal accelerating diagram with a maximum ground acceleration on two perpendicular directions on a horizontal plan of 0.3 g, values of 50% higher than the design basis acceleration for DBE.

Following the numeric analysis performed for the rolling insulators on plain surfaces, a seismic acceleration was obtained which operates over the insulating system of 0.3g, the response in acceleration is of 0.02g, according to Fig. 2.13. and the response in relative movements between insulated infrastructure and supra-structure is of maximum 10mm, as in Fig. 2.14.

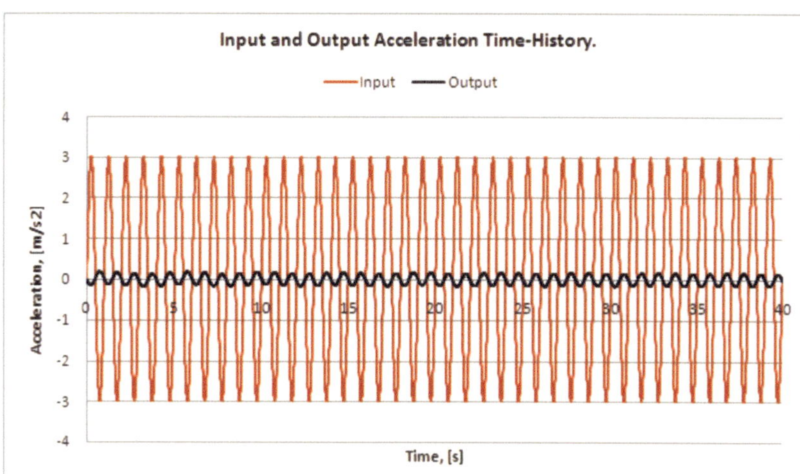

Fig. 2.13. Rolling insulators over plain surfaces. The ground acceleration = 0.3g; Acceleration of insulated supra-structure = 0.02g, for a sinusoidal acceleration diagram

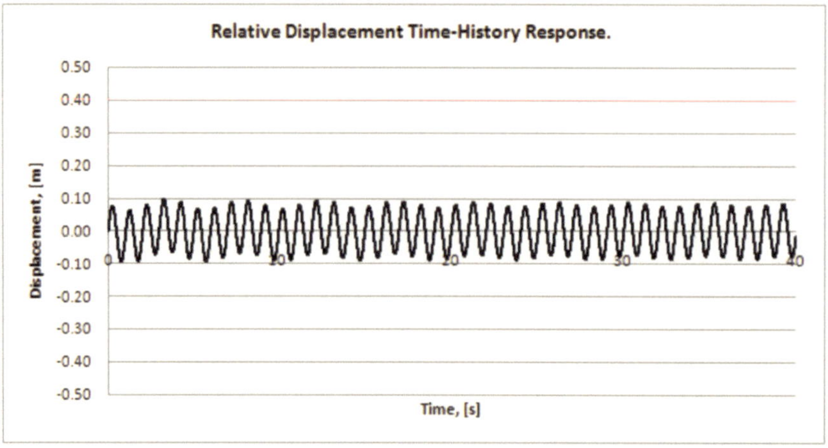

Fig. 2.14. Rolling insulators over plain surfaces. Relative movement (between insulated supra-structure and ground imbedded infrastructure) = 0.1m of the insulating system for a sinusoidal acceleration diagram

2.2 Reduction of the dynamic report of equipments and piping network using SERB-SITON devices

For equipments and piping network, the seismic loads, shocks and vibrations have a larger percentage in load groups than in constructions, (Panait et Serban, 2006). Moreover, in many cases, the qualifying solutions for equipments and piping network for dynamic charge-discharge are in contradiction with their qualifying solutions for charges resulting from thermal expansions, their application being possible by imposing the extra-requirements for the devices used.

Usually, the devices for the reduction of the dynamic response, which for equipment and ducts are called bolsters, must also allow movements in thermal expansion (which besides the takeover and damping of dynamic actions are large because of temperature variations).

SERB-SITON bolsters may absorb the elasticity with pre-settled rigidity, permanent charges (usually from self weight), allow movement in thermal expansions on a direction or in a plan with elastic reactions or constant dynamic charges that they absorb. Depending on the percentage of the dynamic charge, in the group of charges and its cinematic characteristics, a few bolster (devices) options are presented below.

Alternative 1 – *Bolster for heavy equipment support, which undergoes shocks and vibrations*

Usually, heavy equipments are placed on bolsters with horizontal action limiting device. In Fig. 2.15. the cassette bolster type for large loads is presented and in Fig. 2.16. the individual bolster type for large capacities. These bolsters have a non-linear and asymmetric conduct with large absorption. Insulation of heavy equipment like mould hammers lead to the obtaining of a 98 % insulation.

Alternative 2 – For the support of medium or small weight equipment, the bolts in Figs. 2.17. - 2.20. were developed which can also take over the movement in thermal charges of the ducts connected to equipment. The rigidity, the absorption and the movement allowed in any degree of freedom may be reached in the desired values on bolster types. The tunings for equipment horizontality or co-axiality of trees can be done by pre-restraining of bolsters.

Fig. 2.15. Alternative 1 for bolster and cassette equipment

Fig. 2.16. Alternative 2 bolster for heavy equipment

Fig. 2.17. Alternative 1 - isolator for cabinet

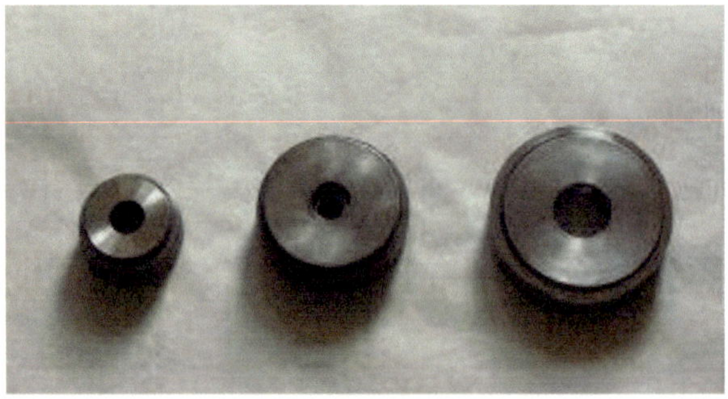

Fig. 2.18. Alternative 1 - light and medium weight equipment

Fig. 2.19. Alternative 2 - medium equipment bolster

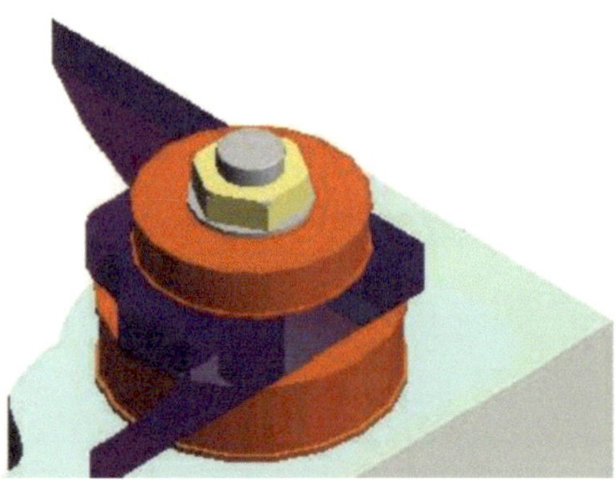

Fig. 2.20. Alternative 1 - catch of the equipment sole

Alternative 3 - For the catch of the pipes and columns to which vibration and thermal expansion movements are imposed in pre-established values for certain directions, the bolsters in Figs. 2.21. - 2.23. are developed. The thermal expansion movement or seismic movements of catch points can be made with constant or elastic reactive forces of pre-established values.

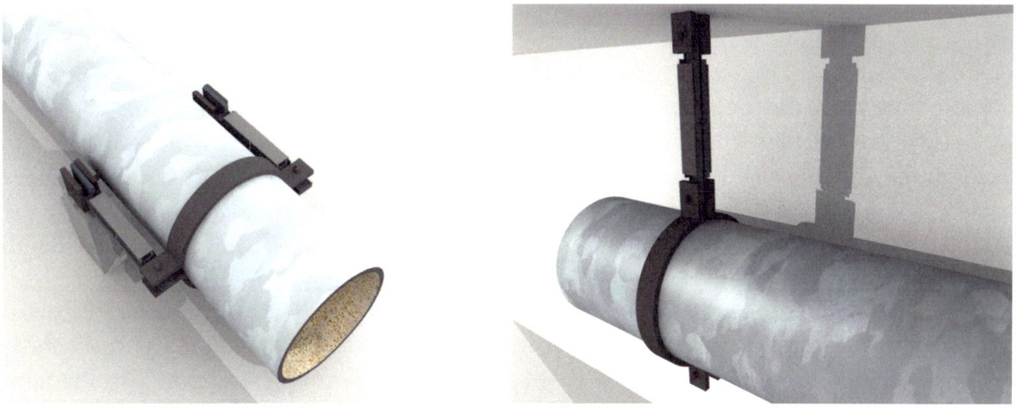

Fig. 2.21. Alternative 1, 2 - Bolster equipment

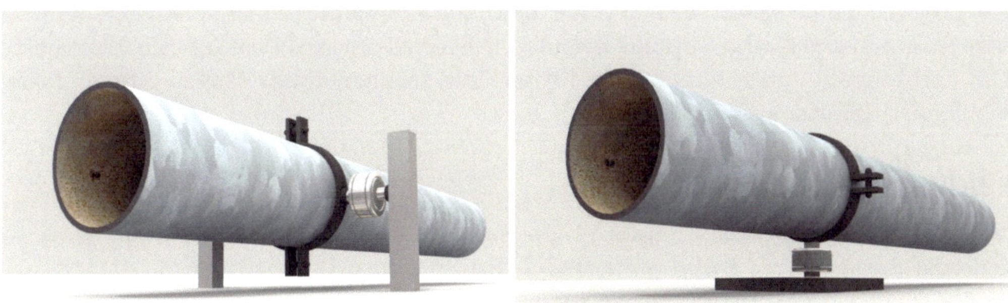

Fig. 2.22. Alternative 3,4 - Bolster equipment

Fig. 2.22. Alternative 5,6 - Bolster equipment

3. Dynamic analysis of structures subjected to Time-History acceleration

The structure in time behavior at dynamic actions, controlled by non-linear mechanical devices with large absorption and to which the shaping force is hystheresis type with consolidation (not degrading), can be done analytically through specific mathematical models, which can take into consideration the increase of rigidity along with an increase in deforming. The present commercial computing programs allow the use of non-linear materials and devices behavior, with only degrading hysthretic characteristics because the mathematical methods used for integration do not allow the use of hystheresis curves with strengthening because these lead to numeric instability.

The new computing method of structures is simple, containing a limited number of masses and elements of rigidity aligned on a vertical line, in which are included the hystheresis curves with rigidity increase, once with an increase in deforming, associated to the non-linear mechanical and absorption devices. Considering the behavior in materials elasticity and the fact that only devices have non-linear behavior, these models are representative for structures.

The proposed mathematical method is the direct and simultaneous integration of non-linear differential equations, which include the uni-dimensional model of structure simultaneously with the hysteresis curves with the rigidity increase, mathematically shaped through Bouc-Wen models, (Sireteanu et al., 2008).

4. BOUC-WEN model

The Bouc-Wen model, widely used in structural and mechanical engineering, gives an analytical description of a smooth hysteretic behavior. It was introduced by Bouc (Bouc, 1967) and extended by Wen (Wen, 1976), who demonstrated its versatility by producing a variety of hysteretic characteristics. The hysteretic behavior of materials, structural elements or vibration isolators is treated in a unified manner by a single nonlinear differential equation with no need to distinguish different phases of the applied loading pattern. In practice, the Bouc-Wen model is mostly used within the following inverse problem approach: given a set of experimental input–output data, how to adjust the Bouc-Wen model parameters so that the output of the model matches the experimental data. Once an identification method has been applied to tune the Bouc-Wen model parameters, the resulting model is considered as a "good" approximation of the true hysteresis when the error between the experimental data and the output of the model is small enough from practical point of view. Usually, the experimental data are obtained by imposing cyclic relative motions between the mounting ends on the testing rig of a sample material, structural element or vibration isolator and by recording the evolution of the developed force versus the imposed displacement. Once the hysteresis model was identified for a specific input, it should be validated for different types of inputs that can be applied on the testing rig, such as to simulate as close as possible the expected real inputs. Then this model can be used to study the dynamic behavior of different systems containing the tested structural elements or devices under different excitations.

Various methods where developed to identify the model parameters from the experimental data of periodic vibration tests. A frequency domain method was employed to model the hysteretic behavior of wire-cable isolators , iterative procedures were proposed for the parametric identification of a smoothed hysteretic model with slip (Li et al., 2004), of a modified Bouc-Wen model to portray the dynamic behavior of magnetorhological dampers, etc. The Genetic Algorithms were widely used for curve fitting the Bouc-Wen model to experimentally obtained hysteresis loops for composite materials (Horning), nonlinear degrading structures (Ajavakom, 2007) or magnetorheological fluid dampers (Giuclea, 2004 et Kwok, 2007).

In the present work, our primary focus is to give closed analytical relationships to determine the parameters of the Bouc-Wen model such as the predicted hysteresis curves and the experimental loops to have same absolute values of the maximum forces and same coordinates of the loop-axes crossing points. The derived equations can be used for fitting the Bouc-Wen model to both symmetric and asymmetric experimental loops. The asymmetry of experimental hysteresis curves is due to the asymmetry of the mechanical properties of the tested element, of the imposed cyclic motion, or of both factors. In most cases, the identified model output turns out to be a "good" approximation of experimental output. When this approximation is not satisfactory, the obtained parameter values can be used as initial values within an iterative algorithm to improve the model accuracy.

4.1 Fitting the Bouc-Wen model to symmetric experimental hysteresis loops

Suppose the experimental hysteretic characteristic is a asymmetric loop, $-F_{\mathrm{m}} \le F(x) \le F_{\mathrm{m}}$, obtained for a periodic motion $-x_{\mathrm{m}} \le x(t) \le x_{\mathrm{m}}$, imposed between the mounting ends of the tested element. The loop-axes crossing points are: A(0, F_0), C(x_0, 0), D(0, -F_0) and E(-x_0, 0). By introducing the dimensionless magnitudes:

$$\tau = t/T,\ \xi(\tau) = x(\tau T)/x_{\mathrm{u}},\ \xi'(\tau) = \mathrm{d}\xi/\mathrm{d}\tau,\ z(\xi) = F(x_{\mathrm{u}}\xi)/F_{\mathrm{u}},$$
$$\xi_{\mathrm{m}} = \max|\xi(\tau)|,\ z_{\mathrm{m}} = \max|z(\xi)|,\ \xi_0 = x_0/x_{\mathrm{u}},\ z_0 = F_0/F_{\mathrm{u}} \tag{4.1}$$

Where, T is the period of the imposed cyclic motion and $x_{\mathrm{u}}, F_{\mathrm{u}}$ are displacement and force reference units such as $\xi_{\mathrm{m}} \le 1,\ z_{\mathrm{m}} \le 1$, a generic plot of the symmetric hysteresis loop $z(\xi)$ can be represented as shown in Fig. 4.1.

The Bouc-Wen model, chosen to fit the hysteresis loop shown in Fig. 4.1., is described by the following non-linear differential equation:

$$\frac{dz}{A - |z|^n\left[\beta + \gamma\,\mathrm{sgn}(\xi' z)\right]} = d\xi \tag{4.2}$$

where A, β, γ, n are loop parameters controlling the shape and magnitude of the hysteresis loop $z(\xi)$. Due to the symmetry of hysteresis curve, only the branches AB, BC and CD, corresponding to positive values of the imposed displacement $\xi(\tau)$, will be considered. The model parameters are to be determined such as the steady-state solution of equation (4.2) under symmetric cyclic excitation to satisfy the following matching conditions:

$$z(0) = z_0 \text{ at A},\ z(\xi_{\mathrm{m}}) = z_{\mathrm{m}},\ z(\xi_0) = 0,\ z(0) = -z_0 \text{ at D} \tag{4.3}$$

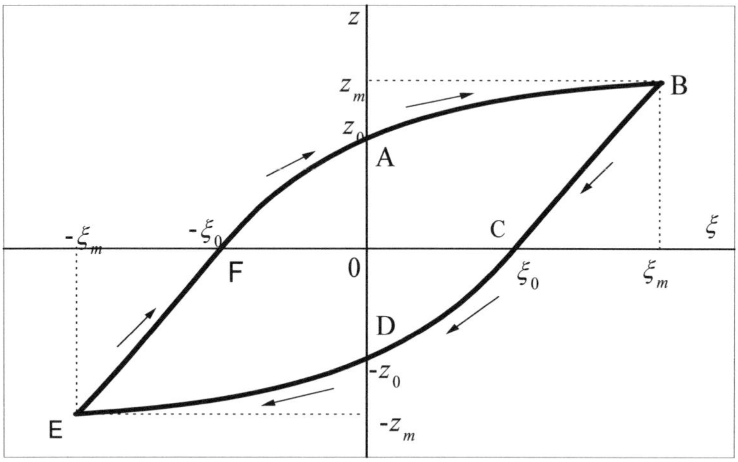

Fig. 4.1. Generic experimental hysteretic loop

Equation (4.2) is solved analytically for $n = 1$ and 2. For arbitrary values of n, the equation can be solved numerically. In the present work, the proposed method for fitting the solution of equation (4.2) to the experimental hysteresis loop shown in Fig. 4.1., is illustrated for $n = 1$. Introducing the notation:

$$\sigma = \beta + \gamma, \ \delta = \beta - \gamma \tag{4.4}$$

the equation (4.2) takes on three different forms for each the three branches AB, BC and CD shown in Fig. 4.1.:

$$\text{AB: } \frac{dz}{A - \sigma z} = d\xi \ , \ \text{BC: } \frac{dz}{A - \delta z} = d\xi \ , \ \text{CD: } \frac{dz}{A + \sigma z} = d\xi \tag{4.5}$$

From equations (4.5) one can calculate straightforward the slopes α_1 and α_2 of AB and BC branches in the point B:

$$\alpha_1 = \left.\frac{dz}{d\xi}\right|_{\substack{\xi \to \xi_m \\ \text{on AB}}} = A - z_m\sigma, \ \alpha_2 = \left.\frac{dz}{d\xi}\right|_{\substack{\xi \to \xi_m \\ \text{on BC}}} = A - z_m\delta \tag{4.6}$$

Since the condition $\alpha_1 < \alpha_2$ holds for any physical hysteresis loop, from equation (4.6) one obtains $\sigma > \delta$. Therefore, the Bouc-Wen model can portray a real hysteretic behavior only for positive values of parameter γ .

Integration of equations (4.5) on each branch yields three different relationships between the parameters ξ_m, ξ_0, z_m, z_0 , measured on the experimental loop, and the Bouc-Wen model parameters A, σ, δ .

4.2 Fitting the Bouc-Wen model to asymmetric experimental hysteresis loops

Suppose the experimental hysteretic characteristic is a asymmetric loop $-F_{m2} \leq F(x) \leq F_{m1}$, obtained for a periodic motion $-x_{m2} \leq x(t) \leq x_{m1}$, imposed between the mounting ends of the tested element. As before, the loop-axes crossing points are: A(0,f_0), C(x_0,0), D(0,-f_0) and E(-x_0,0).

With notations similar to (1), the asymmetric hysteresis loop $z(\xi)$ is modeled by:

$$\begin{aligned} z(\xi) &= \frac{1}{2} z_1(\xi_1)\left[1+\operatorname{sign}\xi_1\right] + \frac{1}{2} z_2(\xi_2)\left[1-\operatorname{sign}\xi_2\right], \\ \xi(\tau) &= \frac{1}{2}\left\{\xi_1(\tau)\left[1+\operatorname{sign}\xi_1\right] + \xi_2(\tau)\left[1-\operatorname{sign}\xi_2\right]\right\} \end{aligned} \tag{4.7}$$

where $z_1(\xi)$ and $z_2(\xi)$, are the solutions of the symmetric Bouc-Wen equations:

$$\frac{dz_1}{A_1 - |z_1|\left[\beta_1 + \gamma_1 \operatorname{sgn}(\xi_1' z_1)\right]} = d\xi\,, \quad \frac{dz_2}{A_2 - |z_2|\left[\beta_2 + \gamma_2 \operatorname{sgn}(\xi_2' z_2)\right]} = d\xi \tag{4.8}$$

For each of these equations, the loop parameters are determined according to the fitting algorithm presented in the previous section. As the loop-force axis crossing points of both branches have same coordinates, $z(\xi)$ is continuous in these points. By using equations (4.8), the continuity conditions of its derivative in these points lead to:

$$A_1 - z_0\sigma_1 = A_2 - z_0\sigma_2\,, \text{ where } \sigma_1 = \beta_1 + \gamma_1,\ \sigma_2 = \beta_2 + \gamma_2 \tag{4.9}$$

As the parameters $A_1, \sigma_1, A_2, \sigma_2$ are uniquely determined such as $z(\xi)$ to have imposed extreme values and axes crossing points, one must take into consideration a trade-off between these requirements and curve smoothness condition (4.9), such as to minimize a given accuracy cost function. This optimization of Bouc-Wen model fitting to asymmetric experimental hysteresis loops can be approached by iterative or Genetic Algorithms methods.

4.3 Application to experimental asymmetric hysteresis loops

The fitting method was applied to identify the differential Bouc-Wen models, which portray the hysteretic behavior of two vibration control devices: SERB-B-194 for earthquake protection of buildings by bracing installation (Serban et al., 2006), and SERB-B 300C for base isolation of forging hammers. The results presented in Figs. 4.2 and 4.3 prove the efficiency of the proposed method for fitting experimental hysteresis loops. Only a few iterative steps were needed in order to obtain a good approximation of experimental data.

5. Application and characteristics

The new technology developed by SITON has been applied by now in classic and nuclear objectives as follows:

- In 2003, the isolation of vibration shocks and seismic actions of a forging hammer located in IUS Brasov-Romania, and having the weight of 360kN which, as per the initial foundation solution, the shocks generated by the hemmer blow were transferred

Fig. 4.2. Vibration control device SERB-B-194

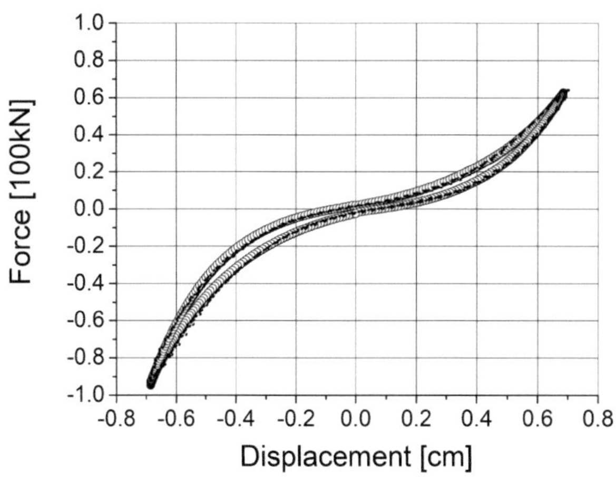

Fig. 4.3. Fitting the hysteretic characteristic of vibration control device SERB-B-194.

— analytical model: $A_1 = 0.22$, $\beta_1 = -3.6$, $\gamma_1 = 0.7$; $A_2 = 0.28$, $\beta_2 = -3.9$, $\gamma_2 = 0.7$

..... experimental data: $\xi_{m1} = \xi_{m2} = 0.69$, $z_{m1} = 0.65$, $z_{m2} = 0.92$, $\xi_0 = 0.07$, $z_0 = 0.02$

Fig. 4.4. Device SERB-B 300C

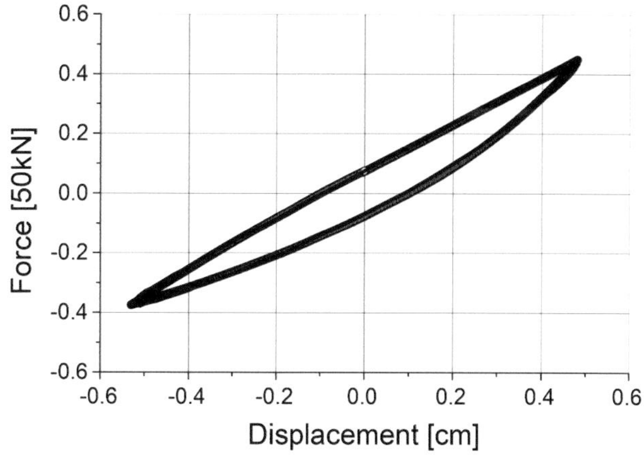

Fig. 4.5. Fitting the hysteretic characteristic of vibration control device SERB-B 300C.

— analytical model: $A_1 = 0.77$, $\beta_1 = -1.1$, $\gamma_1 = 1.05$; $A2 = 0.80$, $\beta2 = 0.3$, $\gamma2 = 0.7$

····· experimental data: $\xi_{m1} = 0.48$, $z_{m1} = 0.45$, $\xi_{m2} = 0.53$, $z_{m2} = 0.37$, $\xi_0 = 0.1$, $z_0 = 0.07$

to the near-by building (300 m and 800 m distance) and were resulting in the vibration of the building floors by a speed up to 52 mm/sec exceeding by 3,5 times the allowable limit of 15 mm/sec. After having installed SERB-SITON isolation devices, the value of the building floor vibration speed was reduced down to 6,75 mm/sec;

- also in 2003, the isolation against shocks, vibrations and seismic actions of pressurized air inlet and outlet pipes to the forging hammer. By the installation of the isolation devices the volume compensator on these pipes with an average service-life of 30 days were eliminated and the costs related to the maintenance and repairs were reduced;
- In 2005 a similar work with the one in 2003 for another forging hammer. The adopted solution was more performant meaning that the values of the building floor vibration speed was reduced to 0.085mm/sec from 52mm/sec. The isolation rate experimentally determined is 89%;
- Between 2005-2006 the strengthening, extension and rehabilitation of an old reinforced concrete framework building in order to withstand violent earthquakes with a 0.29g acceleration on 2 orthogonal directions in horizontal plane. Strengthening was done by inserting a small number of panels braced by SERB type telescopic devices symmetrically arranged as to the building symmetry plane. SERB device are controlling, limiting and damping the relative level displacements of the building. The columns (pillars) and beams of the building have not been strengthened, except those pertaining to the placed panels which have been lined with metal profiles;
- In 2006, installation of a SERB-SITON type of support on the pipe 1056 located in Drobeta-Turnu Severin Factory-Romania. After the installation the amplitude of the pipe vibrations was reduced 6 times;
- In 2007, the isolation of the electric and I&C panels associated to the H_2S compensators in GS3 section in ROMAG PROD against shocks and vibrations and seismic movements

by the use of SERB type sliding supports. After the installation of the seismic isolation devices in the cabinets, the serial components inside the cabinet could be also installed but without verifying the behavior of the cabinet during an earthquake because the seismic acceleration transferred to the cabinet by the isolating;

- In 2008, seismic qualification of COLD-BOX columns for the radioactive tritium separation located in the Cryogenic Research (ICSI) in Ramnicu-Valcea, Romania. The seismic qualification consisted of the installation of 4 SERB supports on each column for to control, limit and damp the swinging movement of the columns during an earthquake;
- In 2009, the isolation of the electric and I&C panels associated to the H_2S compensators in GS4 section ROMAG PROD against shocks and vibrations and seismic movements by the use of SERB type rolling supports;

6. Conclusions

The purpose of this work is to demonstrate that in SITON was developed an innovative method of reduction of dynamic response of structures subject to dynamic actions like shocks, vibrations and seismic movements starting with the new types of devices used for structure isolation and/or dissipation of the energy transferred, up to the use of adequate mathematic models and analysis method that allow a representative evaluation of structures conduct in dynamic actions. The method developed by SITON was checked on physical models, on prototypes on a 1/1 scale and by certain applications in classical and nuclear industry.

This method can be successfully applied to the rehabilitation or achievement/construction of classical and nuclear objectives, because it possesses the following advantages over the current methods:

- Reduces the seismic response of constructions through the enlargement of absorption capacity, including at small and medium deformations as well as through the control and limitation of relative level deformations to small values so as the structure of the building remains in the linear behavior range. Seismic energy is taken over and damped by these devices on a hysthetic cycle;
- The structural element of the construction do not reach local overload which may lead to their degrading and to the appearance of plastic articulations including in case of a violent earthquake, and the construction in functional after an earthquake, exceeding with up to 50% the designed earthquake level;
- The control and limitation of level movements is performed with SERB-SITON mechanical devices, which can be installed either in central and ex-central bracing, or around the nodes of a panel or at the interface between building parts;
- The typical SERB-SITON telescopic devices can take over axial charges of stretch and compression up to 1500 kN and can scatter the seismic energy on a cycle up to 80% of a seismic energy on a cycle. The SERB-SITON isolator devices can take over compression load up to 5000kN and tension load up to 1500kN;
- The consolidation of the constructions can be made without the evacuation of the inhabitants and the time needed for the consolidation is reduces two or three times more than the classic method and the price is lowered with approx. 10-30%;

- The devices for equipment and piping network can be used to isolate these and/or to the reduction of the relative movement; These can take over the movements in thermal expansions with the elastic or constant reaction force;
- The SERB-SITON mechanical devices work efficiently at low speeds as well as high vibrating speed which allows their use without any restrictions, both in case of fast surface earthquakes and slow earthquakes like the intermediary ones or the earthquakes on unconsolidated soft soils;
- The devices are not affected by the aging phenomenon and can be installed including in areas with high radiation flux;
- The devices are safe, cheap, have a low weight and are easy to install unlike the hydraulic absorbers or other devices existent on the market.

7. References

Romanian Seismic Design Code, P100-1/2006;

Panait, A., Serban, V., *Isolation and damping of shocks, vibrations, impact load and seismic movements at buildings, equipment and pipe networks by SERB-SITON method*, International Conference on Nuclear Engineering, 2006, July 17-20, Miami, Florida, USA;

Serban, V., Brevet de Inventie OSIM Nr. 119822 B1/29.04.2005, Structura sandvis si dispozitiv compact, pentru preluarea si controlul incarcarilor statice si dinamice;

Serban, V., Brevet de Inventie OSIM Nr.119845 B1/29.04.2005, *Structura sandvis, dispozitiv avand in componenta aceasta structura si retea de dispozitive pentru preluarea si amortizarea incarcarilor, pentru controlul comportarii, sub sarcina, a constructiilor, sistemelor si echipamentelor*;

Sireteanu, T., Giuclea, M., Serban, V., Mitu, A.M., *On The fitting of experimental hysteretic loops by Bouc -Wen model*, SISOM - 2008, Bucharest, 29 -30 May, 2008;

Bouc, R., *Forced vibration of mechanical systems with hysteresis*, Proceedings of the Fourth Conference on Non-linear oscillation, Prague, Czechoslovakia, 1967;

Wen, Y.K., *Method for random vibration of hysteretic systems*, Journal of the Engin. Mechanics Division 102(2)(1976) 249–263;

Li, S.J., Yu, H., Suzuki, Y., *Identification of non-linear hysteretic systems with slip*, Computers and Structures 82 (2004) 157–165;

Hornig, K. H., *Parameter characterization of the Bouc-Wen mechanical hysteresis model for sandwich composite materials by using Real Coded Genetic Algorithms*, Auburn University, Mechanical Engineering Department, Auburn, AL 36849;

Ajavakom, N., Ng,C.H.,Ma, F., *Performance of nonlinear degrading structures: Identification, validation, and prediction*, Computers and Structures xxx (2007) xxx–xxx (in press);

Giuclea, M., Sireteanu, T., Stancioiu, D., Stammers, C.W., *Model parameter identification for vehicle vibration control with magnetorheological dampers using computational intelligence methods*, Pro. Instn. Mech. Engers. 218 Part I: J. Syst. and Control Engin., no.17, 2004, pp.569-581;

Kwok, N.M., Ha, Q.P., Nguyen, M.T., Li, J., Samali, B., *Bouc–Wen model parameter identification for a MR fluid damper using computationally efficient GA*, ISA Transactions 46 (2007) 167–179;

Serban, V., Androne, M., Sireteanu, T., Chiroiu, V., Stoica, M. *Transfer, control and damping of seismic movements to high-rise buildings,* International Workshop on Base Isolated High-Rise Buildings, Yerevan, Armenia, June 15-17, 2006;

Serban, V., Sireteanu, T., Androne, M., Mitu, A.M., *The effect of structural degradation on the seismic behaviour of structures*, SISOM - 2010, Bucharest, 29 -30 May, 2010;

4

Advances in CAD/CAM Technologies

Mircea Viorel Drăgoi
Transilvania University of Braşov
Romania

1. Introduction

The product development, as a stage of product life cycle, has to meet the requirements enforced by the technological component on one hand, and by the economical one, on the other hand.

The technological component enforces restrictions referring to materials, cutting tools, machine-tools foreseen for processing, and mainly to the possibility to work a certain part by the means available.

The economical component comes with restrictions referring to the costs of developing the product. It is clear that the two types of restrictions work contrary: in the most cases, the more a material, a tool, or a technology is effective, the more expensive it is. On the other hand, the cheap resources cannot always meet all the technical / technological specific requirements of a certain product. Consequently, a permanent optimisation is necessary. Such an optimisation means finding the best compromise between the technological and the economical criteria. This compromise must lead to producing a product that fully meets all the customer's technical specifications, and which is attractive in terms of price, as well.

Optimisation has to be applied in all the stages of product development, consequently in the simultaneous design - constructive and technological - of the product, and its integrated conception with the production.

It is known that the CAD/CAM software is very expensive. Many SME's that do not use sophisticated hardware resources, and do not use to produce very complicated parts are interested to benefit from low-priced pieces of software that can give answers to the most common problems of designing and manufacturing. Some specific tools that ease the designer's job in terms of finding the best solution to technological problems, or automating some specific laborious tasks are presented in this chapter. Note that the solutions here provided do not intend to replace the well-known specific software, but rather to be an alternative to it, there where those are not a real necessity.

Some samples of problems approached here are automatically checking if a certain part can be machined by up to three axes CNC milling machines, optimizing the approximation of the curves in order to engender automatically CNC files, automatically engendering the tool-path for CNC milling in vertical planes (a new method to compute ball nose milling cutter radius compensation in Z axis for CNC), transferring geometrical data from CAD systems to CAM system (in order to engender CNC files) as a mean of integration of CAD and CAM systems.

AutoCAD and AutoLISP were chosen as the support of CAD/CAM systems.

2. 3D modeling of the product

3D modeling is the necessary stage the geometry of the part is designed in. In the modern product engineering, describing in terms of geometry of a part cannot be completely separated from taking into account the technological criteria. Further more, the quality of the outcome of the conception process is strongly influenced by the level of integration of the constructive and technological aspects of the stage of designing. In these terms, at least two aspects should be taken into account:

- using the constructive-technological entities, which integrate geometrical features (dimensions, shape) and technological features (limits or deviations, surface quality), as well;
- the machinability of the shape.

2.1 Using the constructive-technological entities in designing

The constructive-technological entity (Chicoş & Ivan, 2004; Chicoş et al., 2009) has the advantage that the use of data with technological specific feature creates the basic prerequisite for automatically/computer aided engendering the process plan of the surface under debate, as a necessary stage in elaborating the technology of the part.
Binding technological specific features to a geometric entity (i. e. a line, that in shop-floor drawing may represent a surface of the part) can be done in AutoCAD by means of specific user-defined functions.
Such user-defined functions present dialog boxes by means of which the features (either geometrical or technological) are input. The geometrical data are used to define the entity, calling from the function the appropriate AutoCAD command. The technological data are bound to the new entity by means of its associated list, in the area of extended data.
A sample of user-defined function, named EXDATROUG1 (EXtended DATa ROUGhness), that binds to an existing entity the roughness, as extended data, is presented bellow. The entity the extended data is bound to must be selected by user.

Program line	Effect
(defun C: EXDATROUG1 ()	
(setq addrug (entget (car (entsel))))	Gathers the associated list of the entity (the user selects) to which extended data must be attached
(setq RR (getreal "Input roughness:"))	Gathers from the user the value of the roughness, to be attached as extended data
(regapp "Rug")	Creates an application named *Rug* in order to be attach the extended data
(setq data_extinsa_rug	
(list (list -3 (list "Rug" (cons 1040 rr)))))	Builds the list that defines the extended data
(setq addrug (append addrug data_extinsa_rug))	Appends to the associated list of the initial entity the extended data
(entmod addrug))	Attaches the extended data to the associated list of the entity

Table 1. EXDATROUG1 user-defined function. Explanations

In a similar manner new entity having extended data attached can be created by means of a user-defined function. Building the associated list that defines the attached extended data is done as in the sample above (Table 1). All the dotted pairs that define the geometric data and the AutoCAD properties (e. g. layer, thickness, color, linetype, a.s.o.), and which constitute the associated list of the new entity must be composed as well. Assembling the new entity and drawing it is performed using the standard AutoLISP function *ENTMAKE*. Such a function, EXDATROUG2 (EXtended DATa ROUGhness), is described bellow:

Program line	Effect
(defun C: EXDATROUG2 ()	
(setq RR (getreal " Input roughness:"))	Gathers from the user the value of the roughness, to be attached as extended data
(regapp "Rug")	Creates an application named *Rug* in order to be attach the extended data
(setq data_extinsa_rug	
(list (list -3 (list "Rug" (cons 1040 rr)))))	Builds the list that defines the extended data
(setq P1 (getpoint "Input start point:")	Gathers interactively from the user the first point of the line
P2 (getpoint "Input end point:"))	Gathers the second point of the line
(setq l1 (list (cons 0 "LINE"))	Defines the type of the new created entity (*LINE*)
l2 (list (cons 8 "0"))	Selects the layer where the entity to be placed
l3 (list (cons 10 p1))	Defines the start point of the new created line
l4 (list (cons 11 p2)))	Defines the endpoint of the new created line and ends creating the associated list
(setq addrug (append l1 l2 l3 l4 data_extinsa_rug))	Attaches the extended data to the associated list of the new entity
(entmake addrug))	Creates the new entity (line) with extended data

Table 2. EXDATROUG2 user-defined function. Explanations

Remarks:

- if try to attach new extended data to an entity that already has some one, the old extended data is overwritten. **Sample:** if the entity described by the associated list
  ((0 . "LINE") (8 . "0") (10 5.0 5.0 0.0) (11 5.0 15.0 0.0) (-3 ("Rug" (1040 . ***12.5****))))*
  is processed by **EXDATROUG1** user-defined function, binding the value of 6.3 to the roughness, the associated list of the entity becomes
  ((0 . "LINE") (8 . "0") (10 5.0 5.0 0.0) (11 5.0 15.0 0.0) (-3 ("Rug" (1040 . ***6.3****)))),*
  and not
  ((0 . "LINE") (8 . "0") (10 5.0 5.0 0.0) (11 5.0 15.0 0.0) (-3 ("Rug" (1040 . **6.3**))) (-3 ("Rug" (1040 . 3.**2**)))),
  as somebody might think;

- independent of the way the extended data was attached, it can be later gathered by using the standard AutoLISP functions *ENTGET* and *ASSOC*. *ENTGET* must however be called with an additional parameter as against the usual way to use it: this parameter must point to the application that defines the extended data to be gathered. So, the call (ENTGET (entlast)) – the common use – will return the associated list
  ((0 . "LINE") (8 . "0") (10 5.0 5.0 0.0) (11 5.0 10.0 0.0)),
  while the call (ENTGET (entlast) „Rug") will return
  ((0 . "LINE") (8 . "0") (10 5.0 5.0 0.0) (11 5.0 15.0 0.0) (-3 ("Rug" (1040 . **6.3**))))

One might want to use the extended data associated to an entity from within a piece of software that acts as a technological processor (Ivan 1994; Ivan 1996). This is possible by accessing the associated list of the targeted entity and cutting off from the list the wanted data. To automate this task, another user-defined function was created. It is named *GETROUG* and is presented in the following table:

Program line	Effect
(defun C:getroug ()	
(setq l (entget (car (entsel)) (list "Rug"))	Get the associated list of the selected entity
l (cdr (assoc -3 l))	The extended data in the associated list: (("Rug" (1040 . 3.2)))
l (car l)	The first sub-list in the extended data ("Rug" (1040 . 3.2))
l (cdr l)	The right member of the previous list: ((1040 . 3.2))
l (car l)	The first sub-list from the previous: (1040 . 3.2)
Rug (cdr l))	**The roughness**
)	Ending the function

Table 3. GETROUG user-defined function. Explanations

This function may be used either within a technological processor, or in any other piece of software that has to use the value of the roughness as an extended data.

2.2 Machinability of the surfaces

In this context, ***Machinability*** is understood as the potential of a surface to be obtained by certain (simple) means, at the required precision of its shape and dimensions.

With no doubt, any surface of a part has to meet all the requirements in terms of functional and aspect. However, the technological aspect may not be neglected. Some shape details do not affect the good working of the part, but might have major influence on the machinability of a part. A non-technological shape may lead to unjustified expenses related to tool consumption, to necessity of using sophisticated and expensive resources (technologies, machine-tools, fixtures). An appropriate sample in this acceptance is the presence or not of the filleting radius between two adjacent surfaces, as shown in Fig. 1.

A special attention must be given to the grooves at the junction of the perpendicular surfaces that must be grinded, or at the end of internal threads.

Furthermore, a special importance has the shape of the cast parts or of those made by plastic mass injection, in the context of making the cast models and moulds by CNC technologies.

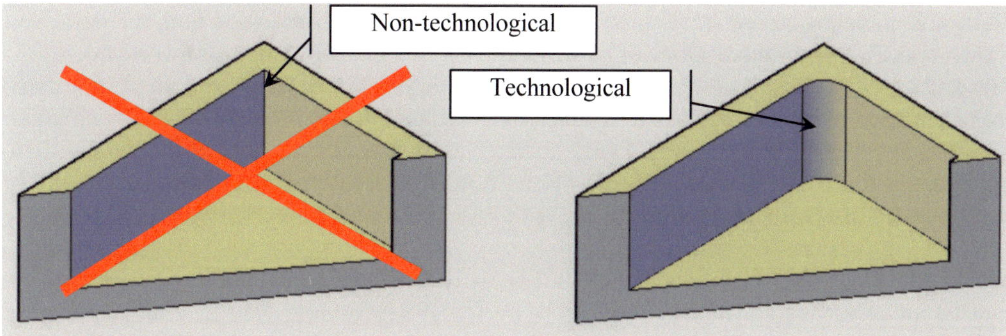

Fig. 1. Filleting radiuses influence the part machinability

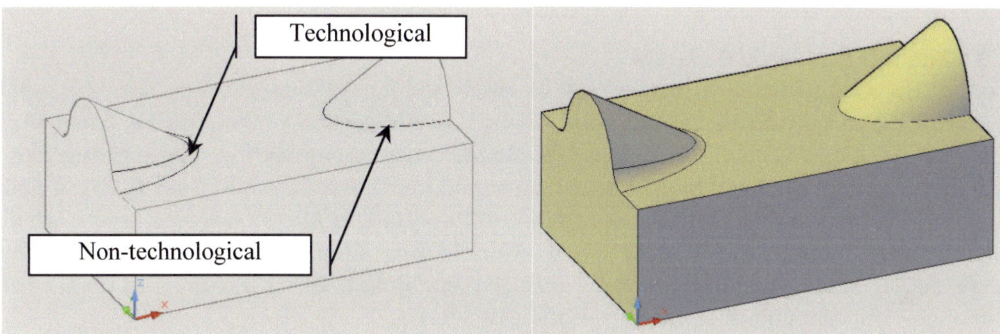

Fig. 2. Filleting radiuses influence the part machinability

2.3 Checking the machinability of the parts in order to be processed by milling on CNC machine-tools

The most CAD/CAM systems suffer from a big deficiency: their machining (CAM) modules engender correctly the tool-paths in terms of collisions, but without reporting if in certain cases the part cannot be completely be worked, that is the shape cannot be completely achieved. For milling processes, this happens if a CNC equipment with no more than three axes is used, and the orientation of a surface is not proper relative to the vertical plane, as can be seen in Fig. 3.

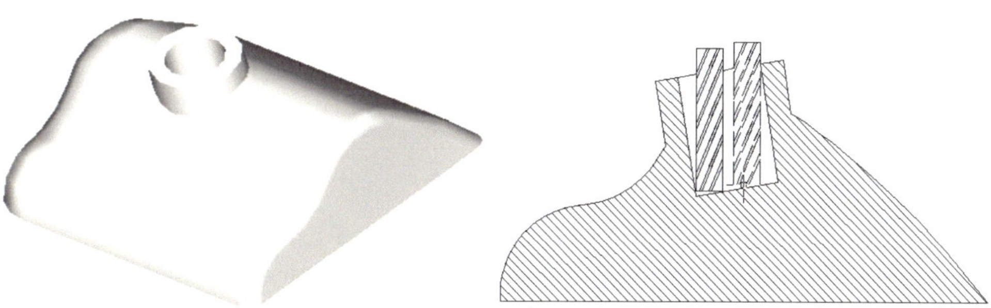

Fig. 3. Part which cannot be processed (the hole because it's inclination) on a 2 ½ or 3 axes CNC machine tool

To be able to be processed (machinable) on a 2 ½ or 3 axes CNC machine tool, the part must meet a specific requirement: none of its surfaces may be bevelled "under the vertical".
The problem that must be solved in such cases is to detect the surface elements (if they exist) that cannot be machined on 2 ½ or 3 axes CNC machine-tools, because of their inclination reported to vertical. Furthermore, if such surfaces exist, must be sought out the possibility to yet process the part by modifying its orientation. If a solution is available, it should be described by the plane and the angle in/with which the part must be rotated to come in a proper position as to be worked.
To identify the surface elements that cannot be worked (in the context of those stated above) may be followed two ways: the study can be performed sectioning the part either by vertical or horizontal planes, each of it having specific advantages and disadvantages.
The solving here proposed assumes the study of the 3D model of the part, either built in AutoCAD, or imported in AutoCAD from other CAD systems, by means of IGES or DXF files.

2.3.1 Scanning in vertical planes

To find the surface elements that cannot be machined for the reason stated above, the 3D model of the part is successively sectioned by equidistant vertical planes, the outcome being a family of regions. Each of the regions is exploded becoming a polyline, whose geometrical features can be investigated by means of data stored in the associated list (Fig. 4). Every pair of points P_i, P_{i+1} is studied, and the occurrence of the situation $X_{i+1}<X_i$ means the existence of a surface element that cannot be machined. We call such a situation *undercut*. In this case the angle the segment P_iP_{i+1} makes with the vertical is computed, taking into account the sign of this angle. The convention of the signs is indicated in Fig. 4.

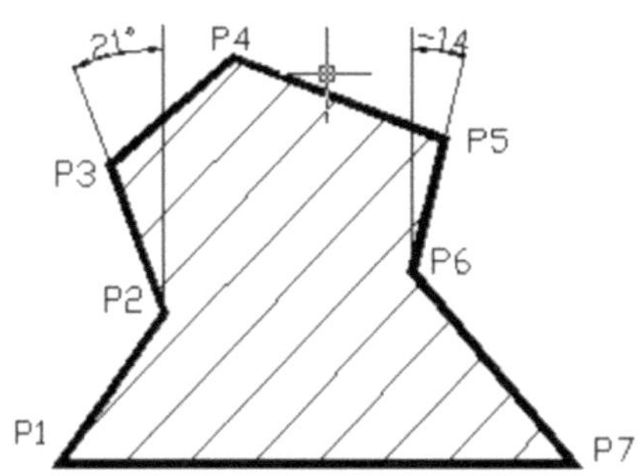

Fig. 4. Contour obtained by sectioning the part with a vertical plane

As more such situations arise, either in the same section, or in different sections, the maximum and the minimum value of the angle are stored. At the end of the scanning process the presence and the type of undercuts are evaluated. Depending on the result of this evaluation, some conclusions are made, as follows:

a. if no undercuts, the part can be machined on a 2½ or 3 axes CNC machine-tool. (final conclusion);
b. if undercuts exist, and they have the same sign, the part must be rotated in vertical plane with a minimum angle equal to the maximum undercut (Fig. 5); Scanning is performed again, until a final conclusion can be made;
c. if at least two undercuts having opposite signs exist, the part cannot be machined on a 2½ or 3 axes CNC machine-tool, even if the part should be rotated (final conclusion).

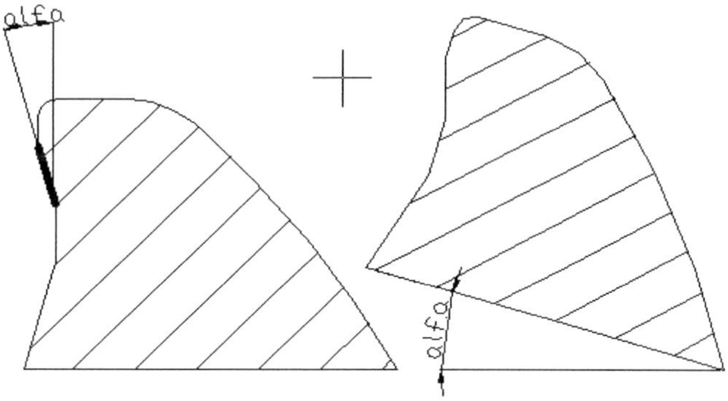

Fig. 5. Rotating the part in order to avoid undercuts

2.3.1 Scanning in horizontal planes

This method is simpler and more reliable than scanning in vertical planes, but doesn't offer solutions in terms of finding an appropriate orientation of the part, as it to be machinable, even when such solutions exist. To describe the algorithm two sample parts are used, those in Fig. 6.

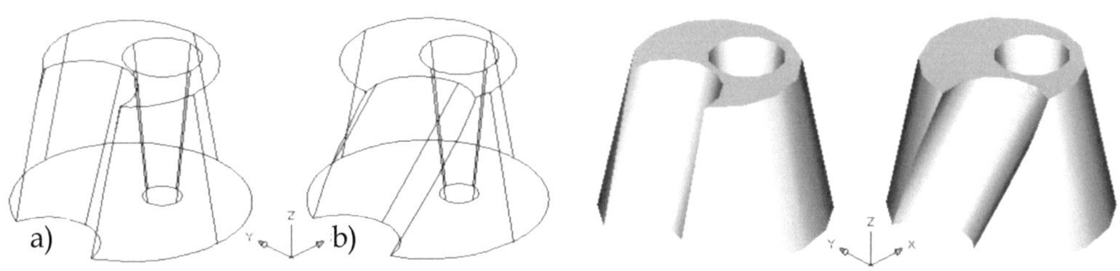

Fig. 6. Parts with similar configuration

The parts are conic trunks from which a cylinder and another conic trunk were extracted. The cylinder was previously rotated 10° about Y axis (Fig. 6a) and 15°and 10° about X and respectively Y axis (Fig. 6b). The rotation of the cylinder 15° about X axis makes machining impossible on a 2½ or 3 axes CNC machine-tool (if reorient the part by rotating it -15° about X axis an undercut occurs on the internal conical surface.

The working principle of the method consists in sectioning successively the 3D model of the part with equidistant horizontal planes, beginning from the bottom (Fig. 7).

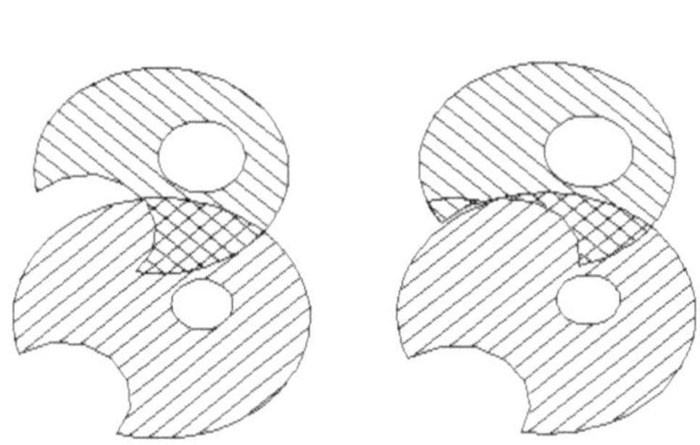

Fig. 7. Horizontal sections through the two sample parts

The outcome of the sectioning is a region. Pairs of regions are studied in the following way: two consecutive regions are copied in the same horizontal plane (the upper region is moved downward to the plane of the lower one), without being moved in horizontal direction. The two objects in the same plane are intersected. If the output is congruent with the region that initially was upper, on the lateral surfaces between the planes of the regions analyzed do not occur undercuts. The mentioned congruence is equivalent with that in a projection on a horizontal plane the upper section is entirely included in the lower one. Such a congruence is emphasized in Fig. 8a), where the smaller region is entirely covered by the larger one. In Fig. 8b) the regions do not overlap integrally, a sector of the smaller region (filled with colour) is not covered by the larger region.

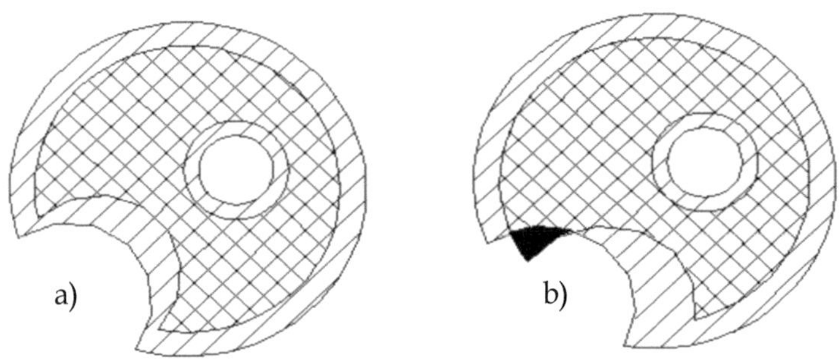

Fig. 8. Intersection (overlaping) of the regions

3. Tools for computer aided engendering of CNC files

3.1 Finding the size of solids in AutoCAD

It is usual to section a 3D model with vertical (YZ or ZX) planes in order to reduce the 3D problem of determining the CNC tool-paths to a set of 2D much easier to be solved problems. "YZ" or "ZX" are not singular planes, but families of parallel planes. To indicate a particular plane of a family, a point the plan goes through must be specified. To get a reasonable effect from sectioning a solid with a plane, of course, the plan must intersect the solid. That is, the sectioning plane must go by a point inside the bounding box of the solid. At a glance, it seems very easy to indicate a point inside a box - that if work interactively. If the job is to be performed from within an AutoLISP program, is not so simple, all the more it has to be done repeatedly exactly from an end of the box to another. So, knowing the data that describe the bounding box of a 3D model is a necessity to handle the solid correctly. The single mean to get the bounding box of a solid is the command MASSPROP. This command returns the wanted data either displayed on the screen, or saved in a file, but doesn't store it in system variables as to be ready to hand for the programmer. A piece of software that gives the user the coordinates of the ends of the bounding box would be very useful. A user-defined function was built, it is named 3DSIZE, and its' return are the minimum and maximum values of X, Y, and Z coordinates of a 3D model in AutoCAD. This function is useful, as well, when designing the billet a part will be cut off from.

The function is based on the fact that MASSPROP command returns all the mass properties of a solid (including its bounding box) in a report having the following format:

```
---------------- SOLIDS ----------------

Mass:              1003.5348
Volume:             1003.5348
Bounding box:        X: -2.1511  --  113.0500
                   Y: -166.0205  --  -156.5898
                   Z: -2.3466  --  -0.1022
Centroid:           X: 62.7399
                   Y: -161.9498
                   Z: -1.2244
Moments of inertia:  X: 26324880.1352
                   Y: 5074009.6821
                   Z: 31395132.5908
Products of inertia: XY: -10197522.0729
                   YZ: 198988.3441
                   ZX: -77088.8178
Radii of gyration:   X: 161.9634
                   Y: 71.1065
                   Z: 176.8744
Principal moments and X-Y-Z directions about centroid:
                   I: 2942.6215 along [1.0000 -0.0008 0.0000]
                   J: 1122292.9852 along [0.0008 1.0000 0.0000]
                   K: 1124487.1726 along [0.0000 0.0000 1.0000]
```

Table 4. The MASSPROP report

If save this report, all the wanted data can be read from that file, following a specific rule. After read the data from file, the text values must be converted in numbers and stored in variables. The first section of the function is presented in Table 5. Only retrieving the values of z_{max} and z_{min} is described, the same algorithm being followed for X and Y co-ordinates.
Remark: the mass properties are computed appertained to the current coordinate system. So, if before calling the function the origin of the coordinate system is placed in the point that is going to become the part datum, the values returned by the function can be used for CNC programming.

Program line	Effect
(defun C:3Dsize ()	
(setq prisma (car (entsel)))	Select the object to be evaluated
(command "massprop" Prisma "" "Y" "c:\\temp.mpr")	Call MASSPROP command and write data to file
(setq fis (open "c:\\temp.mpr" "r"))	Open the data file
(repeat 5 (setq linie (read-line fis)))	Read lines that contain irrelevant data
(setq linx (read-line fis)) ;X	Read the line with data about solid extent along X axis
(setq liny (read-line fis)) ;Y	Read the line with data about solid extent along Y axis
(setq linz (read-line fis)) ;Z	Read the line with data about

	solid extent along Z axis
(setq fis (close fis)	Close file
c %	Binding values to some variables
cifre (list "1" "2" "3" "4" "5" "6" "7" "8" "9" "0" "." "-")	
s linz	
)	
(while (not (member (substr s 1 1) cifre))	Processing the line with data about the height of the solid. Get the data referring the extent of the solid along the vertical, and computing the height of the object.
(setq s (substr s 2))	
)	
(setq zmins "")	
(while (member (substr s 1 1) cifre)	
(setq zmins (strcat zmins (substr s 1 1))	
s (substr s 2)	
)	
)	
(setq s (substr s 5))	
(while (not (member (substr s 1 1) cifre))	
(setq s (substr s 2))	
)	
(setq zmaxs "")	
(while (member (substr s 1 1) cifre)	
(setq zmaxs (strcat zmaxs (substr s 1 1))	
s (substr s 2)	
)	
)	
(setq zmin (atof zmins)	
zmax (atof zmaxs)	
lz (- zmax zmin)	
....	Program sections related to the extents of the solid along X and Y axis. They are similar to Z
....	
)	End the function

Table 5. 3DSIZE user-defined function. Explanations

3.2 Gathering polyline data from DXF files in order to engender CNC files

The DXF files are often used because can be handled by any CAD/CAM system. The function described bellow gathers the data describing polylines from DXF files and use it for automatically engendering the CNC files. Since the polylines in discussion are 2D objects, the application can be used for outlining processes (milling, cutting by laser or water jet). In the DXF files data are always presented in groups of two lines. The value on the first line of the group is always a numeric one, and represents a code that indicates the meaning of the value on the second line. The value on the second line may be a numeric or a text one. The data bundles that describe polylines in DXF files are grouped in four or six lines, as they represent an end of a line or of an arc. The first group of four lines describe the start point of the polyline. The next groups describe the points the polyline passes along. The code 42,

named also *Bulge*, indicates that the next point is reached along an arc. The value of *Bulge* allows computing data that describe the arc. The sign of *Bulge* gives the sense of travel through the arc (positive - counter clockwise, negative - clockwise). The value that follows the code *10* represents the X coordinate of the point, and that follows the code *20* is the Y coordinate of the point. The value following the code *42*, named *Bulge*, describes the curvature of the arc. In fact, *Bulge* represents the tangent of the quarter of the arc's included angle. An arc on the polyline is perfectly defined by three data: startpoint, endpoint, and *Bulge*. Anyway, this minimum set of data is not enough to draw an arc, or to program it in a CNC file. Therefore, *Bulge* must be converted in an angle, as then compute the radius of the arc.
Equations (1) .. (5) are used to compute the data that describe the arc as it to be drawn or programmed in CNC files. The meanings of the notation are according to Fig. 9, as follows: U=the included angle of the arc; d=the length of the arc chord P_1P_2; R= the arc radius.

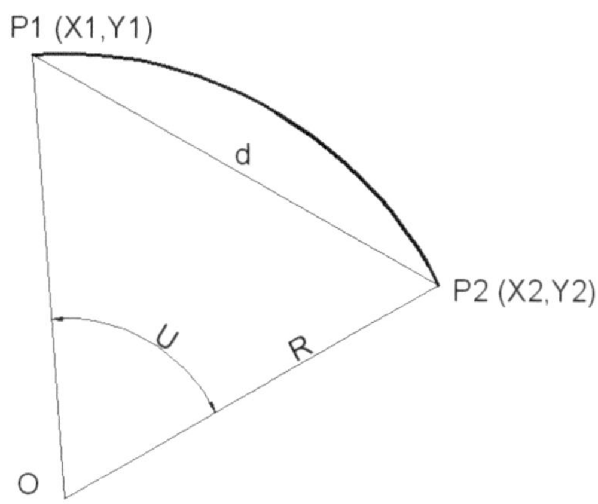

Fig. 9. Arc elements

The included angle of the arc is computed wit equation (1), since bulge is available. The length of the chord, d, is computed with equation (2). Applying in triangle OP_1P_2 the generalized theorem of Pythagoras - equation (3), and furthermore transforming, the value of the radius is found.

$$U=4*\arctan(\text{Bulge}) \tag{1}$$

$$d=\sqrt{(x1-x2)^2+(y1-y2)^2} \tag{2}$$

$$R^2+R^2-2*R*R*\cos(U)=d^2 \tag{3}$$

$$2*R^2*(1-\cos(U))=d^2 \tag{4}$$

$$R=\frac{d}{\sqrt{2*(1-\cos(U))}} \tag{5}$$

The DXF_CNC function opens the DXF file, finds in it data that describe the polyline, and gathers the required pieces of information. For the arc, it computes the included angle and radius. As calculi are made, the segments and the arcs are drawn on the display and the G codes are written to the CNC file. The function is described bellow.

Program line	Effect
(defun C: DXF_CNC ()	
(command "UNDO" "BE")	
....	Prepare input data (names of files)
(write-line "TEST" CNC)	Writes the first lines to CNC file
(write-line "G28 Z30" CNC)	
(write-line "M06 T01" CNC)	
(write-line (strcat "M03 S" (itoa (fix s))) CNC)	
(setq linie "" inceput T)	
(while (not (= linie "LWPOLYLINE"))	Ignores in the DXF files the irrelevant data
(setq linie (read-line fis))	
)	
(while (not (= linie "ENDSEC"))	
(setq linie (read-line fis))	
(if (= linie "LWPOLYLINE") (setq inceput T bulge 0))	
(cond	
((= (atof linie) 10.0)	Gathers the coordinates of the point
(setq x (atof (read-line fis)))	
)	
((= (atof linie) 20.0)	
(setq y (atof (read-line fis)))	
(if inceput (progn	If it is the start point of the polyline, programs the tool start position and entering into the billet
(setq p1 (list x y) inceput nil)	
(write-line (strcat "G00 Z-" (rtos ad)))	
(write-line (strcat "G01 X" (rtos x) " Y" (rtos y) " F" (itoa (fix Avans))) CNC))	
(progn	If it is an arc, computes the radius - equations (1) .. (5), and programs the arc, clockwise, or counter clockwise, as the sign of *Bulge* impose
(setq p2 (list x y))	
(if (/= bulge 0)	
....	
)	
(Progn (command "LINE" p1 p2 "")	Programs linear interpolation when need
(setq p1 p2)	
(write-line (strcat "G01 X" (rtos x) " Y" (rtos y)) CNC)	
)))))	
((= (atof linie) 42.0)	Gets the value of Bulge, computes the included angle in radians, and transforms it in degrees
(setq bulge (atof (read-line fis))	
u (* 4 (atan bulge))	
ugrad (/ (* u 180) pi)))))	
(setq fis (close fis))	Close DXF file
(write-line "G28 Z30" CNC)	Writes the ending sequence to CNC file
(write-line "M02" CNC)	
(write-line "M30" CNC)	
(setq CNC (close CNC))	Close CNC file
(setvar "OSMODE" os)	
(command "UNDO" "E")	
)	

Table 6. DXF_CNC user-defined function. Explanations

3.3 Gathering polyline data from the associated list of AutoCAD entities in order to engender CNC files

In table 6 a simplified form of the function was presented. The function is equipped with specific mechanisms that allow it to know if the polyline is open or not. If the polyline is closed, its first point is stored, as it can be used as the last point of the polyline, as well, even this is not explicitly given by the DXF file.

Through two points, two arcs having the same radius, but different included angle (a smaller than 180° and a larger than 180° - see Fig. 10) can be drawn. To distinguish them, a rule has to be followed: the arc larger than 180° has to be programmed in the CNC file as having a negative radius. This aspect is also solved by the function.

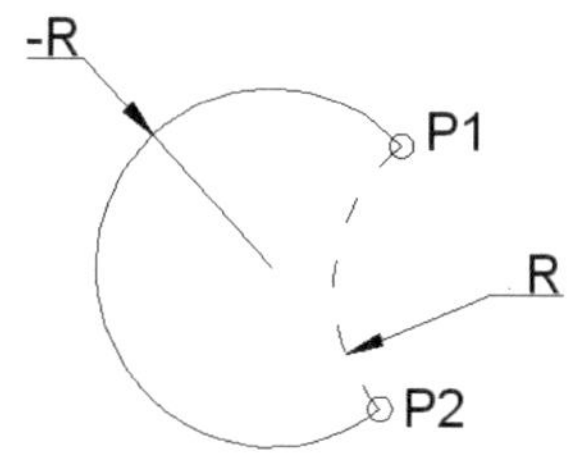

Fig. 10. Two arcs having the same radius and going through the same two points, P_1 and P_2

The function described above was adapted to gather data from the associated list of the polyline in an AutoCAD drawing. The operating and computing algorithm is the same, differs only the way data is gathered.

The list bellow is associated to a pentagon having the edge length of 50, and whose corners are filleted with the radius of 10.

((-1 . <Entity name: 7ef083b8>) (0 . "LWPOLYLINE") (330 . <Entity name: 7ef06cf8>) (5 . "1F7") (100 . "AcDbEntity") (67 . 0) (410 . "Model") (8 . "0") (100 . "AcDbPolyline") (90 . 10) (70 . 1) (43 . 0.0) (38 . 0.0) (39 . 0.0) (10 27.2654 -1.77636e-015) (40 . 0.0) (41 . 0.0) (42 . 0.0)
(10 62.7346 -1.42109e-014) (40 . 0.0) (41 . 0.0) (42 . 0.32492)
(10 72.2451 6.90983) (40 . 0.0) (41 . 0.0) (42 . 0.0)
(10 83.2057 40.643) (40 . 0.0) (41 . 0.0) (42 . 0.32492)
(10 79.573 51.8233) (40 . 0.0) (41 . 0.0) (42 . 0.0)
(10 50.8779 72.6716) (40 . 0.0) (41 . 0.0) (42 . 0.32492)
(10 39.1221 72.6716) (40 . 0.0) (41 . 0.0) (42 . 0.0)
(10 10.427 51.8233) (40 . 0.0) (41 . 0.0) (42 . 0.32492)
(10 6.79429 40.643) (40 . 0.0) (41 . 0.0) (42 . 0.0)
(10 17.7549 6.90983) (40 . 0.0) (41 . 0.0) (42 . 0.32492) (210 0.0 0.0 1.0))

All the dotted pairs of the associated list are successively cut off until the first one to have the code 70 is found. Here is the piece of information that says if the polyline is closed (value 1) or open (value 0). After this data was gathered, the function looks for the first dotted pair which has the code 10. Beginning from this point the geometrical data of the polyline are obtained. They are processed exactly as they were in the **DXF_CNC** function. Optionally, three special program lines may be added at the beginning of the CNC file, to be simulated as it should be run on a 2 ½ axes milling machine from Denford, UK, and having a Fanuc controller. This version of the function is more advantageous than the DXF one because of if in the drawing there are more polylines, one/those wanted to be processed can be selected. The function LAsoc_CNC is described bellow.

Program line	Effect
(defun C:LAsoc_CNC ()	
(command "UNDO" "BE")	
(setq os (getvar "osmode"))	
(setvar "OSMODE" 0)	
....	Prepare input data
(setq nume_fis (getfiled "Open DXF" "c:\\ DXF" "CNC" 1)) (setq CNC (open nume_fis "W"))	Prepare CNC file to be written
(write-line "[BILLET X100 Y100 Z10" CNC) (write-line "[EDGEMOVE X0 Y0 Z0" CNC) (write-line "[TOOLDEF T01 D30" CNC) (write-line "G28 Z30" CNC) (write-line "M06 T01" CNC) (write-line (strcat "M03 S" (itoa (fix s))) CNC)	Write the opening sequence to CNC file
(setq l (entget (car (entsel "Select the polyline:"))))	Gets the associated list of the polyline
(while (/= (caar l) 70) (setq l (cdr l))) (setq inchis (cdar l)) (if (= inchis 1) (setq inchis T) (setq inchis nil))	Learn whether the polyline is closed or not
(while (/= (caar l) 10) (setq l (cdr l)))	Cuts off the irrelevant data
(setq p1 (cdr (nth 0 l)) x1 (car p1) y1 (cadr p1) bulge (cdr (nth 3 l)) xprimul x1 yprimul y1)	Gets the first point and stores it in order to be used as the last one for the closed polylines
(write-line (strcat "G0 X" (rtos x) " Y" (rtos y) " F" (itoa (fix Avans))) CNC) (write-line (strcat "G01 Z-" (rtos ad)) CNC)	Programs the tool start position and entering into the billet
(repeat 4 (setq l (cdr l)))	
(while (> (length l) 1)	
(setq p2 (cdr (nth 0 l)) bulge1 (cdr (nth 3 l)) x2 (car p2) y2 (cadr p2) i 0)	Gets the next point
(setq u (* 4 (atan bulge)) ugrad (/ (* u 180) pi))	Computes the included angle
(if (/= bulge 0)	
(progn (command "ARC" p1 "E" p2 "A" ugrad) (if (> bulge 0) (setq g "G03 ") (setq g "G02 ")) (setq x1 (car p1) y1 (cadr p1) x2 (car p2) y2 (cadr p2) d (sqrt (+ (* (- x1 x2) (- x1 x2)) (* (- y1 y2) (- y1 y2)))) r (sqrt (/ (* d d) 2 (- 1 (cos u))))	If it is an arc, computes the radius and programs the arc, clockwise, or counter clockwise, as the sign of *Bulge* impose
(if (> (abs ugrad) 180) (setq r (- r)))	Changes the sign of R if the included angle of the arc is larger than 180°
(write-line (strcat g "X" (rtos x2) " Y" (rtos y2) " R" (rtos r)) CNC) (setq p1 p2 bulge 0))	
(Progn (command "LINE" p1 p2 "") (write-line (strcat "G01 X" (rtos x2) " Y" (rtos y2)) CNC)))	Programs linear interpolation when need
(repeat 4 (setq l (cdr l)))	Removes from the list the sub-lists of the current point
(setq p1 p2 bulge bulge1)	
(write-line "G28 Z30" CNC) (write-line "M02" CNC) (write-line "M30" CNC)	Writes the ending sequence to CNC file
(setq CNC (close CNC))	Close CNC file
(setvar "OSMODE" os)	
(command "UNDO" "E")	
)	

Table 7. LAsoc_CNC user-defined function. Explanations

Bellow is shown the CNC file as an output of the function, run for the pentagon described by the associated list presented above.

```
[BILLET X100 Y100 Z10
[EDGEMOVE X0 Y0 Z0
[TOOLDEF T01 D30
G28 Z30
M06 T01
M03 S3000
G00 X27.2654 Y0 F125
G01 Z-2
G01 X62.7346 Y0
G03 X72.2451 Y6.9098 R10
G01 X83.2057 Y40.643
G03 X79.573 Y51.8233 R10
G01 X50.8779 Y72.6716
G03 X39.1221 Y72.6716 R10
G01 X10.427 Y51.8233
G03 X6.7943 Y40.643 R10
G01 X17.7549 Y6.9098
G03 X27.2654 Y0 R10
G00 Z10
G28 Z30
M02
M30
```

3.4 Computer aided engendering of CNC files for milling the complex surfaces

The function named VertMill is used to engender the CNC files for finishing by milling following a strategy in vertical planes. To gather the geometrical data in order to process the part, its 3D model is sectioned by vertical planes (YZ or ZX, by the choice of the user). The regions so obtained are exploded, getting first polylines, and then, by a new explosion elementary 2D entities are separated: lines, arcs, ellipse arcs or splines. From these are kept only those are useful for the freeform shapes (thick in Fig. 11), not the sides or the bottom of the outline.

When gathering and processing the geometrical data, in order to define the tool-path, some aspects must be taken into account:

- most of the 2 ½ and 3 axes CNC equipment can perform only linear and circular interpolation;
- the tool-path is obtained most frequent starting from a polyline;
- the polylines cannot include among their elements ellipse arcs or splines.

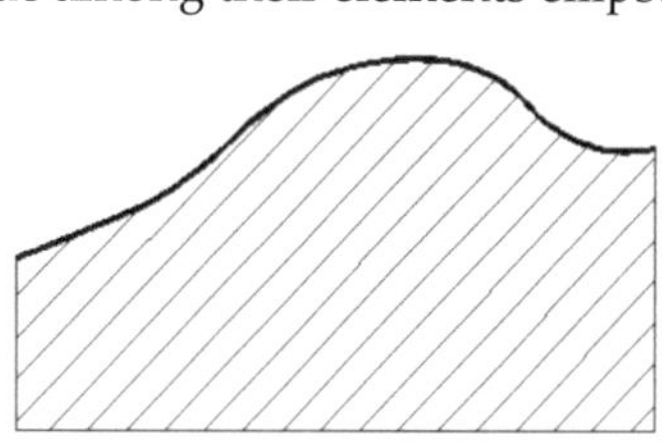

Fig. 11. Section through a freeformed part. Useful elements

Keeping in mind those stated above, the following problems have to be solved:

- approximate the ellipse arcs and the spline curves with sequences of lines and/or arcs;
- assembling these approximations into polylines;
- finally, determining and programming the tool-path.

3.4.1 Approximation the arc of ellipse by lines

A curve, no matter what type it is, can be approximated more or less precisely by a sequence of lines. How many lines should replace a curve? This is a question hard to be answered, because in measuring the quality of the output, two criteria operate: the more segments, the better approximation is, and fewer segments means less tool-path direction changes, that is better dynamic conditions for the machine-tool. Of course, the most important criterion is that of geometrical precision of the machined surface. This must not lead to the impression that a however big number of segments is acceptable, because anyway the computations are easily and fastly performed by the computer. Always a certain admissible deviation from the nominal shape is foreseen. This is the main input that must be taken into account when determining the number and length of the segments that approximate a curve: the maximum distance between the curve and any of the segments (chords of the curve) may not exceed the admitted deviation from the nominal shape.
Determining the set of segments that approximates an arc of circle is simple, because of the curvature of the circle is constant, and consequently all the segments are equal in length. Finding the substitute of an ellipse is more complicated because of the variable curvature of this curve. This means that to maintain the same distance between the curve and any of the segments, these must differ in length (Fig. 12).

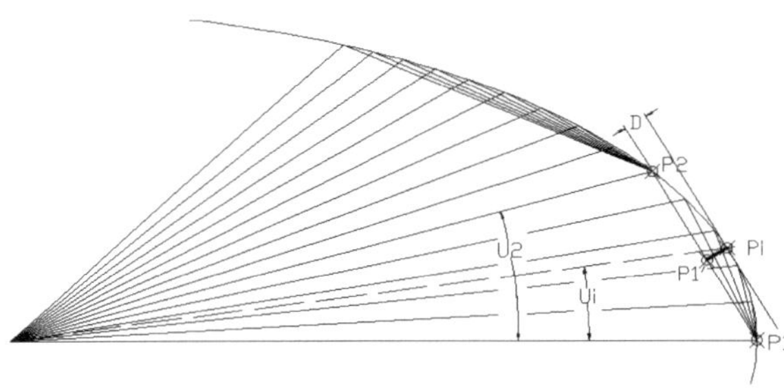

Fig. 12. Data required to compute the distance between the curve and chord

The computing algorithm is as follows: an included angle is such chosen, as it is small enough to ensure that the distance (D) between the ellipse and chord is for sure less that the admitted deviation. This angle is progressively increased towards point P_2 and compute again the distance (D) between the curve and chord. Always the previous point is stored. The procedure is repeated until the distance (D) overcomes the admitted deviation. When this happens the previous point is put in the list of those that describe the approximation. This point becomes P_1 for the next arc. Computing the distance between the curve and the arc is performed as follows: compute the distance in ten intermediary points and store maximum value. This is the distance (D). To perform these calculi the following stages are passed:

- the coordinates of points P_1 and P_2 that end the arc are computed using the parametric coordinates of the ellipse (6);

$$x_i = x_c + a * \cos U_i ; y_i = y_c + b * \sin U_i \tag{6}$$

Where x_i, y_i, x_c, y_c are the coordinates of the current point and of the centre of ellipse, respectively

- write the equation of the chord determined by points P_1 and P_2 (7);

$$y - y_1 = \frac{y_2 - y_1}{x_2 - x_1} * (x - x_1) \tag{7}$$

- to be used later, equation (7) is put in its general form Ax+By+C=0, which has the quotients (8);

$$A = \frac{y_2 - y_1}{x_2 - x_1}; B = -1; C = y_1 - \frac{y_2 - y_1}{x_2 - x_1} x_1 \tag{8}$$

- the intermediary points P_i are determined using equations (6);
- the distance d is computed using equation (9)

$$d = \frac{|Ax_0 + By_0 + C|}{\sqrt{A^2 + B^2}} \tag{9}$$

Some supplementary calculi have to be performed if the ellipse is rotated (has not the axes parallel to those of the coordinate system).

3.4.2 Approximation the spline curves by lines

Approximation the spline curves by segments is performed with a user defined function - TranSpline - using the data in the associated list of the entity. For that is applied the hypothesis that the spline curves obtained by sectioning solids are smooth and have not big curvature. A spline is generally defined by control points (code 10 in the associated list) and knots (code 11 in the associated list). Spline curves result when sectioning cones, cylinders or thoruses, if the sectioning plane is not perpendicular or parallel to the solid's axis. Such curves have not control points, but knots only. In Fig. 13 a spline is shown and then it's associated list.

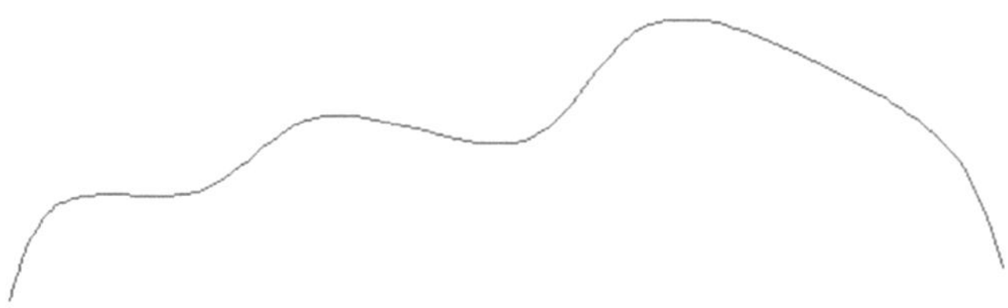

Fig. 13. Spline curve

(44 . 1.0e-010) (40 . 0.0) (40 . 0.0) (40 . 0.0) (40 . 0.0) (40 . 5.14612) (40 . 12.6226) (40 . 18.5784) (40 . 30.3778) (40 . 37.0406) (40 . 47.4409) (40 . 55.131) (40 . 60.5652) (40 . 60.5652) (40 . 60.5652) (40 . 60.5652) (10 45.72 8.55 0.0) (10 46.31 10.42 0.0) (10 47.77 15.01 0.0) (10 55.81 12.13 0.0) (10 61.47 20.31 0.0) (10 72.13 12.74 0.0) (10 77.06 24.35 0.0) (10 86.71 19.79 0.0) (10 93.77 16.32 0.0) (10 95.17 11.98 0.0) (10 95.75 10.18 0.0) (11 45.72 8.55 0.0) (11 48.17 13.08 0.0) (11 55.603 13.89 0.0) (11 60.58 17.15 0.0) (11 72.37 16.52 0.0) (11 76.99 21.32 0.0) (11 87.23 19.51 0.0) (11 93.57 15.16 0.0) (11 95.75 10.18 0.0))

The function for approximation the spline is presented bellow.

Program line	Effect
(defun C:TranSpline ()	
(command "UNDO" "BE")	
(setq osvechi (getvar "OSMODE")	
curba (car (entsel))	Select the spline to be processed
lista (entget curba)	Get the associated list
l lista	
lp1 (list) lp2 (list))	
(while (/= (caar lista) 10)	Remove from the associated list the irrelevant elements for the geometrical data
(setq lista (cdr lista))	
)	
(while (= (caar lista) 10)	Gather the control points of the spline
(setq lp1 (cons (cdr (assoc 10 lista)) lp1)	
lista (cdr lista)))	
)	
)	
(while (= (caar lista) 11)	Gather the knots of the spline
(setq lp2 (cons (cdr (assoc 11 lista)) lp2)	
lista (cdr lista)))	
(setq lp1 (reverse lp1)	
lp2 (reverse lp2)	
i1 (length lp1)	
i2 (length lp2)	
i 0	
)	
(command "line"	Draw a line through the control points (if they exist)
(repeat (1+ i1) (command (nth i lp1))	
(setq i (1+ i))	
)	
(command "")	
)	
(setq i 0)	
(command "color" 5)	
(command "line"	Draw a line through the knots
(repeat (1+ i2) (command (nth i lp2))	
(setq i (1+ i))	
)	
(command "")	
)	
(command "color" "bylayer")	
(command "Erase" curba "")	
(command "Pedit" "l" "Y" "J" "all" "" "")	Join the segment drawn through the knots into a single polyline that approximates the spline
(setvar "OSMODE" osvechi)	
(command "UNDO" "E")	
)	

Table 8. TranSpline user-defined function. Explanations

In Fig. 14 two splines are shown together with their approximations. The curve a) has big inflexions; consequently the polyline drawn through the control points is distant from the curve itself. The polyline drawn through the knots has large deviations from the original object (emphasized in circles), as well. The curve b) obtained by sectioning a 3D object is smooth, has not control points, and the polyline drawn through the knots has not visible deviations from the original curve.

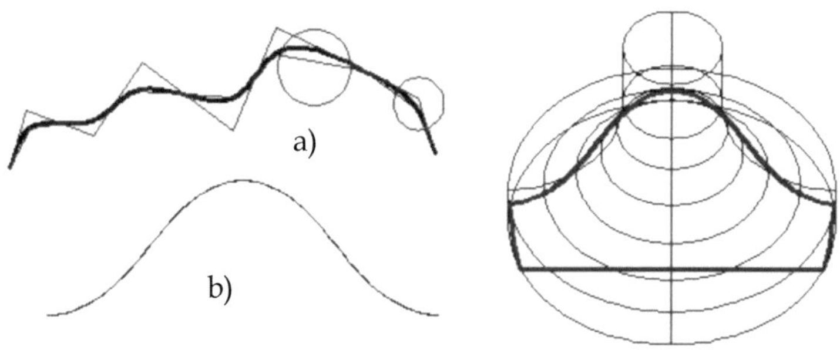

Fig. 13. Aproximation of the curves by polylines a) some spline, b) smooth spline

Assembling the entities that constitute the polyline that approximates the spline is performed using the PEDIT command of AutoCAD. To ensure a correct selection of those entities, the section they come from is placed in a special layer, the single one thaw at the moment of the processing.

3.4.3 Finding the tool-path for milling finishing in vertical planes

Finding the tool-path in vertical planes that is going to be programmed raises some problems if use a ball nosed milling cutter (the most used for finishing). The difficulties occur if the slope of the profile varies, because the contact point of the tool with the surface machined migrates along the tool profile, as shown in Fig. 14. Other approaches are presented in (Bo.& Bangyan, 2006; Kim & Yang, 2007).

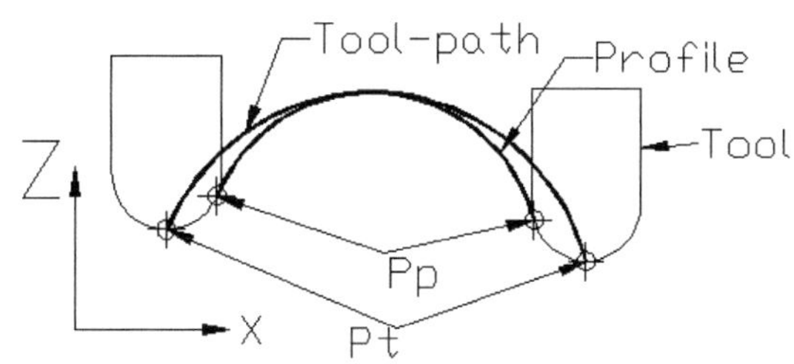

Fig. 14. The contact point Pp migrates along the cutting edge

Because of the tool radius compensation mechanism does not work in a vertical plane, the position of the programmed point must be permanently evaluated. The way the profile slope influences the relative position of the contact point (P_c) between tool and part, on one hand, and the programmed point on another hand is depicted in Fig. 15 (Drăgoi, 2009). Furthermore, in Fig. 15 can be seen the distortion of the tool-path reported to the machined profile.

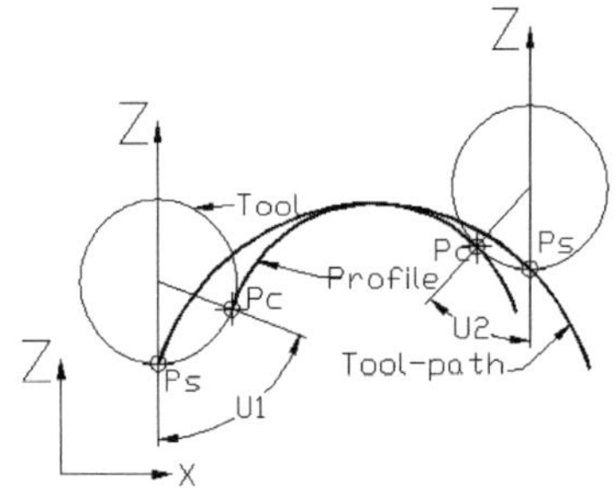

Fig. 15. The relative position of the contact point (P_c) and the programmed one (P_s)

In Fig. 16 are presented the data required to compute the coordinates of the programmed point as a function of the point on the part profile.

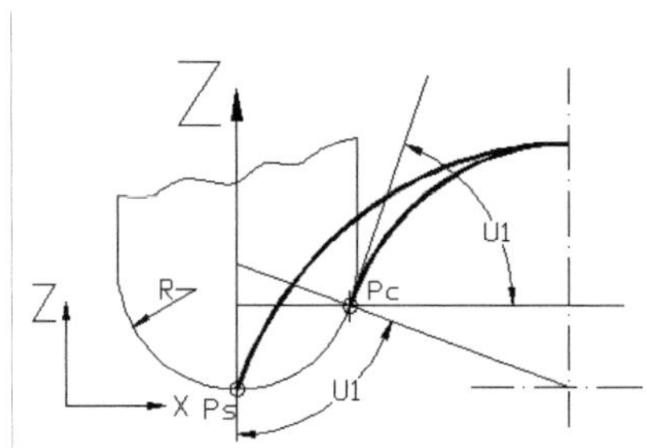

Fig. 16. Computing the position of the programmed point

The coordinates of the programmed point are computed with equations (10) and (11) related the coordinates of the point on the profile, tool radius, and the slope of the profile.

$$X_{Ps}=X_{Pc}-R*\sin(U1); \quad (10)$$

$$Y_{Ps}=Y_{Pc}-R*(1-\cos(U1)); \quad (11)$$

Where R=tool radius, U1=the contact angle, equal to the profile slope, P_c=contact point, and P_s=point to be programmed.

Although computing the coordinates of the programmed point is not difficult, it becomes a problem in the case of curved profiles, where the slope varies continuously along the profile. Because of computing the tool-path point by point cannot be a solution, an AutoLISP user-defined function was designed to solve this problem by means of AutoCAD. The stages of the algorithm are the following:

1. section the 3D model of the part by vertical planes;
2. the region obtained is exploded into the constituent entities;
3. keep only those entities that describe the profile to be machined, not the sides or the bottom of the former region;
4. join the entities retained at the previous stage into a polyline;
5. apply *Offset* command on the polyline, above, at a distance equal to the tool radius;
6. move the output of *Offset* command downward with a displacement equal to the tool radius.

The curve obtained at the stage 6 is the tool-path that must be programmed in the CNC file. The new offset seems to be (and it is) distorted from the part profile. This curve is not an equidistant to the profile, since a point of it (P1) lies on the profile, and another point (P2) is at a certain distance in any other position. Furthermore, that distance varies permanently, related to the slope of the profile, as can be seen in Fig. 17.

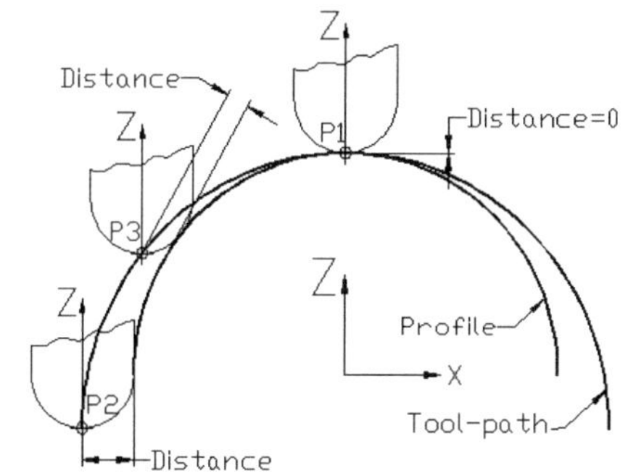

Fig. 17. The tool-path is not equidistant to part profile

This manner to determine the tool-path has another major advantage against computing the coordinates of the programmed point: the equations for computing the coordinates must be modified if the surface to be machined is concave instead of convex. In other words, the equations are different if machining inside or outside the profile.
In practice, problems may occur if some sectors of the profile are curves that cannot be joined to a polyline (i. e. ellipse arcs or splines). In such cases, if every entity is offseted individually the output cannot be used as a tool-path because of the discontinuities (Fig. 18a).

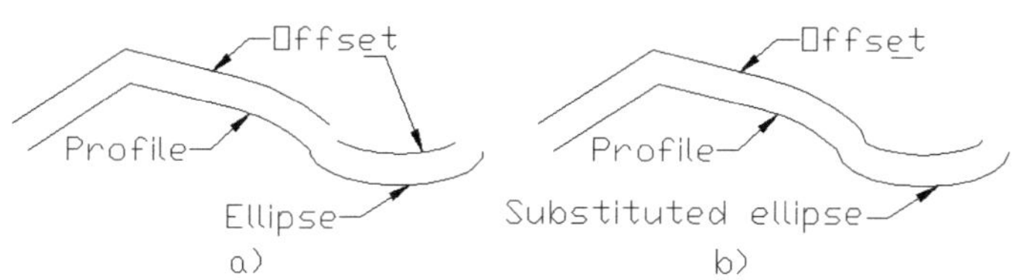

Fig. 18. Offsetting entities individually is not a solution, because of the discontinuity of the output

To overcome this disadvantage, the arcs of ellipse and splines must be substituted - as have been described above in this chapter - by sequences of lines, which can be joined to a polyline. Such a polyline can now be offseted, and the output can act as tool-path (Fig. 18b).
Everything has been presented so far related to tool-path finding for milling in vertical planes is only the theoretical support of the problem. Solving the problem is the output of the AutoLISP user-defined function called VertMill, which steps through the following stages:

1. opens the file where the CNC program will be saved;
2. gets by means of 3Dsize function the sizes of the 3D model;
3. adopts the sizes of the billet and generates the billet;
4. builds an axial section of the milling cutter that will be used later;
5. sections successively the part with vertical planes, YZ or ZX as the user chooses;

For each section the following tasks are performed:

6. align the UCS to the region obtained by sectioning. Aligning is necessary in order to perform later some specific 2D operation that can only be done in the XY plane;
7. explode the region in the current XY plane;

8. every entity obtained by explosion is examined by means of the associated list.
9. the arcs of ellipse and splines are substituted by sequences of lines;
10. the old arcs of ellipse and splines are deleted;
11. all the segments and arcs (if they exist) are joined into a polyline;
12. an offset is drawn above the profile at a distance equal to the tool radius;
13. the offset is moved downwards with a displacement equal to the tool radius - this is the tool-path;
14. gather from the associated list of the offset its' geometrical data;
15. coordinate transformations are applied to the points on the offset (they are in the current XY plane, and must be transferred to the YZ or ZX plane of the part);
16. The new coordinates are written to the CNC file;
17. sweep the section of the tool (obtained in stage 4) along the tool-path;
18. the solid obtained at the previous stage is subtracted from the billet. Stages 16 and 17 constitute the components of the machining simulation module;
19. the vertical plane is moved with an increment equal to the transversal feed-rate;
20. stages 6 to 18 are repeated until the 3D model is entirely processed;
21. close the CNC file.

Only the main stages have been here presented. Many other auxiliary actions are necessary to perform all the transformations: managing the layers and UCSs, moving entities from a layer to another to facilitate the required entities selections, operating with OSNAP modes, saving some system variables at the beginning of the program and restoring it at the end of the program, many computations.

4. CAD for Rapid tooling

Rapid tooling is, in the widest sense, the fast making of tools, in the context of manufacturing engineering. Particularly the *Rapid tooling* syntagma is associated with rapidly producing mould inserts for injection moulding of plastics and casting of metals. As a step towards producing the tools mentioned before, CAD of inserts for injection of plastic mass parts is very important. Some aspects related to the design of inserts, having as data input the 3D model of the part itself are discussed bellow. For illustration, take as sample a mould to obtain the plastic shell for packing a tooth-brush and a mould for plastic injection of the tooth-brush's handle.

4.1 3D modelling of the product

As already has been mentioned, in the context of simultaneous design, even in the CAD stage must be taken into account the technological aspects. That is, the shape of the part has to be adequate to the technology it will be made by. In order to have the input for designing the plastic shell for packing, not the material and inner details are important, but the outer shape of the part. The shape of the proper tooth-brush is obtained by extruding a 2D contour. The handle is built using the LOFT command of AutoCAD applied on a set of elliptical sections aligned to the same axis (Fig. 19).

Taking into account of the shape of the handle (the cross section is elliptical), the plastic shell will not follow exactly the aspect of the product; that is the shell does not mould the handle under it's the horizontal medial plane, noted with P-P in Fig. 20.

Consequently, in order to design the mould, the model of the product must be adjusted. For that, the handle is sectioned by plane P-P and the region obtained is extruded downward (Fig 21).

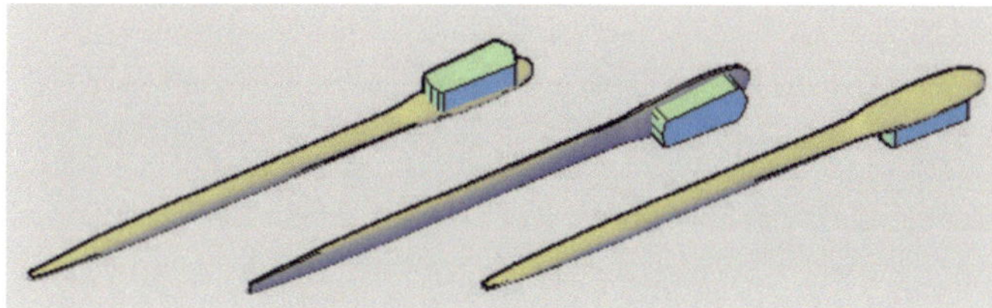

Fig. 19. The 3D model of the product

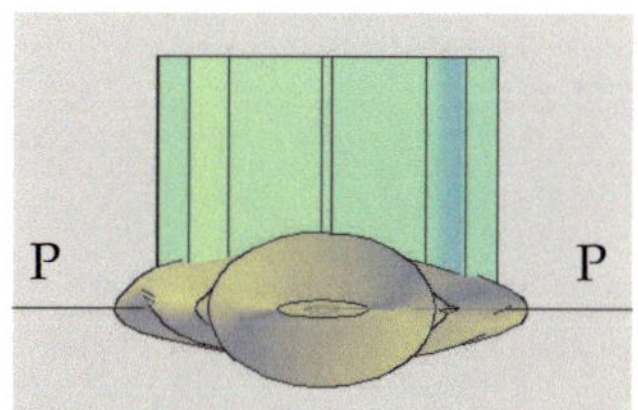

Fig. 20. The 3D model of the product. Detail

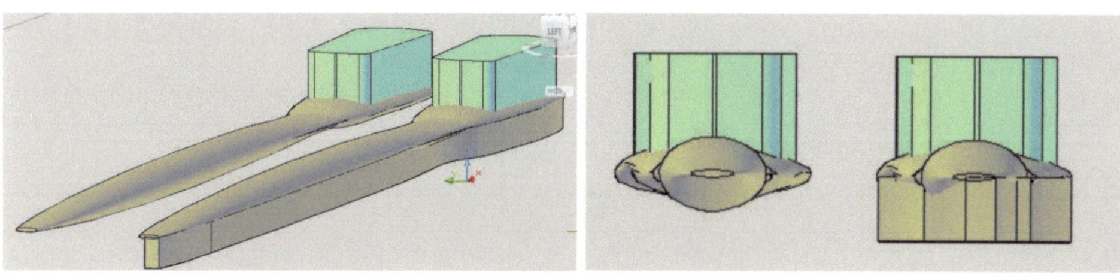

Fig. 21. The 3D model adjusted in order to design the mould

4.2 Design of the mould used to shape the plastic shell

Modelling the mould supposes building the 3D models of the inserts. For this, first of all a model of the shell itself must be created. To do that the command of AutoCAD SOLIDEDIT/Body/Shell is used. To get the shell outside the model of the product, a negative offset distance is used. The lower face of the object is removed. The output is shown in Fig. 22.

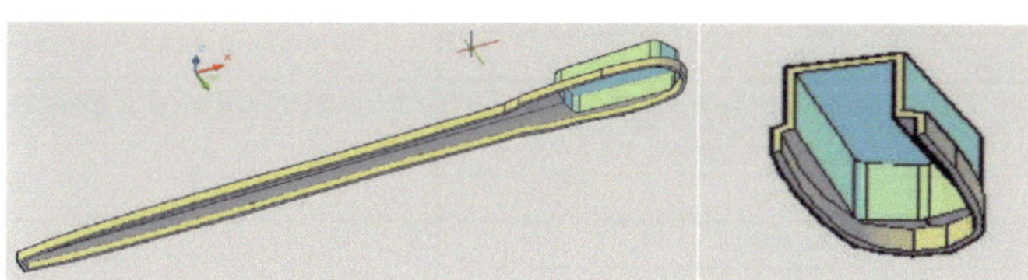

Fig. 22. The model of the plastic shell

Later, this shell is subtracted together with the model of the product from a box-shaped body, to get the upper insert of the mould (Fig.23a). The lower insert is obtained unioning the product 3D model to a box-shaped object (Fig.23b). Of course, in this way, only the active parts of the mould were obtained, forwards all the other details being designed (determination if part retainer or "chase" is required, incorporation of "locks" to prevent shifting of core and cavity, specification of the ejection required, if any, finding the size and location of spruem runners, and gates, determination of the size, number, and location of water lines for temperature control, if required).

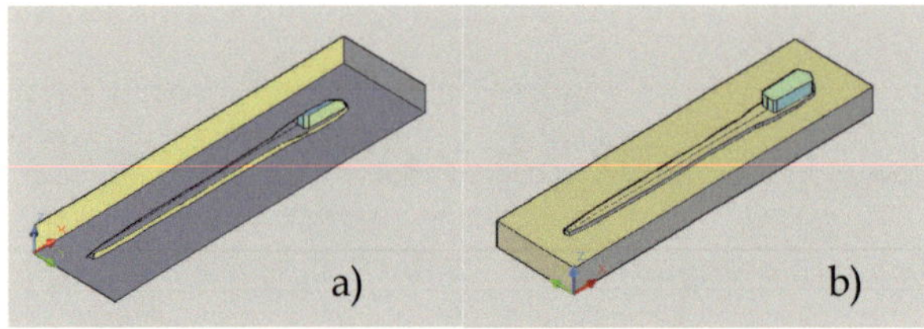

Fig. 23. The inserts. The upper one – a), the lower one – b)

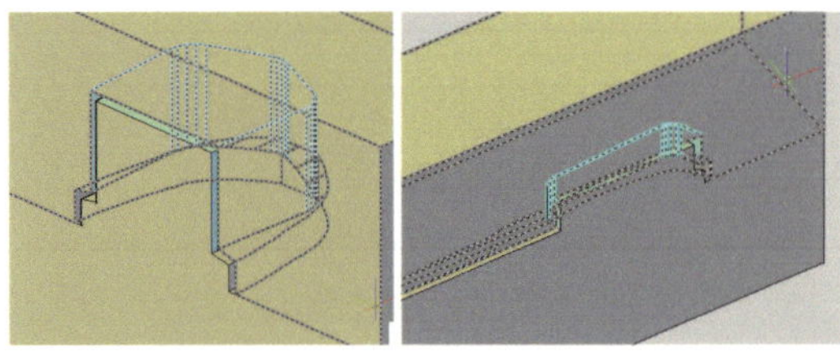

Fig. 24. The two inserts assembled (two sections - details)

The sequence of operations presented was turned into an AutoLISP user-defined function named Mould1. Its important stages are presented in Table 9.

Program line	Effect
(defun C:Mould1 ()	
....	
(setq produs (car (entsel "\n Select the object to be processed ")))	Select the object to be processed
(setq g (getreal "\The thicknes of the shell "))	Input data
(command "copy" produs "" "0,0,0" "")	Make a copy of the object
(setq cs_prod (entlast))	Give a name to the object
(command "copy" produs "" "0,0,0" "")	Make a new copy of the object
....	
(command "Solidedit" "B" "S" copie "" (- g) "" "")	Create the shell
(Princ"\nBuild the two inserts: ")	
(setq P1 (getpoint "\nFirst corner of the insert:")	Input the sizes of the inserts
P2 (getpoint "\nOther corner ")	
h1 (getreal "\nUpper insert height:")	
h2 (getreal "\nLower insert height::")	
)	
(command "BOX" P1 P2 h1)	Make the rough inserts
(setq sup (entlast))	
(command "BOX" P1 P2 (- h1))	
(setq inf (entlast))	
(command "-layer" "N" "Produs" "")	Manage the layers
(command "CHPROP" produs "" "LA" "Produs" "")	
(command "-layer" "F" "Produs" "")	
(command "Subtract" Sup "" copie cs_prod "")	Make the active parts of the inserts
(command "Union" inf ci_prod "")	
)	

Table 9. Mould1 user-defined function. Explanations

4.3 Design of the mould used to make the product by plastic injection

Modelling the inserts of the mould supposes making the cavities in the inserts, the plastic mass is going to be injected in. The sample is the tooth-brush handle. The main stages of the process are as follows:

- scaling the model with the thermal dilatation quotient;
- defining the billet;
- subtracting the product 3D model from the billet;
- separating the two inserts by slicing the billet with a horizontal plane at he level of the separating plane

The sequence of operations presented was turned into an AutoLISP user-defined function named Mould2. The main stages are presented in Table 10.

Program line	Effect
(defun C:Mould2 ()	
(command "Undo" "BE")	
(setq os (getvar "osmode"))	
(setvar "OSMODE" 0)	
(setq produs (car (entsel "\ Select the object to be processed ")))	Select the object to be processed
(setq P1 (getpoint "\nFirst corner of the insert:")	
P2 (getpoint "\nOther corner ")	Input data to describe the sizes of the inserts
h1 (getreal "\nUpper insert height:")	
h2 (getreal "\nLower insert height:")	
Sep (getpoint "\nThe level of sepparating plane ")	Input the separating plane
)	
(command "UCS" "O" Sep)	
(setq coef (getreal "/n thermal dilatation quotient:"))	
(command "Scale" produs "" "0,0,0" coef)	
(command "BOX" P1 P2 (+ h1 h2))	Make the rough insert (unique for now), and name it
(setq pastile (entlast))	
(Command "move" pastile "" (list 0 0 (- h2)) "")	
(Command "Subtract" pastile "" produs"")	Make the active part of the insert
(command "Slice" pastile "" "XY" "0,0,0" "B")	Separate the inserts
(command "-layer" "N" "Produs" "")	
(command "CHPROP" produs "" "LA" "Produs" "")	
(command "-layer" "F" "Produs" "")	
(setvar "OSMODE" os)	
(command "Undo" "E")	
)	

Table 10. Mould2 user-defined function. Explanations

After run the function, the ensemble in Fig. 25 is obtained (In Fig. 25 slices of the inserts are presented, as the inner details to be visible).

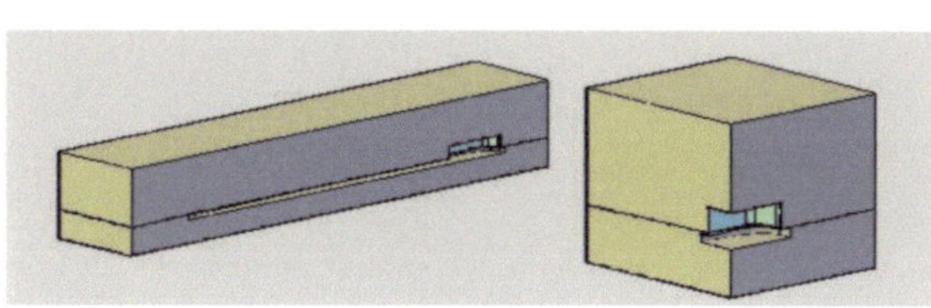

Fig. 25. The mould inserts

4.4 Application. Modelling the mould inserts for plastic mass injection

The application uses the function Mould2 to design the active part of the inserts of a mould for plastic mass injection. In Fig. 26 the part to be produced is presented in two 3D views. In Fig. 27 a slice through the two inserts knocked together, and in Fig. 28 the two inserts individually are depicted.

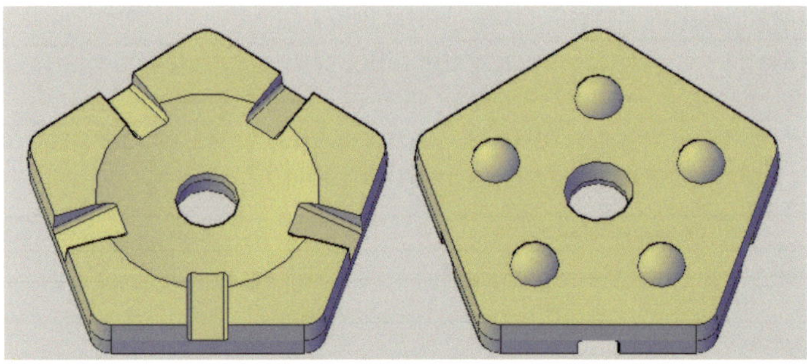

Fig. 26. The part to be processed. Special roundel. Up and down view

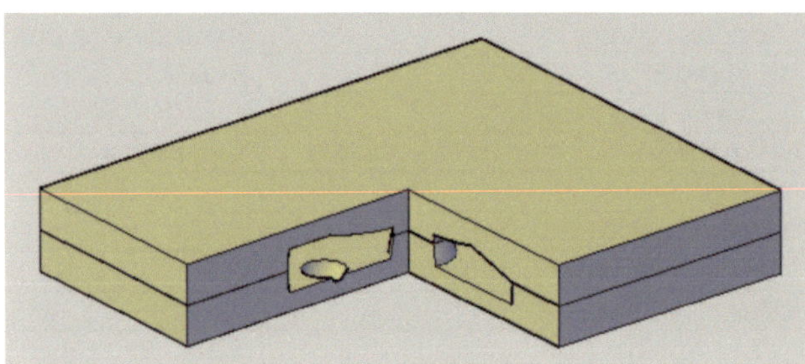

Fig. 27. The mould inserts. Slice to view inside details

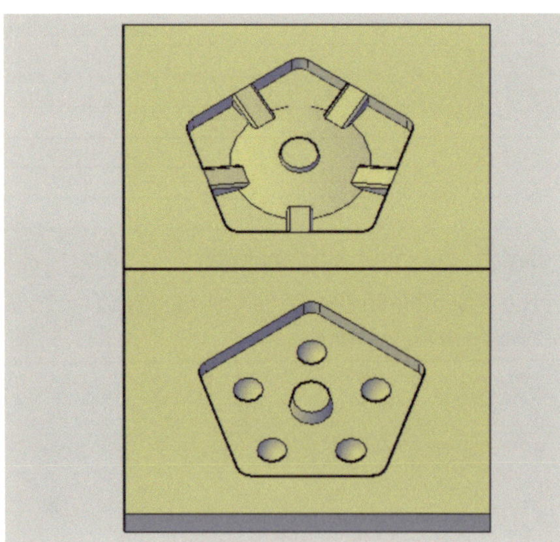

Fig. 28. The pair of inserts

5. Conclusion

The approach of the subject here presented is a unitary one, all the solutions proposed being obtained by means of AutoCAD and AutoLISP. The set of AutoLISP functions cover several specific tasks of the manufacturing engineer. The manufacturing engineer is thought as a specialist that acts both in the constructive and the technological sides of design, being able to integrate them. The integration goes now further, being applied to design and manufacturing by means of CAD/CAM systems. Two new user-defined functions for the constructive-technological design were described (EXDATROUG1 and EXDATROUG2). New techniques of scanning 3D models in horizontal and vertical planes are set up, as means of checking the machinability of the parts by 2 ½ or 3 axes CNC machine-tools.
2D CNC milling processing is targeted by the new user-defined functions DXF_CNC and LAsoc_CNC. Means to aid the engineer in designing the active inserts of some moulds are also presented in this chapter.

5.1 Future research

After presenting the new tools offered to manufacturing engineers, some questions might rise: are these tools good as they are? Are they too specialized and too focused to narrow problems, or are they too general? The answers can be given only by those interested to use it, after they will get benefits from it, or will find certain drawbacks.
Future researches are foreseen to improve the tools as potential users would suggest. Here could be mentioned extending the set of functions consecrated to mould design with new ones that make easier the job of designing it in terms of determination if part retainer is required, incorporation of "locks" to prevent shifting of core and cavity, specification of the ejection required, if any, dimensioning and placing the water lines for temperature control, if they are necessary, and others.
New useful tools are intended to be provided in the area of reverse engineering, namely reconstituting the 3D model of a part, having as input the CNC file used to produce the object.

6. Acknowledgement

Some achievements presented in this chapter (2.1, 3.2, 3.3, 4.1, 4.2, and 4.3) were obtained within the research project Sisteme Expert de Optimizare a Proceselor Tehnologice (Expert System for Optimisation of Technological Processes) - ESOP, PN II, 2007-2010; nr. 71-133/18.09.2007 funded by the Romanian Ministry of Education and Research.

7. References

Bo, Z. & Bangyan, Y. (2006). Study on 3D tool compensation algorithm for NC milling, Proceedings of *International Technology and Innovation Conference 2006 (ITIC 2006)*, pp. 290 -294, ISBN: 0 86341 696 9, Hangzhou, China, 6-7 November 2006

Chicoş, L.A. & Ivan N.V.,(2004), Programmes Package Regarding Concurrent Engineering for Product Development - New Applications, *Academic Journal of Manufacturing Engineering (AJME)*, Vol.2, Nr.2, (June 2004), pg.31-37, ISSN: 1583-7904,

Chicoş, L.A., et al., (2009)., Simultaneous approach of CAD and CAM technologies using constructive-technological entities, *Annals of DAAAM for 2009*, Volume 20, No.1 (November 2009), pp.377-378, ISSN 1726-9679.

Drăgoi, M. V., (2009,) Ball Nose Milling Cutter Radius Compensation in Z Axis for CNC.. Proceedings of *the 8th WSEAS International Conference on Recent Advances in Software Engineering and Parallel and Distributed Systems (SEPADS '09)*, pp. 57-60, ISBN 978-960-474-052-9 Cambridge, UK, February 21-23 2009.

Ivan, N. V. (1994), Prozessoren für geometrische und technologische Beschreibung der Werkstücke. *Proceedings of 5th Internationales DAAAM Sympozium*, pp. 173-174, ISBN 86-435-0084-4, Maribor, Slovenia, October 1994, University of Maribor, Maribor

Ivan, N. V., et al., (1996) Technological Processor Structure. *Buletin of Transilvania University of Brasov, New Series*, A, Vol 3 No 38, (October 1997) p.129-134. ISSN 1223-9631

Kim, S.-J. & Yang, M.-Y.(2007), Incomplete mesh offset for NC machining", *Journal of Materials Processing Technology* Vol. 194 No. 1-3 (November 2007), pp. 110–120, ISSN: 0924-0136

5

Computer Simulation of Plaque Formation and Development in the Cardiovascular Vessels

Nenad Filipovic
University of Kragujevac
Serbia

1. Introduction

Atherosclerosis is a progressive disease characterized in particular by the accumulation of lipids and fibrous elements in artery walls. Over the past decade, scientists come to appreciate a prominent role for inflammation in atherosclerosis.

Atherosclerosis is characterized by dysfunction of endothelium, vasculitis and accumulation of lipid, cholesterol and cell elements inside blood vessel wall. This process develops in

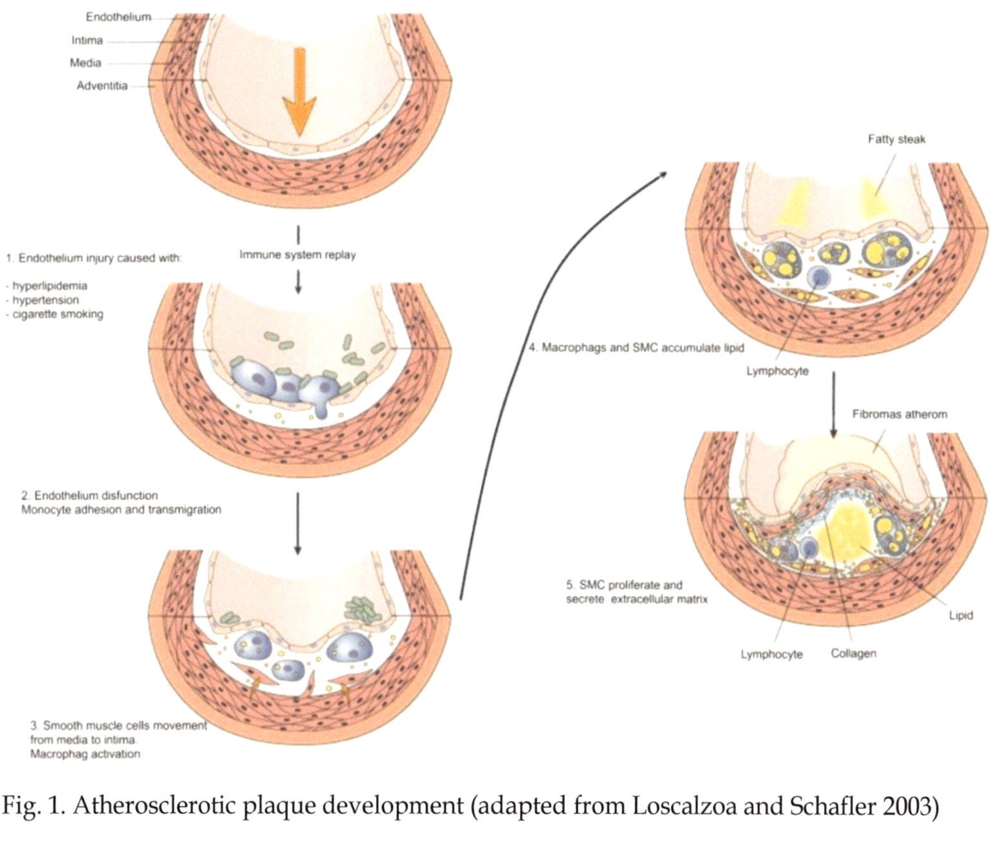

Fig. 1. Atherosclerotic plaque development (adapted from Loscalzoa and Schafler 2003)

arterial walls. Atherosclerosis develops from oxidized low-density lipoprotein molecules (LDL). When oxidized LDL evolves in plaque formations within an artery wall, a series of reactions occur to repair the damage to the artery wall caused by oxidized LDL. The body's immune system responds to the damage to the artery wall caused by oxidized LDL by sending specialized white blood cells-macrophages (Mphs) to absorb the oxidized-LDL forming specialized foam cells. Macrophages accumulate inside arterial intima. Also Smooth Muscle Cells (SMC) accumulate in the atherosclerotic arterial intima, where they proliferate and secrete extracellular matrix to form a fibrous cap (Loscalzo & Schafer, 2003). Unfortunately, macrophages are not able to process the oxidized-LDL, and ultimately grow and rupture, depositing a larger amount of oxidized cholesterol into the artery wall. The atherosclerosis process is shown in Fig. 1.
This chapter describes a completely new computer model for plaque formation and development. The first section is devoted to the LDL model of transport from the lumen to intima and detailed three-dimensional model for inflammatory and plaque progression process. The next section describes some of the benchmark examples for 2D, 2D axi-symmetric and 3D model of plaque formation and development. At the end a complex specific-patient 3D model is given. Finally the main conclusions of the work are addressed.

2. Methods

In this section a continuum based approach for plaque formation and development in three-dimension is presented. All algorithms are incorporated in program PAK-Athero from University of Kragujevac (Filipovic et al., 2010).

2.1 Governing equations for modeling of LDL transport through the arterial wall

The governing equations and numerical procedures are given. The blood flow is simulated by the three-dimensional Navier-Stokes equations, together with the continuity equation

$$-\mu\nabla^2 u_l + \rho\left(u_l \cdot \nabla\right)u_l + \nabla p_l = 0 \tag{1}$$

$$\nabla u_l = 0 \tag{2}$$

where $\boldsymbol{u}_l$ is blood velocity in the lumen, p_l is the pressure, μ is the dynamic viscosity of the blood, and ρ is the density of the blood.
Mass transfer in the blood lumen is coupled with the blood flow and modelled by the convection-diffusion equation as follows

$$\nabla \cdot \left(-D_l \nabla c_l + c_l u_l\right) = 0 \tag{3}$$

in the fluid domain, where c_l is the solute concentration in the blood lumen, and D_l is the solute diffusivity in the lumen.
Mass transfer in the arterial wall is coupled with the transmural flow and modelled by the convection-diffusion-reaction equation as follows

$$\nabla \cdot (-D_w \nabla c_w + k c_w u_w) = r_w c_w \tag{4}$$

in the wall domain, where c_w is the solute concentration in the arterial wall, D_w is the solute diffusivity in the arterial wall, K is the solute lag coefficient, and r_w is the consumption rate constant.

LDL transport in lumen of the vessel is coupled with Kedem-Katchalsky equations:

$$J_v = L_p\left(\Delta p - \sigma_d \Delta\pi\right) \tag{5}$$

$$J_s = P\Delta c + \left(1-\sigma_f\right)J_v\bar{c} \tag{6}$$

where L_p is the hydraulic conductivity of the endothelium, Δc is the solute concentration difference across the endothelium, Δp is the pressure drop across the endothelium, $\Delta\pi$ is the oncotic pressure difference across the endothelium, σ_d is the osmotic reflection coefficient, σ_f is the solvent reflection coefficient, P is the solute endothelial permeability, and $\bar{c}$ is the mean endothelial concentration.

The basic relations for mass transport in the artery. The metabolism of the artery wall is critically dependent upon its nutrient supply governed by transport processes within the blood. A two different mass transport processes in large arteries are addressed. One of them is the oxygen transport and the other is LDL transport. Blood flow through the arteries is usually described as motion of a fluid-type continuum, with the wall surfaces treated as impermeable (hard) boundaries. However, transport of gases (e.g. O_2, CO_2) or macromolecules (albumin, globumin, LDL) represents a convection-diffusion physical process with permeable boundaries through which the diffusion occurs. In the analysis presented further, the assumption is that the concentration of the transported matter does not affect the blood flow (i.e. a diluted mixture is considered). The mass transport process is governed by convection-diffusion equation,

$$\frac{\partial c}{\partial t} + v_x\frac{\partial c}{\partial x} + v_y\frac{\partial c}{\partial y} + v_z\frac{\partial c}{\partial z} = D\left(\frac{\partial^2 c}{\partial x^2} + \frac{\partial^2 c}{\partial y^2} + \frac{\partial^2 c}{\partial z^2}\right) \tag{7}$$

where c denotes the macromolecule or gas concentration; v_x, v_y and v_z are the blood velocity components in the coordinate system x,y,z; and D is the diffusion coefficient, assumed constant, of the transported material.

Boundary conditions for transport of the LDL. A macromolecule directly responsible for the process of atherosclerosis is LDL which is well known as atherogenic molecule. It is also known that LDL can go through the endothelium at least by three different mechanisms, namely, receptor-mediated endocytosis, pinocytotic vesicular transport, and phagocytosis (Goldstein et al., 1979). The permeability coefficient of an intact arterial wall to LDL has been reported to be of the order of 10^{-8} $[cm/s]$ (Bratzler et al., 1977). The conversion of the mass among the LDL passing through a semipermeable wall, moving toward the vessel wall by a filtration flow and diffusing back to the mainstream at the vessel wall, is described by the relation

$$c_w v_w - D\frac{\partial c}{\partial n} = Kc_w \tag{8}$$

where c_w is the surface concentration of LDL, v_w is the filtration velocity of LDL transport through the wall, n is coordinate normal to the wall, D is the diffusivity of LDL, and K is the overall mass transfer coefficient of LDL at the vessel wall. A uniform constant concentration C_0 of LDL is applied at the artery tree inlet as classical inlet boundary condition for eq. (7).

2.2 Finite element modeling of diffusion-transport equations

In the case of blood flow with mass transport we have domination of the convection terms due to the low diffusion coefficient (Kojic et al., 2008). Then it is necessary to employ special stabilizing techniques in order to obtain a stable numerical solution. The streamline upwind/Petrov-Galerkin stabilizing technique (SUPG) (Brooks & Hughes, 1982) within a standard numerical integration scheme is implemented. The incremental-iterative form of finite element equations of balance are obtained by including the diffusion equations and transforming them into incremental form. The final equations are

$$\begin{bmatrix} \frac{1}{\Delta t}\mathbf{M}_{\mathbf{v}} + {}^{n+1}\mathbf{K}_{\mathbf{vv}}^{(i-1)} + {}^{n+1}\mathbf{K}_{\boldsymbol{\mu}\mathbf{v}}^{(i-1)} + {}^{n+1}\mathbf{J}_{\mathbf{vv}}^{(i-1)} & {}^{n+1}\mathbf{K}_{\mathbf{vp}}^{(i-1)} & \mathbf{0} \\ \mathbf{K}_{\mathbf{vp}}^{T} & \mathbf{0} & \mathbf{0} \\ {}^{n+1}\mathbf{K}_{\mathbf{cv}}^{(i-1)} & \mathbf{0} & \frac{1}{\Delta t}\mathbf{M}_{\mathbf{c}} + {}^{n+1}\mathbf{K}_{\mathbf{cc}}^{(i-1)} + {}^{n+1}\mathbf{J}_{\mathbf{cc}}^{(i-1)} \end{bmatrix} \times \tag{9}$$

$$\begin{Bmatrix} \Delta\mathbf{V}^{(i)} \\ \Delta\mathbf{P}^{(i)} \\ \Delta\mathbf{C}^{(i)} \end{Bmatrix} = \begin{Bmatrix} {}^{n+1}\mathbf{F}_{\mathbf{v}}^{(i-1)} \\ {}^{n+1}\mathbf{F}_{\mathbf{p}}^{(i-1)} \\ {}^{n+1}\mathbf{F}_{c}^{(i-1)} \end{Bmatrix}$$

where the matrices are

$$\begin{aligned}
&(\mathbf{M}_{\mathbf{v}})_{jjKJ} = \int_V \rho N_K N_J dV, && (\mathbf{M}_{\mathbf{c}})_{jjKJ} = \int_V N_K N_J dV \\
&\left({}^{n+1}\mathbf{K}_{\mathbf{cc}}^{(i-1)}\right)_{jjKJ} = \int_V D N_{K,j} N_{J,j} dV && \left({}^{n+1}\mathbf{K}_{\mu v}^{(i-1)}\right)_{jjKJ} = \int_V \mu N_{K,j} N_{J,j} dV \\
&\left({}^{n+1}\mathbf{K}_{\mathbf{cv}}^{(i-1)}\right)_{jjKJ} = \int_V N_K\, {}^{n+1}c_{,j}^{(i-1)} N_J dV && \left({}^{n+1}\mathbf{K}_{vv}^{(i-1)}\right)_{jjKJ} = \int_V \rho N_K\, {}^{n+1}v_j^{(i-1)} N_{J,j} dV \\
&\left({}^{n+1}\mathbf{J}_{\mathbf{cc}}^{(i-1)}\right)_{jjKJ} = \int_V \rho N_K\, {}^{n+1}v_j^{(i-1)} N_{J,j} dV && \left({}^{n+1}\mathbf{K}_{\mathbf{vp}}^{(i-1)}\right)_{jjKJ} = \int_V \rho N_{K,j} \hat{N}_J dV \\
&\left({}^{n+1}\mathbf{J}_{\mathbf{vv}}^{(i-1)}\right)_{jkKJ} = \int_V \rho N_K\, {}^{n+1}v_{j,k} N_J dV
\end{aligned} \tag{10}$$

and the vectors are

$$\begin{aligned}
{}^{n+1}\mathbf{F}_{c}^{(i-1)} = {}^{n+1}\mathbf{F}_{q} + {}^{n+1}\mathbf{F}_{sc}^{(i-1)} - \frac{1}{\Delta t}\mathbf{M}_{c}\left\{{}^{n+1}\mathbf{C}^{(i-1)} - {}^{n}\mathbf{C}\right\} - \\
{}^{n+1}\mathbf{K}_{cv}^{(i-1)}\left\{{}^{n+1}\mathbf{V}^{(i-1)}\right\} - {}^{n+1}\mathbf{K}_{cc}^{(i-1)}\left\{{}^{n+1}\mathbf{C}^{(i-1)}\right\}
\end{aligned} \tag{11}$$

$$\left({}^{n+1}\mathbf{F}_{q}\right)_K = \int_V N_K q^B dV \qquad {}^{n+1}\mathbf{F}_{\mathbf{sc}}^{(i-1)} = \int_S D N_K \nabla\, {}^{n+1}\mathbf{c}^{(i-1)} \cdot \mathbf{n} dS$$

Note that $\hat{N}_J$ are the interpolation functions for pressure (which are taken to be for one order of magnitude lower then interpolation functions N_I for velocities). The matrices $\mathbf{M}_{cc}$ and $\mathbf{K}_{cc}$ are the 'mass' and convection matrices; $\mathbf{K}_{cv}$ and $\mathbf{J}_{cc}$ correspond to the convective terms of equation (7); and $\mathbf{F}_c$ is the force vector which follows from the convection-diffusion equation in (7) and linearization of the governing equations.

2.3 Modeling of plaque formation and development

The inflammatory process was solved using three additional reaction-diffusion partial differential equations (Calvez et al., 2008; Boynard et al., 2009):

$$\begin{aligned}
&\partial_t Ox = d_2 \Delta Ox - k_1 Ox \cdot M \\
&\partial_t M + div(v_w M) = d_1 \Delta M - k_1 Ox \cdot M + S/(1-S) \\
&\partial_t S = d_3 \Delta S - \lambda S + k_1 Ox \cdot M + \gamma(Ox - Ox^{thr})
\end{aligned} \tag{12}$$

where Ox is the oxidized LDL or c_w - the solute concentration in the wall from eq. (7); M and S are concentrations in the intima of macrophages and cytokines, respectively; d_1, d_2, d_3 are the corresponding diffusion coefficients; λ and γ are degradation and LDL oxidized detection coefficients; and v_w is the inflammatory velocity of plaque growth, which satisfies Darcy's law and continuity equation (Kojic et al., 2008; Filipovic et al., 2004, 2006a, 2006b):

$$v_w - \nabla \cdot (p_w) = 0 \tag{13}$$

$$\nabla v_w = 0 \tag{14}$$

in the wall domain. Here, p_w is the pressure in the arterial wall.

3. Results

3.1 2D model of plaque formation and development

For the first example of a two-dimensional model of the mild stenosis, a fully developed parabolic steady velocity profile was assumed at the lumen inlet boundary

$$u(r) = 2U_0 \left(1 - \left(\frac{2r}{D}\right)^2\right) \tag{15}$$

where $u(r)$ is the velocity in the axial direction at radial position r; D is the inlet diameter; and U_0 is the mean inlet velocity. At the lumen side of the endothelial boundary, a lumen-to-wall transmural velocity in the normal direction was specified:

$$t_l^T \cdot u_l = 0 \ , \ \ u_l n_l = J_v \tag{16}$$

where $\boldsymbol{t}_l^T$ and $\boldsymbol{n}_l$ are the tangential and normal unit vectors of fluid subdomain, respectively. Oxidized LDL distribution for a mild stenosis is shown in Fig. 2.

3.2 2D axi-symmetric model of plaque formation and development

The plaque formation and development is modeled through an initial straight artery in 2D axi-symmetric model with mild constriction of 30%. The inlet artery diameter d_0=0.4 [cm]. Blood was modeled as a Newtonian fluid with density $\rho = 1.0$ [g/cm^3] and viscosity $\mu = 0.0334$ [P]. The steady state conditions for fluid flow and mass transport are assumed. The entering blood velocity is defined by the Reynolds number Re (calculated using the mean blood velocity and the artery diameter).

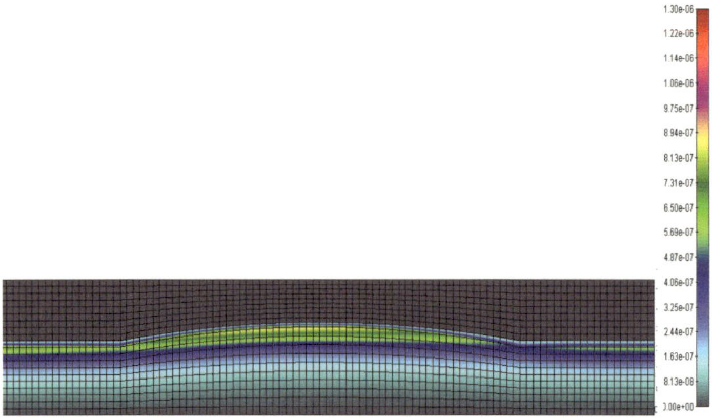

Fig. 2. Oxidized LDL distribution for a mild stenosis (30% constriction by area)

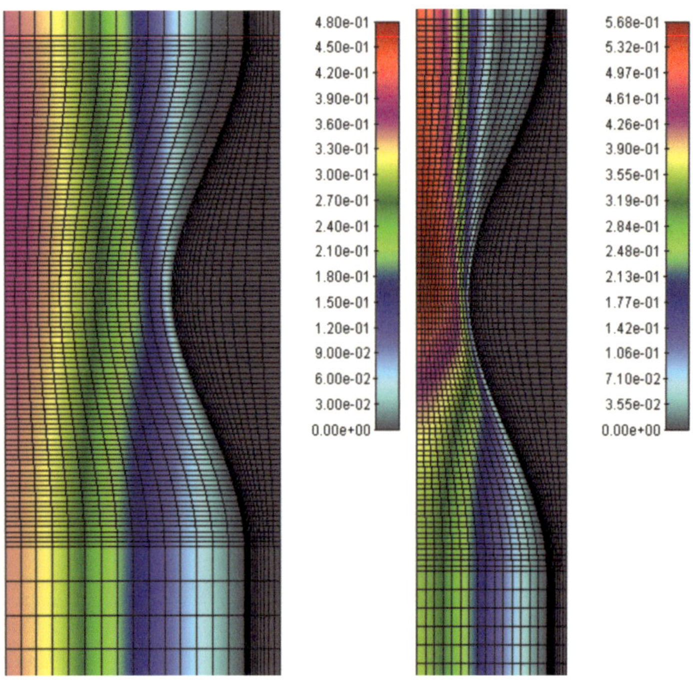

Fig. 3. a) Velocity distribution for an initial mild stenosis 30% constriction by area
b) Velocity distribution at the end of stenosis process after 10^7 sec [units m/s]

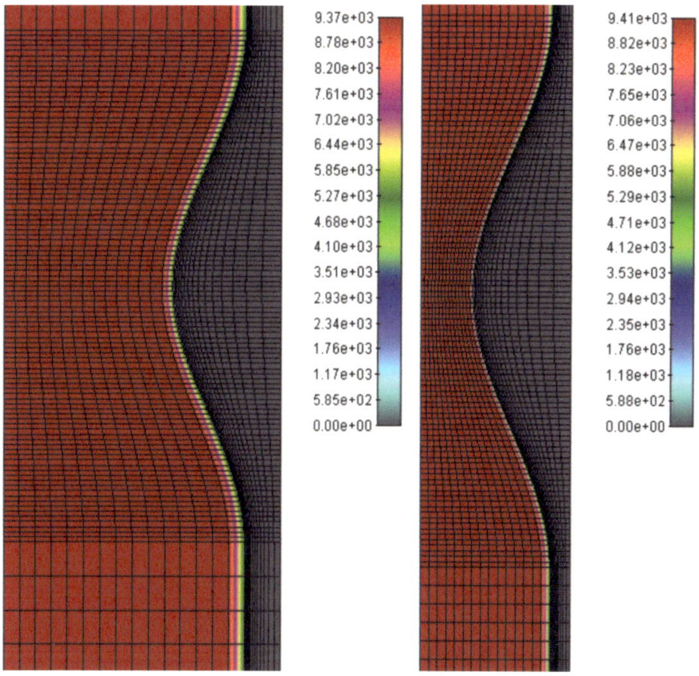

Fig. 4. a) Pressure distribution for an initial mild stenosis 30% constriction by area
b) Pressure distribution at the end of stenosis process after 10^7 sec[units Pa]

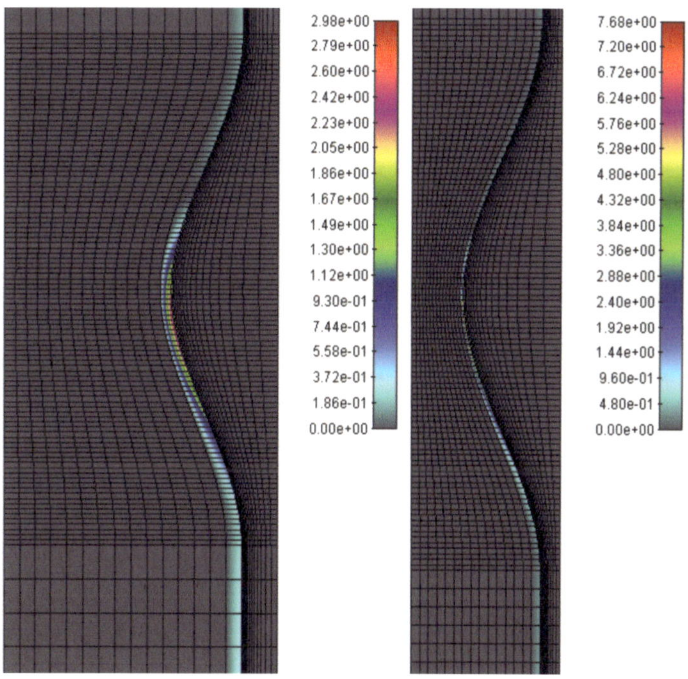

Fig. 5. a) Shear stress distribution for an initial mild stenosis 30% constriction by area
b) Shear stress distribution at the end of stenosis process after 10^7 sec[unitsdyn/cm^2]

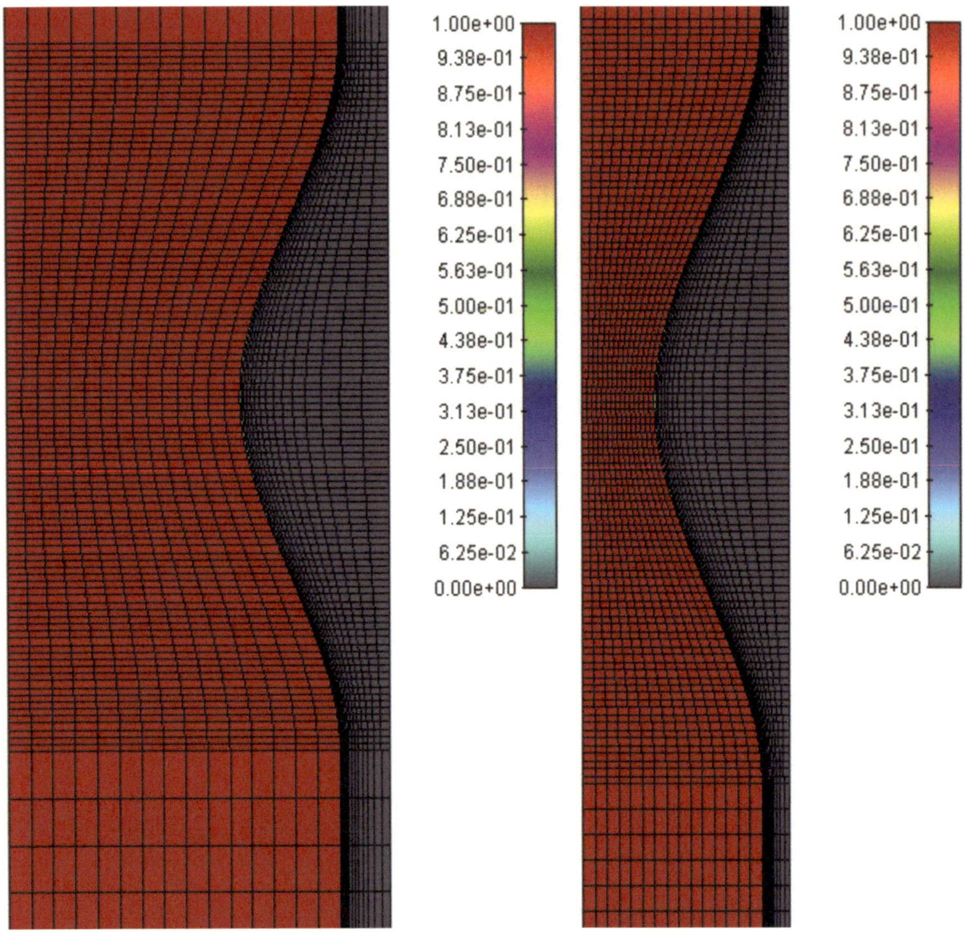

Fig. 6. a) Lumen LDL distribution for an initial mild stenosis 30% constriction by area b) Lumen LDL distribution at the end of stenosis process after 10^7 sec[units mg/mL]

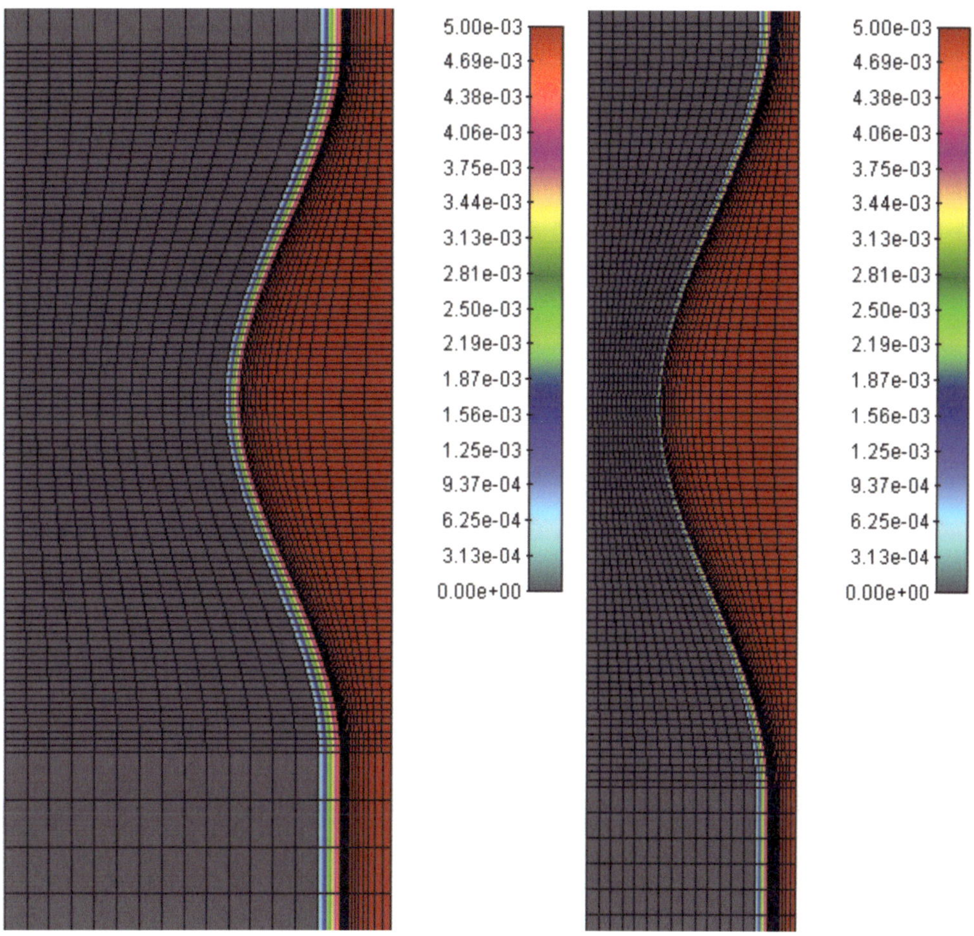

Fig. 7. a) Oxidized LDL distribution in the intima for an initial mild stenosis 30% constriction by area b) Oxidized LDL distribution in the intima at the end of stenosis process after 10^7sec[units mg/mL]

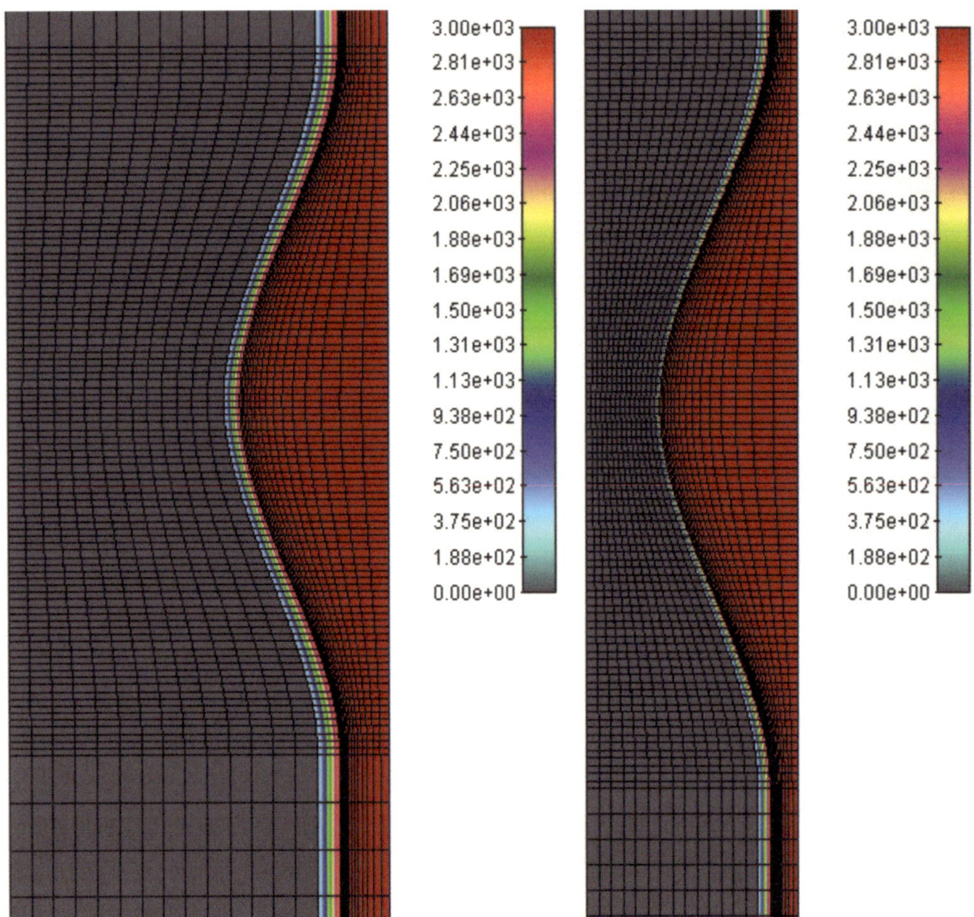

Fig. 8. a) Intima wall pressure distribution for an initial mild stenosis 30% constriction by area b) Intima wall pressure distribution at the end of stenosis process after 10^7sec[units Pa]

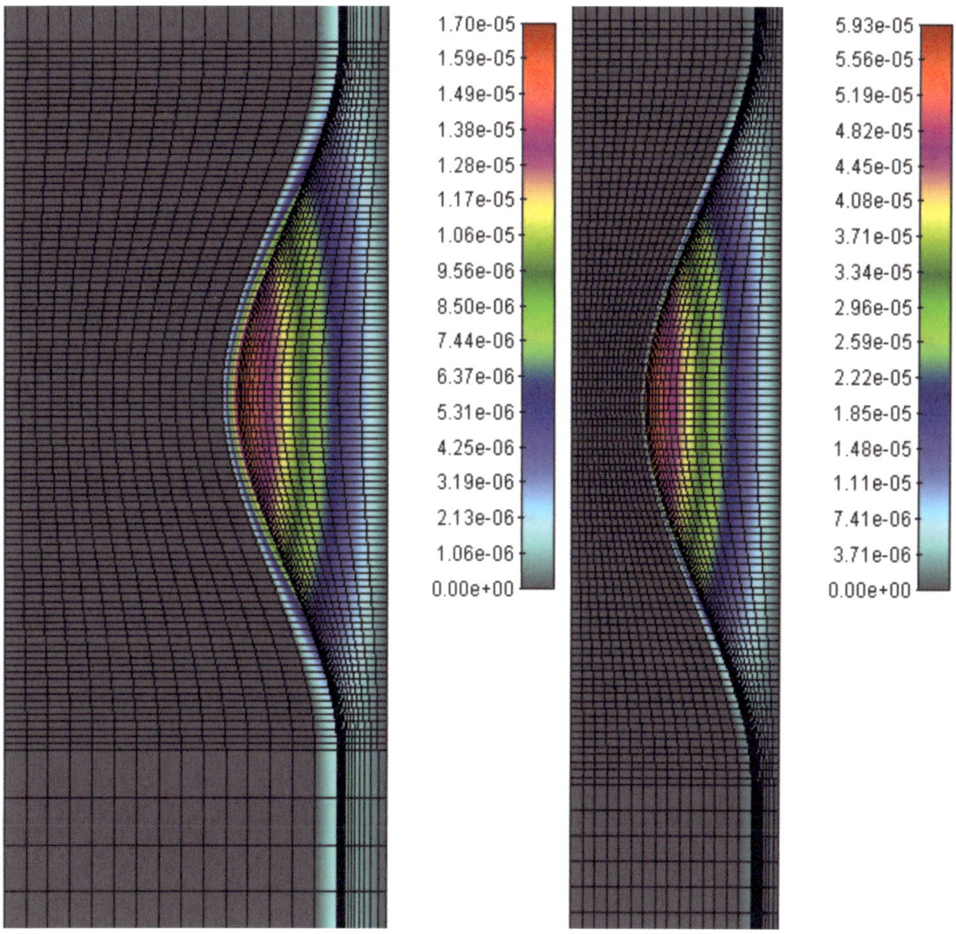

Fig. 9. a) Macrophages distribution in the intima for an initial mild stenosis 30% constriction by area b) Macrophages distribution in the intima at the end of stenosis process after 10^7 sec[units mg/mL]

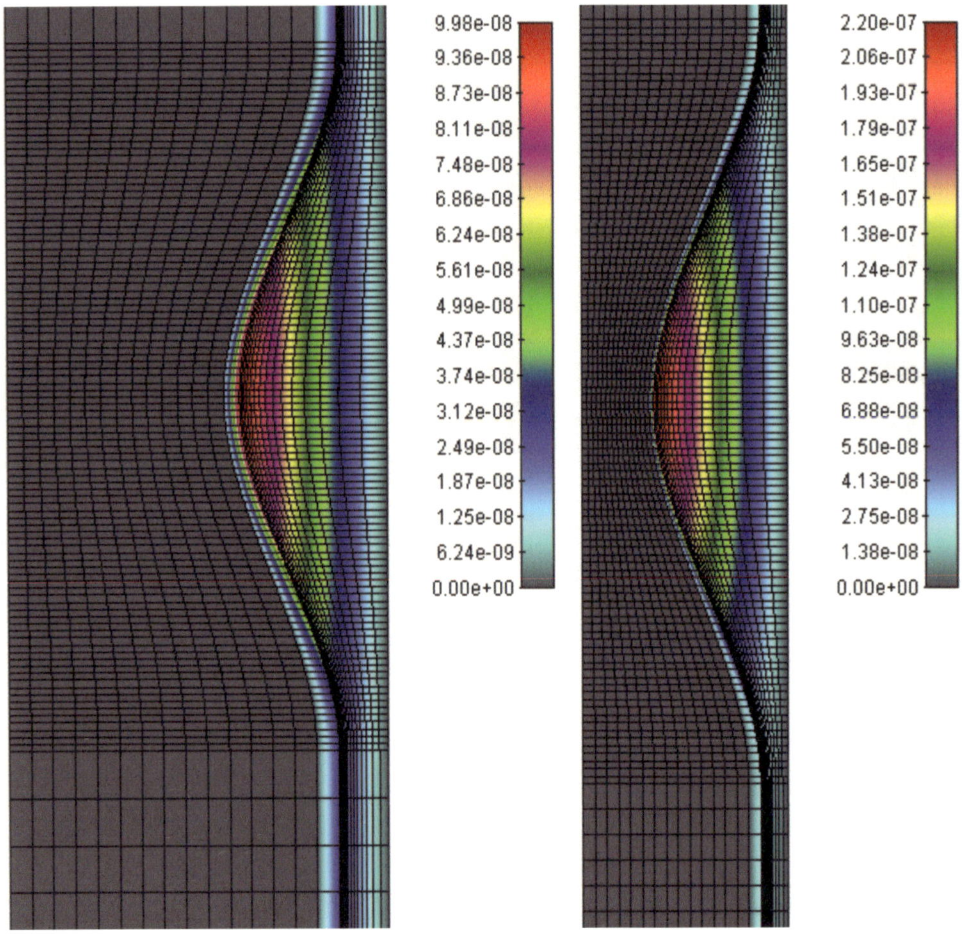

Fig. 10. a) Cytokines distribution in the intima for an initial mild stenosis 30% constriction by area b) Cytokines distribution in the intima at the end of stenosis process after 10^7 sec[units mg/mL]

3.3 3D model of plaque formation and development

In order to make benchmark example for three-dimensional simulation we tested simple middle stenosis with initial 30% constriction for time period of t=10^7 sec (approximately 7 years) and compare results with 2D axi-symmetric model. The results for velocity distribution for initial and end stage of simulations are presented in Fig. 11a and Fig. 11b.

The pressure and shear stress distributions for start and end time are given in Fig. 12 and Fig. 13. Concentration distribution of LDL inside the lumen domain and oxidized LDL inside the intima are presented in Fig. 14 and Fig. 15.The transmural wall pressure is presented in Fig. 16. Macrophages and cytokines distributions are shown in Fig. 17 and Fig. 18.The diagram of three-dimensional plaque volume growing during time is given in Fig. 19. It can be seen that time period for developing of stenosis corresponds to data available in the literature (Goh et al., 2010).

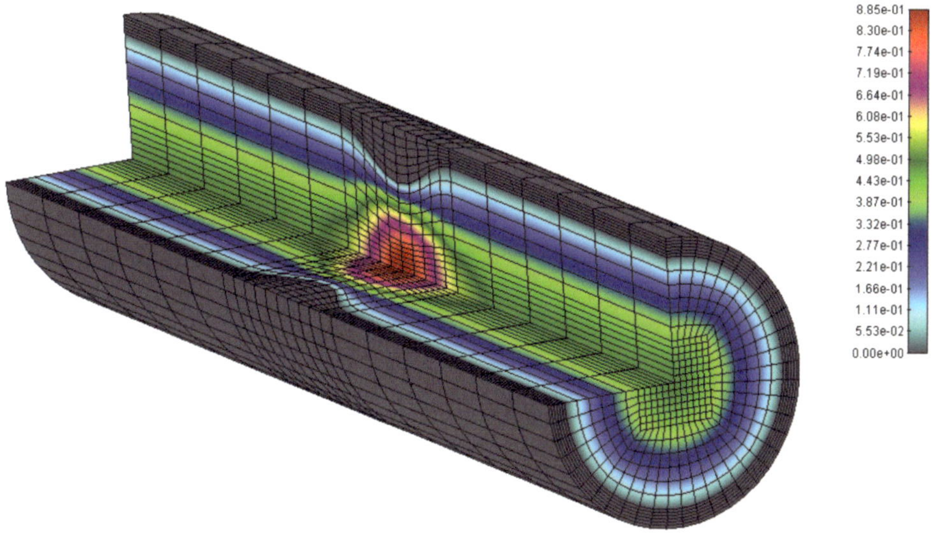

a)

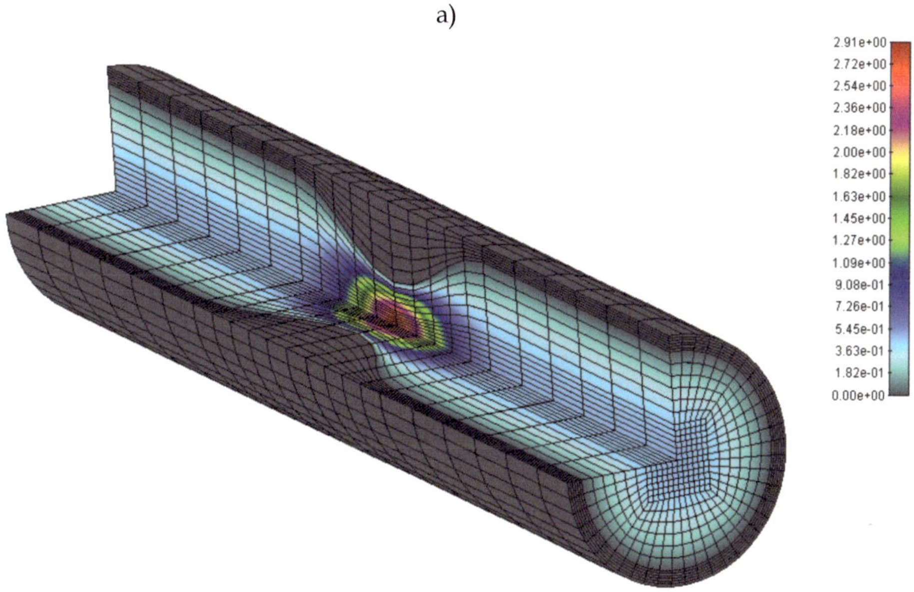

b)

Fig. 11. a) Velocity distribution for an initial mild stenosis 30% constriction by area
b) Velocity distribution at the end of stenosis process after 10^7 sec[unitsm/s]

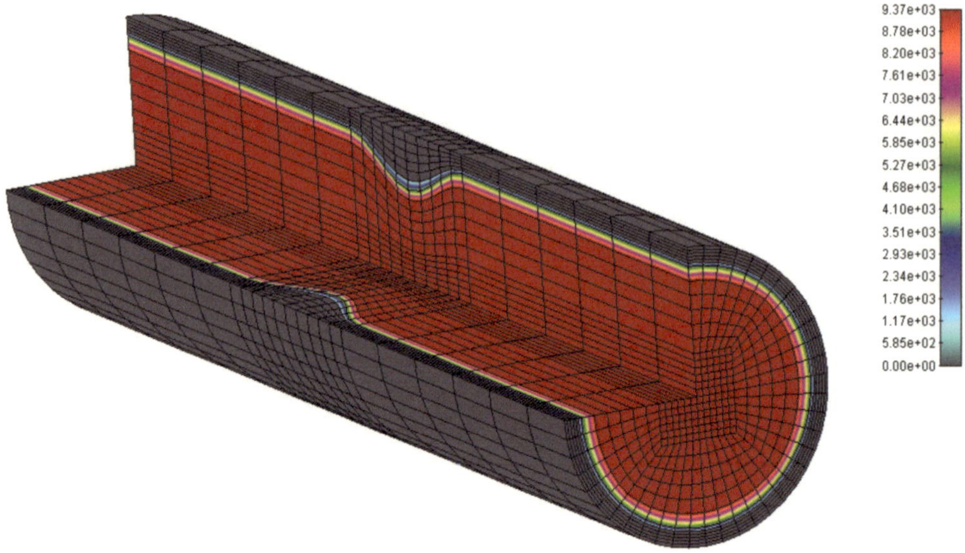

a)

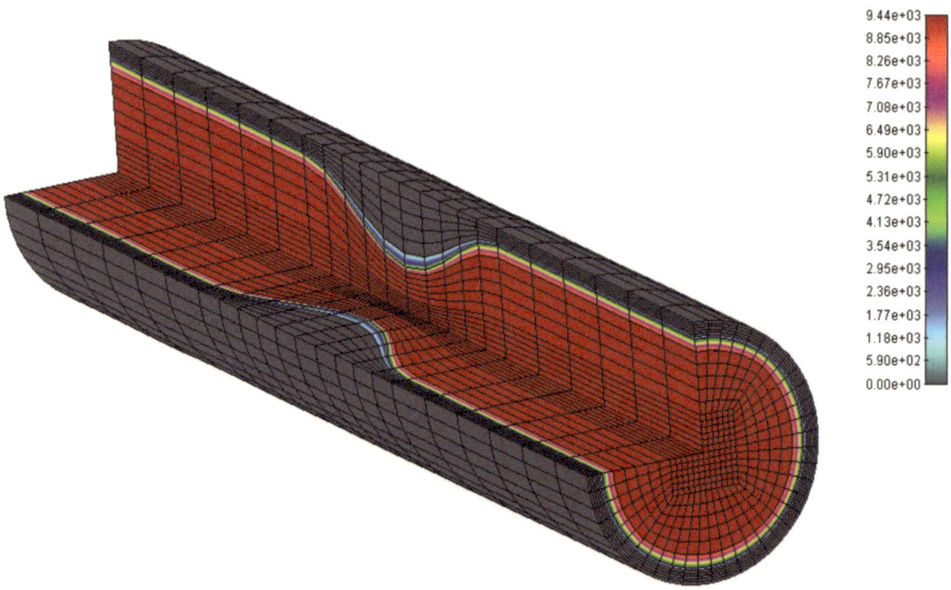

b)

Fig. 12. a) Pressure distribution for an initial mild stenosis 30% constriction by area
b) Pressure distribution at the end of stenosis process after 10^7 sec[units Pa]

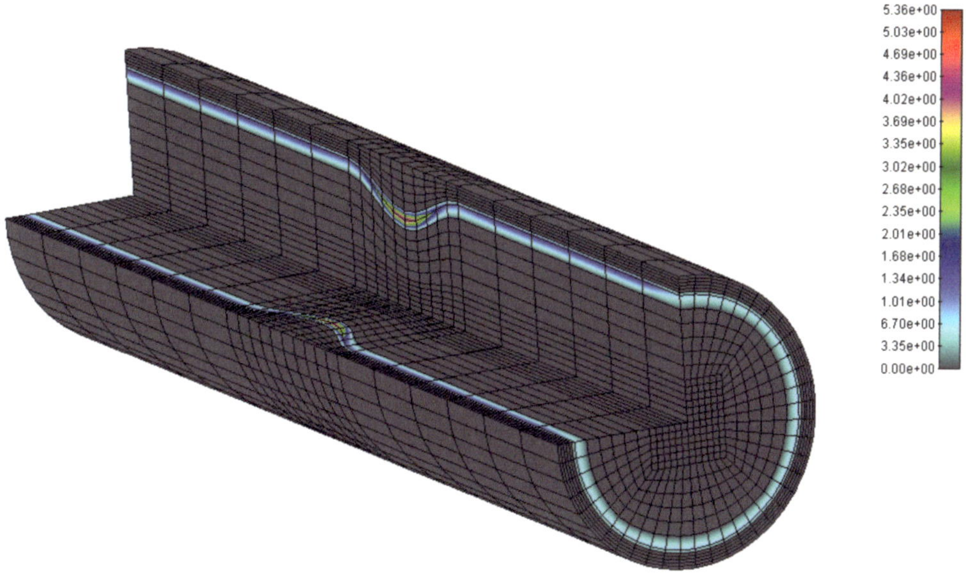

a)

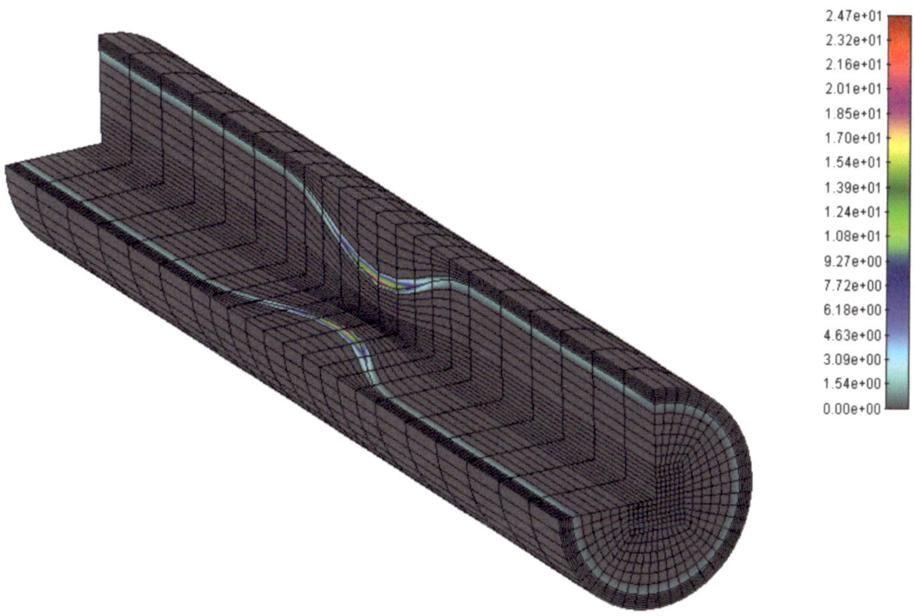

b)

Fig. 13. a) Shear stress distribution for an initial mild stenosis 30% constriction by area
b) Shear stress distribution at the end of stenosis process after 10^7 sec[unitsdyn/cm^2]

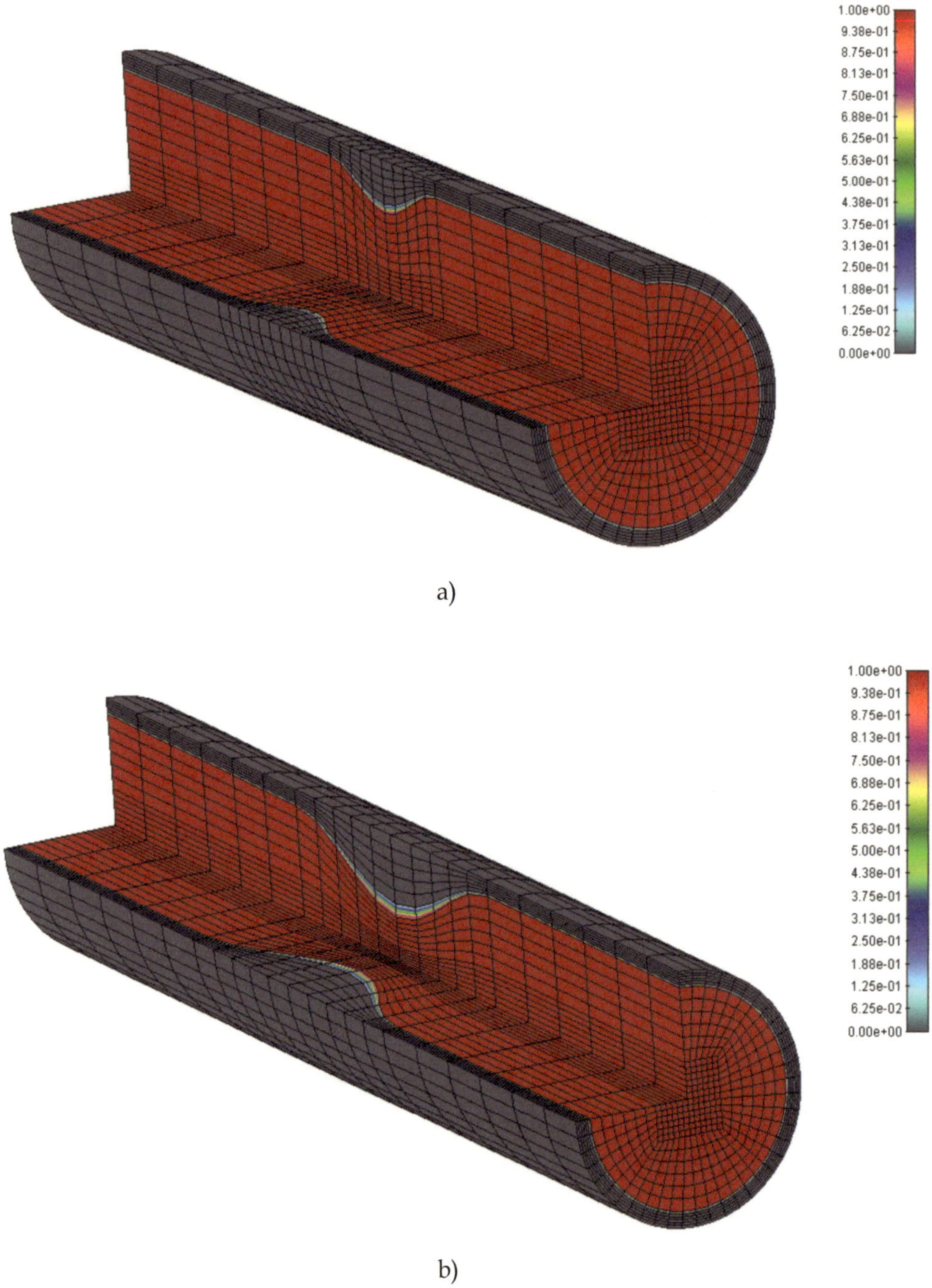

Fig. 14. a) Lumen LDL distribution for an initial mild stenosis 30% constriction by area b) Lumen LDL distribution at the end of stenosis process after 10^7 sec[units mg/mL]

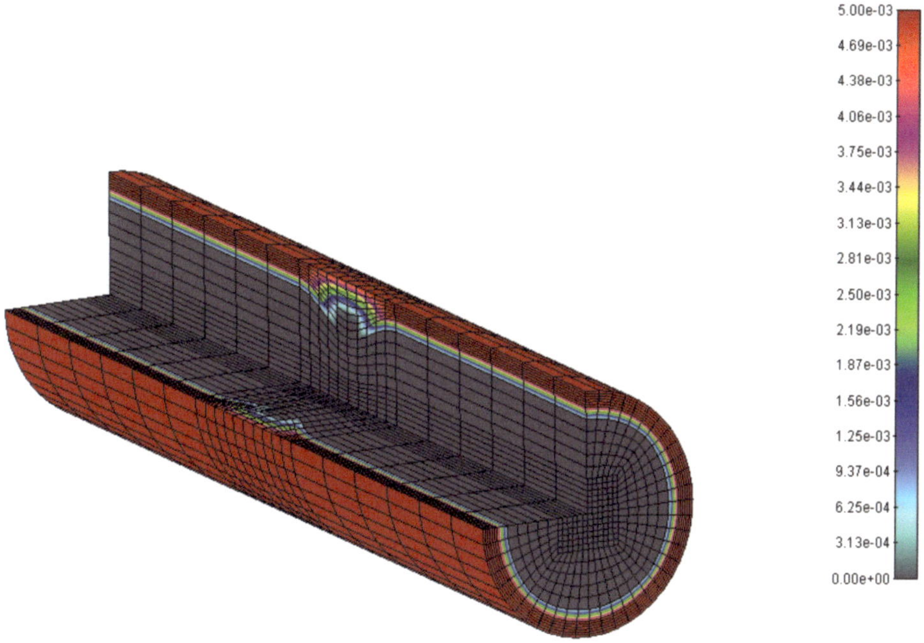

a)

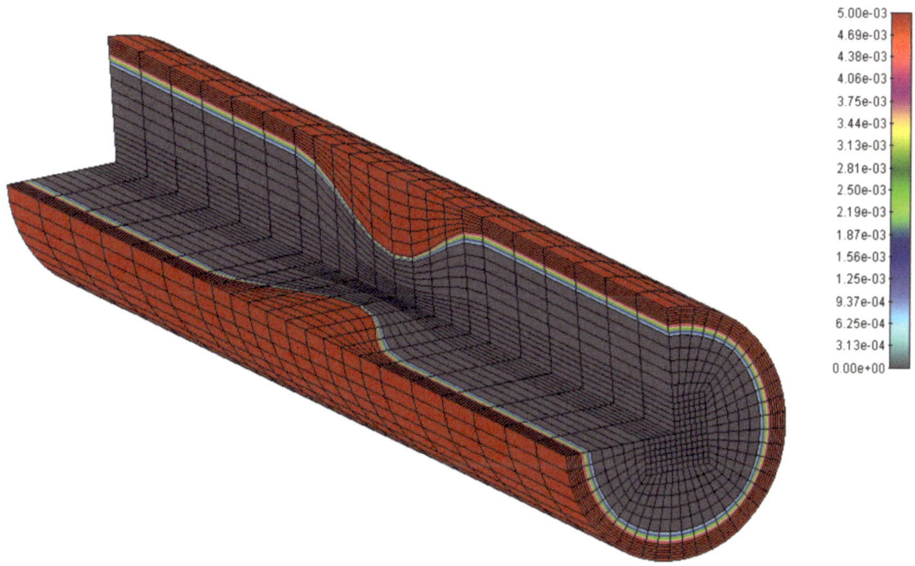

b)

Fig. 15. a) Oxidized LDL distribution in the intima for an initial mild stenosis 30% constriction by area b) Oxidized LDL distribution in the intima at the end of stenosis process after 10^7 sec[units mg/mL]

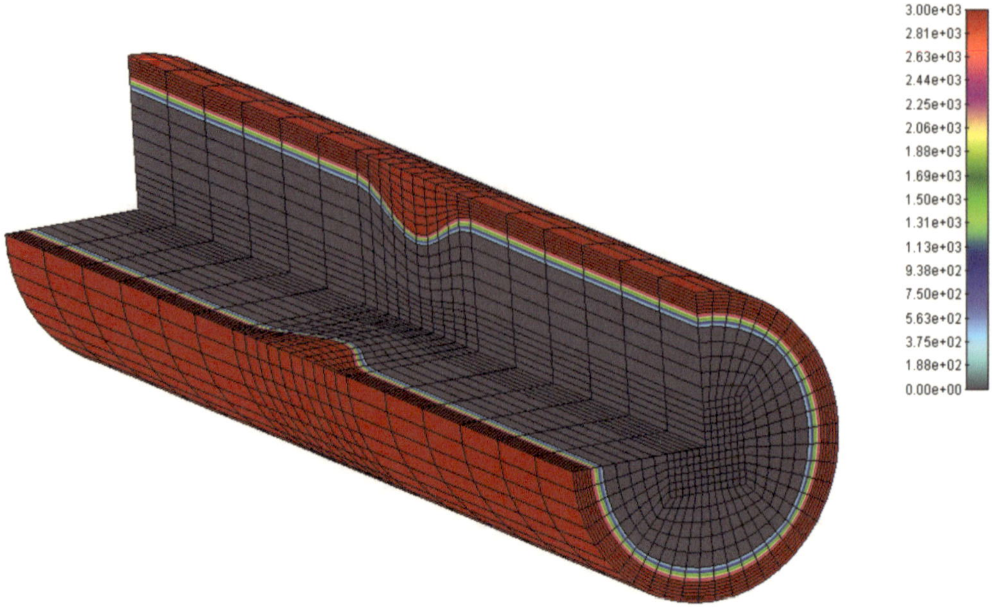

a)

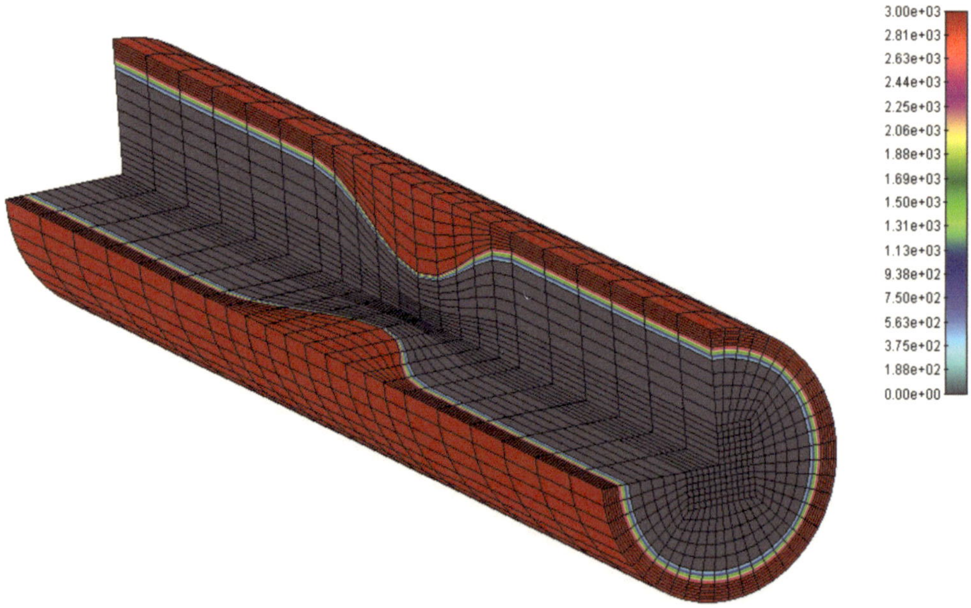

b)

Fig. 16. a) Intima wall pressure distribution for an initial mild stenosis 30% constriction by area b) Intima wall pressure distribution at the end of stenosis process after 10^7 sec[units Pa]

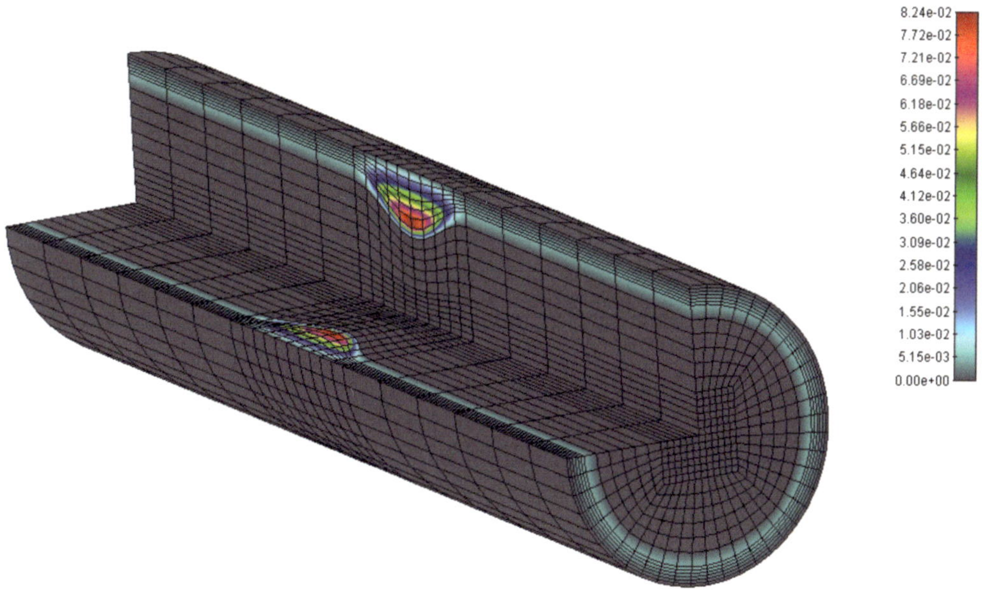

b)

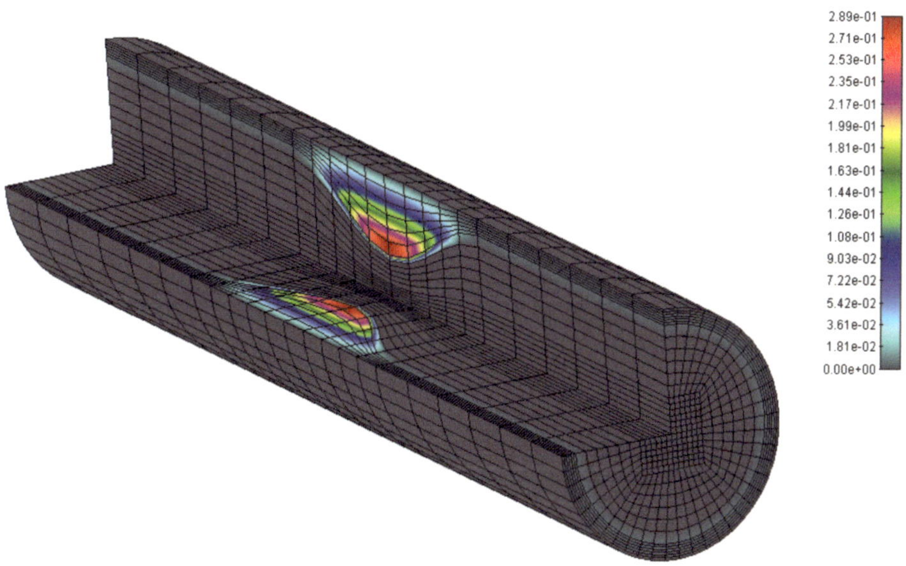

b)

Fig. 17. a) Macrophages distribution in the intima for an initial mild stenosis 30% constriction by area b) Macrophages distribution in the intima at the end of stenosis process after 10^7 sec[units mg/mL]

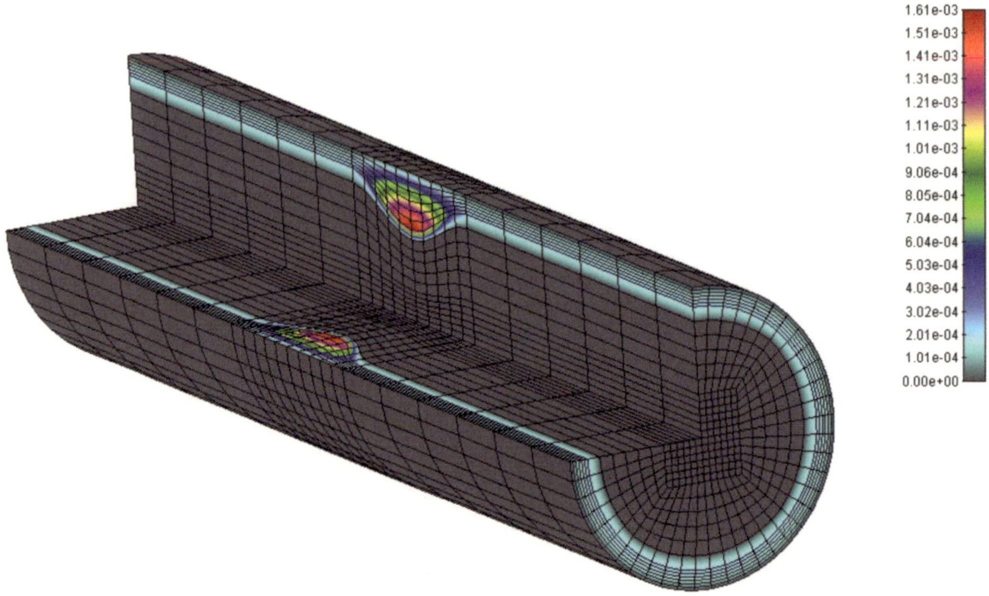

a)

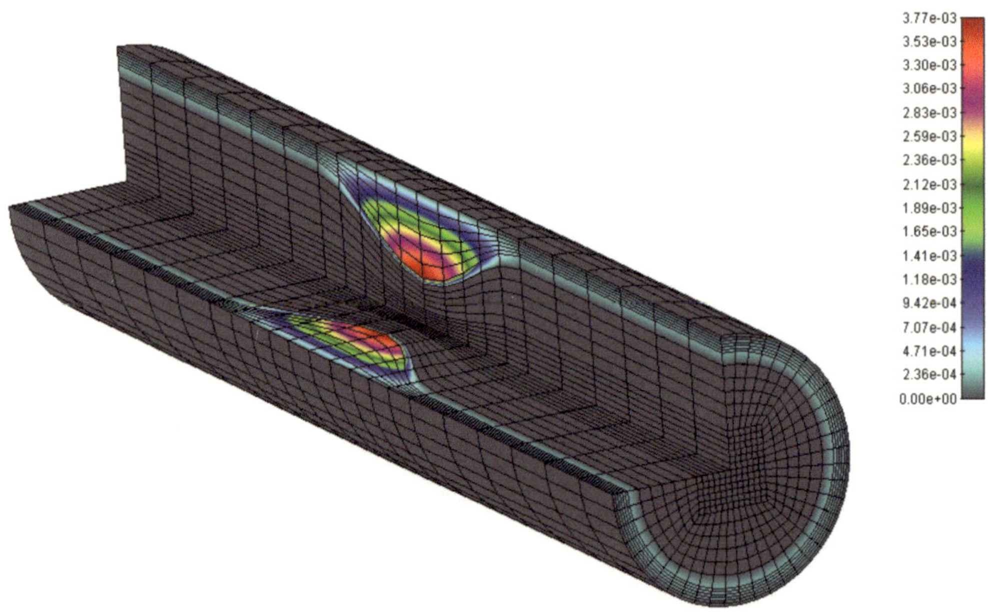

b)

Fig. 18. a) Cytokines distribution in the intima for an initial mild stenosis 30% constriction by area b) Cytokines distribution in the intima at the end of stenosis process after 10^7 sec[units mg/mL]

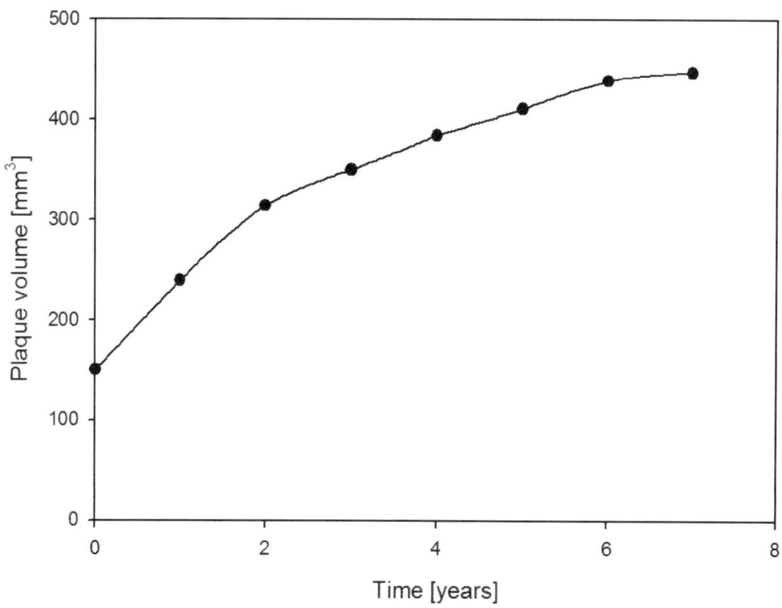

Fig. 19. Plaque progression during time (computer simulation)

From above figures it can be observed that during time plaque is progressing and all the variables as velocity distribution, shear stress, macrophages, cytokines are increasing. Also from Fig. 19 it can be seen that plaque progression in volume during time corresponds to clinical findings (Verstraete et al., 1998).

The last example is a model of the patient specific Left Anterior Descending (LAD) coronary artery for steady flow conditions. Computed concentration of LDL indicates that there is a newly formed matter in the intima, especially in the flow separation region in the LAD artery (Fig. 20).

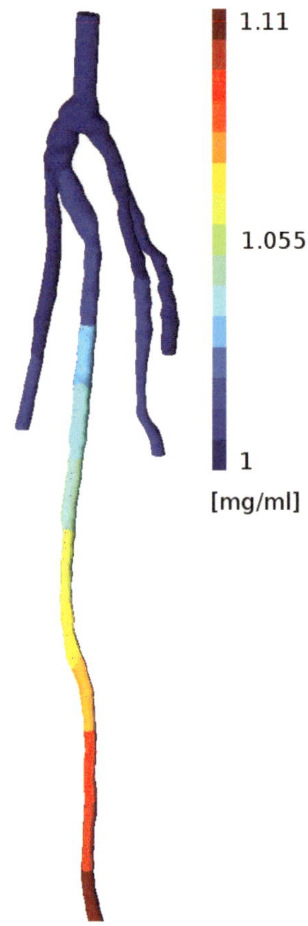

Fig. 20. LDL concentration distribution in the left anterior descending coronary artery

4. Discussion and conclusions

Full three-dimensional model was created for plaque formation and development, coupled with blood flow and LDL concentration in blood. The models for plaque initiation and plaque progression are developed. These two models are based on partial differential equations with space and times variables and they describe the biomolecular process that takes place in the intima during the initiation and the progression of the plaque. The model for plaque formation and plaque progression despite some difficulties concerning the different time scales that are involved and the different blood velocities in the lumen and in the intima, its numerical treatment is developed by using decomposition techniques together with finite elements methods and by splitting the numerical scheme into three independent parts: blood flow and LDL transfer, inflammatory process and atheromatous plaque evolution.

Determination of plaque location and progression in time for a specific patient shows a potential benefit for future prediction of this vascular decease using computer simulation. The understanding and the prediction of the evolution of atherosclerotic plaques either into vulnerable plaques or into stable plaques are major tasks for the medical community.

5. Acknowledgments

This work is part funded by European Commission (Project ARTREAT, ICT 224297).

6. References

Boynard, M.; Calvez, V.; Hamraoui, A.; Meunier. N.& Raoult, A. (2009). Mathematical modelling of earliest stage of atherosclerosis, *Proceedings of* COMPDYN 2009 - SEECCM 2009,Rhodes, Jun 2009.

Bratzler, R. L.; Chisolm, G.M.; Colton, C. K.; Smith, K. A.& Lees, R. S. (1977).The distribution of labeled low-density lipoproteins across the rabbit thoracic aorta in vivo. *Atherosclerosis*, Vol. 28, No. 3, (November 1977), pp. 289–307.

Brooks, A. N. & Hughes, T. J. R. (1982).Streamline upwind/Petrov-Galerkin formulations for convection dominated flows with particular emphasis on the incompressible Navier-Stokes equations. *Comput. Meths. Appl. Mech. Engrg.*, Vol. 32, No. 1-3, (September 1982),pp. 199-259.

Calvez, V.; Ebde, A.; Meunier, N. & Raoult, A. (2008). Mathematical modelling of the atherosclerotic plaque formation, *Proceedings of ESAIM*, Vol. 28, pp. 1-12.

Caro, C. G.; Fitz-Gerald, J. M. & Schroter, R. C. (1971). Atheroma and Arterial Wall Shear. Observation, Correlation and Proposal of a Shear Dependent Mass Transfer Mechanism for Atherogenesis. *Proc. R. Soc. London*, Vol. 177, No. 46, pp. 109–159.

Filipovic, N. & Kojic, M. (2004). Computer simulations of blood flow with mass transport through the carotid artery bifurcation. *Theoret. Appl. Mech. (Serbian)*, Vol. 31, No. 1,pp. 1-33.

Filipovic, N.; Mijailovic, S.; Tsuda, A. & Kojic, M. (2006a). An Implicit Algorithm Within The Arbitrary Lagrangian-Eulerian Formulation for Solving Incompressible Fluid Flow With Large Boundary Motions.*Comp. Meth. Appl. Mech. Eng.*, Vol. 195, No. 44-47, (September 2006), pp. 6347-6361.

Filipovic, N.; Kojic, M.; Ivanovic, M.; Stojanovic, B.; Otasevic, L. & Rankovic, V. (2006b). MedCFD, Specialized CFD software for simulation of blood flow through arteries, University of Kragujevac, 34000 Kragujevac, Serbia

Filipovic, N.; Meunier, N.&Kojic, M. (2010).PAK-Athero, Specialized three-dimensional PDE software for simulation of plaque formation and development inside the arteries, University of Kragujevac, 34000 Kragujevac, Serbia.

Goh, V. K.; Lau, C. P.; Mohlenkamp, S.; Rumberger, J. A.; Achenbach A. & Budoff, M. J. (2010).*Cardiovascular Ultrasound*, 8:5.

Goldstein, J.; Anderson, R. & Brown, M. (1979). Coated pits, coated vesicles, and receptor-mediated endocytosis. *Nature*, Vol. 279, (Jun 1979), pp. 679-684.

Kojic, M.; Filipovic, N.; Stojanovic, B. & Kojic, N. (2008). *Computer Modeling in Bioengineering – Theoretical Background, Examples and Software*. John Wiley and Sons, 978-0-470-06035-3, England.

Loscalzo, J. & Schafer, A. I. (2003).*Thrombosis and Hemorrhage,* Third edition, Lippincott Williams & Wilkins, 978-0781730662, Philadelphia.

Verstraete, M.; Fuster, V. & Topol, E. J. (1998).*Cardiovascular Thrombosis: Thrombocardiology and Thromboneurology,* Second Edition, Lippincot-Raven Publishers, 978-0397587728, Philadelphia.

6

Economic Impact of Information and Communications Technology – Identifying Its Restraints and Exploring Its Potential

Ksenija Vuković and Zoran Kovačević
University of Zagreb
Croatia

1. Introduction

Information and communications technologies (ICT) facilitate the introduction of new relationships, interactions, distribution of forces and novel moments into the classic economic settings. The prominent role of these technologies explains why they provide a challenging topic in the context of the newly arisen economic, technological and even social relationships. Whereas companies are adapting organizationally and are seeking new business models, customers are changing their customer habits and purchasing modes. The state, on its part, is opening up new channels of communication with citizens in the attempt to improve the quality of its services. Regardless of whether the economy in this new context is entitled as the "Knowledge Economy" or the "New Economy", it is profoundly related to ICT that enable networking, knowledge spillover, reduction of costs and allow for new entrepreneurial opportunities. The issues that economic experts need to address are whether economy has actually changed with the advent of ICT and how this new input can be exploited. They also need to identify challenges that have arisen with ICT, including the technological limitations and ways of overcoming them. From the macroeconomic aspect, their impact on productivity and growth as well as on labour market and employment should be considered. On the other hand, from the microeconomic aspect, the impact of ICT concerns the issues of efficiency, competitiveness, changes in market structure, business strategies, emerging industries and the effect of those technologies on opening up of new entrepreneurial opportunities. This overview of impacts of ICT on economy focuses on microaspects, particularly regarding the dynamic processes among companies in the ICT sector. In that respect, it is notable that macro- and micro-level are interdependent and intertwined, which explains why it may sometimes be difficult to draw a clear line between the two. Nevertheless, the truth remains that phenomena of major importance occur in the microeconomic spheres, in other words, in companies and industries.

2. Macroeconomic aspects

2.1 Growth and productivity

Generally speaking, when economy is concerned, productivity growth is imperative. More specifically, discussions related to the impact of ICT on growth and productivity have commonly been reminiscent of the statement made by Solow: "We see computers everywhere except in the productivity statistics." Since the 1950s, the models of growth have considered the impacts that work, capital (human and physical) and the so-called residuals upon growth. The question arises of how these production factors can be combined to achieve optimal efficiency, in other words, total factor productivity. The latter refers to various improvements in the area of efficiency, such as improvements in the managerial practice, organizational changes and, in most general terms, innovative approaches to production of goods and services. It is this segment, also referred to as a residual, that has come into spotlight with the arrival of the New Economy. ICT form part of that residual, although they are by no means their sole component. These technologies are a powerful tool that is penetrating all the segments of production, allows for a more efficient allocation of resources, creates new needs, generates the demand, and is the driving force behind new industries and jobs. At the same time, the vast contribution of ICT to the development of scientific disciplines and other industries as well as their efficiency should be emphasized. The research into the impact of ICT on the increase of productivity has so far focused on three mechanisms. The first of them is their impact through production in the ICT sector. In most countries, the sector that produces ICT has a minor role in respective economies. Nevertheless, that minor sector can have a relatively significant contribution to productivity and growth if it increases faster than the rest of the economy. There are simple statistical analyses that reveal a positive correlation between the segment of the sector that pertains to manufacturing and the increase of the total factor productivity, although they refer to a fairly small number of countries, notably Finland and Ireland (OECD, 2001). Such a positive correlation is to be expected on account of the high rate of technological progress and the total factor productivity growth in ICT manufacturing. However, certain countries whose ICT sector is relatively small, such as Australia, are also experiencing a total factor productivity growth, which entails that the size of the ICT sector is not a prerequisite for total factor productivity growth (OECD, 2002). The second mechanism is the ICT implementation in business activities, which will be further discussed in the section on ICT and entrepreneurship. The third mechanism is technological spillovers. The impact of ICT is visible if ICT investment is accompanied by other changes and investments (Pilat, 2004). ICT primarily affect companies that tend to promote their expertise and know-how and introduce organizational changes. Yet another key factor are innovations since users facilitate that investments into technologies such as ICT become more worthwhile through experimentation and invention. Without the process of complementary innovations, economic impact of ICT would be limited. Research also shows that the adoption and impact of ICT varies from company to company, depending on their size, type of activity, the company's founding year etc. The studies based on longitudinal data emphasize the interaction between ICT and human capital. (Bartelsman & Doms, 2000). Although some longitudinal data bases used in those studies contain data on the employees' knowledge, skills and occupations, most of them interpret human capital in terms of wages claiming that positive correlation exists between them and the employees' knowledge and skills. Companies that adopt advanced technologies increase their expenditure on education and training. Managerial teams that focus on perfecting the quality of their companies' products by adopting an aggressive human resources strategy achieve faster growth by continually improving their employees' knowledge and skills through training and hiring of new staff (Baldwin et al., 2004). All the aforementioned studies refer to developed countries. In the 1990s and early 2000s their economic experts were intensively engaged in elaborating and researching the economic aspects of growth, production and ICT implementation. For other groups of countries that are not so well developed the essential aim is to achieve convergence with the developed countries. One of the shortcomings of transition countries, among which is Croatia, is insufficient exploitation of opportunities that ICT provide for the economy. On the other hand, these countries are in a position to apply the lessons learned by the developed countries (such as USA, the developed EU members etc.) and adapt their own policies supporting the ICT sector development and their implementation in business. In the period 2001-2004 the share of ICT sector in the Croatian economy in terms of the total revenue amounted to 3%, without significant fluctuations over the mentioned period (Kovačević & Vuković, 2007). Its comparison with the size of the ICT sector (measured by its share in the economic added value) for EU countries, Norway, USA, Canada, Japan, Australia, Korea and Taiwan shows that the size of the ICT sector ranges between 3% (Greece) to just short of 12% (Ireland) of the total output, whereas the EU average is 5% of the total production (European Commission, 2006). In the context of the developed countries the need for expanding the ICT sector in Croatia is further emphasized, especially when its extraordinary growth is considered. Namely, the total revenue growth in the ICT sector has exceeded that in the economy as a whole, although it has paralleled the dynamics of the increase, that is, the decrease in the total revenue of the entire economy. ICT services have experienced a more rapid growth than the average of the ICT sector, which makes them the engine of the ICT sector growth. The productivity coefficient[1] of the ICT sector in Croatia grew steadily between 1997 and 2004 (see Figure 1).

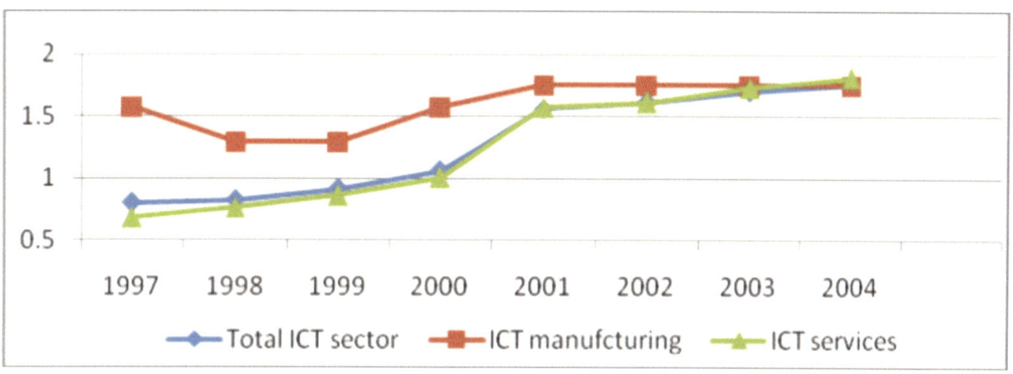

Fig. 1. Productivity coefficient trends in the total ICT sector, ICT manufacturing and ICT services (1997-2004)

In the same period (1997-2004), the ICT manufacturing displayed above-average productivity. ICT services industries in Croatia in that period also increased their productivity. Croatia has a potential for the development of the ICT sector, but companies need institutional support if that potential is to be exploited.

[1] Productivity coefficients are calculated as the share of the total revenue of a sector or its part (in this case, ICT manufacturing and ICT services) in the total revenue of the entire economy divided by the share of employees in that particular sector or its part in the total number of employees in a given economy. Values above 1 indicate productivity that is above the average of that of a given economy, whereas values below 1 indicate productivity below the average of that particular economy.

2.2 Labour market

In the time when creating new jobs is becoming more challenging than ever, the question of the impact of ICT on employment needs to be addressed. The impact of new technologies on employment and wages has been a subject of numerous discussions for a few centuries (Kalmbach & Kurz, 1986). Freeman & Soete (1997) mention the Physiocrats.[2] In the 1940s, following the Second World War, Norbert Wiener forecast that the invention of the computer would result in such huge job losses "that the depression era unemployment rates would seem like a picnic". Consequently, the question arises of the actual potential of ICT in employment, which is highlighted by the fear that certain groups within the workforce will have increasingly smaller employment prospects owing to a lack of qualifications needed. From the historical perspective, taking a long-term view, it could be said that the job creation effect has exceeded the job destruction effect. Nevertheless, it was Ricardo who brought to attention the idea that new jobs will not be analogous to the old jobs in terms of the required skills. This problem of incompatibility between skills and qualifications represents structural unemployment and entails the problem of structural adaptation. In that sense, both direct and indirect effects of ICT should be considered. The indirect effects of the ICT revolution are particularly visible, in the light of Schumpeter's analysis of the "bandwagon effect" following which the opening up of new markets is generated and numerous opportunities for profitable investments are created (Freeman & Soete, 1997). Regardless of whether the questions of the number of employees, sector composition or job structure are concerned, it is hard to accurately forecast the impact of ICT on employment. Several factors determine whether the job creation effect or the job destruction effect will prevail, including: the individuals' behaviour and adaptation to new circumstances and their ability to position themselves in the labour market; macroeconomic policy; employment policy; educational policy. The impact of ICT on employment and labour market should be viewed through the prism of several different questions, including labour demand, skill requirements, wage differentials and geographical dimension. The first question is that of labour demand. Owing to the rise of new industries, vast employment opportunities have already been introduced, most notably in the software industry, industries producing digital goods (films, music), service industries with an intensive use of ICT (banking, commerce). It is very likely that some other service industries in the area of information and communications will enable new employment opportunities. There is a notable impact of the new ICT on employment in the service sector, especially with regard to the Internet, where the speed of changes occurs outside of the controlled liberalization of processes and entails a far more dramatic process of "creative destruction" with a new price structure and changes in the market structure. The second question concerns skill requirements. Economic debates on employment are focused on the skill-biased nature of the more recent technological change. The skill-biased technical change hypothesis deals with creating the need for a highly-qualified workforce arising from ICT implementation, which in turn leads to the rise in wage differentials. ICT both presuppose and cause

[2] The Physiocrats, whose theories in 1760 claimed that land agriculture was the only truly productive economic activity and expressed their concern that the redistribution of French workforce across other activities would reduce national wealth. In the five decades that followed the share of agricultural workers in the French workforce gradually declined. At the same time, owing to productivity growth, the average wages increased by one quarter, while the unemployment did not rise significantly. Similar pessimistic outlooks were expressed by Ricardo and Marx.

structural adaptations of individuals, companies, industries, governments and other institutions that are learning, mainly through trial and error, how to optimally harness these new technologies. Workers tend to become more productive once they have acquired experience in using them. One of the essential problems at the microlevel is that of distribution of skills and incomes caused by structural changes. These changes have given rise to new challenges for policy makers. On the one hand, adaptation to information society requires significant changes in the demand for various types of education and skills. On the other hand, it is very likely that a considerable segment of the unskilled workforce will be excluded. ICT enable the codification of a major part of human skills due to the ability to memorize, store, quickly manipulate and interpret data and information, which is inherent in these technologies. With respect to their specific nature, different types of knowledge distinguish themselves according to tacitness (that is, the presence of tacit knowledge), perceivability, complexity and systemic nature (Winter, 1987). Knowledge can range widely, from highly tacit to completely articulated, depending on the ease with which it can be transferred. Perceivability denotes the amount of knowledge that is disclosed upon the very implementation of knowledge. The complexity degree refers to the quanitity of information that is needed for a problem to be properly defined and the knowledge required for creating alternative solutions. The systemic nature of knowledge indicates whether certain knowledge is completely self-contained and usable in its own right or presents an element of an interdependent system and as such only has significance and value if considered in a specific context. Tacit, imperceptible and complex knowledge that is an element of a larger system is hard to transfer. On the other hand, articulated, perceivable, simple and self-contained knowledge can be easily transferred. Nevertheless, the larger the proportion of knowledge that can be codified, the more significant the remaining knowledge that does not lend itself to codification will be. Since most routinized skills are becoming codified, their importance is diminishing. The required society-wide response to this trend in the context of employment would be to provide education and training for all social groups to ensure that some of them, such as unskilled individuals and the segment of the workforce that possesses routinized skills, are not excluded from economic activity. Apart from the issues of economic and social exclusion resulting from a lack of appropriate skills, the problem of the rise in wage differentials needs to be considered. This problem, which has been a prominent topic of investigation in the field of labour economics over the last thirty years, is mainly dealt with from the perspective of skill requirements, although its interpretations are sometimes also related to trade union erosion and minimal wages that protect low-skill workforce (Morettti, 2008). When it comes to employment in Croatia, it is notable that, in spite of the growing unemployment resulting, on the one hand, from global recession and, on the other, from structural changes in the economy, the demand for ICT experts is not diminishing. Differences between the average wage in Croatia and those in the ICT sector should not be disregarded either. ICT enable a global approach to information and knowledge, and provide companies with the opportunity to reallocate routinized activities that can be traded at the international level. These technologies contribute to economic transparency and generally imply cost reduction achievements resulting from the usage of alternative locations and international outsourcing of particular jobs. However, the question persists of the global distribution of benefits yielded by ICT. Similar to the trend described at the national level, two possible scenarios can be envisaged at the global level. According to the first scenario, ICT and corresponding policies will enable everyone to access labour market and thus reduce the wage differentials. According to the second, the winner-take-all-races principle will prevail. Undoubtedly, the ICT impacts on employment will depend on macroeconomic policy, institutional and legislative reform of the labour and commodity market, technological policy, new distribution policy, the policy that aspires to productivity growth, as well as one that will be aimed at integrating ICT in the society. Compatibility between technology, politics and institutions over a longer period of time may eventually result in full employment.

3. Microeconomic aspects

3.1 Impact on competitiveness and market structure

ICT are significantly contributing to the process of creative destruction through the birth of new companies and industries and the mortality of the less successful enterprises. Such developments have a visible effect on industrial structure as well as implications on employment. Directly or indirectly, ICT can reduce the transaction cost and market friction thus affecting competitive positioning. In high-technology industries, such as ICT, technological changes occur rapidly and competition is based on Schumpeterian innovation. The essential feature of new industries is the process of competition dominated by efforts for creating intellectual ownership through research and development. This often results in a rapid technological change that entails changes in market structures. The widely used term "New economy" corresponds to the spirit of Schumpeterian creative destruction in which innovations are destroying old industries and creating new ones (Schumpeter, 1942). Nowadays, creative destruction has been embraced as a concept beyond economic theory itself. Institutions, whether regional, national, or global, use this term as the point of departure in their policies or recommendations for facilitating growth. The goal is to enable new entrants to more easily gain access to markets and aid the most successful ones to grow. Schumpeter considered the entrepreneur to be the agent of change. However, depending on whether he considered the origin of progress to lie in small or big enterprises, he himself would change his mind thus giving rise to two innovativeness models. He proposed his first model, entitled Schumpeter Mark I, in his book *The Theory of Economic Development* (1934), where he investigated a typical European industrial structure at the end of the 19th century characterized by a large number of small companies. According to Schumpeter, the key features of the model of innovative activity at the time were the technological ease of entry into an industry and a major role of small enterprises in performing innovative activity. The second model, entitled Schumpeter Mark II, discussed in the book *Capitalism, Socialism and Democracy* (1942), was inspired by characteristics of the American industry in the first half of the 20th century. In this book Schumpeter emphasized the relevance of industrial research and development (R&D) laboratories for technological innovation as well as the key role of big enterprises. This view suggests that the model of innovative activity is characterized by the dominance of the existing big enterprises and relevant entry barriers for new innovators. Big enterprises have institutionalized the innovative process by creating research and development laboratories inhabited by researchers, technicians and engineers. Nowadays it has been widely accepted that enterprises are heterogeneous. Consequently, the Theory of the Representative Agent has been abandoned. The heterogeneity of enterprises is manifested in two dimensions: enterprise features (size, technology, behaviour) and performance (competitiveness, profitability etc.). Enterprises differ in their size, level of education of their employees, wages and investment in staff training. Within the population of enterprises, survival chances of many small enterprises, and occasionally big ones as well, are being changed. Economic experts generally agree that "destruction, however painful, is the necessary price of creative progress toward a better material life" (McCraw, 2007). Theoretical discussions and empirical research emphasize "production efficiency" and "dynamic efficiency" which can be broadly defined as productivity growth through innovations (Evans & Schmalensee, 2001). Production (or technical) efficiency arises from innovations through which new or better production methods are introduced that, in the long run, will lead to a higher productivity rate, which means achieving the "dynamic efficiency", most commonly through research and development. Intensive investment into creating intellectual ownership results in a substantial scale economy, which results in seller concentration. These phenomena, although typically associated with endogenous sunk costs industries, in which costs related to advertising, research, product design and development are important aspects of competition (Sutton, 1991), are particularly visible in high-technology industries. Such investment is directed at increasing the demand by creating new markets or increasing the buyers' readiness to pay for the existing products or services (Kaniovski & Peneder, 2002). Each selective environment that encourages competition for the required product quality generates investments through research and development, advertising or human resources. Provided that technological potential exists and that the customers are ready to accept new combinations on the side of the supply, the nature of differences in the growth at the industry level becomes endogenous with regard to entrepreneurial activity (Peneder, 2001). A higher ratio of entrepreneurial type of industry would imply a higher capacity of an economy for generating income and growth. However, market leadership cannot be considered to be a steady condition as there is a constant threat of drastic innovations on the part of competition. Still, certain authors suggest the possibility of creative construction (Agarwal et al., 2007) and express a fairly optimistic attitude on construction as an alternative to destruction. They exemplify this alternative process by means of the knowledge spillover mechanism leading to creation of new ventures and, ultimately, industrial and regional growth.

3.2 Entrepreneurial characteristics of the ICT sector in Croatia

Industries in the ICT sector are young and feature a large number of small enterprises. It is on examples of enterprises in the ICT sector that the relative innovative advantage of small enterprises in highly innovative industries (Acs & Audretsch, 1987) is particularly notable. For instance, a research on innovativeness in Croatia (Račić, 2005) shows that service enterprises in the ICT sector are among the three most innovative companies in the entire service sector. Generally speaking, new technologies have reduced the significance of scale economies so that even small enterprises that successfully implement innovations can thrive in the growing market. Structural changes, which account for a higher share of services in economies, have opened up new possibilities for entrepreneurship in the service sector. For instance, out of 6.6 million employees in the ICT sector in EU, 27.5 million (75.5%) are employed in the service sector. That sector is characterized by a low level of initial capital investment, whereby entry barriers are minimized and the beginning of functioning of a new small enterprise is facilitated (Verheul et al.). Europe has been experiencing an increase in research and development expenditure in the service segment of the ICT sector, especially in the software industry, which has compensated for the decrease in the research and development expenditure in ICT manufacturing (European Commission, 2009). The statistical data obtained in our research on the contribution of the ICT sector to Croatian economy showed that the share of the Croatian ICT sector in the economy is relatively small but is growing faster than the rest of the economy. The research also revealed above-average dynamic processes in industries manifested through above-average growth and productivity with respect to the economy as a whole. The ICT sector itself is heterogeneous concerning the differences in structural and dynamic features of particular industries (Kovačević & Vuković, 2006). The heterogeneity was determined on the basis of observing the following characteristics of industries: enterprise size distribution, distribution of employees across small, medium and big enterprises, market shares, productivity, profit and profit rates, entry barriers (capital intensity and minimum efficient size), entry rates, exit rates, enterprise survival. This heterogeneity is already evident at the moment of dividing the sector into the manufacturing segment and services. When the Croatian ICT sector is concerned, the service industry outperforms manufacturing. ICT services have been growing faster than the average of the ICT sector, and the dominant number of employees in the ICT sector belongs to services. The minimum efficient size, which is taken in research as a measure of scale economies or a measure of sunk cost, can negatively affect entry (entry barrier) if a huge output is required for potential entrants to reap the benefits from economies of scale. The minimum efficient size in ICT sector industries (calculated as the average of the total revenue per company over a four-year period and divided by one thousand for easier readability) shows that this determinant is not an obstacle for new enterprises to enter the ICT sector. There are exceptions (see Table 1), in which the minimum efficient size is fairly large (telecommunications, manufacture of insulated wire and cable, manufacture of television and radio transmitters and apparatus for line telephony and line telegraphy). A small minimum efficient size in most industries in the ICT sector speaks in favour of the conclusion that a smaller minimum efficient size of an industry increases the survival chances of the enterprises in that particular industry. In markets in which companies smaller than the minimum efficient size constitute a major share in the industrial population, new companies may stand better chances for survival. Industries in the ICT sector are not capital intensive. Instead, if we consider new investment concepts, such as the so-called extended investment concept that also includes human capital, it can be said that ICT industries are human capital intensive.

For an approximate illustration of human capital intensiveness (for lack of other data) we used the data on average wages in certain industries in the ICT sector. The average wage in the ICT sector is above the Croatian average; in 2000 it was nominally 20% above the average. In 2002 this difference was 19%, in 2003 it was 15% and in 2004 it amounted to 19% (Kovačević & Vuković, 2006). The lack of entry barriers has resulted in high entry rates to the ICT industry. Net entry rates provide an insight into general tendencies in demographic trends in the industrial population on the entry side. A somewhat vague impression made by net entry rates arises from the fact that the net rate conceals the actual scope of gross entries into each industry. The extent to which net entries differ from the actual number of entries depends on the scope of exits from the industry. Net entry rates represent a simpler method of calculating entry than calculating gross entry rates since it is not necessary to track data for each entrant and incumbent company separately. Instead, the total number of enterprises within a given period is considered. Net entry rates are calculated as follows:

$$\text{Net entry rate} = \frac{N_t - N_{t-1}}{N_{t-1}} \tag{1}$$

Where N_t refers to the total number of enterprises in year *t*, and N_{t-1} to the total number of enterprises in year *t-1*. According to research in developed countries and in Croatia, high entry rates in ICT industries support the product life cycle theory (Vuković, 2006). That theory suggests that entry rates are particularly high in young industries (see Figure 2). High entry rates in this sector can also be interpreted as support to vintage models and models of economic growth which emphasize the importance of creative destruction as a prerequisite for innovations. All the aforementioned theories imply that innovative activity and the adoption of new technologies are related to the process in which new and innovative companies replace the old, less productive ones. The structure of the economy is also changed in the process.

Industry	Minimum efficient size
Manufacture of office machinery	4,2
Manufacture of accounting and computing machinery	8,2
Manufacture of insulated wire and cable	80,3
Manufacture of electronic valves and other electronic components	3,8
Manufacture of television and radio transmitters and apparatus for line telephony and line telegraphy	32,7
Manufacture of television and radio receivers, sound or video recording or reproducing apparatus, and associated goods	2,0
Manufacture of instruments and appliances for measuring, checking, testing, navigating and other purposed, except industrial process control equipment	2,0
Manufacture of industrial process control equipment	2,4
Wholesale of computers, computer peripherial equipment and software	9,9
Wholesale of other electronic parts and equipment	4,3
Renting of office machinery and equipment, including computers	3,6
Telecomunications	215,8
Hardware consulting	3,0
Software supply and software consultancy	1,6
Dana processing	1,5
Dana base activities	2,1
Maintenance and repairing of office and accounting machinery, and computers	3,0
Other computer related services	2,0

Table 1. Minimum efficent size in ICT industries in Croatia

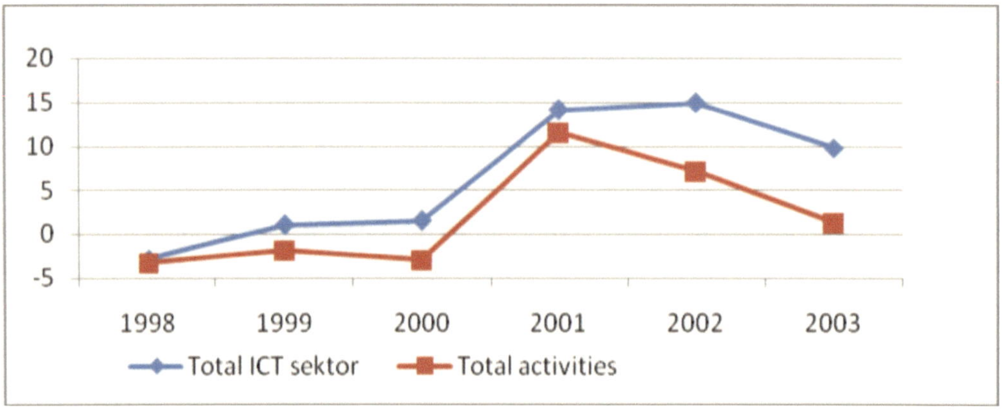

Fig. 2. Net entry rates of companies in the ICT sector and the total of all the activities in Croatia

High net entry rates, considerably above the average of a given economy, are indicative of an entrepreneurial regime or expansion model, which is characterized by huge innovation and imitation potential, both of which allow for a continuous entry of new companies into the industry. The technological regime concept provides a description of the technological environment in which a company performs its business activity. The technological regime identifies the characteristics of the learning process, sources of knowledge and the nature of knowledge bases. In literature two technological regimes are distinguished: the "entrepreneurial" regime facilitates an innovative entry, while the "routinized" regime makes innovation easier for the existing incumbent firms (Winter, 1984). Research in Croatian ICT sector confirms the hypothesis stated in the literature according to which the entry of small enterprises originates from three principal sources: positive impact of fast-growing industries and high technological opportunity, lack of entry barriers and companies' reliance on strategies which facilitate market entry. According to the lifecycle theory, entry rates will decrease (which is evident in the figure above), and the character of technological regimes is also likely to change with time owing to mobility among enterprises (that is, with small enterprises becoming medium and big enterprises). Every year major companies in the ICT sector improve their ranking. Some of them are among the leading businesses in the entire economy. It is the largest enterprises that have contributed to the growth of the ICT sector. As in the economy on the whole, in the ICT sector big companies are gaining importance, occupying an ever larger share in the total revenue and profit after taxes at the expense of small enterprises. The existing manufacturers will accumulate knowledge through learning by doing process. A wider knowledge base will probably lead to a larger minimum enterprise size and mobility of incumbent companies. Enterprise turnover rates (that is, sums of entries and exits) in certain industries are very high owing to high entry rates : telecommunications 30.6%, computer equipment consulting 20.2%, database development and management 19.66%, and other computing activities 20.48% (Vuković, 2006). The average turnover rate in the service segment of the ICT sector is considerably higher than that in the manufacturing segment of the ICT sector. The majority of companies in the ICT sector belong to the category of small enterprises. The fact that the largest part of the ICT sector is constituted by small enterprises highlights the entrepreneurial character of the sector. A fairly modest ratio of employees in the total number of employees within the economy as well as the modest employment growth in the ICT sector in this case indicate a neutral influence of entrepreneurship on employment. In general, it can be said that the beginning of the transition process in Croatia was recognized as an opportunity for the establishment of independent businesses, which is documented by the number of newly founded small enterprises in Croatia by 1995 (Kovačević, 2001). In the ICT sector a large number of small enterprises also emerged. This coincided with the disintegration of big computing centres, whose engineers and ICT experts would start up their own businesses. Popular entrepreneurship theories emphasize motives that serve as an impetus for starting an entrepreneurial venture: risk taking, managerial ability, wealth, preferences for control, flexibility and other job attributes that come with being one's own boss (Ohyama et al., 2009). Other tendencies we have noticed by monitoring industrial dynamics and evolution of ICT industries in Croatia have also been pursued in the recent entrepreneurship theory: successful new start-ups are founded by gifted individuals whose business is based on ideas generated in their previous workplace (Klepper & Thompson, 2009). As our research has shown, small enterprises in the ICT sector show better performance when compared to small enterprises in other industries. The research conducted among the leading start-ups (Bhidé, 2000; Kaplan et al., 2005; Klepper & Thompson, 2009) showed that two features are essential for the success of a small enterprise. One is that it belongs to high-growth industries: software, semiconductors, lasers, biotechnology. The other is that the founders of a new small enterprise are highly-educated persons. A large number of them are reputable experts in a particular technical field with appropriate experience in management, quite often occupying senior positions. It is evident that ICT industries employ and require highly educated experts while opening up possibilities for entrepreneurial ventures and self-employment. Based on the conducted research, the following conclusions can be made: 1) industrial dynamics (entries, exits, enterprise mobility) affects the performance of the sector (the growth of the sector measured in terms of total revenue, profitability, productivity); 2) service industries, with respect to their performance, represent the engine of the ICT sector growth; 3) comparison of performances of companies pertaining to different categories indicates that medium and big enterprises are more successful than small enterprises; 4) the relation between the performance of the sector and dynamic features of the industry highlights the potential role of institutions and economic policy. The results of the research suggest that the role of regulation should be directed at creating circumstances that would encourage survival and growth of profitable enterprises while presenting a minimum hindrance to experimentation related to companies' market entries and exits. Possible areas in which institutions and legislation could be actively involved and have an impact on industrial dynamics (which in turn affects productivity, profitability and other aspects of economic performance) are: administrative barriers and barriers to entrepreneurship, availability of capital (level of capital market development), education system, freedom of exit out of the market (bankruptcy law and employee protection regulation). The availability of capital for new small enterprises is generally problematic due to the commonly risky character of projects involved and modest means that such companies have at their disposal. In Croatia there are no special funds for innovative projects intended for small enterprises. Moreover, the existing traditional ways of financing are hardly available considering the insufficiently developed capital market.

3.3 Implementation of ICT in small and medium enterprises

In a competitive environment, ICT, when used efficiently, can aid successful enterprises in increasing their market share at the expense of less productive enterprises. This can be achieved by increasing productivity. The increase in the usage of ICT is mainly motivated by a company's internal growth and restructuring (the growth of certain companies and a decline in the business activity of others). This is particularly true of young companies, some of which thrive and continue to grow, whereas the others cannot manage to survive in the market. Technological development, especially implementation of ICT, is beneficial for small-scale production, which was established in our research on static features of industries in the ICT sector in Croatia. Cheaper capital goods are becoming available, the minimum efficient scale is being reduced and flexible specialization is possible (Verheul et al., 2002). Nevertheless, although the usage of ICT enables a relatively easy entry of small and medium enterprises into the market and their expansion, big enterprises can also appear as competition in areas dominated by small and medium businesses. Research brings to attention problems related to adoption and usage of ICT in small and medium enterprises (Harindranath et al., 2008.). In an extremely competitive environment and under financial restrictions entrepreneurs fear problems related to technology obsolescence and long-term investment into external consultants and suppliers arising from the lack of one's own skills in using and assessing the potential of ICT investment. For small and medium enterprises the implementation and usage of ICT is by no means simple, as it presupposes complementary expenses related to training and organizational changes as well as the direct costs of investing into hardware and software (OECD, 2004). In their research into small and medium enterprises in the USA, Wymer and Regan (2005) identified factors that impact on the adoption and usage of ICT. Some of these factors are related to technology, whereas the other part belongs to business activity (external and internal factors - internal knowledge and expertise, financing). Rashid (2001) proposes four categories of factors that affect the ICT adoption: technological (relative advantage, complexity, compatibility, cost), environmental (competition, supplier/buyer pressure, public policy), organizational (size, quality of systems, information intensity, specialization) and individual factors (managers' innovativeness and knowledge). Yet costs and financing of material factors are clearly only one dimension of the problem. The other concerns the complementariness with other factors, primarily the human factor - managers-owners and employees. Arendt (2008) presents the results of the research into barriers to ICT adoption in SMEs comparing selected regions in Spain, Portugal and Poland with a similar research conducted in the USA. According to his research, the major issue is not the access to ICT but rather the lack of appropriate education, knowledge and skills on the part of managers and employees. It is evident that small and medium enterprises suffer from the problem of a lack of human resources needed for using ICT. Their occupation with numerous operational and financing problems prevents them from making the assessment of benefits of using ICT one of their priorities. Research conducted among small enterprises with 3-80 employees reveals that the adoption of the Internet depends on the existence of technology-savvy employees (Mehrtens et al., 2001). It is not necessary that such persons are ICT professionals. Instead, their interest in technology is crucial. This is reminiscent of the problem highlighted in the passage on the impact of ICT on labour market; namely, there is no doubt that the demand for ICT experts as well as for individuals inclined toward using the new technologies will persist. The higher the number of ICT users and ICT producers, as well as of individual users, the higher the possibility of spillovers. There are two conditions that determine spillovers or external economies. The first condition is the interdependence of economic subjects, whereas the second condition is a lack of market compensation for the effect caused by one of the subjects on another. Producer-related spillovers arise from easier diffusion of production knowledge within certain spatial limits or institutional frameworks. User related spillovers refer to positive externalities arising from the implementation of certain products and services. ICT spillover effects have been recognized in theory although empirical research on that topic is scarce. This may result from the fact that spillovers are difficult to measure and their long-term effects have to be examined.

4. Conclusion

This review, based on the extensive literature on the impact of ICT on economy, encompasses both theoretical foundations and empirical research in the field. It was our goal to bring to attention certain elements that may be particularly relevant in determining the path of the development of national economies and regions. The review also focuses on issues that deserve particular attention by institutions responsible for passing measures concerning development support. The part dealing with the microeconomic domain presents the research into the contribution of the ICT sector to Croatian economy by using statistical methods. In our research on the microdynamic processes in ICT industries we used the tools for industrial demography analysis. We noticed the benefits of the impact of the ICT sector: more rapid growth and above-average performance in comparison to other industries, opening up of new entrepreneurial opportunities owing to a lack of sunk costs in service industries (that is, exploration instead of exploitation). Strong spatial concentration of ICT industries suggests potential transformation of regions through ICT industries and ICT entrepreneurship. Future research should aim to identify the way ICT are being used to improve local and regional economic efficiency, innovation and entrepreneurship. Nevertheless, at this point we should take into consideration that, although theory is more inclined toward researching dynamic industries, in practice we should depart from the evident differences among regions in terms of resources and tradition of economic activity. In that sense the potential role of ICT should be highlighted again, as even less attractive industries that are implementing ICT and complementary innovations and changes can become attractive. In all research the importance of human capital and high potential of ICT services is notable. This could provide the less developed countries with an opportunity to converge with their developed counterparts, provided they exploit their human potential by investing into education. ICT industries are by no means immune to business cycles, which explains the decrease in activity and employment shown by statistics. Nevertheless, recession can be considered as a "filtering period" in which restructuring occurs at relatively lower costs, although the social dimension of the problem manifested in the loss of the existing jobs or inability of creating new ones should not be disregarded. Research into enterprise survival indicates an increase in exit rates (with the majority of unsuccessful companies very likely to experience collapse) that discourage potential entrants. As economic experts, we believe that the role of ICT should not be mystified, although we are aware of the profound and far-reaching economic and social changes introduced by them. Instead of focusing on short-term effects, it is necessary to consider how the described phenomena will evolve in the long run, which is the approach commonly practiced in economics. Although it is very hard to differentiate "short-run" from "long-run", we are certain that in the long run everything is subject to change. From the perspective of digitization of economy, taking into consideration time and space (which indicate huge differences between the developed and the undeveloped), it is evident that this process is analogous to transition. In ensuring that the world is not reduced to polarization of winners and losers, that is, to the winner–take-all-races landscape, the role of institutions should be highlighted. It is through them that a more even distribution of knowledge, skills and income should be achieved.

5. References

Acs, Z. & Audretsch, D. (1987). Innovation, market structure and firm size. *Review of Economics and Statistics,* Vol. 69, No. 4, pp. 567-574, ISSN 0034-6535

Agarwal, R.; Audretsch, D. & Sarkar, M.B. (2007). The process of creative construction: knowledge spillovers, entrepreneurship, and economic growth. *Strategic Entrepreneurship Journal,* Vol. 1, No. 3-4, pp. 263-286, ISSN 1932-4391

Arendt, L. (2008). Barriers to ICT adoption in SMEs: how to bridge the digital divide. *Journal of Systems and Information Technology,* Vol. 10, No. 2, pp. 93-108, ISSN 1328-7265

Baldwin, J.R., D. Sabourin & Smith, D. (2004). Firm Performance in the Canadian Food Processing Sector: the Interaction between ICT Advanced Technology Use and Human Resource Competencies, In: *The Economic Impact of ICT – Measurement, Evidence and Implications,* pp. 153-181, OECD, ISBN 9789264021037, Paris

Bartelsman, E.J. & Doms , M. (2000). Understanding Productivity: Lessons from Longitudinal Micro Datasets. *Journal of Economic Literature,* Vol. 38, No. 3, pp. 569-594, ISSN 0022-0515

Bhidé, A.V. (2000). The Origin and Evolution of New Business, Oxford University Press, ISBN 9780195131444, Oxford

European Commisssion (2006). Effects of ICT production on aggregate labour productivity growth, Enterprise and Industry Directorate General, Unit: Technology for Innovation, ICT industries and e-Business, Brusells. (Available at http://ec.europa.eu/geninfo/legal notices en.htm)

European Commisssion (2009).European Industry in a Changing World – Updated Sectoral Overview 2009, Commission Staff Working Document SEC (2009)1111, Brusells

Evans, D.S. & Schmalensee, R. (2001). Some economic aspects of antitrust analysis in dynamically competitive industries, *NBER Working Paper Series,* Working Paper No. 8268

Freeman, C. & Soete, L. (1997). *The economics of industrial innovation,* Third edition, MIT, ISBN 0-262-56113-1, Cambridge, Massachusetts

Harindranath, G., Dyerson, R. & Barnes, D. (2008). ICT Adoption and Use in UK SMEs: a Failure of Initiatives. *Electronic Journal of Information Systems Evaluation,* Vol.11, No. 2, pp. 91-96, ISSN 1566-6379

Kalmbach, P. & Kurz, H.D. (1986). Economic Dynamics and Innovation: Ricardo, Marx and Schumpeter on Technological Change and Unemployment, In: *The Economic law of Motion of Modern Society – A Marx-Keynes-Schumpeter Centennial,* Wagener, H.-J. & Drukker, J.W. (eds), pp. 71-92, Cambridge University Press, ISBN 0521300924, Cambridge

Kaniovski, S. & Peneder, M. (2002). On the structural dimension of competitive strategy, Industrial and Corporate Change, Vol.11, No. 3, pp. 557-579, ISSN 0960-6491

Kaplan, S.N., Sensoy B.A. & Stroemberg (2005). What are Firms? Evolution from EarlyBusiness Plans to Public Companies, *NBER Working Paper* No. 11581

Klepper, S. & Thompson, P. (2009). Who Founds the Best Firms and Why? Intra-industry Spinoffs and Disagreements, *Working Paper*, Carnegie Mellon University

Kovačević, Z. (2001). *Restrukturiranje hrvatskih poduzeća*, Politička kultura, ISBN 953-6213-33-8, Zagreb.

Kovačević, Z. & Vuković, K. (2006). Performanse poduzeća u hrvatskom sektoru informacijsko-komunikacijske tehnologije (ICT), Ekonomska misao i praksa, Vol. 15, No. 2, pp. 217-240, ISSN 1330-1039

Kovačević, Z. & Vuković, K. (2007). Industrija informacijsko-komunikacijske tehnologije (ICT) u Hrvatskoj, Poslovna izvrsnost/Business Excellence, Vol.1, No. 1, pp. 97-112, ISSN 1846-3355

McCraw, T.K. (2007). *Prophet of Innovation: Joseph Schumpeter and Creative Destruction.* Belknapp Press and Harvard University Press, ISBN 978-0-674-02523-3719, Cambridge, MA

Mehrtens, J., Cragg, P.B. & Mills, A.M. (2001). A Model of Internet Adoption by SMEs, *Information & Management*, Vol. 39, No. 3, pp. 165-176, ISSN 0378-7206

Moretti, E. (2008). Real Wage Inequality, *NBER Working Papers* 14370, National Bureau of Economic Research, Inc.

Ohyama, A. & Braguinsky, S. (2009). Schumpeterian entrepreneurship, *DRUID Summer Conference*, Copenhagen Business School, Frederiksberg, Denmark, June 17-19, 2009

OECD (2001). *Science, Technology and Industry Outlook 2001*, Drivers of Growth: Information Technology, Innovation and Entrepreneurship, OECD, Paris, ISBN 9789264195554

OECD (2002). Measuring the Information Economy, *OECD Working Papers*, OECD, Paris. (Available at http://www.oecd.org/dataoecd/16/14/1835738.pdf)

OECD (2004). *The Economic Impact of ICT – Measurement, Evidence and Implications*, OECD, Paris. (Available at http://www.oecd.org/publications/e-book/9204051E.PDF)

Peneder, M. (2001). *Entrepreneurial Competition and Industrial Location*, Edward Elgar, ISBN 18406443, Cheltenham

Pilat, D. (2004). The ICT productivity paradox: insights from microdata, *OECD Economic Studies*, No. 38, 2004/1

Račić, D. (2005). Inovacije u hrvatskim poduzećima. *Inovacijsko žarište*, Vol. 2, No. 6, pp. 4-5, ISSN 1334-8663

Rashid, M.A. & Al-Qirim, N.A. (2001). E-Commerce Technology Adoption Framework by New Zeland Small to Medium Enterprises. *Research Letters Information Mathematical Science*, Vol. 2, No. 1, pp. 63-70, ISSN 1073- 2780

Schumpeter, J. (1934). *The theory of economic development*, Cambridge, MA.

Schumpeter, J. (1942). *Capitalism, socialism and democracy*, Harper, New York.

Sutton, J. (1991), *Sunk Costs and Market Structure: Price Competition, Advertising, and the Evolution of Concentration*, MIT Press, ISBN 0-262-19305-1, Cambridge

Verheul, I. , Wennekers, S., Audretsch, D. & Thurik, R. (2002). An *Eclectic Theory* of *Entrepreneurship: Policies, Institutions and Culture*, In: Entrepreneurship: Determinants and Policy in a European-US Comparison, Springer Netherlands, Audretsch, D.B, Thurik, A.R., Verheul, I. and Wennekers, A.R.M. (Eds), pp. 11-81, Kluwer Academic Publishers, ISBN 978-0-306-47556-6, Boston/Dordrecht

Vuković, K. (2006). *Industrijska dinamika u sektoru informacijskih i komunikacijskih tehnologija*. Ph.D., University of Zagreb

Wiener, N. (1948). *Cybernetics: or, Control and Communication int he Animal and the Machine,* MIT Press, ISBN 026273009, Massachusetts

Winter, S. (1984), Schumpeterian competition in alternative technological regimes, *Journal of Economic Behavior Organization,* Vol. 4, No. 1, pp. 287-320, ISSN 0167-2681

Wymer, S.A., Regan, E.A. (2005). Factors influencing e-commerce adoption and use by small and medium businesses, *Electronic Markets,* Vol. 15, No. 4, pp. 438-453, ISSN 1422-8890

7

Operating System Kernel Coprocessor for Embedded Applications

Domen Verber and Matjaž Colnarič
University of Maribor, Faculty of Electrical Engineering and Computer Science
Slovenia

1. Introduction

The silicon evolution yields advances in contemporary processor architecture. As a result of the ever-increasing number of components in a chip, multi-core solutions have emerged. In general computing systems, their goal is to accommodate the parallel execution of processes, tasks or threads. Apart from general computing, the parallel execution of tasks is characteristic of asynchronous and dynamic embedded applications like automotive systems, process control, multimedia processing, security systems, etc., for which, in recent times, multi-core architecture has also raised interest [Lee (2010)].
However, in the case of processors for embedded systems and their ultimate requirement being predictability of temporal behaviour, the implementation of traditional multiprocessing is not straightforward. Their advanced architecture features (pipelines, cache, etc.), which are devised to improve average computing speed, may introduce severe sources of non determinism and unpredictability.
Instead of symmetrical multiprocessing, it is more adequate to employ multi-core processors for specialised operations. One of these is the execution of operating system services with a goal to deal with the nondeterministic and unpredictable time delays caused by the very nature of asynchronous events by separating the execution of process control tasks from real time operating system (RTOS) kernel routines. This approach is similar to the idea of math coprocessors, graphical accelerators, intelligent peripherals, etc. These specialized units are able to perform operations much faster than general processors that implement them as software programs.
The idea of migrating scheduling out of the main processor is already old [Halang (1988); Cooling (1993); Lindh et. all. (1998), etc.] However, with the advent of multi-core processors on one hand, and programmable hardware devices for prototyping on the other, its implementation has become much more feasible and realistic. In this contribution we are presenting a prototype of a separate Application specific integrated circuits (ASIC) implemented coprocessor performing operating system kernel functionalities.
First, some background regarding the real-time properties of embedded systems is given, and some of the most characteristic solutions of real-time operating systems which jeopardise predictability are pointed out. Then, an architectural solution to the problem is proposed and validated with the prototype.

2. Real-time properties of the embedded system

An embedded system is a special-purpose computer system designed to control or support the operation of a larger technical system which usually has mechanical and electrical components in which the embedded system is encapsulated. Unlike a general-purpose computer, it only performs a few specific, more or less complex pre-defined tasks. It is expected to function without human interaction and therefore, it usually has sensors and actuators, but no peripheral interfaces like keyboards or monitors, except if the latter are required to operate the embedding system. Often, it functions under real-time constraints, which means that service requests must be handled within pre-defined time intervals.

Embedded systems are composed of hardware and corresponding software parts. The complexity of the hardware ranges from very simple programmable chips (like field programmable gate arrays or FPGAs) over single micro-controller boards to complex distributed computer systems. In simpler cases, the software consists of a single program running in a loop, which is started on power-on, and which responds to certain events in the environment. In more complex cases, operating systems are employed. The application for the embedded system (and the others) usually consists of several processes or tasks that must be executed more or less simultaneously. The operating system (OS) provides features like multitasking and scheduling to allocate the active tasks to limited processing resources by means of different scheduling policies. The OS also provides task synchronisation, resource management, etc. [Silberschatz et. all. (2009)].

The main focus of this paper is the scheduling of processes operating under hard real-time constraints as a basis for other operating system kernel services (event management, synchronisation, etc.). For such systems, the essential and characteristic requirement is that each task, regardless of circumstance, must finish its work prior to the predefined deadline. Here, obviously, task scheduling is the critical operation. Some functionalities of operating systems (e.g., virtual memory, mass storage device management, etc.) are rarely relevant for the embedded system and are not considered here.

Although the discipline of real-time research was established thirty years ago, even now inappropriate scheduling policies (e.g., fixed priority) are very often employed. For singleprocessor systems operating in the real time regime, theoretical aspects of task scheduling have been acknowledged at least since the well-known paper [Liu-Layland (1973)]. The advantage of the often used, although inadequate, fixed priorities-based scheduling is that in most cases, it is built into the processors themselves in the form of priority interrupt handling systems. Thus, implementation is fast and simple. However, it is difficult to assign adequate priorities to tasks, which leads to the starvation of other tasks which are waiting for blocked resources. Usually, priorities are not flexible enough and cannot adapt to the current behaviour of systems. With rate-monotonic scheduling, a set of periodic tasks is considered. In this case, the tasks are scheduled according to their periods. In the paper mentioned above, the scheduling of such a task set is proven to be feasible, however, only if the utilisation of the processor is less than approximately 70%.

It is widely accepted that in a general case, the deadline-driven scheduling policy is the most appropriate, more specifically, the Earliest Deadline First (or EDF). In this case, the priority of the task is determined by its deadline. The task with the closer deadline has higher priority than the task with the more distant one.

When the deadline-driven scheduling is employed, the actual schedule can and should be tested for feasibility during run-time (schedulability check). Each time a new task is added to the system, a test must assess whether the deadlines of all active tasks will be met. To perform this test, the sum of the (remaining) execution times of each task and the tasks that will be executed before it must be smaller than or equal to its designated deadline. The schedulability check depends on the accurate estimation of the execution times of tasks. To calculate this properly, all aspects of the embedded systems (hardware, operating systems and application) must behave with temporal predictability.

Typical embedded systems are expected to react to events from the environment. Traditionally, this is implemented by means of interrupts that signal the main processor when a specific event occurs. The problem with this method is that the interrupts and interrupts handling also introduce sporadic delays asynchronous to the execution of the running processes, and this jeopardises the temporal predictability of the latter.

Another problem facing the real-time behaviour of embedded applications is the operating system itself. Traditionally, operating systems are software services running on the same processor along with the application software, with the goal to support the application execution on the target hardware systems. Each call of the operating system routines prolongs the execution of the application. It is usually very difficult to get adequate execution times for these routines because the calculation depends on the number of active tasks currently running.

3. Outline of the architectural solution

Embedded applications usually consist of several tasks or processes, and the OS is responsible for scheduling these for execution on devices with limited processing resources. In addition, the OS is responsible for proper task synchronisation, inter-task communication, reaction to events in the system, etc. The reason adequate OS operations for real-time systems are seldom supported is that their implementation is difficult and impractical due to their complexity and often unacceptable overhead. By implementing the scheduling in hardware operating in parallel, complexity is not an issue any more, and the overhead becomes negligible. First, the hardware implementation usually outperforms any software execution. Second, in hardware, many operations can be executed in parallel, further speeding up the execution. In addition, the ever-decreasing cost of hardware devices on one hand, and a steadily increasing degree of integration on the other, justifies the use of a hardware approach even for complex solutions.

The outlook of the hardware architecture is shown in Figure 1. The main processor, where the tasks' code is executed, accesses the operating system services via its system bus. The set of registers implemented within the coprocessor are thus addressable within its memory space, providing communication with the OS kernel functions. Instead of executing the specific function on its own, the OS system routines set the appropriate parameters in the coprocessor's registers and issue a specific command. After that, the OS responses are read from the coprocessor's registers.

Task administration is split into two parts. The internal states (contexts) of tasks are kept at the main processor. The coprocessor only maintains the statuses and the essential parameters of tasks, and determines which task must be executed next. Furthermore, the coprocessor also maintains synchronisers, shared variables, etc., and is responsible for controlled system event management.

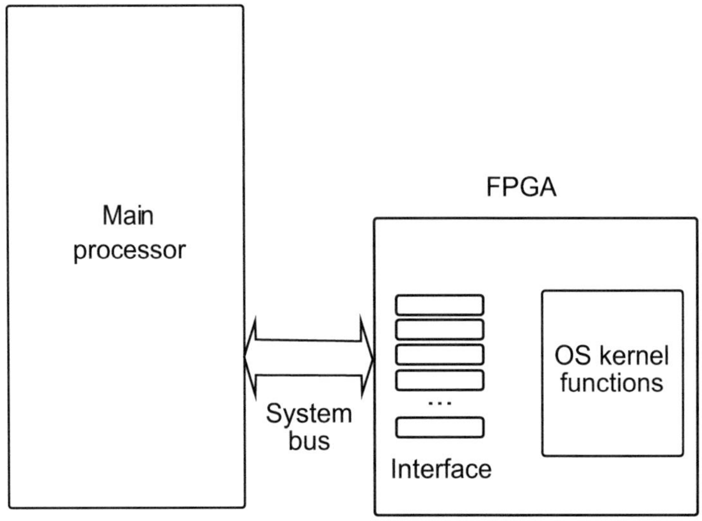

Fig. 1. General scheme of the implementation

4. Coprocessor instruction execution

The coprocessor operates by means of instructions (requests for operations) issued by the main processor (the host). Each instruction consists of an operation code and associated parameters (operands). For example, in the case of task activation, these parameters are the task identification number and the task scheduling constraints. Usually, the instruction is executed immediately after it is put into the interface registers. In addition, the coprocessor can store several instructions for future execution.

Such instructions are triggered by certain conditions that are also set by the host. When these conditions are satisfied, the instruction is issued to the instruction execution unit. There the operation code is decoded and an appropriate set of signals is generated to carry out the required operation. The process of instruction execution and its implementation is presented in Figure 2.

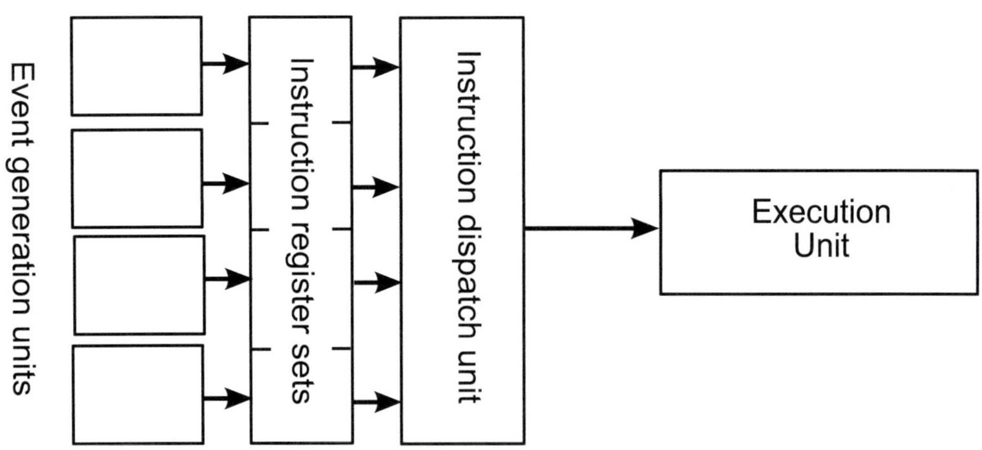

Fig. 2. The instruction execution implementation

Several instruction register sets hold the operation code and parameters of the specific instructions. The instruction dispatch unit is constantly monitoring which of the stored instructions is ready and forwards them further for execution. When the conditions of several instructions are fulfilled at the same time, the instruction register with the lowest number takes precedence.
The readiness of the instruction for execution is determined by one of the event generation units (event generators). These units generate signals when specific conditions are met, and these signals in turn may then initiate the instruction execution. Which event generation unit is connected to a specific instruction register set is determined by the application. Each register set may also be configured for immediate instruction execution. In this case, the instruction is issued immediately when the operation code is placed (i.e. the application first sets the instruction parameters and then triggers the instruction execution by writing the operation code). In this way, parameters for several instructions may be pre-set in advance. For command execution, only the operation code must be set.
There are different kinds of event generation units. The so-called external event reaction unit reacts to the events generated outside of the system. This is similar to interrupt and event handling in traditional microprocessors. The events from different external and internal sources can be combined, and some events may be masked.
Another event generation unit is the periodic event generator. This unit generates signals for periodic instruction execution. The main application may set the time of the instruction's first occurrence, the period of the repetition and the overall duration interval.
An additional unit observes the shared variable registers (described in more detail later). When a new value is written into a shared variable, a signal is generated, which can be used for instruction execution. In addition, the relevance range of the value can be associated with each shared variable separately. In this case, the signal is generated only if the value written into the register is outside of the predetermined range. In this way, different message-passing algorithms for inter-task communication may be implemented, and the system may react to some conditions which are related to the values of some parameter (e.g., temperature too high).
As will be described later, the event generators can also be used as a part of the task synchronization mechanisms. Furthermore, the event generator can be configured to generate interrupts to the main processor.

5. Operating system functions implementation

The execution unit implements specific operating system functionalities. One of the primary goals of this research was to eliminate operating system temporal interference. The main processor should read the results of the operation as soon as the operation is written into the coprocessor. When implemented in software, the execution times of most OS instructions depend on the number of tasks that are currently active. With the appropriate approach it was possible to achieve a constant execution time for each OS instruction, independent of the number of tasks (i.e., time complexity of O(1)). The consequence of this is increased silicon consumption, which is not an issue anymore. There are several groups of OS instructions that were implemented with the coprocessor.

5.1 Task scheduling

For the implementation of task scheduling, a sorted list of tasks is maintained. Each element of the list holds relevant parameters of an active task: the task ID number and the parameters related to scheduling. For EDF scheduling, the latter would be the deadline of the task, remaining execution time, etc. For other scheduling policies, it would be the priority or period of a task. Each time the task information is added to, or removed from the list, the parameters are updated accordingly. Some parameters are also updated periodically during this time. For example, the remaining execution time of the task that is currently running must be periodically decreased. Other parameters related to task synchronization and other operations, is described in the next sections. Other parameters of tasks, not relevant to the OS coprocessor (e.g., the context of the task), are kept in the main processor. Data in the list are sorted according to the scheduling policy. In the case of EDF scheduling, this is done based on their increasing deadlines. For priority-based systems, it would depend on priorities. For rate monotonic scheduling, the sorting criterion would be the period of tasks. The list can be observed as a set of independent cells or components with the same functionalities. This is illustrated in Figure 3.

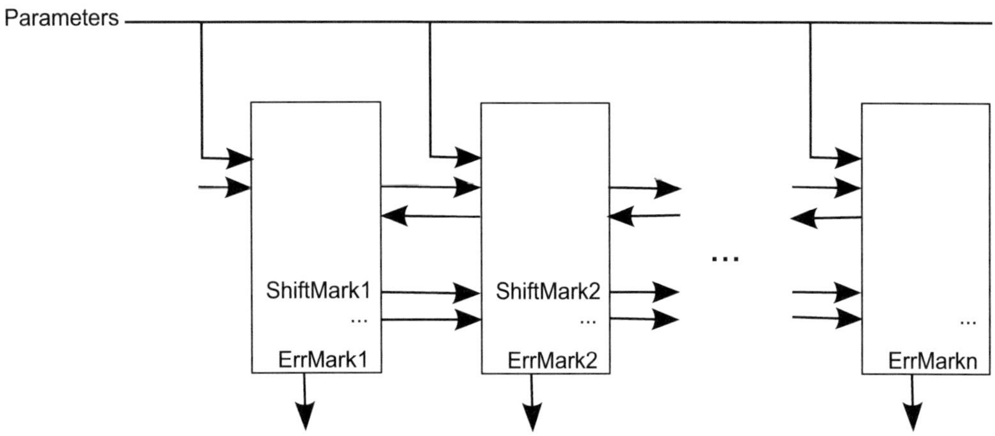

Fig. 3. Implementation of task scheduling

The parameters of the current OS instruction (such as the index of the task, its deadline and remaining execution time) are put on the common bus. Then, a series of control signals (not shown in the picture) are generated to execute different steps of the specific instruction. Each cell contains a set of registers that hold the task's parameters. Apart from the registers, each cell has two sets of inputs and outputs that are used during different phases of some OS instructions. The data from a single cell may be shifted into the next or into the previous cell. Several logical signals are used to synchronize these shift operations (ShiftMark) or to signal if there is a deadline violation or some other error (ErrMark). In addition, each cell consists of digital logic divided into several parts, which are responsible for executing different OS operations. Such division allows for parallel execution during the completion of OS instruction. One part of this digital logic is responsible for the identification of the cell by means of its task ID, the second part is responsible for the deadline comparison, and yet another part performs the arithmetic for cumulative execution time calculation, etc. In this way, it is possible to achieve the same execution time for each instruction. For example, when a new task is being added to the list, a proper position is determined: First, the deadline for the arriving task is compared with the deadlines for all tasks already in the list. Then, room is made by shifting the proper set of cells to the end of the list and finally, the new task information is put into place. The removal of a task from the list is performed in a similar fashion. A more detailed elaboration of the procedure is given in Verber (2009). A dedicated logic within the cell also signals the current state of the task (i.e., if it is ready for execution or if it is suspended for some reason). In parallel to the list of tasks, an additional component determines which task must be executed next by detecting the first task in the list that is not in a suspended state. This variable may change every time an instruction is executed. An instruction may be executed independently from the main processors by means of instruction register sets and event generators as described above. In this case, when the new task must be dispatched for execution, an interrupt signal to the main processor is generated. The interrupt service routine may access the ID of the executing task through the coprocessor interface. This component also incorporates support for the so-called non-interruptible critical sections. A task, executing in a non-interruptible critical section, cannot be replaced by another one even if there is an active task with a shorter deadline. There are two instructions for entering and leaving such a section.

When using the EDF policy, another important part of task scheduling is the schedulability check. Each time a new task is put into the list, it must be proven that the deadlines of all active tasks will be met. To this end, each element of the list maintains the remaining execution time of the current task, as well as the cumulative execution time of the tasks to be executed prior to and including the current one. For the schedule to be successful, the latter must be smaller than or equal to the designated deadline of the current task. To maintain the sum of execution times, when a task is added to the list, its execution time is added to all cumulative execution times of the elements which come after the newly arrived one. Similarly, when the task is removed, its remaining execution time is subtracted from sums in subsequent elements of the list.

5.2 Task synchronisation

Occasionally, an active task may not be in a position to continue the execution due to the unavailability of exclusive resources, because it must wait for another task to complete its job, etc. In such cases, the OS puts the task in a suspended state. When, for example, the exclusive resource becomes available, one of the tasks waiting for it is removed from the suspended state. The main difficulty in this execution is that some sort of queue of waiting tasks must be implemented. This is easily done in the software, however, maintaining several queues in hardware would consume too many resources. Instead, each element of the sorted list contains a set of bits that represent the various synchronisers for which the current task is waiting. These bits are shifted together with every task addition/removal operation and are maintained by the synchroniser control circuits. Each sychronisation unit can be associated with the specific synchronisation bit in each cell. The task is suspended if either of these synchronisation bits in the cell is set to one. This is achieved with a simple logical operation of disjunction (or). Different synhronisation control units can be used to implement different synchronization mechanisms. In these experiments, the binary semaphore primitives Lock and Unlock have been implemented. When a Lock instruction is executed for a semaphore for a certain task, the control unit checks to see if the semaphore is already locked. If it is, a corresponding bit in the cell is set and the task becomes suspended. In other cases, if the semaphore is unlocked, the control logic marks it locked and the task remains non suspended. Upon the Unlock operation, when several tasks are waiting for the same synchroniser, the left-most one in the list becomes ready. In the case of the EDF scheduling algorithm, this is the task with the shortest deadline. In this way, the possibility of deadline violation is minimized. The binary semaphore is implemented with simple flip-flop logic. The control logic for other synchronization mechanisms may be easily implemented.

5.3 Inter task communication

To serialize the data-dependent operation between tasks or to employ inter-task communication in general, a set of common shared variables is used. A value, written into a shared variable by one task, may be read by the others. Using the common shared variables, tasks may also be synchronised. For example, one task is waiting until another one changes a value of a variable. This can be implemented by combining the synchronization control logic with shared variable event generators. The same method is used for the implementation of traditional OS signals. Shared variables are mapped into the memory space of the main processor. The shared variables may also have a very important role in distributed embedded systems. In a previous research [Colnaric and Verber (2004)], the hardware support for transparent interprocessor communication in distributed environments was studied and implemented. In order to accomplish this, a new value's contents, when put into the shared variable, is distributed (replicated) to the other nodes in the system. In this way, inter-task communication and synchronization may be implemented between tasks running on different processors. In the current work, those mechanisms are not yet implemented.

5.4 Real-time clock

Although a typical processor may have implemented a real-time clock by other means, its integration into the kernel coprocessor may allow other operations to use and react to the same absolute time source. However, a proper real-time clock must operate even when the system is switched off. This requires battery-powered circuits. Currently, it is not possible to put part of an FPGA device into an operational state during the shutdown of the system Therefore, it is the responsibility of the main processor to set the proper time of the real-time clock at startup. Implementation of the reading and maintenance of the precise real-time clock by means of a dedicated battery-powered real-time clock chip is under development. For an even more precise clock source, the use of a GPS receiver may be considered.

5.5 Support for fault tolerance

Apart from operating within real-time constraints, the embedded systems are frequently used in situations where faults may result in large material losses or even the endangerment of human safety. There are several aspects of fault tolerance that may be incorporated into the coprocessor. For example, in the case of event generators, different self-monitoring circuits may be implemented in hardware in order to detect hardware-related faults. The event generators related to the shared variables can be used to detect abnormal values of a certain variable. The task scheduler is also capable of detecting deadline violation errors. However, for more subtle fault detection and fault management, the coprocessor is usually not adequate. If a fault is detected, a contingency plan must be employed and a new set of tasks should usually be introduced. This can only be done by the main processor.

6. Results of the experiments

To support the theoretical research, studies on an experimental hardware platform were conducted. The main processor is Texas Instruments' digital signal processor TMS320C6771 running at 150 Mhz [Texas Instruments (2010)]. The coprocessor is implemented with Xilinx FPGA device Spartan2E xc2s300e running at 50 Mhz [Xilinx (2010)]. This device consists of 1536 so-called Configurable Logic Blocks (CLBs). Each CLB is capable of performing simple logical functions and/or to be used as a memory element. This is a relatively low performance and low-cost device. In the experiments, four event generators, four synchronisers, eight shared variables and eight task scheduling cells were implemented. By this method, approximately half of the available silicon resources were used. Another half, it is planned, will be used in future work. The newest FPGA devices and dedicated ASIC chips may have hundreds of times more silicon resources and are much faster. On the main processor the artificial tasks were used for the test bed. The tasks were created with a proprietary realtime operating system on the evaluation board. Nevertheless, the OS operations were issued through the coprocessor. The task IDs and the operation codes are one byte in size. All other parameters require 16 bits. All temporal values and constraints are represented in a relative fashion (i.e., as a number of basic clock cycles relative to the current moment in time).

Each instruction is executed in four basic clock cycles. This is 80 ns at 50 Mhz. Some instructions could be executed in fewer clock cycles, however, we found that it is much easier to implement the instruction execution unit if the same four cycles are used every time. For simple instructions during some execution cycles, the instruction execution unit is idle. In any case, the execution time of a single instruction is shorter than the memory access time of the main processor. I.e., the main processor may read the results as soon as the operation code is provided.

7. Conclusion

With the ever-increasing density of silicon chips, it is possible to dedicate some areas on the chip to the implementation of operating system functionality. The situation is similar to that of the early 1990s. In the beginning, floating-point operations were implemented in software, then math coprocessors replaced software routines and execution times shrunk to only a small portion of their original. Later on, with miniaturization, the coprocessors were integrated into the processor cores. In the research described here, it was shown that the same scenario may be applied to the implementation of OS functionalities. If consumption of silicon is not an issue, the functionalities of the coprocessor may be executed in constant time (i.e., with O(1) time complexity). Hardware implementation of the operating system's functionalities has little impact on application development. Within traditional development tools, the OS support is usually considered on the application programming interface (API) level. If only the inner parts of the API to OS routines are modified, no change in the development tools is required.

Although the number of components in the experiment were limited, the proposed implementation is modular enough to be easily expanded and modified to manage a different number of tasks, synchronisers, shared variables, etc.

The main focus of our research is real-time systems. However, the principle of the OS coprocessor can be effectively used with any kind of operating system. For example, priority-based scheduling can be implemented much easier than EDF. In this case, the number of parameters in each element of the task list is greatly reduced and there is no need for the feasibility check of the schedules. The synchronisation mechanisms, shared variables and other elements of the coprocessor may remain the same.

8. References

Edward Lee (2010). Design Challenges for Cyber Physical Systems, In: *Strategies for Embedded Computing Research*, Eutema, Vienna, March 2010.

Halang,W. A. (1988). *Parallel Administration of Events in Real-Time Systems, Microprocessing and Microprogramming*, Vol. 24, Jan 1988, pp. 678 - 692, North–Holland.

Cooling, J. (1993). Task Scheduler for Hard Real-Time Embedded Systems, *Proceedings of Int'l Workshop on Systems Engineering for Real-Time Applications*, pp. 196–201, Cirencester, 1993, IEE, London.

Lindh, L., Strner, J., Furuns, J., Adomat, J. and Shobaki M. E. (1998), Hardware Accelerator for Single and Multiprocessor Real-Time Operating Systems, *In Seventh Swedish Workshop on Computer Systems Architecture*, Sweden, 1998.

Silberschatz, A., Galvin, P. B., Gagne, G. (2009). Operating System Concepts. Wiley, 2009.

Liu, C.L. and Layland, J.W. (1973). *Scheduling algorithms for multiprogramming in a hard real-time environment, Journal of the ACM*, Vol. 20, No. 1, 1973, pp. 46–61, IEE, London.

Verber, D. (2009). Attaining EDF Task Scheduling with O(1) Time Complexity *Proceedings of 7th Workshop on Advanced Control and Diagnosis*, November 2009, Zielona G´ ora, Poland.

Colnarič, M., Verber, D. (2004). Communication infrastructure for IFATIS distributed embedded control application *PRTN 2004 : proceedings of 3rd intl. workshop on real-time networks*, pp. 7-10. June 2004, University of Catania, Italy.

Texas Instruments (2010). *http://www.ti.com.*

Xilinx (2010). *http://www.xilinx.com.*

8

Modern Internet Based Production Technology

Dobroslav Kováč, Tibor Vince, Ján Molnár and Irena Kováčová
Technical University of Košice
Slovak Republic

1. Introduction

Recent a globalisation of technology and economics world causes that possibility for Internet applications are moved from off-line data processing to on-line data streaming. Internet is often becoming a part of real production processes that thanks to Internet can grow into more efficient and better distributed. A typical example of an implementation of modern production technologies using modern information and communication technologies (ICTs) can be a company with headquarters in one country but having chain of production sites all over the other countries. Internet in this case can be used not only for data transfer from e.g. economics, management, logistics and marketing, but also for data flow regarded to quality of production, failures and others. It enables to achieve recent information about technology used. Moreover, it allows interfering with production system, e.g. in case of need it is possible to actively interact into production system.

2. Modern production topology – recent development and future directions

Higher demands on quality growth and productivity of production processes lead to necessity of integration within information and control systems. It tackles whole production system from top management to the most modern technology, sensor systems and integrated production unit. This is called also Computer Integrated Manufacturing (CIM). In order to create CIM, to provide only simple connection of the all technical devices using communication interfaces to connect them mutually is not sufficient. However, it is necessary to compose reasonable control and information architecture. In general the architecture is divided into 4 levels based on 4R principle as it is displayed on Fig. 1 (Yang et al., 2002).

The principle of 4R is an abbreviation of 4 particular levels of the information pyramid: *Response time*, *Resolution*, *Reliability* and *Reparability*.

Response time.

In direction to top part of the information architecture requirements on fast time response decrease. While on the level of sensors and activators it is necessary to update the information in milliseconds, on the level of management the average update made in hours is sufficient.

Resolution.

It covers level of data abstraction that occurs on each of the levels of the information pyramid. The highest level of the pyramid, management and planning, works with the most abstract data.

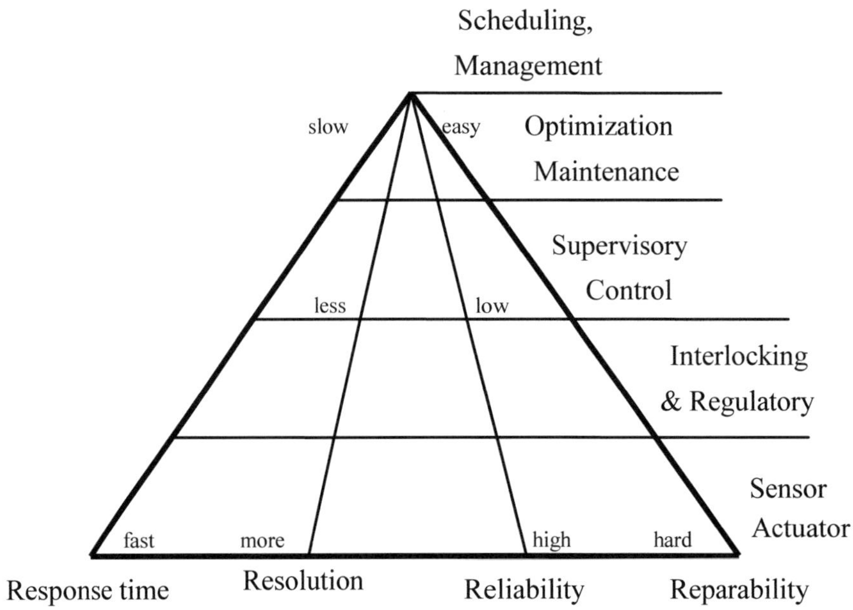

Fig. 1. An information pyramid

Reliability.

In direction of top to basis of the information pyramid, not only the requirements for fast time response of the system are decreasing. Also the requirements on reliability of system decrease. For example in case of computer blackout on the level of management and planning, system can be safely turned off in the time interval of hours but even days with slight loss only. However, if the network blackout occurs for few minutes on the level of control and/or on lower level, this can lead to breakdown, emergency situation and/or blackout of the entire production system.

Reparability.

This term describes the ordinary process to conduct the repairs and maintenance of computer devices on the particular levels of information pyramid. Technical devices on the level of the production line are mostly more complicated from the maintenance and repair point of view that those applied on the level of management and planning.

2.1 Internet applications in production technologies

For technical design of the information pyramid it is necessary to use provide appropriate type of the communication bus besides the particular technical and information devices with suitable communication inputs and outputs.

2.1.1 Strong and weak points of Internet usage

An analysis of strong and weak points of Internet as bus that connects particular levels of information pyramid can be seen in Table 1.

From the analysis in the table above the text it can be concluded that in the technological production chain it is possible to use Internet as bus for data (data bus line).

The most significant problem that occurs is the need of data flow in real-time especially in case of transferred data have transfer time period shorter than hundreds of milliseconds. It occurs due to fact that the delay in data transfer via network can fluctuate from few milliseconds to hundreds of milliseconds. In case of periodic data transfer with periodicity of higher than few seconds this problem can be neglected.

Existing information level	Information exchange	Strong points	Weak points
Management level	System of commercial data	Enables to share and exchange information between management and clients /customers/	Not suitable for real-time monitoring and control
Level of plant-wide optimization	Global database	Enables to gather information about processes in the company easier	Not suitable for real-time monitoring and control
Level of supervisor	Process database	More comfortable to reach the real-time control, suitable for extended control implementation	Does nor dispose with information from management
Level of regulation	PLC, control units	Enables connection of control units directly to Internet	High risk in case of hackers' attacks
Level of sensors and actuators	Intelligent devices	Possibility of direct monitoring and control of intelligent devices via Internet	High risk in case of hackers' attacks, high demands on speed and security of connection

Table 1. Options of interconnection of control levels via Internet

Another, relatively recent issue in this field is the security of the transfer, authorized access to information (Tomčík & Tomčíková, 2008). In case of the connection of technological production devices to global network - Internet (not only on isolated LAN network) there is a danger of unauthorized access. Therefore, people not authorized to have an access to the data could reach information not for their usage.

In spite of the mentioned problems, the advantages that Internet offers are far more dominant that the disadvantages.

One of the most important advantages is the possibility of huge distance data exchange. Internet is the most wide-spread computer network and it enables users to connect to it from any place on the world. Nowadays, speed of Internet connection is so advanced that the data transfer delay between two continents is hundreds of milliseconds. Because of the fact that Internet is generated by license-free protocols, its usage is very strong and economically lucrative tool in process of data transfer.

Another strong advantage of Internet is its versality and availability of Internet protocols that are applicable as for small closed networks as for large-scale networks. Hence, it is possible to find application for the same data acquisition chain and control process for technologies in distance of meters or thousands of kilometres (theoretically it can be applied for unlimited distances). However, it had to b taken into an account that longer the distance, higher the data transfer delay and also greater the uncertainty, unpredicted fluctuations in timing of data transfer).

One of the advantages is the number of nods connectable to the network. While in case of the other bus types the number of the nods that are able to be connected is limited, e.g. for RS-232, Profibus, MPI and others where the number of nods is ranging from 32 to 128 depending on the bus type, in case of Internet the number of nods is restricted with number of IP addresses. When the Ipv4 (Internet Protocol version 4) protocol is used, theoretically 2^{32} addresses can be created and therefore, 4. 10^9 nods can be connected. In the case of IP protocol of newer generation, e.g. IPv6 it is possible to use 2^{128} addresses, which enables to connect more than 3. 10^{38} devices to the network. Therefore, Internet has no competition in the situation when it is necessary to connect higher number of devices to one bus.

One of the practical advantages is the transmission capacity of the Internet. Most of the local networks nowadays operate with data transfer rate of 10Mb/s or 100Mb/s, respectively. However, in case of higher bitrate demands, it can be increased to reach value of bus data transfer rate up to 10Gb/s.

Last advantage to be mentioned, but not the last one in general, is the independence on the type of the transmission medium. It means that data can be transmissed no matter the type of the transmission is using wire-based or wireless connection or even using the combination of both. When the system for control and monitoring production technology is developed based on the Internet, it can be equally reliably functioning with no regards on the type of the transmission channel. The condition is that data transfer rate of the slowest way of connection was sufficient to requirements of the control system. This advantageous feature of the Internet in case of both types of busses is very complex and difficult for realisation. In this way control system based on the Internet can be fully adjustable to the environment, in which it is implemented.

Nowadays, Ethernet communication is a worldwide standard. Therefore, many of technical devices produced recently have already the interface for network connection integrated. It enables direct communication via Internet with central control system without need of the indirect connection to computer.

2.1.2 Methods of application of Internet into production technologies

There are several architectures that can be applied in the technological chain based on Internet. To answer the question which of the architectures is the most suitable depends on the particular situation.

Regarding the connection of particular device to control and information technological network 2 main ways of connection can be distinguished:

- indirect connection - using PC (Fig. 2)
- direct connection - without using PC (Fig. 3)

In case of indirect connection is the Internet connection realized by local computer that communicates with technical device via interface, e.g. RS-232, LPT, USB and others. Local computer is functioning as converter of the interfaces. It also can carry function of security firewall that controls authorization of the particular access to data, enables data encryption, data compression, filtration or pre-processing of data. It can serve as database of history of the processed data. In case it is required, it is possible to connect several devices to one nod in the same moment while computer has function of multiplexer. Indirect connection is preferred mostly in controlling of devices on greater distances.

Second type of the connection requires that there is interface for direct connection to network implemented in the technical device. Except of the various programmable logic controllers (PLC), oscilloscopes and other multifunction devices, there are nowadays also intelligent sensors equipped with this interface.

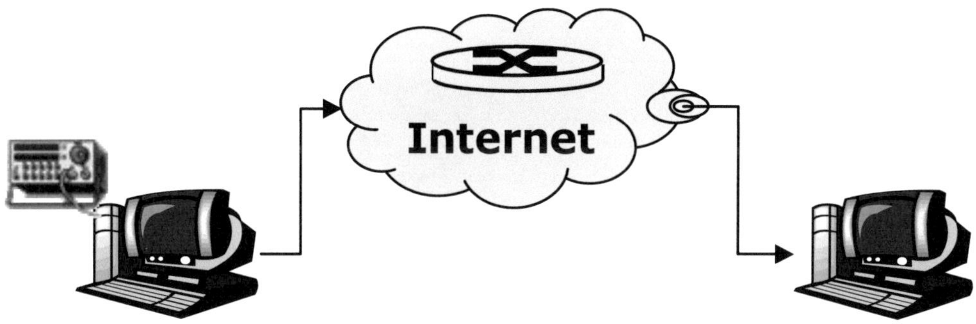

Fig. 2. Indirect connection to Internet

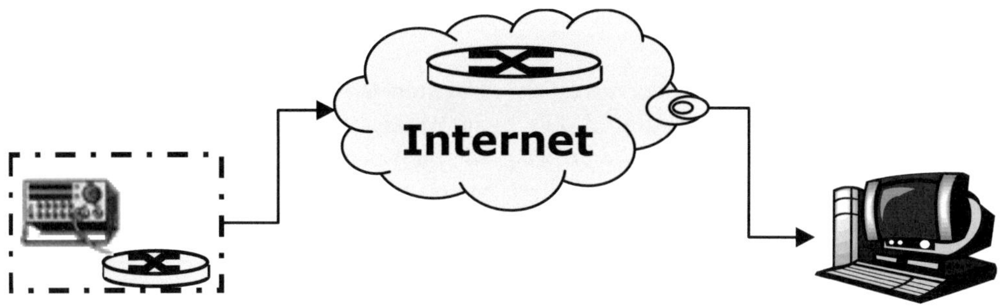

Fig. 3. Direct connection to Internet

The most important advantage of this type of connection is no need of another computer included in the control chain. In many devices there is next to hardware interface also software interface implemented. For example it is web server, ftp server, OPC server and others. Direct connection is financially and spatially not demanding, also technically less difficult to operate for end user. It is more suitable for distant control of systems in the LAN networks (for shorter distances).

According to the interaction between (client and server) computer and controlled device we can classify:

a) Device in the role of server (Fig. 4)

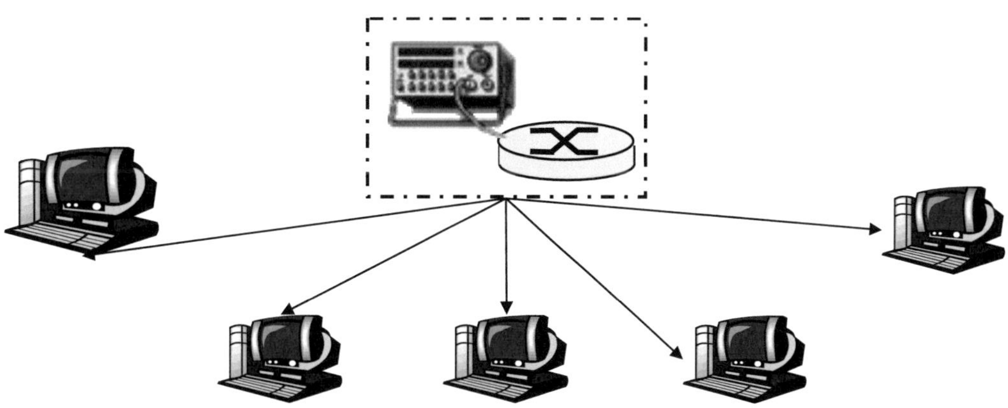

Fig. 4. Device works as server

b) Device in the role of client (Fig. 5)

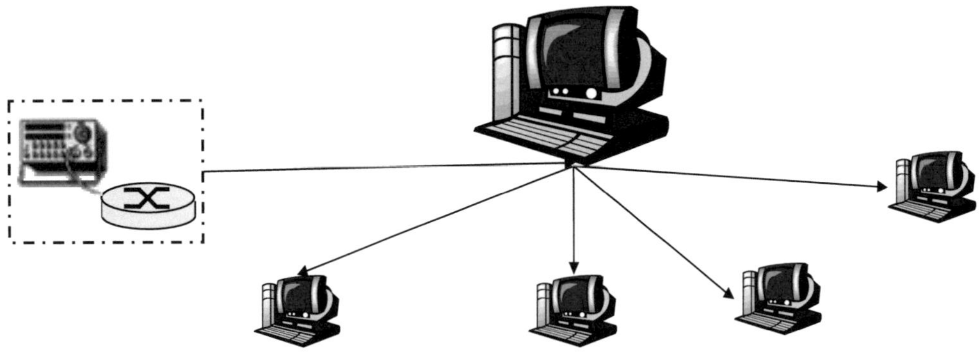

Fig. 5. Device works as client

When device has function is the server, all the computers that require technical data connect to this device that is sending required data using multicast messages (from on or several sources the message is sent to several targets). The biggest advantage of this architecture is that it is not necessary to have powerful server and users have the most recent updated information. This type of the architecture is suitable especially for decentralized systems. A disadvantage is that the faster data transfer is necessary on the line on which the device is connected and it is also higher intelligence of device necessary in order to manage and fulfil several tasks in the same moment. Other disadvantage is the time of the response that can vary depending on the number of the requests.

When device has a function of the client, there are data required that are transferred to one computer, which provides the further distribution on demand. In case of this architecture it is necessary to have computer in role of server. However, there are not high requirements on data link transfer rate, on which technical devices are connected to (because the data are provided to one target). Thanks to written about system has relatively constant time response. Server provides better security and data administration. This type of architecture is suitable for centralized systems.

2.2 Computer network

Computer network creates a connection of several computers, technical devices and nods of then network. It enables transfer of the information among of them. There are various types of networks: Token Ring, Novell, Ethernet etc. Interconnecting those particular local networks composes Internet.

2.2.1 Network layers classification

In consequence of the fact it would be practical if several devices from different producers can communicate with each other via Internet connection and it would be necessary to create certain communication standard. With intention to build up structured communication model in computer networks, an international organisation ISO (International Organization for Standardization) established system OSI/RM (Open Systems Interconnection Reference Model). The aim of the OSI/RM was to design how the different parts of the network communication system should cooperate and interact. This model determines what is necessary to do generally, not what has to be done particularly during realization of the tasks. Model OSI divides communication services to 7 categories. Generally, system OSI/RM creates a complete model with 7 layers of network communication (Seven-Layer Model - SLM):

1. *Application layer* - consists of a set of the different protocols for upper level of application, specifies background in which network application communicate with network services as electronic mail, file transfer, connecting distant terminals and others.

2. *Presentation layer* - it is also theoretical and not used widely. It is devoted to semantics of particular bits and controls formatting of the data transfer. It consists of specification for encryption and decryption of symbols' sets. It enables communication of computers with different internal representation. This layer is the most suitable for occasional implementation of cryptography.

3. *Relation layer* - is the most spread transport layer. It provides services that control data and synchronization. It processes errors of transmission and transfer and stores reports about the transfers made. This layer is purely theoretical and it is used only exceptionally and with few applications. The set of Internet protocols does not support this layer.

4. *Transport layer* - provides functions for creating particular connections, initialize data transfer and release the connection after end of the transfer. It does not deal with lost messages. Transport layer ensures reliable network services. It divides the message (from relation layer) to smaller particles (packets) and match them sequence numbers (for purposes of repeatable composition of the message) and send it over. The most used protocols of transport layer are protocol X.25 and protocol IP. In case IP protocol the packets must be repeatedly put in order according to the sequence because it could happen that due to delayed of the packets the order of them could change.

5. *Network layer* - is devoted transferring of the packets. It describes processes that divide data into network addresses and control whether the messages were send entirely and in time. Network layer finds the best path for sending the packet, it is described as process called routing, and that can be rather complexion large network as Internet. There are several types of routing: short or fast (it means whether the packet will go to the shortest or the fastest way), static or dynamic routing and others. The most used protocols at the secure transfer connection are X.25 (contact oriented - example of fax line) or IP (contract-free oriented).

6. *Contact layer* - is correcting errors that occur in physical layer, describes processes that detect and correct errors on the level of data during data transfer between physical and layer and layers above physical layer. It configures bits into the frames and controls its correct transfer. Contact layer adds certain bits plus called check sum in each frame. On the side of the receiver the check sum is verified whether damage of the frame occurred during the transfer. If check sum does not match, contact layer requires a retransmission of packet. It sends control message to the source of the damaged packet. Contact layer consists of sublayers:

Logic Link Control (LLC) - defines how the data are transferred using medium and provides services of data connection to higher layers.

Medium Access Control (MAC) - defines who can use the line where there are several computers attempting to connect in the same time (for example Token passing, Ethernet [CSMA/CD] etc.).

7. *Physical layer* - is aimed to transfer the particular bits, it describes electrical, mechanical and functional requirements on processing transfer data. It characterizes processes that control data as bits passing through the hardware. There are topics of voltage levels of logic zero and logic one, transfer bitrate, one-way and two-way transfer, protocols of data transmission (for example X.21, RS-232 for serial communication and others), topics solved

on the level of physical layer. Because of the fact that relation and presentation layer are only theoretical and they are used not very frequently, we will focus more on the % layers (5-layer model). The activity of the particular layers is displayed on Fig. 6.

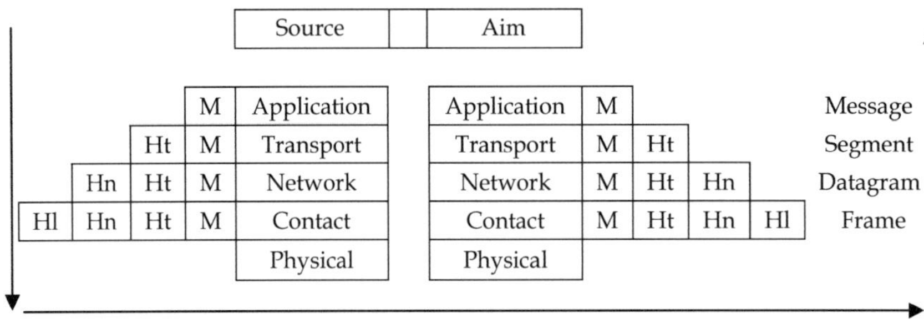

Fig. 6. Activities of the particular layers OSI/RM

Message is sent from the application layer of the source to the transport layer. In the transport layer the message is divided into the segments of appropriate size and each part of the message will be matched with Transport layer Head (Ht) by transport layer. This is the process how segment is made. Segment is removed to network layer. In the network layer the most optimal route to send the message is selected according to the target address, this calls routing. In the network layer the Net layer Head (Hn) is added to segment in order to create datagram. Datagram is moved from the network layer to contact layer, where Link layer Head (Hl) is paired with datagram in order to create frame. Frame is decomposed into particular bits, which are transported with physical layer using suitable medium (coaxial, optical fibre or using microwaves). After the frame is transported to the target, message M is forwarded from physical layer back to application layer by inverse functions. It means integrity of transferred frame is inspected, instead of the decomposition the message is recomposed once again. In each layer head is removed and information is passed to higher level. For example in the transport layer net layer head is removed from the particular segments. They are composed into one compact message and then the message is send to application layer. Application layer transmit the message to particular application for further processing (Tanenbaum, 2003).

2.2.2 Network components

For exact and reliable functioning of the network, there are besides the transfer medium and computers also large number of other components included in the computer network. Correct recognition and application of these components and these features allows composing better computer network. Because Internet as global network made by local networks, it means it is necessary to take into the account communication requirements of other types of networks with their different structure and topology.

The network includes these components from physical point of view:

- nods
- transfer routes

Basic nods of network

<u>*Network adapter*</u> – it is an interface between computer's motherboard and transfer medium. Each technical device connected to network must dispose with appropriate network adapter.

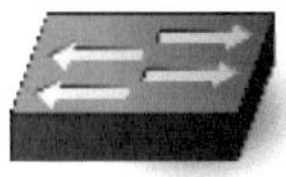

Repeater – 2-port electronic device that reproduces only what it receives on input port into the output port. It is used for instance when the data are transported on bigger distances.

Bridge – more sophisticated repeater with option to filter the packets (it is disposing with contact layer - OSI level 2). It means that except of the sending packets from one port to the other, it enables to inspect the packets and allows transporting only undamaged packets (frames). In case of damaged packets bridge send a request to source of damaged packets to send the undamaged packet.

Hub – is a multiport repeater.

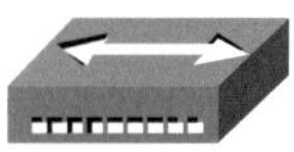

Switch – is a multiport bridge.

Router – connects two or more networks (also different types of networks), conducts references with appropriate "routing" information. It operates on the level of network layer (OSI level 3). It has to include wide range of information about internal network (routing table). It is usually used in distant networks that interconnect networks with different communication protocols. Router stores the routing table, routes between the nods and it is able to select optimal route for data transfer. If the error occurs in one-way data transfer, it will try to find an alternative. There are static and dynamic routers. Static router includes table of directions that is static. Only network administrator can change it, e.g. when the structure or network topology is changed. Dynamic router can find on its own how the particular branches of the network loaded and adjust the table of directions in order to send packets in the most optimal (fastest) route.

Gateway – similarly as the router, it connects two networks of different type. It can operate on OSI level higher than 3rd. it enables to convert instruction file of the sending network to corresponding instruction file of the receiving network. These instructions about the way of data processing are required by some of the network systems, mostly mainframe and minicomputer systems. For example gateway converts the request for accepting the file form the connected computers into the command that is readable by mainframe computer. In praxis for gateway can be computer with more network cards where each card serves for different type of the network. It also can provide function of bridge and router.

Data flow routes

Data can be transferred by three main ways:

- telephone line
- wired media
- wireless media

Telephone line

It belongs to the oldest way to connect computer to network. In spite of this way of connection is still wide spread and used, it is slowly in regression. There are several types of the telephone connection. The most wide-spread are following:

Classic modem connection - direct connection on router using telephone modem. The data transfer rate is up to 56 Kbps (kilobits per second).

ISDN connection (Integrated Services Digital Network) - digital connection with router. The transfer rate can be up to 128Kbps for both-way digital connection.

ADSL connection (Asymmetric Digital Subscriber Line) - is asymmetric digital connection. It enables connection with bitrate 1Mbps in direction PC -> Router and bandwidth 8Mbps in direction Router -> PC.

Wired media

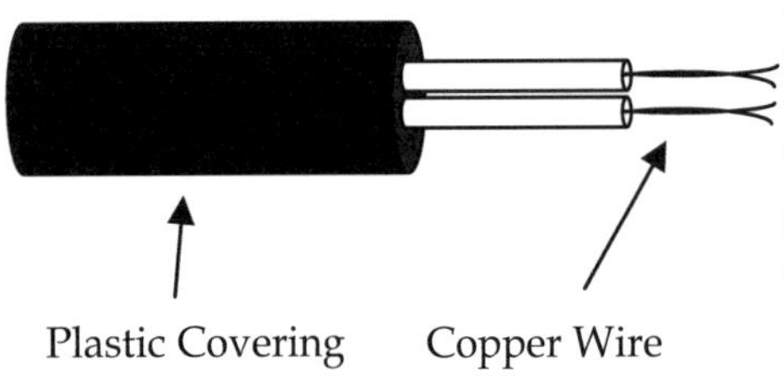

Twisted pair: is used often in telephone wiring where several fibres are twisted together. Twist pair is characterized by low bitrate: 250 KHz and low relation of signal and noise. It can cause signal skip, which results to low data rate. This type of the medium is suitable for short distance communication. Therefore, they are used especially in the LAN networks.

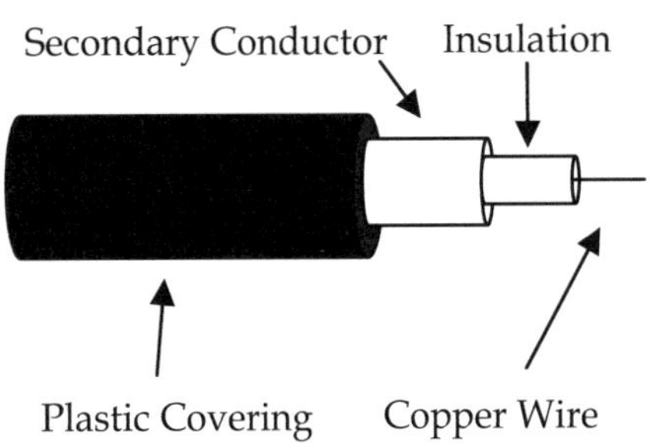

Coaxial cable: is used especially in LAN networks. It is simple central line surrounded round isolation layer and conductive shadow. In comparison of twisted line wire coaxial cable has higher data rate: up to 400 MHz, higher quality of transfer (higher relation of signal and noise distance) and maximal bandwidth - 100 Mbit/s. When higher frequencies are reached, loss of signal can occur.

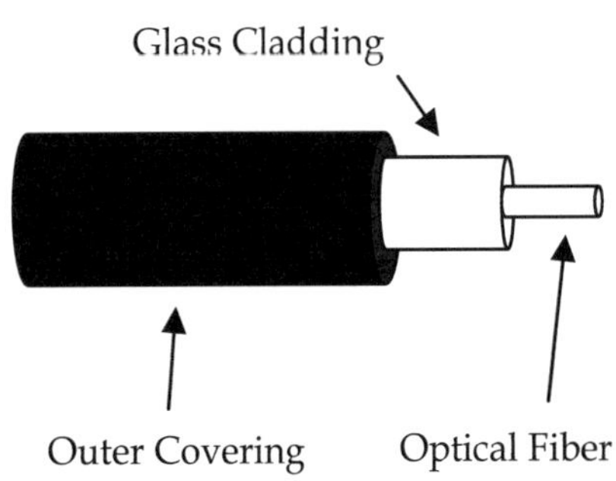

Optical cable: is suitable for high quality and high-speed data transfer. Optical impulses are used for transfer of particular bits instead of the electrical impulses. This enables hide range of frequencies (20,000 MHz) and data rate: 400 Mbit/s and more. Optical cable is used for intercontinental lines. It becomes more and more often popular connection for LAN networks. Extremely low noise and very problematic tapping the line (high level of security) increases advantages of this medium. Main recent disadvantage is relatively high costs.

In project 802 IEEE the specification for cables for transferring Ethernet signals was established. IEEE - Institute of Electrical and Electronic Engineering is American organization that defines standards relating the networks and other areas. IEEE 802.x are probably the most known standards in the field of computer networks. There are series of the standards, recommendations and information documents related to computer networks and communication. Names 10Base5, 10Base2, 10BaseT and 10BaseF (FOIRL [Fibre Optic Inter Repeater Link]) label thick coaxial, think coaxial, unshielded twisted double line, cables with optical wire, respectively. Number "10" means Ethernet data transfer rate - 10Mbit/s. "Base" describes basic band, 1 communication channel for each cable. Last number was originally naming maximal length of cable in hundreds of meters. After implementation of codes 10BaseT and 10BaseF, the last letter is used to distinguish type of the wiring: T is used for twisted double line (twisted pair) and F indicates optical fibre (fibre-optic).

Table 2 shows names and characteristics of some selected medium types. Above-mentioned wiring (twisted pair, coaxial cable and optical fibre) belong to the most wide-spread and most often used.

Cable label	Type of medium	Maximal segment length	Maximal number of nods
10Base5	Thick coaxial	500 m	100
10Base2	Thin coaxial	185 m	30
10BaseT	Twist pair	100 m	1024
10BaseF	Optical fibre	2.000 m	1024 k

Table 2. Table of features of wire medium

Wireless media

In case of the medium without line wire the signals are transferred using electromagnetic spectrum. To ensure the signal transfer there is no physical wire or cable used and the signal is transferred both-ways. The main disadvantages of the media without wire line are the effects related to electromagnetic field and signals: reflection, blocking the signal by objects, intereference.

Nowadays, there are several ways of wire free line connection used. The basic types of the wireless connection are:

Shared microwave network - it connects router with end-users. Bitrate of particular channels is up to 54 Mbps.

Wireless local network- physical fibres are replaced by radio spectrum. Data rate e.g. in network Lucent Wavelan that can represent this type of wireless network is 10Mbps.

Wider wireless network - e.g. DCPD (Cellular Digital Packet Data) is wireless access to ISP (International Standardized Profile) connecting router via cellular network. Bitrate of this connection reached tenths of kbps.

Satellite connection - enables wireless connection in the channels up to 50 Mbps, or more channels with lower permeability. Final delay is maximally 270ms. The greatest advantage of satellite connection is that end-user does not need to be close to basis connected to network using cables, wires and in spite of it the connection can be made anywhere on the globe. Extremely high costs are the main disadvantage.

Nowadays, it is possible to reach relatively high speed of the data transfer using physical media. However, those higher transfer rate causes lower quality of transferred data. In case of the analogue image or sound transfer it can cause only graininess or noise, in case of packets it would mean devaluation of the whole packet if only one of the bit is changed within the packet. It would result to situation that when using high-speed data transfer, the ratio of number of successfully received packets and entire number of sent packets would decrease.

Therefore, very high data transfer rate can cause connection failure. It is very important to realize that transfer parameters from producer are conditional by "ideal conditions". To reach them is practically impossible in praxis. It is the reason why while designing the network it is necessary to take into the account relatively big difference between data transfer rate given from producer and real speeds that are reached in practical usage.

2.2.3 Internet protocols

Internet protocols are among the network communication protocols used worldwide the most. They are "open-system" protocols and they also are universal. They are used in the local (LAN) but also wide (WAN) networks and they can be applied as well as independently on the network type (Ethernet, Token Ring, etc.). They consist of complex of protocols, the most known and most often used are TCP (Transmission Control Protocol) and IP (Internet Protocol). They include not only protocols of lower layer as TCP a IP but also protocols of common applications as electronic mail, protocols for file transfer, hypertext data transfer and others.

First protocols were developed in half of seventies by DARPA (Defence Advanced Research Projects Agency). They were interested in the network based on the packets' exchange that enabled communication among various computer systems in the research agencies. With the aim to create heterogenic interconnection DARPA funded the research at Stanford University, Bolt, Beranek a Newman (BBN). As a result of this research was set of the Internet protocols finished at the end of seventies.

TCP/IP was later connected with Berkley Software Distribution (BSD) UNIX and since then it became a background on which Internet and World Wide Web are based on. Documentation of Internet protocols, including new and sophisticated protocols and their policies are specified in the technical documentation, technical reports called Request for Comments (RFC). They are reviewed and published by Internet community. Any improvement of protocols is then published in the new RFC.

Fig. 7 displays an overview of the most spread protocols classified according to the OSI reference model, depending on the OSI level, in which they operate.

Physical layer: on the level of physical layer there are not specified any protocols yet. Transfer of particular bits is purely task of the network layer and transfer media.

Contact layer: on the contact layer there are mainly 2 protocols active: ARP and RARP. In order to ensure that 2 devices could communicate, the physical address of the paired devices has to be known (so called MAC address).

- Address Resolution Protocols (ARP) is a protocol that help host to find out MAC address corresponding the IP address. After accepting MAC address, IP device remembers temporarily and stores cache in order to avoid sending ARPs once again in case of restoring the connection. If the contacted device does not react for certain time interval, cache record is deleted.
- Reverse Address Resolution Protocols (RARP) has reverse function, finds IP address that correspondents with particular MAC address. RARP protocols represent logical inversion of ARP are used at workstations without disks that reply on RARP server with mapping table MAC-IP because they find out IP address while booting.

Network layer: disposes mainly with 3 protocols: ICMP, Routing protocols and IP protocol.

- ICMP protocol provides information packets back to the source, reports about errors and other information related to the IP packets processing. Packets of protocol ICMP are generated in several situations, including inaccessibility of the target, echo requirement and reply, diverting, time-out etc. The typical feature of the ICMP protocols is that if ICMP packet does not reach the target, other one is not generated in order to avoid an incomplete cycle of sending the packets.

OSI reference model

<table>
<tr><td>Application layer</td><td rowspan="3">FTP, Telnet,
SMTP, SNMP, HTTP</td><td>NSF</td></tr>
<tr><td>Presentation layer</td><td>XDR</td></tr>
<tr><td>Session layer</td><td>RPC</td></tr>
<tr><td>Transport layer</td><td colspan="2">TCP, UDP</td></tr>
<tr><td rowspan="2">Network layer</td><td colspan="2">IP, Routing Protocols, ICMP</td></tr>
<tr><td colspan="2" rowspan="2">ARP, RARP</td></tr>
<tr><td rowspan="2">Datalink layer</td></tr>
<tr><td colspan="2" rowspan="2">not specified</td></tr>
<tr><td>Physical layer</td></tr>
</table>

Fig. 7. Overview of different protocols classified according to OSI model

- Routing protocols – routers inside the Internet are organized hierarchically. Routers used in the autonomous systems are called internal routers. The use Interior Gateway Protocols (IGPs), e.g. Routing Information Protocol (RIP). Routers that transfer information among autonomous systems are called external routers. External routers use for information exchange among autonomous systems Exterior Gateway Protocols (EGPs), e.g. Border Gateway Protocol (BGP).
- IP protocol – includes address information and some of the control information that enables alignment of the packets. IP is main protocol in complex of Internet protocols. Together with TCP protocol it presents main keystone of Internet protocols. IP has two primary functions: provide the most powerful delivery of datagram via computer network and provide a fragmentation and rebuilding datagram to support data connection with different Maximum Transmission Unit (MTU).

Transport layer: uses one of the most known Internet protocols TCP and UDP. These protocols play key role in the control and regulation processes via Internet. Both mentioned protocols, TCP and UDP, use the most known and most often used protocol IP for their operations. Protocol TCP transports data using TCP segments that are addresses to particular applications. Protocol UDP transports data using so called UDP datagram. Protocols TCP and UDP ensure connection between application running on distant computers. Difference between protocols TCP and UDP is in fact that protocol TCP is linking service, receiver confirms the received data. In case of data loss, loss of TCP segment, receiver requires resending the message. Protocol UDP transfers the data using datagram, sender sends datagram but is not interested in the information whether it was successfully received or not.

All the protocols of transport layer, including TCP and UDP, use IP protocol for their basic functions. IP is on itself not reliable protocol. It does not provide any guarantee that datagram or packet will be successfully delivered. It happens that packet is lost or damaged, e.g. due to overload of the network. Also a situation that packets are not received in correct order can occur. For applications that require certain level of guarantee of successful delivery of undamaged packets, e.g. Telnet, Email and others, is the uncertainty of delivery unacceptable. These applications need the guarantee that sent data will reach target in initial, unchanged form. This reliability is assured using virtual circuit, when two applications based on TCP need to communicate with each other. In this circuit every data exchange is monitored between the particular applications. Receiver confirms sender that each message was received. Message that was sent and there was no delivery report about will be resent again.

- Fig. 8 displays particular fields of packet in TCP protocol. The meaning of the particular fields is following:
  Source port is 16-bit number that indicates number of sender's port.
  Target port is 16-bit number that indicates number of receiver's port.
  Sequence number helps receiver to restore the received data from the segments in case they arrived not in initial order.
  Acknowledgement code is byte that sender of the message expects as confirmation that packet was successfully received.
  Data offset is number of 32-bits words in the header (place where data begin).
  Reserve is made for usage in the future (recently it does not have any function).
  Flag means control bits that are used e.g. when initializing the connection.
  Window specifies size of window of sender (size of buffer that should be in disposition for incoming data).
  Urgent is index that indicates the 1st urgent data byte in the packet.
  Option specifies different TCP settings.
  Data includes data from higher layer.
  Header of TCP packet is mostly in size of 20 bytes. However, maximal size of header is up to 60 bytes.
  TCP provides 5 key services: virtual circuit, managing input/output (I/O) applications, managing I/O networks, controlling flow and reliability of delivery.
  Every time two applications need to communicate one with the other using protocol TCP, there is a virtual circuit created between them. If it is necessary to provide communication among more computers in the same time, there is virtual circuit created for each coupled pair of computers separately. Virtual circuit is basic for all other services provided based on TCP.

<table>
<tr><td colspan="3">Source port</td><td>Target port</td></tr>
<tr><td colspan="4">Sequence number</td></tr>
<tr><td colspan="4">Acknowledgement number</td></tr>
<tr><td>Data offset</td><td>Reserve</td><td>Flag</td><td>Window</td></tr>
<tr><td colspan="3">Check sum</td><td>Urgent index</td></tr>
<tr><td colspan="4">Options</td></tr>
<tr><td colspan="4">Data</td></tr>
</table>

Fig. 8. Fields of TCP packets

Applications communicate among each other sending data to local TCP that transfer data using virtual circuit on the other side where they are delivered to the target applications. TCP provides I/O buffer for applications. It enables them to send and receive data as fluent data flow. TCP converts data to individually monitored segments and it sends them using IP protocol of network layer.

TCP provides also managing I/O of network for IP, creating segments that can effectively flow through the IP network and particular segments repeatedly transfer to data flow appropriate for applications.

Different hosts in the network have different characteristics, including memory, width of network band, capability and other. For this reason not all the hosts are able to send data with the same data transfer rate and TCP have to be able to cope with those differences. TCP has to conduct all these services continuously without performing any activities form the side of applications.

TCP provides reliable transfer by monitoring of data that have been sent. TCP uses sequence numbers for monitoring of particular bytes of data, acknowledgement numbers in case of loss of any packet and acknowledgement numbers for own control of the data. All the services make protocol TCP robust transport protocol.

- UDP packet is displayed on Fig. 9.

Source	**Target**
Length of UDP	UDP control sum
Data	

Fig. 9. Fields of UDP packets

As it is visible from Fig. 9, packet UDP is much simpler in comparison with TCP packet. Protocol UDP is often named as simple interconnection of network layer (IP protocol) with higher layers, e.g. with application layer. UDP provides transfer services for applications that don't need to or cannot use services connection oriented services provided by TCP. The term connection oriented services is mutual dialogue of sender and receiver in data transfer. It is characterized by successful delivery of each one of the packets. UDP is the most used by applications that often send packets with updated information or in cases when it is not necessary to deliver each one of the packets.

At UDP protocol that sends UDP packets, immediately after sending protocol "forgets" the packet and does not care of it anymore. While TCP provides high level of reliability using virtual circuit, UDP offers high level of network performance (Hall, 2000).

Application layer: (together with Presentation and Relation layer) uses protocols to support particular application e.g. for applications of electronic mail, protocol SMTP and so on.

- Network File System (NFS) - distributed file system developed by Sun Microsystem for simplification of work with files on distant system.
- External Data Representation (XDR) - abstract (not depending on machines) syntax for data structures description. XDR was developed as part of NFS.
- Remote Procedure Call (RPC) – transparent mechanism. Using this mechanism one procedure called on one computer by program running on other computer. This mechanism provides simple way how to implement relation client-server.
- File Transfer Protocol (FTP) – is a protocol that allows user transfer of files from one computer to the other. For this purpose it is using protocols of lower layer of TCP/IP.

- Simple Mail Transfer Protocol (SMTP) - provides simple service of electronic mail. It is using protocol TCP for sending and receiving messages.
- Simple Network Management Protocol (SNMP) - is applied to control services of network management and data transfer connected to network management.
- Telnet - has ability of emulation of the terminal and enables log in from distant network.
- Hypertext Transfer Protocol (HTTP) - fast and object oriented protocol used mainly in World Wide Web (WWW). HTTP enables agreement between Web client and server.

2.2.4 Latency in network

Power of computer network is characterized by several existing and mutually interconnected parameters. One of parameters is latency in the network. Latency is time that is needed for transfer of the empty message (a packet without any data) between two computers. It is entire sum of the latency due to software of the sender, software of the receiver, latency of the access to the network and latency of the network transfer.
Other parameter is data transfer rate. It is a speed that data are transferred between sender and receiver via network. Size of the parameter is given in Bits/s.
Time required for transfer of packet then can be calculated:

$$MTT = L + ML / TS \tag{1}$$

where: *MTT* - message transfer time, *L* - latency, *ML* - message length, *TS* - transfer speed
Bandwidth is also very important parameter. Bandwidth is entire volume of the transfer that can be transported through the net. Maximal speed of data transfer can be then calculated:

$$MTS = BW * \log_2 (1 + (signal/noise)) \tag{2}$$

where: *MTS* - Maximal Transfer Speed, *BW* - Bandwidth of transfer medium
The maximal transfer speed calculated in this way is purely theoretical and it cannot be practically used. It can be regarded as higher limit for transfer speed of given network.
All the parameters describing power of the network show that the main disadvantage of Internet when the point of view is an application of Internet in the technological chain is latency. Certain latency occurs in every type of remote control (RS-232, Profibus and others). However, latency that occurs in Internet connection is very dynamic parameter that can change its value in wide range (from few milliseconds to seconds). While packet is passing through the network there are 4 types of latency in each nod:

- Latency that results from processing packet in the nod (control whether the packet arrived without damage and according to the target address forwarding to appropriate output channel),
- Latency due to queue (waiting while transmission line is occupied),
- Latency due to transfer (time from transmitting packet to medium),
- Latency due to dataflow (time that signal needs in order to change position from one to next nod using medium).

The greatest latency is caused by latency due to queuing that is becoming significant especially at more loaded networks. How the network is loaded it can be calculated:

$$TI = L * A / W \tag{3}$$

where: *TI* - Traffic Intensity, *L* - average length of network, *A* - average number of incoming packets, *W* - link bandwidth (bps).

When relation $L*A/RP$ is close to zero, latency from queuing is very small. If it reaches value 1, latency due to queuing increases exponentially. If the value is greater than 1, latency due to queuing is becoming "infinite", as there are more packets incoming to the nod than it is able to process (Tanenbaum, 2003).

3. Modern production topology – future challenges

All mentioned circumstances are nowadays only a recent premise for designing future automated technological complexes that will be reconfigured via Internet (Fig. 10). The main advantage of such modern solutions is their adaptability and variability that will assure higher efficiency of production and competitiveness. Herewith, other activities connected to production process and its modifications can be also automated and so eliminate the subjective human factor that was responsible for performing mentioned activities.

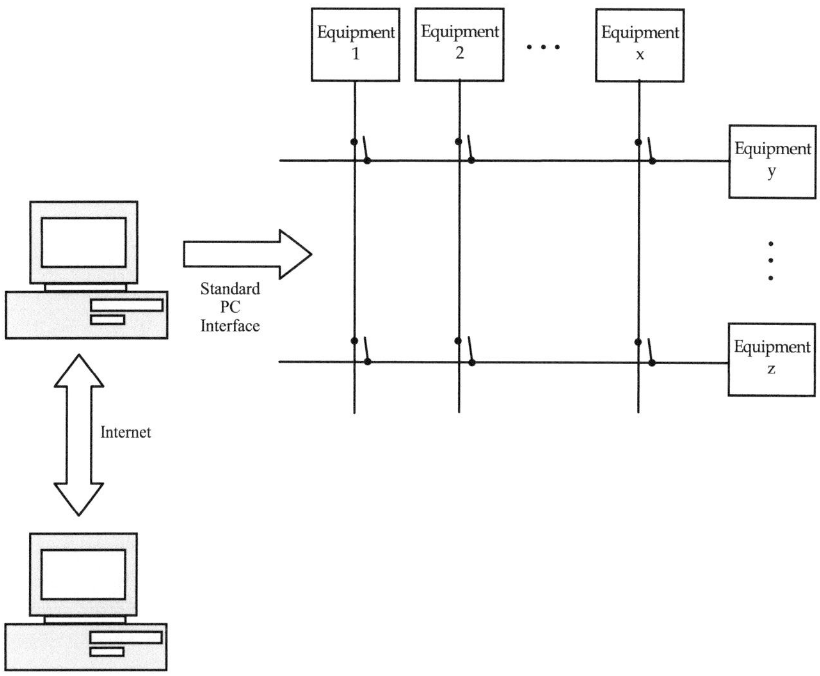

Fig. 10. Production line based on matrix topology and configurable via Internet

Basic assumption of the matrix production line configurable via Internet is fact that all or most of parts of production technology must be designed in advance and constructed as mobile or universal devices. In that case it is possible to control change of topology of production line, and so also production character and its quality and productivity via Internet. For this purpose it is possible to use reconfigurable mobile transport systems remotely controlled via Internet. If they exist, particular parts of the production line can be stationary.

There is detailed example of Virtual Measurement Laboratory (VML) realised on matrix topology basis. It is also the first workplace designed in this manner. It serves for expert measurements of various applications; at the moment Slovak patent application (Kováč et al., 2009) is pending. It is necessary to underline that VML (Fig. 11, Fig. 12, and Fig. 13) is not only a virtual workplace.

Fig. 11. Real part of VML

However, it is a real laboratory equipped with excellent measurement technologies based on above-mentioned principle of complexes that can be reconfigured via Internet. It is available for wide range of clients, users with no regard on their location and time. Due to these facts the expensive expert measuring systems enabled to utilize the equipment effectively and provide better access to technology. Therefore, this laboratory became a background for building a centre of excellence in the field of electrical engineering in the wide scope of application.

Work with the system starts in program VM LAB, in which user chooses in the roll-down menu "Type of element" from the catalogue of devices (those that are available in given laboratory). After selecting device user places it on the right side of the window of the program. After selecting all the devices that user is planning to work with and the ones that user wants to use for measurements. Then these devices are connected according to the electric circuit designed by user. In order to compose this circuit, user can choose all necessary components from the catalogue, e.g. cable, crossing, nod etc. in this way user can create scheme of electrical circuit that user has intention to perform practical measurement. In the moment user finished editing of circuit topology, user can send it to server using main menu. After connection with server, server accepts the request and interconnects particular measuring system according to the circuit scheme designed by user. If this is finished, left lower part of the program window becomes accessible for user. There user can change settings for required parameters of particular sources used in the measurement system in order to be able to modify parameters of the circuit according to user's ideas and the technical description of the devices. Output measured parameters are displayed in form of the graphical output as it is displayed on the particular measured instrument.

Real-time work with system is finished in the moment user disconnect from server and closing the program VM LAB, in case of working in off-line mode. User can save all edited schemes on disk of computer for purpose of the future use.

Design of measuring system begins in the program VM Lab developed at department of Theoretical Electrical Engineering and Electrical Measurement, Technical University of Kosice (Fig. 12). Left part of the program window serves for selection of component from the catalogue. The component is then added to designed system as part of it, e.g. direct current or alternating current power source, resistance, coil, voltmeter etc. After selecting category of the component, program offers user list of the components from selected type

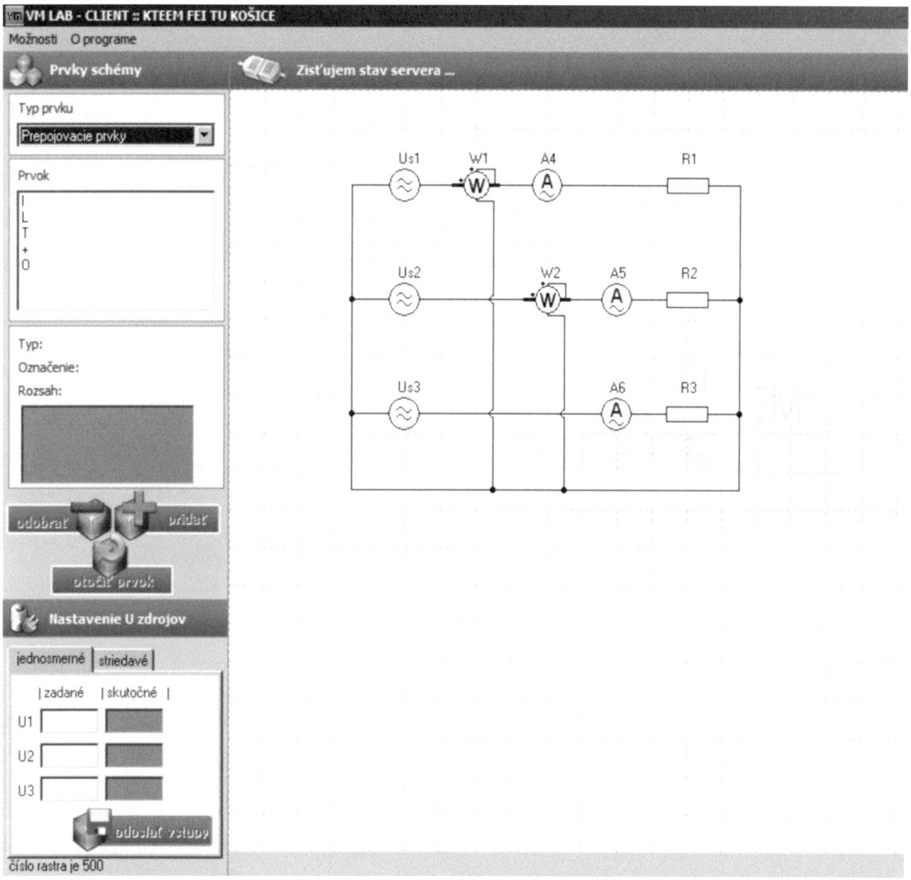

Fig. 12. Virtual part of VML

including the parameter's description. For example, if the selected component is resistance, then program will offer certain number of different resistances with their parameters and additional information, e.g. value of resistance, dimensioning power, technological form (ceramic, resistance wire etc.). If the selected resistance meets requirements, user will add it to right side of the program window. After all the components are selected, user connects them with lines. In this was the entire system is designed. For example Fig. 12 shows Aron's set/up for three-phase power measurement. After the scheme of the circuit is created, program connects to the server where there is real hardware as a part of the measurement laboratory (Fig. 11). In this way all selected components are physically interconnected according to the design in the program VM Lab. In this moment it is possible to change the input settings for adjustable sources and to see values achieved from measuring devices. Designed measurement circuit can be saved into the files and they can be later reloaded in case of need. In this way it is possible to create whole library with the circuits and edit schemes that have already been created.

It is necessary to emphasize several important facts related to unique workplace:

- Program does not serve for simulating of measurement systems. Displayed output from measurements does not result from calculations but they are resulting from real measured data.

- Hardware does not work as switch between limited set of pre-defined connection but it enables creating any real circuits with components available in VML. Circuits can be created from any computer connected on Internet.

For these features developed system is unique.

Fig. 13. Communication board

4. Conclusion

All the information included in the chapter can be implemented also in the area of manufacturing technologies. For this purpose it will be necessary, besides the modern technical devices, to provide also sufficient human resources. Design, realization and administration of these technologies will require wide multidisciplinary knowledge in the field production technologies, mechanical engineering, electrical engineering, automation, informatics, artificial intelligence, management and logistics, but also mathematics, physics, material engineering. Due to this fact their practical applications are not easy and they become a challenge for coming decades.

All results presented in this chapter has been prepared under the support of Slovak grant projects VEGA No.1/0660/08, KEGA No. 003-003TUKE-4/2010, KEGA No. 3/6386/08, KEGA No. 3/6388/08.

5. References

Yang, S. H.; Tan, L. S. & Chen, X. (2002). Requirements Specification and Architecture Design for Internet-based Control Systems, *Proceedings of COMPSAC'02 26th IEEE Annual International Computer Software and Applications Conference*, pp. 75-81, Oxford, 2002, UK.

Tanenbaum, A. C. (2003). *Computer Networks*, Prentice Hall, ISBN-10: 0130661023, New Jersey, USA

Hall, E. (2000). *Internet Core Protocols: The Definitive Guide*, O'Reilly & Associates, ISBN 1-56592-572-6, USA.

Kováč, D.; Vince, T.; Kováčová, I. & Molnár, J. (2009). Scheme for automated and variable wiring of electrical components and equipments into electrical circuits, *Patent application No. 00023-2009*, Košice, Slovak Republic

Tomčík, J.; Tomčíková, I. (2008). Safety policy of automation and SCADA systems. *Journal of Electrical Engineering and Energetics*. Vol. 14, No. 1 (2008), pp. 9-11, ISSN 1335-2547

9

eLearning and Phantasms

M.sc. Ivan Pogarcic and B.sc. Maja Gligora Markovic
Polytechnic of Rijeka, Business Department
51000 Rijeka,
Croatia

1. Introduction

None of the phenomena created within human activities cannot be qualitatively analysed if isolated from its environment. Reason of its appearance is regularly a consequence of influencing the observed segment of universal system upon its environment and/or inner interaction between its own segments. Further on, every phenomenon is regularly positioned in space and time that determine its dimensions as basis for measuring its influence over the environment. The process rarely includes timely and spatially independent phenomena. Naturally, the component of time can be observed and comprehend in different manners. The same can be applied on component of space especially if space indicates its location concentration and specificity. The third dimension of analysis is pragmatic nature of phenomena, respectively absolute and relative capacity of its usability observed from user's standing point.

Considering the above stated, learning as human activity or a specific phenomenon also requires detailed and comprehensive valorisation. From historical point of view, learning has been changed and developed thorough time in different forms, applicable according to momentarily conditions, circumstances and disposition of elements that define learning at specific moment of time. From primitive modes when student simply imitates teacher or gains knowledge and skills upon his own experiences, learning has been modified and developed to modern forms recognisable upon class/grade articulation, connections and relationships between elements. The basic elements that make a crucial axis are nevertheless student and teacher. All other elements are determined and subordinated to them. These other elements are still important up to a certain level, especially learning contents. Together with learning contents, student and teacher also determine learning goals. System defined in such mode simultaneously determines and is determined or interdependent upon environment in which learning is being performed. Only with this environment respectively context in which learning is being performed the concrete forms of learning's realisation will be defined. Learning can be differently treated dependently upon angle of its analysis.

However, every frame of analysing learning as phenomena has necessarily included validation of all factors due to creating a comprehensive overview of relationships, activities or any other initiating elements. Holistic approach to learning as phenomena is a combination of both verification and validation of learning as phenomena, but also of the individual elements that comprise learning. Absence of validation of any element that comprises learning or factors which define its realisation cause incomplete picture and basis for invalid comprehension of learning as a complex phenomena. This paper defines all inaccurate conceptions and thesis of learning and its determinants as phantasms. Accordingly, a phantasm connotes desirable, but from certain reasons, unrealised form of certain phenomenon/a. (wordnetweb.princeton.edu/perl/webwn, visited 30.VI.2010.). It is important to mention that desirable form of realisation can exist only when circumstances which enable its full realisation are being removed or neutralised. Therefore, a learning phantasm cannot be constrained only to a wrong conception respectively mental illusion of realistic circumstances. Though Platon's definition of phantasm is more strict and restricted to sensibly distorted perceptions of phenomena, for this paper's purpose other impassive factors have been embedded as well.

2. Learning, paradigms and phantasms

Learning itself is a complex phenomenon. Term phenomenon is here used as preliminary term since learning can be observed from different point of views. Alteration of attitudes indicates different characteristics of learning with emphasis put on some of them. Further analysis will try to define and describe learning in different ways. Intention is to form an integral approach respectively consideration of more relevant factors which define learning regardless to stronger attendance of traditional or modern doctrine such as eLearning or mLearning. Hence, by changing approach to observing and analysing, learning can be treated as:

- System
- Process
- Method
- Technique
- Model

Can one of these approaches be neglected and what are the consequences? Holistic approach requires consideration of all relevant facts due to insuring the highest level of learning quality. Insisting on integration and polyvalence in analysis is basis used in this paper for observing wrong anticipations about learning, especially eLearning. As phenomenon learning has been changing during the time and has developed to different forms by tracking changes and growth of general human knowledge through usage of accomplishments in technique and technology whether as supporting instruments or basic objects of interest. Technology and technique consequently regained the more important position in learning's realisation. Consequences of including technology in the process have besides advantages and benefits also brought specific problems that can influence learning's quality.

Another important fact should also be taken into consideration. In general, learning is a formal realisation of educational process respectively gaining knowledge and skills. Articulation of such realisation has special importance both for individual and whole society where it has been realised. With this in mind, it should be said that governments or administrative bodies regulate learning as activity through their legislation. This assumes certain consensus of whole society considering content, form and mode of realisation, and especially consensus in defining goals that should be reached. Aims set whether by

individual or society are mutually supplementary. This process at the same time requires institutionalism of learning and education in shapes that ask for such a consensus.
Time period of considerable development of computer technology and applicable software is short and abound with line of important accomplishments. A need for automation of simple tasks which frequently repeat at same pattern is strong reason for fasten development of automation and computer technology in combination with need for processing a huge quantity of data or complicating the processing and calculation in a short period of time. Speed of changes and frequency of accomplishment in this field require appropriate moderations in all activities that have implemented computers in their business. More frequent ICT application in learning requests alterations that will change learning considerably, but these changes should not harm the quality of learning. Quite contrary, these alterations should increase a quality of learning. ICT usage demands their understanding and possession of specific information competencies that indirectly puts learning participants into unfavourable position and sometimes it complicates learning. ICT application and change of learning's realisation modes demand readiness of all subjects included in learning to willingly accept that mode. In other terms, a critical mass should accept new mode of learning and validate respectively confirm all characteristics of such organised learning. A structure of individuals who will accept this moderation is naturally heterogeneous and they can be classified according to their nature and willingness to accept eLearning. They can usually be distributed in five groups: innovators, early supports, early majority, late majority and sceptics or indecisive (Zemsky, 2009). Each of mentioned groups own in a certain way defined and accepted mode of thinking and anticipating. That is the most important characteristic of a group. The recent literature frequently, though without a precise determination, uses a term paradigm. Though the concept is not new just Thomas Kuhn in 1962 used term paradigm in theoretical discussions about development of science and technology. Literature provides many different interpretations of this term. The most common definitions imply paradigm as frame of thinking and believing that interpret phenomenon or reality. Mentioned definitions have nevertheless general character. More detailed definition of a paradigm implies: „A paradigm is a constellation of concepts, values, perceptions and practices shared by a community which forms a particular vision of reality that is the basis of the way a community organises itself."(Kuhn, 2009) Kuhn used in his definition a general term of society though after he had insisted on differentiating normal and revolutionary science. This attitude was confirmed within ICT field since speed and development moderate a classic meaning of term evolution and revolution.
Acceptance of paradigm and its application is not just declarative though the realisation is usually kept on a level that makes no difference. Interpretation of paradigm and its acceptance and application also imply a subjective aspect. Individual will usually interpret it the way it suits him in certain circumstances when trying to apply generally accepted rules. However, eLearning requires also a different approach in all phases starting with planning up to realisation, so the problems must emerge. If an emphasis is put on individual approach to student and teacher respectively learning contents, then their specificities are important and should be taken into consideration. If one considers all other specificities of eLearning, divergence from concrete paradigm leads to distorted conceptions and mistakes that could have irreparable consequences. Every such deviation is in this paper imposed as phantasm. The fact that is perceptible in realisation is a possibility of conflict between

paradigms. (Hesselgrave, 2006) Conflict can be initiated by different factors. Without detailed definitions of individual paradigms indicated by (Hesselgrave, 2006), paper further on considers a projection of specific paradigm on eLearning and it will try to describe connection between conflicts of paradigms and wrong perception of eLearning that can refer to phantasm.

Which proclaimed characteristics of eLearning can be basis for wrong perception or wrong definition of eLearning itself? Conditionally more important characteristics of eLearning can be defined since their inaccurate perception and treatment can lead to wrong perception or even more rigid, wrong foundations for application of eLearning. These are:

- Phantasm of time independence
- Phantasm of location/physical independence
- Phantasm of transferring quantity into quality
- Phantasm of communication integrity
- Phantasm of availability
- Phantasm of evolution
- Phantasm of holistic approach
- Phantasm of general usage

Each of these mentioned concepts requires careful analysis of connections to general form of eLearning's realisation and elaboration of characteristics that determine or can determine specific forms of realisation. Each of mentioned concepts determines definition according to eLearning as phenomenon since they are interconnected.

3. eLearning is/not ...... !?

Recent practice has proved that introduction and implementation of ICT in learning is a complex process that requires team approach. At the same time, the process treats ICT as technology that supports educational process. However, technology as such alternates the structure of process. Form and intensity of influence and modification are evident both in learning and learning environment. A carrier of learning activity - teacher, in opposition to traditional role is put in moderated position that requires additional or modified activities. Barycentre of a system is moved from his position to student's or content's position. Besides standard methodical, didactic and pedagogic knowledge and skills, teacher should also possess technological knowledge and specific skills in ICT application. Since teacher is a part of team, other team members should as well be trained in ICT application.

Usually team includes three to five persons who in cooperation with teacher coordinate preparation of individual eLearning elements (s11). These are instructional and graphical designers and different field's programmers (Caplan, 2005). Number of team members is not fixed and usually depends upon course itself and institution where it is being prepared. For example, graphic designer and programmer - multimedia author can be the same person in an institution, if that person owns required knowledge and skills (Caplan, 2005. According to Caplan (Caplan, 2005) a team leader should be instructional designer.

Teacher is considered a field expert or content expert. His jurisdiction includes definition of learning outcomes, preparation of learning materials and responsibility on author's level. Instruction designer should have knowledge relevant for learning methodical important in choosing the strategy of performing and applying educational technology. Graphical

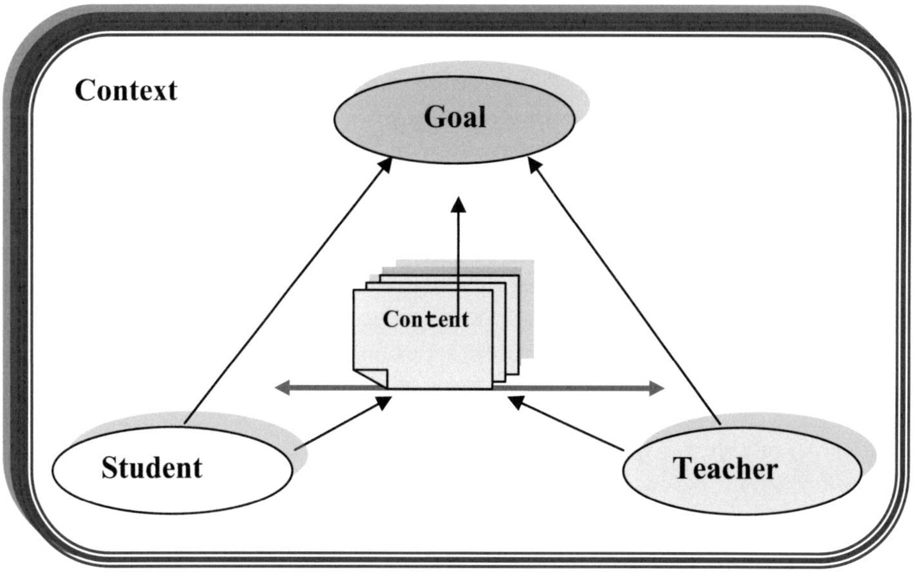

Fig. 1. Modified Pratt's model of learning (Authors)

designer, programmer and web designer are authorised for realisation of users' interface in compliance with good practice and necessary conditions.

Learning supported by ICT can be realised as hybrid learning and pure online learning. Both forms should support different learning and teaching styles. User's interface is realised in web environment so the creation of patterns for learning's realisation is one of the most important eLearning's team assignments. Different applicative software is used in designing the patterns and complete courses. Realisation of eCourse demands specialised applicative solution known as Learning Management System. For example, LMS Moodle is open type software that can be upgraded since author's additions and solutions can be implemented. Moodle is an acronym of Modularly object-oriented dynamic educational interface and is considered ad open code project, so the insight and moderation of application's original code moderation is possible in order to adjust system to own needs (Bosnić, 2006).

3.1 eLearning as system

System is a general concept and has an abstractive definition that can be concretised in specific frames. Generally system is defined as an assembly of interrelated elements or segments that make unity trying to realise its activities. Some authors emphasize a need for some sort of surveillance or feedback connection. Real system is part of actuality and its reality is usually determined by certain level of its importance for users. At the same time, eLearning represents a system that has a tendency of realising a-priori defined goals. According to one of common classifications, learning in general and eLearning as well can be observed as dynamic system. Specificity of dynamic systems is resources' control that enables monitoring the system's condition and more importantly, the existence of feedback relationships as mode of controlling the behaviour of system's elements and controlling the performances of the complete system. All described elements of eLearning are mutually connected and control of learning goals' realisation is ensured through observing the conditions of individual elements and usage of available resources. Learning can be defined as business system as well. Exact definition of business system puts emphasis on final business products, goods and services that bring profit. From that aspect learning can be observed as system, respectively a business system. Basic elements of learning and location of its realisation can be determined by vectors E = (S, T, Co, G, Cx, L, Tm) (Pogarčić at al., 2008). The first five components of vectors are entities of Pratt's learning model (Pratt, 2003 while last two define time and location of realisation. Interdependence of elements in time and space is obvious and it defines structure and quality of learning. System's functioning is defined by system's structure and its position in environment.

However, business function can connote performing professional duties and activities within a certain profession, but also it may relate to occupation and usage of services. That's one reason more to observe learning as a completely specific business system. Teacher is performing role of a service provider, students becomes service' user, while at the same time roles are not strictly divided. When the position of learning content is observed, it is easily to notice its bivalency. Learning content respectively learning itself is manufacturing resource, though it can also be a final product. Within a context of specificity, picture becomes clearer. Albeit differences in learning performed in kindergarten and at doctoral study are significant, both activities succeed learning as abstractive form of the same system. Importance of learning as business system becomes even more highlighted when giving importance to a lifetime education. All the above mentioned refers also to eLearning as a specific form of learning.

3.2 eLearning as information system

It is not difficult to prove that learning is an information system. According to definition, information system is a system that assembles, handles, saves and distributes necessary information to all business system's users who have corresponding rights of an organisation that requires that information together with appropriate authorisation. Usually terms information system and digitally supported information system are used wrongly as synonyms. Another opinion is that of information system as data picture of a business system's process realised through models of data, processes and users. This definition is pragmatic and includes the most factors necessary for a valid valorisation of information system.

Usually when defining information system one claims that information system is a part of business system. Relationship between these two terms is actually symbolic. Information system cannot exist independently, but is connected to business system instead. Purpose of information system is to support a realisation of business processes and a complete business function. Within eLearning frames this connection is obvious: eLearning owns information dimension integrated in realisation of educational process as business activity.

To what measure is learning an information system and to which business system it belongs? Learning is a system that all required information divides into two groups: information subjected to process and resources for shaping the final product of learning – certified attendant/student/pupil and information essential for functioning of learning as a system (in certain way additional or service information). The usual process of collecting and saving required information, which represent learning contents, also has its own specificities. Learning contents and their adjustment to applicable forms are prerequisite to functioning of learning as a system. Processing the mentioned information is also specific in time and space. In traditional learning the organisation of learning/information processes requires spatial and time articulation. Simultaneously time doesn't have a same meaning for teacher and student, respectively time as a concrete moment of realisation in some institutional frames such as classroom has the same meaning for teacher and student, while time differentiates by its meaning from the aspect of age. Special specificity refers to saving and preserving information since the final goal is their implementation by students respectively student's understanding and remembering. More important is active application of exploitations in everyday working and life activities. Formal preservation of learning contents is an important activity but it represents a minor problem.

3.3 eLearning as process

By definition process is an assembly of connected activities used for realisation of certain goal in combination with usage of specific resources (material, financial, machines, employees, information) during a certain time interval (e.g. studying, listening the lectures, preparing the graduate thesis, elaborating the same etc.) Learning as process is defined by its basic elements (Pratt, 2008): student and teacher gathered around content, motivated and directed towards the same cause. The goal is to adopt content through different functions and interaction of basic elements as fast and efficient as possible respectively to adopt necessary knowledge and to gain certain skills. Even though the goal is mutual, both teacher and student approach it from different aspects. Learning is a complex process that embeds several simple processes, phases and actions. Structure of a learning process can be observed on micro and macro level. To macro structure of learning process important are basic (or general) components and determinants of typical structures of learning and teaching situations (adopting the knowledge, practicing, evaluation etc.) that depends upon legitimacy and logic of learning and teaching processes. Micro structure is determined by concrete situation, content, goals and assignments. ICT implementation moderates learning as process. Traditional learning for elementary unit and concrete situation occupies learning hour. Within these frames learning as process is being structured and monitored while resources are at the same time temporally immanent and location-concentrated. In eLearning this organisational term does not represent a basic situation.

3.4 eLearning as technology, technique or method

Learning is realized through processes of learning and teaching. Every such process implies communication between teacher, attendant and content. Articulation of communication requires application of certain techniques and methods. Alteration or highlighting of one element's role or change of their relationship can essentially change mode of teaching/learning. Since these two processes comprise learning, their change in approach to learning or teaching changes learning itself. Term technology is wide and dependable upon a concrete situation. Usually technology refers to usage of different tools and procedures that improve certain process and ensure more qualitative results and complete goals. If a teacher includes educational technology in the process of learning, then other elements of educational system will not change. Shape of connections between elements will change as well as their mutual interaction. By introducing the ICT into educational system basic terms of education usually get prefix "e" - electronic or more recently "m" - mobile. Actually, an assembly of electronic tools, methods and algorithms is directly connected to learning as process within new technological frames. eLearning recontextualizes learning process. Recontextualization is discernible already at the level of programs and it becomes more developed when it approximates more complex shapes. Complete process of learning is realized through new forms and it loses its former static characteristic.

Learning can also be analyzed as assembly of techniques respectively methods that are used in realization of a learning process. ICT implementation, especially from aspect of different program solutions makes eLearning, from methodological point of view, more interesting and acceptable then line of traditional methodic approaches and solutions.

3.5 eLearning as model

Education represents a specific form of human activity, so consequently, learning as a realisation of same activity also becomes specific. This is even more obvious when considering specificity of frames within which learning is being realised. Perhaps not the most sensible, but certainly most detailed form of analysing learning's characteristics is treating it as a model.

Specification according to model in compliance with any model's definition requires neglecting its determined characteristics or at least abstracting or simplifying system up to a level where it can be independently run. Basic characteristic of a model is its capability of imitating the real-time system in different situations. Its basic purpose is to simulate realistic situations possible in system's functioning and making decisions that are applicable in concrete situations. Simulation is a process that enables imitation of a real-time system regardless to time or location of its realisation.

Learning plan and program can be observed as a specific model of an educational system which has to satisfy the same basic postulates as any other model. Same conditions could be defined for eLearning as a form of education. Traditional form of learning plan and program doesn't consider completely all factors which can emerge during realisation.

Basic characteristics of a qualitative model are comprised in its ability to replicate a real-time system at any possible modification of executive factors so to ensure an overview of system's structure respectively the conception of its structure. However, the most important characteristic of any model is its anticipation or ability to predict situations that already haven't been realised. Since eLearning or to be more precise, eCourse is designed as ICT "product" it has certain static characteristics which a-priori restrict realisation. When these boundaries are set by content, goals or restrictions which refer to student or teacher, then they should be considered as specificities. However, qualitatively prepared eCourse has a purpose of realisation, even in unpredictable situations that can, along with small modifications, be accepted as a qualitative realisation.

eCourse can also be treated as a software solution. In that case, its testing as a final product is essential due to verifying all three above mentioned characteristics of a model. Students' success defined as a final goal of course mainly depends upon quality of a model that is aligned with described characteristics. ELearning is an exclusive area where model has to be used as a real-time system in all of his performances.

4. Paradigm as cause of phantasm

A book entitled "Paradigms in conflict" (Hesselgrave, 2006) confronts paradigms which author recognizes within modern social frames. Though book considers reflections of Christian doctrines up to a present moment that also implies certain restriction of applicability, frames can also be applied to other areas which register any kind of human mind's activities. Learning is, without any doubt, a field where individual supported by other persons and specific means shapes himself for the future. This moderation also comprehends cognitive aspect respectively moderating the awareness of oneself and one's possibilities. In momentarily circumstances of accomplishments and fast changes this can no longer be analysed as one-time basis process, but instead as continuing process defined by individual's needs. The process is today usually categorized as lifetime learning.

Such process connotes a possibility of education and qualification without the usual boundaries. ICT application and possibilities that follow it through web and similar forms mainly remove such boundaries. Possibility of using the computer network implies a possibility of approaching knowledge sources embedded in specific forms adjustable to individual's needs or those of a group with similar demands.

Within these circumstances learning transforms from traditional forms to forms adjusted to conditions mutual to all activities that have implemented ICT. To put in usage the applicative software solutions alludes much more than using the application as learning medium or tool. Web as a global "class" puts eLearning to a position it belongs to - place of productive activity which can define its value as any other product placed on market. This is where the real problems emerge and they require a complete definition respectively a holistic approach to eLearning as realisation of education respectively as a system, process, method or technology. Placing the individual towards such forms of learning requires his acceptance of paradigm already supported by other students who share activities defined in lectures.

eLearning and possibilities that it offers puts in front of individual a dilemma of paradigm's independence and respecting tight rules set by one who plans learning and its goals. Web, on the other hand, neutralizes power of independence in a way that it allows an individual a possibility of choosing online courses which mostly suit and in the best possible way ensure goals aligned with his needs. Misunderstanding of such possibilities leads to wrong conceptions about eLearning that will be elaborated in a paragraph referring to phantasms.

Possibilities of web and eLearning provide one more advantage in comparison to traditional forms of learning. For example, hybrid learning also includes ICT so restrictiveness of traditional learning becomes reduced. Important advantage of eLearning is probability of so-called peer-to-peer communication that enables abandoning "eClass" no matter how much it can be virtualised by dimensions. However, paradigm of "common land" - peer-to-peer environment imposes a caution that has to follow qualitatively organise eLearning. As a product formed and put within frames of eBusiness every eCourse or other eForm demands protection of copyrights that belong to the author.

Holistic approach to preparation and realisation of eLearning or some of its forms is the most important precondition of its success. However, at such planning of eLearning and treating its particular elements that comprise its essence, sometimes it is necessary to determine sequence and/or priorities of certain activities. Though it is necessary to determine sequence or priorities for competitive activities in their realisation, still holistic approach should be represented regardless to significant divergences.

4.1 Paradigm as cause of phantasm

Are phantasms mistakes, misconceptions or just wrong conceptions?

4.1.1 Phantasm of time independence

eLearning is proclaimed as timely independent process. Term time independence should be considered extremely carefully. When within eCourses attendants are advised that learning process is not timely defined it basically means an attendant can freely decide when he will learn. Individual component is nevertheless expressed in a bit "distorted" light. Truth is that eCourse's realisation plan has to precisely emphasize how much time is required for certain content, but student is not imposed with schedule which determines a sequence of adopting the content. This trap is usually noticed in high number of attendants who give up courses or achieve bad results in the end.

Time in eLearning should be treated as polyvalence. The first treatment requires a definition of time's quantity necessary for adoption and realisation of content. That period has to be static and appropriate to predictably average attendant. The second treatment is more complex and includes time of communication on all communication levels that exist in eLearning (see table). This time has also static character. Finally, real time independence of eLearning is restricted by already described conditions and it has a relative character. Simply be said, "one can learn whenever he wants to in comparison to eCourse' s duration, but in line with schedule of one's other assignments". Other characteristics of time such as age of attendants are not analysed in this paper.

4.1.2 Phantasm of location/physical independence

Conception of physical independence is extremely phantasm-characteristic. Namely, slogan of location or localization independence of eLearning is strictly connected to existence of network infrastructure, web and Internet approach. True, there is a possibility of Intranet and creation of locally restricted networks, but even then realisation is conditioned by different factors.

Assumption is that certain society owns arranged infrastructure with tendency of global networking. Moment that is exceptionally important in this treatment is safety of network and communication. To describe it scenically "the crush of Internet within eLearning and earthquake within traditional class do not have the same circumstances". Phantasm of location independence should still be observed with much caution.

4.1.3 Phantasm of transferring quantity into quality

This kind of articulation overwhelmingly reminds of some Marx's philosophical approaches but it exceeds those frames. Hypothetically it can be foreseen that specific area or specific subject in a certain period of time can be supported by relevant number of qualitative eCourses when browsing web. Certainly, the choice of courses belongs to individual/attendant and is strictly determined by his possibilities. Since eLearning becomes eBusiness it can be assumed that financial preconditions will determine criteria and range of acceptable forms. In conditions of levelled prices this certainly can impose a problem, if no coordination on a higher level of decision making.

4.1.4 Phantasm of communication integrity

The next assumption that frequently leads to wrong conceptions of eLearning is conditionally said "assumption of communication integrity". This phantasm is a direct consequence of wrong assumption considering location independence.
Communication integrity or better yet completeness connotes undisturbed communication between elements of eLearning. Stricter form implies existence of willingly communication between elements shown in Fig. 2.. From psychological point of view willingly communication can be treated as necessary communication that underlines phantasm of complete communication in eLearning.

	Teacher	Student	Content	Goals	Context
Teacher	indirectly directly actively /passive	directly actively	directly	indirectly actively	indirectly actively
Student	directly actively	indirectly directly actively /passive	directly	indirectly actively	indirectly passively
Content	directly actively	directly actively	directly indirectly	indirectly	indirectly
Goals	indirectly actively	indirectly actively	indirectly actively	indirectly conditionally	indirectly likely
Context	indirectly directly	indirectly indirectly	indirectly	indirectly	indirectly

Fig. 2. Determination of eLearning according to communication between elements(Auth)

4.1.5 Phantasm of accessibility

Accessibility connotes two dimensions: time and content accessibility. Time accessibility is conditioned by reliability of network in local and international level. Both reliabilities are conditioned while level of reliability is inversely relative to phantasm's level in both situations.

4.1.6 Phantasm of revolution

Trend of ICT development and involvement in all activities, learning as well, leaves an impression of necessity for technical facilities, especially in computers. ICT are essential facility in learning, but form of learning that will be applied should be precisely defined according to specificities of certain situation. ELearning can be preferred as executive form due to line of its own characteristics that are advantage in comparison to other forms of learning. Location independence of eLearning allows attendant to perform his assignments at home. That possibility drastically reduces costs of learning both for institution and attendants, so the continuance of eLearning's application can be anticipated. The process will probably lead to disappearance of other learning forms. Therefore eLearning should be observed as evolutionary form of traditional learning. Hybrid form of learning is the most expected form in future. Regardless to velocity of changes, eLearning cannot be regarded as revolution in education. Especially since changes in all aspects of traditional learning do not emerge simultaneously and at same pace, especially in fields of methodic, didactic, pedagogy and psychology.

4.1.7 Phantasm of general applicability

Though computers as ICT segment have found a wide application in all activities, it is not possible in learning programs of certain courses or parts of those courses. Where manual skills or motor capabilities are required, contents of that type are hard to qualitatively express by some sort of eForm. Also, when physical model is required or simulation by physical model, eLearning is not appropriate. Some contents require detailed and complex preparation so they are demanding in time and finances. Wrong concept of general applicability therefore becomes a phantasm that leads to extensive deadlines of preparation or spending more financial means than anticipated.

4.1.8 Phantasm of holistic approach

Necessity for team work and team approach demands certain level of common knowledge and common concept about goals of specific activity. Even if there is a consensus between team members regarding the concept of final goals that are to be realised through preparation of eLearning, still misunderstandings can emerge in comparison to attendants. Since eCourse is prepared carefully and planned as reusable model, its static segment should be carefully determined. On the other hand, it is necessary to upgrade contents and adjust them with modern cognitions. Holism assumes comprehensive attitude towards phenomenon and treating all its parts as integrity. Treating some part or element separately, without observing connections and influences to other elements and over the other elements, almost always leads to mistakes. Mistakes can be overlooking or neglecting important facts which can condition realisation.

Position of teacher in eLearning has been significantly modified since he is required additional skills and capabilities concerning ICT. Already a wrong judgement or self-evaluation of his capabilities can diminish chances for success of eCourse. On the other hand, a part of responsibilities that can be left to other team members frequently loose pedagogic-psychological support. Valorisation and validation of content which will be included in course represent sensitive assignment due to a huge choice of disposable prepared forms, so this can lead to undesirable results.

Still, the most sensitive part remains attendants' evaluation of success since basic goal is certification of users and evaluation of their success in course. Since through these eCourse activities authority is being returned to teacher or a content expert, preparation of tools and modes for evaluation of attendants once again requires comprehensive approach where one shouldn't neglect postulates of good pedagogic-psychological practice.

5. Conclusion

ICT application in learning has a vital importance for all human activities. The final implementation of computers, whether as learning subjects in learning or education in general is timely dissonant. Preparation and execution of eCourse assumes certain level of information-science education that can be provided through specific form of learning. Considering their high financial value and general treatment of formal learning as activity, computers are relatively late included into the learning process. Only the first level of software's development, liberalisation and conscience of necessity and importance of open platforms has enabled learning to possibly apply ICT in preparation and realisation. Velocity of changes and achievements in ICT field open a wide range for possibility of their implementation. Possibilities provided by web, globalisation of business and market connectivity provide a possibility for placing eLearning as a product perhaps for the first time in a way that proves truly economic value of that activity.

However, velocity also brings a risk that can be noticed in superficiality or in "market language" education is put in a situation where it is observed as "bofl" - goods of bad quality. Quality of learning content has a crucial importance in achieving a goal - good educated and certified attendant. Certainly, one shouldn't forget the value and importance of other elements, persons and learning forms in the process.

Insuring quality of (e)Learning is precondition of a final goal - qualitatively educated attendants - and consequently ensuring all benefits that can use to individual or a society as whole. Therefore, in the complete process of preparing and realising eLearning as educational form special attention should be given to treating all facts that could in any way influence quality. The final goal isn't, though it is not forbidden, to create a brand in educational eContents. Finally, all national universities have tried to ensure a top quality in education. If in global frames global eInstitutions of higher education do emerge, one should consider a quality in all levels of realisation. Phantasm is not necessarily negativity if treated as measure of imperfections in planning and/or realising the education through eLearning as one form of it.

6. References

***; wordnetweb.princeton.edu/perl/webwn

Zemsky, R.; (2009). Making Reform Work: The Case for Transforming American Higher Education, Rutgers University Press, ISBN-10: 0813545919

Kuhn, R.; (2009). The Structure of Scientific Revolutions, Books LLC, ISBN-10: 1443255440

Hesselgrave, D.,J.; (2006), Paradigms in Conflict, Kregel Academic&Professional, USA, ISBN 978-0-8254-2770-1

Caplan, D. (2005); Development of online courses, *Edupoint*, 40(V), downloaded on 20th April 2010 from http://www.carnet.hr/casopis/40/clanci/2

Bosnić, I. (2006); Moodle: Seminar textbook. downloaded on 20th April 2010 from http://www.open.hr/e107_files/downloads/Moodle_prirucnik.pdf

Pogarčić, I.; Šuman, S.; Žiljak, I. ; (2008) eLearning as an Information Process // *RECENT ADVANCES IN EDUCATION and EDUCATIONAL TECHNOLOGY, Proceedings of the 7th WSEAS International Conference on EDUCATION and EDUCATIONAL TECHNOLOGY (EDU'08)* / Subhas C. Misra, Harvard

University, USA (ur.). Venice, Italy, November, 2008: WSEAS Press, 2008. 283-289

Pratt, D. D. (2003): Good teaching: one size fits all? U: An Up-date on Teaching Theory, Jovita Ross-Gordon (Ed.), San Francisco: Jossey-Bass, Publishers.

10

Model of the New LMS Generation with User-Created Content

Dragan Perakovic, Vladimir Remenar and Ivan Jovovic
University of Zagreb, Faculty of Transport and Traffic Sciences
Croatia

1. Introduction

The Learning Management System (LMS), designed by the Faculty of Transport and Traffic Sciences, has been in use since 2004, when it was developed according to both the German Fachhochschule concept and specific requirements of studying in the Republic of Croatia at that time. The system consists of 5 individual modules (DMS, SAN, e-Student, SMSCentar i FPZ*mobile*) which in mutual synergy represent the LMS of the Faculty. The Document Management System (DMS) is a system used for managing documents and processes used by the Faculty staff for authorized access to the modules for monitoring work in computer labs, e-Learning system administration (publication of instructional materials, checking and evaluating of seminar papers, etc.), and the module for managing documents and processes within the Faculty (equipment orders, malfunction reporting, updating of the online directory, etc.). The Authorization and Control System (SAN) is a combination of technologies and applications which enable monitoring of students working in the computer labs of the Faculty and personalization of working environment in the Windows operating system, regardless of which computer a student is working on. The e-Student System enables students to have authorized access to instructional materials, exercises and instructions. Furthermore, the system provides support to registration and preparation of seminar papers as well as other forms of testing (e-Blitz, e-Test, e-Quiz).

The SMSCentre and SAN system, supported by the smsCRM application (Customer Relationship Management via SMS), present possibilities for a wide range of information services, based on interactive communication through text messages, available to both students and the Faculty staff alike. Within the last three years the m-Learning module has been added to the e-Learning System of the Faculty in the form of an interactive application, called the FPZmobile, which provides real time information to students through mobile terminal devices. The basic idea was to develop an application which would offer accurate and up-to-date information about the Faculty and students' achivements (seminars, practical courses, preliminary exams, etc.) to all students while studying at the Faculty in the form adapted for viewing on interfaces of mobile terminal devices such as mobile phones, PDA devices and Smartphones. The modularity of this system enables quick implementation of new services developed in accordance with the ideas and suggestions provided by the teaching and non-teaching staff as well as students.

The e-Student module, as part of the Faculty's LMS, has been in use for five academic years. Although there have been improvements in the quality of the studies, the module has also indicated the existence of problems arising from the system processes, technologically obsolete equipment as well as the inability to support the Bologna process and the "Outcomes of Studying at the University of Zagreb" program ("Ishodi učenja na Sveučilištu u Zagrebu" - SveZaIU, 2009). The system is currently used by all 509 courses at the Faculty, serving 7,489 students and 362 staff. Over the past few years the system has been used almost 850,000 times by students and almost 50,000 by instructors. The number of published materials, altogether 5,645 data files, is still significantly smaller than the number of data files created and published by the most important users of the system: the students! Students have published 16,000 topics for seminar papers and more than 15,000 seminar papers. The LMS gives students freedom in working which they gradually accept. It has been noticed that the system is increasingly used in the period between 24:00 and 08:00, whereas working in the period between 08:00 and 16:00 is becoming less and less popular, as shown in Fig. 1.

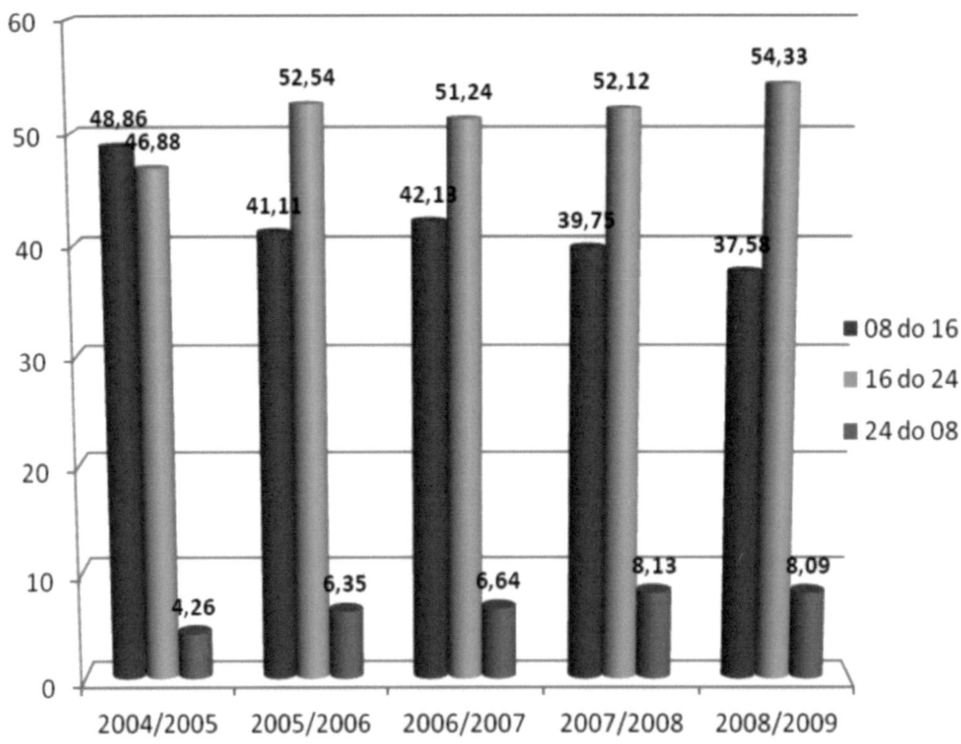

Fig. 1. Percentages of accessing the LMS according to the periods of the day

It is necessary to minimize or completely eliminate the negative aspects of the LMS usage as well as the shortcomings in the operation of the existing system in order to reach the desired level of the quality of studies. For this purpose, an entirely new system will be developed, based on the model described in this paper. The new system will be adapted to the needs of studying according to the Bologna process and in accordance with the "Outcomes of Studying at the University of Zagreb" program ("Ishodi učenja na Sveučilištu u Zagrebu" - SveZaIU, 2009), the ISVU system (Infromation System of Higher Education Institutions) as well as the current trends in learning technology and communication both between students and between students and teachers. Furthermore, the model proposed in this paper will provide guidelines for developing the system according to the e-Learning 2.0 paradigm, which makes use all the Web.2.0 tools, such as Wiki, Blog, forums, social networking and others, in its operation.

2. Initial development

The e-Learning System of the Faculty of Transport and Traffic Sciences was designed, developed and implemented in 2004 to satisfy the needs of studying according to the German Fachhochschule concept as well as the specific requirements for studying in the Republic of Croatia. There are five modules currently in function (DMS, SAN, e-Student, SMSCentar and FPZmobile) which together represent the Learning Management System (LMS) of the Faculty of Transport and Traffic Sciences.

2.1 Why we started

As information and communication technologies and accompanying services developed, so did the idea of having their own distance learning system at the Faculty of Transport and Traffic Sciences. The idea was to develop a distance learning system which would become an indispensable tool in an electronic interactive communication in the teaching process and student welfare. That would make the Faculty of Transport and Traffic Sciences one of the first faculties in the Republic of Croatia to help students acquire and test their current knowledge more easily by using new technologies. Keeping that in mind, new hardware and software conditions were created in order to develop the Faculty's own LMS, which has been constantly improved and used since 2004.

2.2 The system currently in function

The DMS (Document Management System) is a system used for managing documents and processes used by the Faculty staff for authorized access to the modules: monitoring of work in the computer labs (checking of the access to exercises, etc.), system of e-Learning administration (publication of teaching materials, checking and evaluating of seminar papers, etc.), and the module for managing documents and processes within the Faculty (equipment orders, malfunction reporting, updating of the online directory, etc.).

The SAN (Croatian: Sustav autorizacije i nadzora, English: The authorization and control system) is a combination of technologies and applications which enable monitoring of students working in the computer labs of the Faculty and personalization of working environment in the Windows operating system, regardless of which computer a student is working on. The SAN system consists of 6 modules (SAN Server, SAN Application, SAN Client, SAN Administration, SAN Web and FPZBrowser), which are currently used to monitor three computer labs, and public computers located in the Faculty building.

The e-Student system, as shown in Fig. 2, enables authorized access to the teaching materials, tasks, exercises and instructions to students. Furthermore, the system provides support for the registration and creation of seminar papers and other forms of knowledge testing (e-Blitz, e-Test, e-Quiz).

The SMSCentar and SAN system, supported by the smsCRM application (Customer Relationship Management via SMS), present possibilities for a wide range of information services, based on interactive communication through text messages, available to both

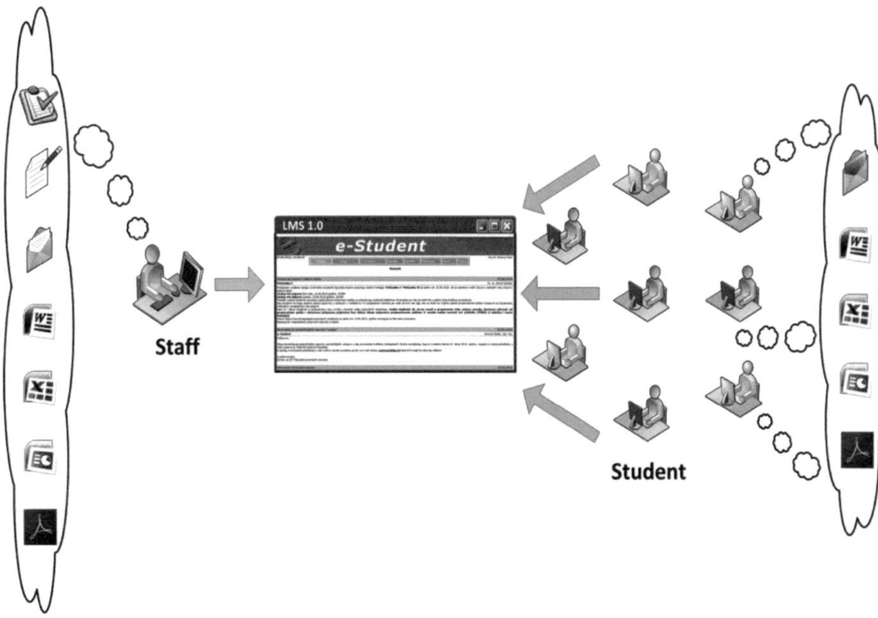

Fig. 2. LMS in use at the Faculty of Transport and Traffic Sciences

students and the Faculty staff alike. The current interactive services available to sudents, which are connected to the PCLab, are the services of obtaining the status of computers and booking of the same in the PCLab (Peraković, Remenar, & Jovović, 2010). By sending an apropriate text message to the SMSCentre phone number, the SMSCentre will forward the request to the SAN Server module, which in turn sends a message containing the information about the status of the computer in the desired computer lab to the SMSCenter. The message also contains infomation on the location of the PC Lab, date and time of the message creation, working hours for that particular day, total number of available computers, number of the computers students are working on, number of computers available for use, as well as the number of booked computers.

There is a whole range of possible additional usages of SMS with the aim of improving the quality and speed of informing the participants of the teaching process about changes in the schedule of lectures and practical courses, the current status of a seminar paper, exam dates and consultation hours of the teachers, exam results, reporting of computer malfunctions etc. The modularity of this system enables quick implementation of new services developed in accordance with the ideas and suggestions provided by the teaching and non-teaching staff as well as students.

Within the last three years yet another module has been added to the e-Learning system of the Faculty in the form of an interactive application, called FPZmobile, which provides real time information to students through mobile terminal devices. The basic idea was to develop an application which would offer accurate and up-to-date information about the Faculty and students' achivements (seminars, practical courses, preliminary exams, etc.) to all students while studying at the Faculty in the form adapted for viewing on interfaces of mobile terminal devices such as mobile phones, PDA devices and Smartphones. As the target users of the application are students, it was thought best to develop an application which could be used with as many mobile terminal devices as possible. By using this application, users would have access to the relevant information at any given time, regardless of their location or terminal equipment they are using. The FPBmobile application was developed within the Java 2 Micro Edition (J2ME) environment intended for mobile terminal devices. The reasons behind this choice arise from the fact that there are numerous different operational systems on the mobile terminal device market and they all support applications developed in J2ME environment (Peraković, Jovović, & Forenbacher, 2009).

All systems are based on communication realized through SQL database as well. This enhances the modularity of the system, i.e. upgrading and system changes are made easy, which in turn makes it possible to adapt the system for operation at other faculties.

3. Implementation results

The modularity of this system enables quick implementation of new services developed in accordance with the ideas and suggestions provided by the teaching and non-teaching staff as well as students. The LMS system of the Faculty of Transport and Traffic Sciences has been in function for 6 calendar years, i.e. 5 full academic years, and has been used by 362 users for 509 courses. The entire system is used by 7,489 students which represents 100% of the courses in the undergraduate, pre-graduate and graduate studies.

3.1 Positive implementation results

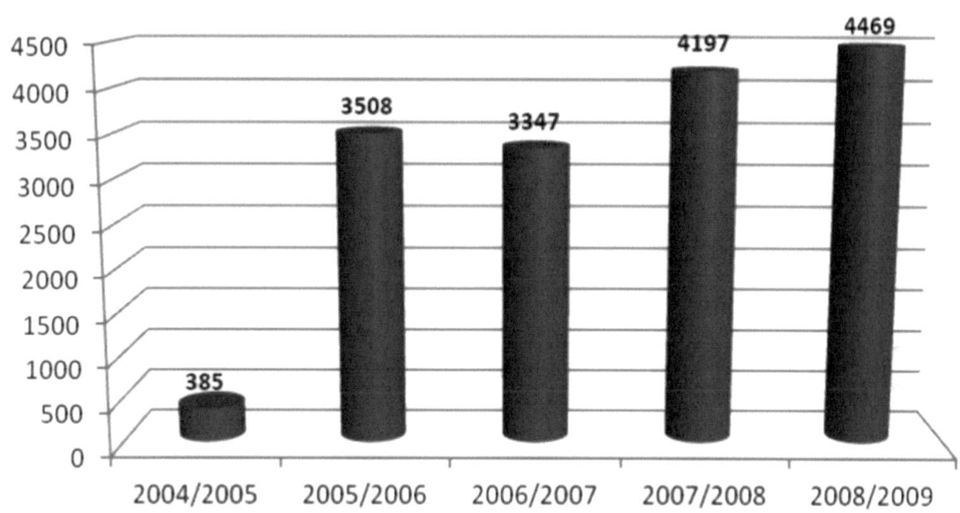

Fig. 3. Submitted topics for seminar papers

Within 5 academic years, the system has been used 836,870 times by students when submitting 15,906 topics for seminar papers (Fig. 3) for as many as 15,033 seminar papers. At the same time the teaching staff used the system 48,444 times to give out 2,742 announcements for their students and publish 5,645 data files of teaching materials as shown in Fig. 4.

The implementation of the system indicated the positive aspects of studying at the Faculty of Transport and Traffic Sciences brought about by the LMS system. Students have accepted the freedom of studying with the support that the LMS system provides extremely well.

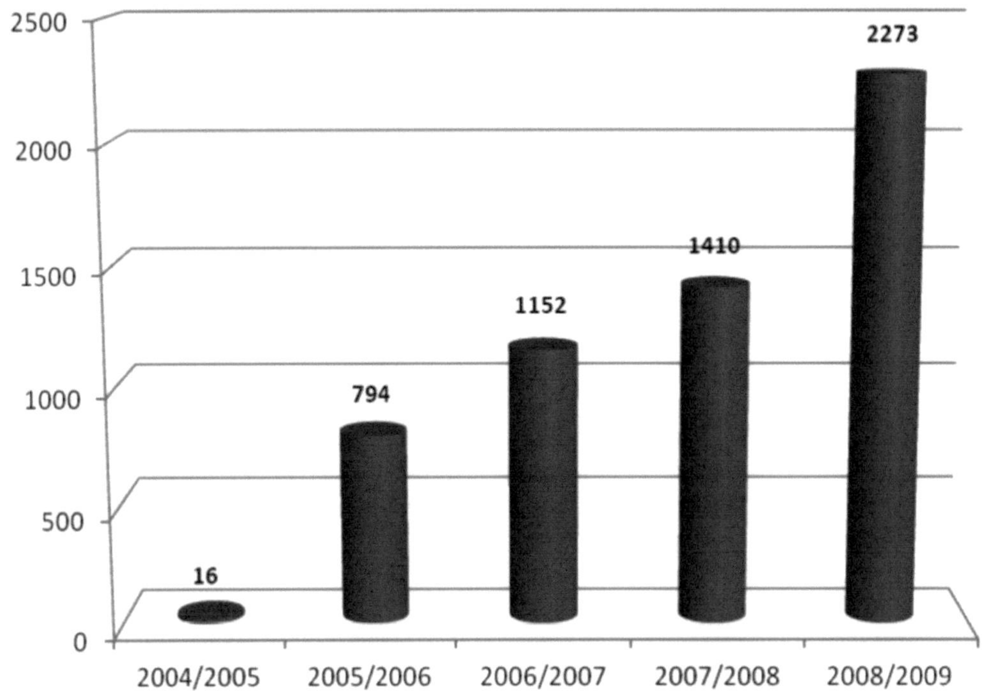

Fig. 4. Published teaching materials

Throughout academic years the number of students working on their tasks in the period between midnight and 8 am has been increasing gradually, whereas the number of students working between 8 am and 4 pm has been falling constantly.

It is also important to notice that the number of submitted seminar papers grows with each new academic year and in the academic year 2008/2009, compared with the year 2007/2008, the number of submitted seminar papers rose by 27%, as shown in Fig. 5. In order to maintain the quality level of the study program , the number of refused seminar papers and seminar papers graded with a D increased significantly in the academic year 2008/2009, whereas the number of seminar papers graded with a B or A dropped.

Fig. 6 shows continuous improvement of the teaching staff in using e-Test knowledge testing which resulted in a 200% higher number of published e-Tests in the academic year 2008/2009 compared with the academic year 2007/2008.

3.2 Negative implementation results

Although there are many advatages to using e-Learning, and surveys and research present e-Learning as a very positive way of learning and one that will be used in the future, drawbacks are often neglected and need to be looked into in order to find a solution.

The basic drawback to any e-Learning system lies in the IT equipment every student needs to possess. A fast Internet connection is very often necessary to manage almost real time communication between either instructors and students or students themselves, to download teaching materials or take exams. Although the e-Learning system in educational institutions is frequently set up as an addition to the traditional way of learning, regular

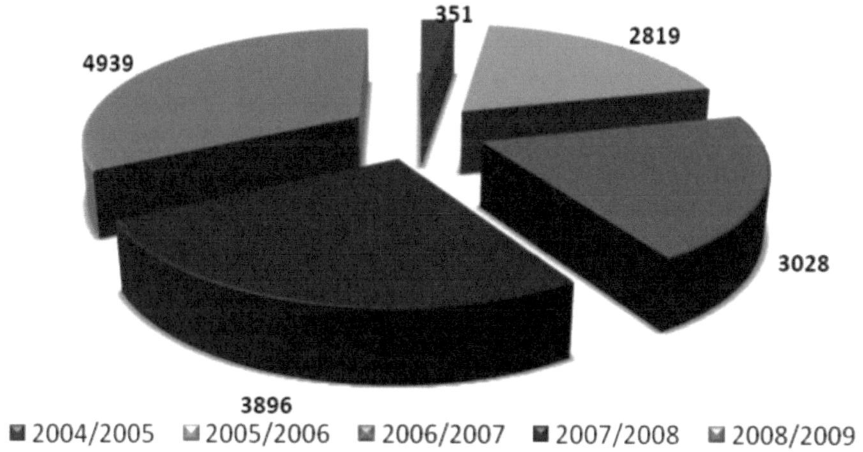

Fig. 5. Posted seminar papers

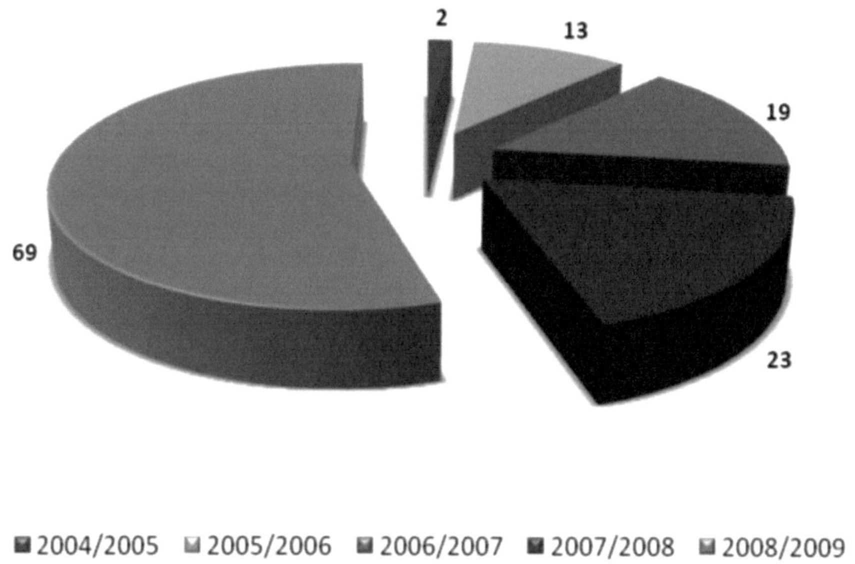

Fig. 6. Published e-Tests per year

maintenance of the equipment at the Faculty is often necessary to support this way of teaching. In order to reach the full potential of e-Learning it is necessary for students to have computers with Internet access. (Peraković, Kavran, & Remenar, 2006)

As the number of e-Tests grew (Fig. 6), it became obvious that a certain number of students were always trying to cheat, whether they were taking exams in a traditional way or by using the e-Learning system for testing of their knowledge. The solution to this problem is complex and starts with giving students authorisation, then monitoring the way students research teaching materials, and finally defining the time for taking exams and randomly choosing questions when applying e-Testing. The availability of materials enables students to cheat when writing seminar papers by submitting someone else's and/or publicly available papers as their own. This problem can be partially eliminated by monitoring the way papers are created and submitted and by browsing the titles and contents of the papers. The downside to this is that students are required to use the e-Learning system right from the beginning of their studies, while working on their seminar papers, taking exams and finally writing their final thesis. Creating an electronic file containing all relevant information on a student may significantly contribute to minimizing the chances of deceiving the system.

The lack of interaction between both instructors and students and students themselves leads to the feeling of isolation. It is more difficult to do group work when using e-Learning system than when using traditional methods with a group of students in the same classroom. By developing social networks and implementing collaboration systems like forums and exchanges of real time e-mail messages, as well as video conferencing systems, it is possible to mitigate the feeling of a computer being an instructor and students being isolated and dislocated. There are indicators which show that people with lower self-esteem manage to improve their interaction when using e-Learning as they can voice their attitudes and opinions without revealing their true identity.

4. Development of a new LMS system

Many advantages as well as disadvantages and technological obsoleteness have been noticed in the last 6 calendar years, i.e. 5 academinc years, during which the current LMS has been in function. Furthermore, while developing and using the current LMS, new insights and knowledge have been gained, which will be used in designing a new LMS.

It will be possible to design the new system according to the model proposed in this paper. The model suggests the use of up-to-date Web 2.0 technologies and principles, compliance with the European Commission guidelines, adaptation to the Bologna education model, ISVU system and the "Outcomes of Studying at the University of Zagreb" program ("Ishodi učenja na Sveučilištu u Zagrebu" - SveZaIU, 2009). Furthermore, the proposed LMS 2.0 model needs to enable efficient and simple use of the system to people with disabilities.

4.1 The European Commission Guidelines

When developing a new LMS, special attention will be paid to the relevant European Commission guidelins, starting with the e-Learning initiative proposed by the European Commission, and aiming at the realization set by the European Council in Lisbon. Ways to mobilize education and train communities, as well as economic, social and cultural development have been devised, all with the prupose of accelerating the process of introducing knowledge society. (COMMISSION OF THE EUROPEAN COMMUNITIES, 2000)

As there is a cause-and-effect link between the development of information and communication technologies (ICT) and LMS, a question arises about defining and separating of the mentioned concepts since learning in a technology enriched environment (e-Learning) cannot be mistaken for learning about technology (learning about 'e'). For this reason, while developing a new LMS, special attention will be paid to the conclusions of the European Commission which proposes distancing initial and expert educational policies for teachers and instructors from thechnical skills while empowering all participants in educational process to develop new competencies and master media and digital literacy as well as ingrate the above into the daily educational environment. The modular framework for ICT use in education and its integration into education spurs better understanding of the changes in educational processes and educational goals whose fulfillment and implementation have been supported by ICT. (EUROPEAN COMMISSION, 2003)

4.2 The old system of studying

By signing the Bologna Declaration, the Republic of Croatia abandoned the old system of studying which had be fashioned according to the German concept of studying, so called Fachhochschule.

The current LMS in function at the Faculty of Tansport and Traffic Sciences was designed and developed for the purposes of studying according to the old system of studying. As such, it could not fully support the priciples of studying according to the Bologna Process. Due to extreme complexity of the current LMS in function, it is not possible to adapt the system to the Bologna Process. The curent LMS will have to be shut down completely and a new one will be developed in accordance with the conceptual model proposed here.

4.3 The Bologna Declaration and process

The Republic of Croatia signed the Bologna Declaration in 2001. Its purpose is to create a universal higher education body and enable a range of advantages for students such as:

- accepting the system of easily recognizable and comparable levels, among other by implementing the Diploma Supplement in order to promote employment of European citizens and international competitiveness of the European higher education system.
- Accepting the system based on two main cycles, undergraduate and postgraduate. In order to enter the second cycle, it is necessary to complete the first cycle lasting at least 3 years. The level achieved after the first cycle needs to correspond to the qualification requirements of the European labout market. The second cycle will enable obtaining of the Master's and/or Doctorate degree, as is the case in many European countries.
- The implementation of a credit system, such as ECTS, as a suitable means of promoting widespread student exchange. Credits may be gained outside the higer education system, including life long learning, provided they are recognized by the university which is accepting the student.
- Promoting mobility by overcoming obstacles to free movement, particularly: giving students a chance to learn, granting them access to the studies and relevant services; acknowledging and evaluating the time teachers, researchers and administrative staff have spent researching, leacturing or studying, without prejudging their statutory rights.
- Promoting European cooperation in ensuring quality with the aim of developing comparable criteria and methodologies.
- Promoting the necessary European dimension of higher education, esapecially in developing curricula, inter-institutional cooperation, mobility schemes and integrated programs of studies, training and research.

Furthermore, the Bologna model of studying introduces two important concepts: life-long learning with continuous monitoring of a student during the course of studies.

The model of the proposed LMS will include a possibility of providing continuous access to the system to the students who have already graduated. Former students will therefore be able to access all the materials they may need in their work. Furthermore, by being granted continuous access, the students who have already graduated are able to contact their teachers, former colleagues as well as graduate students. The goal of life long education can thus be achieved, and new possibilities of cooperation between former students and their teachers, current students and the institution itself present themselves.

Continuous monitoring of student acheivements throughout their studies is implemented in order to achieve higher quality of education. An electronic dossier (e-Dossier) has been developed, which both students and instructors have access to. The E-Dossier is a collection of relevant data on a student from the time of enrollment to graduation. It contains, among other information, basic information about a student and information on the taken e-Tests, topics of seminar papers, data about seminar papers such as the title, description and grade, information about the student's use of computer labs, information about practical work and similar. As students progress through their studies, their e-Dosiers are automatically updated with all the activities connected to the studies.

E-Dossiers provide instructors with an insight into the students' progress and possible adaptations to the students' needs. By using e-Dossier, it can be become apparent that a certain student, throughout his studies, shows great interest in working with information systems as he/she uses them when writing seminar papers and exercises. Based on this information, the student may be offered to write his/her final paper or thesis on the topic he/she shows interest in, i.e. in the field of information systems.

The most significant advantage for students is the possibility of having their e-Dossiers, which show their interests and their work throughout their studies, printed out. A printed copy of the e-Dossier may be attached to letters of application when looking for a job as a proof of their interest in a certain field or the like.

4.4 Adaptation to specific requirements of studying in the Republic of Croatia

Although the suggested model is universally applicable, specific requirements of studying in the Republic of Croatia make it necessary to design parts of the system according to those requirements. Possible problems which would warant deviations from the universal system include student mentality, simultaneous execution of the studies according to the old program and to the Bologna process, usage of the Information System of Higher Education Institutions (Informacijski sustav visokih učilišta - ISVU) which is applied at universities in the Republic of Croatia, as well as the "Results of Studying at the University of Zagreb" program.

Unfortunately, student mentality in the Republic of Croatia still implies, for some students, attempts of cheating at exams and plagiarizing of seminar papers. A fuzzy logic model for browsing the LMS was developed, among other things, for this purpose (Peraković, Remenar, & Grgurević, 2008). It allows instructors to browse easily through papers and discover plagiarized seminar papers. In order to prevent students from copying and sharing answers while taking e-Tests, a logic of assigning exam questions and showing multiple answers was devised. There is also a possibility of restricting the choice of location where students can take e-Tests. It is possible, for example, to limit e-Test taking only to the computer labs at the Faculty within a strictly defined time frame.

The universities in the Republic of Croatia are still in a transitional period between the old system of studying and the Bologna process. As older students are still active in taking exams and will be in the foreseeable future, the conceptual model proposed here implies providing support for those students. Although the majority of functionalities is the same for both "old" and "new" students, there are certain differences and it is necessary to automatically adapt the LMS to the way of studying of each individual student. The LMS, therefore, enables the students studying according to the Bologna process to access all functionalities demanded by the Bologna Declaration, whereas for the old students these functionalities are invisible and they are not able to use them.

At universities in the Republic of Croatia the ISVU system is used for monitoring student achievements and other organizational tasks. As in the future the LMS model suggested here should be able to fully exchange information with the ISVU system, special attention was paid to the manner of formatting data for information exchange. The ISVU system has options for importing and exporting data and the LMS was designed in such a way to half-automatically or automatically generate and manipulate XML and CSV data files. Even though the ISVU system is still relatively closed and it is not possible to access data directly, the authors of this chapter hope that it will be possible in the near future.

4.5 Adaptation for students with disabilities

Although there are very few students with disabilities at the Faculty of Transport and Traffic Sciences, all students need to be able to use the LMS without difficulty. In order to enable students with certain disabilities to use the system, it will be developed in full compliance with all HTML and CSS standards, as well as in accordance with 508 Accessibility Guidelines. Furthermore, in order to facilitate ease of use for visually impaired students, the LMS design will incorporate the option of high contrast colors and text magnification. It will also include applications which enable blind persons to use the Internet (web readers).

4.6 LMS and Web 2.0 – LMS 2.0

Over the last several years, the Web 2.0 concept has significantly altered the paradigm of user participation on the Internet, shifting from the passive, "read only" kind of participation to full, active participation of all users in content creation. The definitions of the Web 2.0 paradigm vary considerably from author to author; however, they are not mutually exclusive, since each definition includes the concept of active user participation in content creation, collaboration, as well as knowledge and information sharing (Grosseck, 2009). As such, the Web 2.0 constitutes the ideal platform for LMSs within which students become active participants in content creation, thus changing learning paradigms. As can be seen from the overview provided in this paper, students already play a significant role in creating content within the LMS. With the application of Web 2.0 paradigms, services and technologies in instructional processes and LMSs most drawbacks can be overcome (Peraković, Remenar, & Kavran, 2008), which will contribute to the quality of instructional processes as well as enable students to participate more fully in content creation. Virtually any Web 2.0 technology or service may be applied in the LMS 2.0 model. Most authors agree that blogs, microblogs, wikis, RSS feeds, tag-based folksonomies, social bookmarking, multimedia content sharing, forums and social networking sites are becoming a crucial part of the tertiary level educational process, as presented in Table 1, (Grosseck, 2009).

Although Web 2.0 technologies and services provide a number of advantages, such as allowing students freedom and flexibility, some of those listed may have negative effects on student education. For instance, the Wikipedia exerts practically no control over the information which is entered. The information is often incorrect or has not been verified, which is very difficult for instructors to monitor and respond to in timely fashion. For this reason, the new generation model, LMS 2.0, proposes the incorporation of Web 2.0 technologies (wiki, forum, tag-based folksonomies, social bookmarking, RSS, social networking) within the LMS in order to maintain the quality, accuracy and up-to-date character of important instructional materials and information. In contrast, some Web 2.0 services are used as an external addition to the LMS, (Facebook, Twitter, Academia.edu, LinkedIn, etc.), as can be seen in Fig. 7.

According to this conceptual model, the basic principles on which the e-Student system operates, described earlier on in this paper, remain practically the same. Students will continue to access instructional materials added by instructors, register topics for seminar

Web 2.0 technology	Educational application
Blogging	- use blogs for real-world writing experiences - pull class blogs together into one area for easy tracking - quickly give feedback to students, and students to each other - students use peer networks to develop their own knowledge - update new information such as homework and assignments - using comments in blogs can encourage students to help each other with their writing, and get responses to a question without getting the same answer twenty times etc.
Microblogging	- classroom community, exploring collaborative writing, reader response, collaboration across schools, countries, project management, assessing opinion, platform for metacognition, conference or as part of a presentation or workshop, for reference or research, facilitating virtual classroom discussion, creating a learning experience, a Personal Learning Network - use for dissemination of teachers' publications and materials, locating original sources of ideas, quotes, allows for very focused and concrete feedback to students to refine their thinking and improve their skills, fostering professional connections, informal research, for storytelling, follow a professional, get feedback on ideas, event updates, live coverage of events, build trust, build a community etc.
Wiki	- use for student projects; use for collaborating on ideas and organizing documents and resources from individuals and groups of students - use as a presentation tool (as e-portfolios); as a group research project for a specific idea; manage school and classroom documents; use as a collaborative handout for students; writing: student created books and journaling - create and maintain a classroom FAQ; as a classroom discussion and debate area; a place to aggregate web resources; supporting committees, working parties and university projects etc.
Photo / Slides sharing	- share, comment, and add notes to photos or images to be used in the classroom - inspire writing and creativity; create a presentation using the photos - use tags to find photos of areas and events around the world for use in the classroom. - post student presentations to an authentic audience and get feedback from around the world; share professional development materials and have it available anywhere, anytime, to anyone; post presentations of special events
Video sharing	- video professional development on own terms; create an own subject specific videos with students; use video sharing sites to find videos on current issues etc.

Syndication of content through RSS	- professional development, time saving; updated information in teaching area - information coming from constraining sources; sharing work with other educators - RSS feeds can potentially replace traditional email lists, reducing email overload - RSS feeds can be used to keep course specific webpages current and relevant etc.
Social bookmarking	- create a set of resources that can be accessed on any computer connected to the internet; conduct research and share that research with peers - track author and book updates; groups of students doing a classroom project sharing their bookmarks; rate and review bookmarks to help with students decide on usefulness of resources; setup a group tag in order to share educational resources - share one del.icio.us account between a number of different subject specific educators in order to share resources with each other etc.
Social Networking	- event support and continuation, team and community support, aggregation of social media applications, personal learning environments etc.
Other tools	- instant messaging increase the sense of community and accessibility which is required for collaborative learning; VoIP can promote international collaborations and understanding; calendars make calendar events, homework, anything you want available on mobile devices connected to the Internet - survey and polls, online diagrams and web-based word processor, on-line spreadsheet, social search, mind mapping; virtual worlds - virtual conferences and seminars, team meetings and collaboration spaces, simulations etc.

Table 1. Models of integrating Web 2.0 technologies in higher education (Grosseck, 2009)

papers as well as submit them, take e-tests and the like. Adding Web 2.0 functionalities will introduce a range of new services and benefits.

Wiki: According to the creators of the original wiki concept, a wiki is a collection of interlinked web pages which can be freely expanded, i.e. a database which any user may edit with ease. Users may visit, read and easily add or alter the content. Wikis have attracted the attention of educational processes because of the advantages they offer, their simplicity and ease of use. Studies show an increase in use for educational purposes. Wikis are considered to be effective learning and teaching tools since they facilitate collaborative learning and collaborative writing, support project-based learning as well as promote creativity and critical thinking (Usluel & Mazman, 2009). In this proposed model, the wiki system has been implemented in the LMS in order to control the content, which would allow the quality to remain constant. Each student has the opportunity to create, edit and add to the wiki content; however, instructors can easily correct errors and deal with possible omissions, all in one place. It is possible to create wiki content with every entry, whether it is

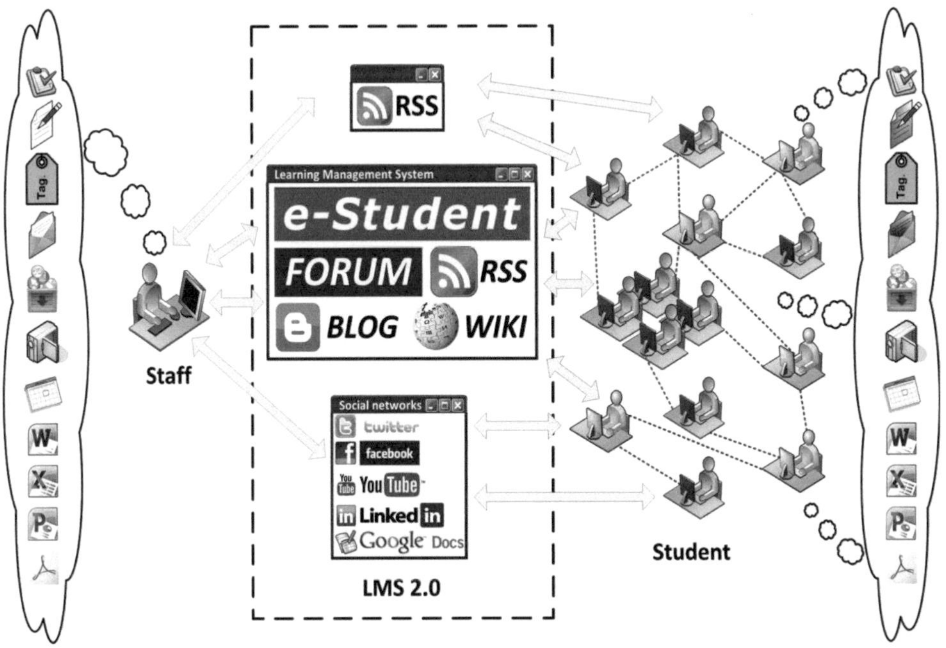

Fig. 7. Conceptual Model of LMS 2.0 at the Faculty of Transport and Traffic Sciences

related to an institute, department, instructor, or a course, teaching unit, lecture, seminar paper or e-test, which would complement the available instructional materials.

Blog: Blogs are also called online diaries. These allow users who possess no technical skills to create, publish and organize their web pages, which contain personal views and opinions presented in chronological order. Studies indicate that blogs develop writing skills, encourage critical thinking and collaborative learning, as well as provide feedback. Blogs may effectively serve as personal online diaries which allow students to share files, resources and knowledge. Furthermore, blogs may function as personal e-portfolios which allow students to track their personal successes, thoughts, achievements and development, (Usluel & Mazman, 2009). For these reasons, the blog constitutes one of the essential Web 2.0 technologies in this conceptual model. Each student will be able to maintain their own blog as well as create blog entries for particular elements within the LMS. For instance, every student will be able to create a blog entry containing his/her review of a particular teaching unit or lecture, which will contribute to the development of their critical thinking and investigative spirit.

RSS: RSS (most commonly expanded as Really Simple Syndication) is a family of web feed formats used to publish frequently updated works—such as blog entries, news headlines, audio, and video—in a standardized format. An RSS document (which is called a "feed", "web feed", or "channel") includes full or summarized text, plus metadata such as publishing dates and authorship. Web feeds benefit publishers by letting them syndicate content automatically. They benefit readers who want to subscribe to timely updates from favored websites or to aggregate feeds from many sites into one place. RSS feeds can be read using software called an "RSS reader", "feed reader", or "aggregator", which can be web-based, desktop-based, or mobile-device-based. (RSS, 2010). The new LMS generation, according to the concept proposed in this paper, will offer the option of distributing multiple RSS feeds, whether these are RSS feeds of particular courses, institutes, departments, instructors or any other entity within the LMS, such as instructional materials and seminar papers. In this way, students can keep up with new information on an almost real-time basis. Furthermore, this model enables each student or instructor to use the LMS as an RSS aggregator, i.e. to aggregate and read their favorite RSS feeds within the LMS, as well as share them with user groups within their social networks or with individual users.

Forum: Online or threaded discussions have been in use for a decade, and therefore a larger body of research literature exists that examines how to use them, what they accomplish, and how to evaluate them. The research on online discussions soon claimed a number of advantages to their use in online and hybrid courses. They increased collaboration, a sense of community, depth and higher-order thinking, interaction, think time, reflection and time on task (Meyer, 2010). In addition to a typical forum which allows students and instructors to post unconnected topics, the new LMS generation will offer the option of starting a discussion topic for each entity within the LMS. For instance, it will be possible to start a discussion on any lecture or file, which will stimulate students' critical thinking, active participation and investigative spirit. On the other hand, instructors will be able to see what their teaching or a particular teaching unit may be lacking, which will enable continuous development and an increase in the quality of instruction and instructional materials.

Social networks: Social networks are applications which facilitate collaboration, sharing of knowledge, interaction and communication between users who are based in different locations but have common interests, needs or goals. Social networks may be viewed as pedagogical tools which encourage the discovery and sharing of information, bring together and maintain networks of people as well as aid the preservation of relationships between members, (Usluel & Mazman, 2009). The application of social network principles within the LMS serves to increase communication between students. The proposed concept specifies the necessity of creating several types of social networks within the LMS. Two networks are predefined: one is a network of students in the same department and the other a network of students taking the same course. The third network would be created personally by each student according to his/her wishes and preferences. In addition to the internal social network, there will also be indirect communication (via RSS feeds) towards independent social networks such as Facebook, LinkedIn, Twitter and so on.

Implementing the above will enable fully flexible creation and editing of content, as well as tagging, storing and sharing of content between students, instructors and previously described social networks (Fig. 8.). Since all these modules make up the LMS of the institution, the entire content may be verified and its quality kept constant.

4.7 Integration with other Faculty systems

Further development of the system will lead towards integrating the e-Student, DMS, SAN, SMSCentar and FPZ*mobile* systems with the other systems at the Faculty: Studomat and ISVU. Integration of all Faculty systems into a single system would simplify the work of both instructors and students. Integration would result in a single set of access data, i.e. a single username and password which would be used to access any system. It would also result in the exchange of student data across systems and a single electronic file for each student. Adapting the visual identity to the needs of system users will make the system significantly easier to use. (Peraković, Remenar, & Jovović, 2010)

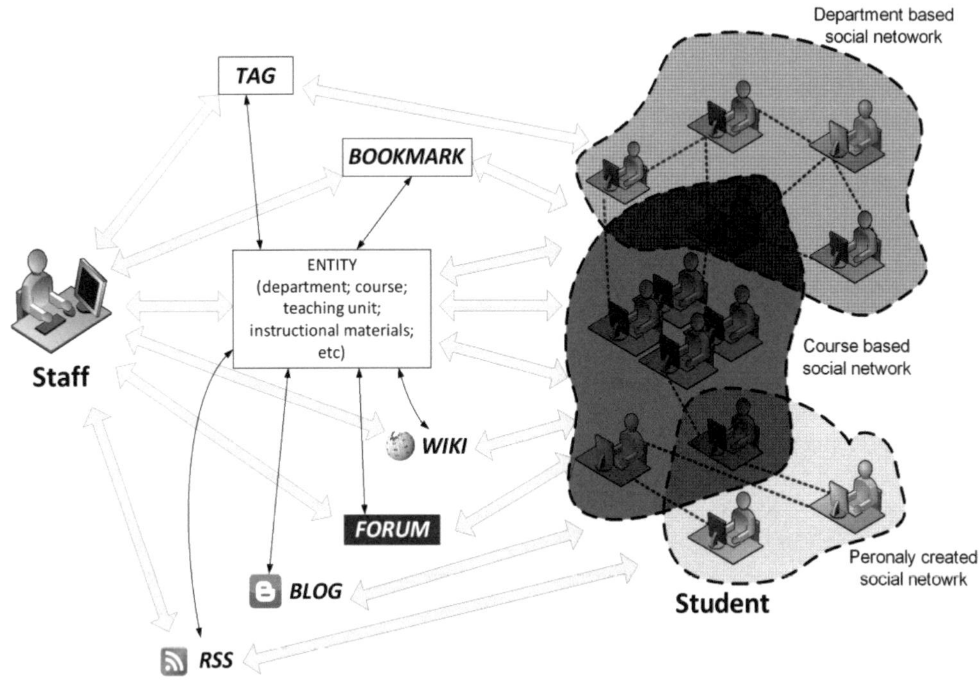

Fig. 8. Creating, editing, tagging, storing and sharing of content

4.8 Security

The fact that the LMS is based on web application modules developed within the institution emphasizes security issues apparent in easy accessibility and a large number of users. It is a priority - not simply an incidental benefit - to maintain a reliable level of functionality and a high level of data security while using and maintaining the LMS, as well as developing new modules. Security testing for existing types of attack is continuously carried out in order to remove hidden flaws, oversights and shortcomings in timely fashion. This serves to protect the distance learning application modules, users and user data stored in the databases used by the LSM. It is crucial to extend the protection beyond the LMS itself, so that organizational, programmatic and structural security mechanisms are implemented and applied to the entire information and communication system of the Faculty. Other services which are in use must also be included, such as the ISVU or the accounting system, as well as specific applications used by the faculty staff. In order to achieve a secure and reliable integral information system of the Faculty of Transport and Traffic Sciences, it is crucial to provide both teaching and non-teaching staff with continuous training on security issues and the risks of transacting business electronically, as well as using a digital signature in the instructional process. It hardly requires explicit mention that the adopted security policy of applying the information and communication infrastructure of the Faculty of Transport and Traffic Sciences must be adhered to fully and unconditionally (Peraković & Remenar, 2007).

5. Conclusion

An analysis of the e-Learning system developed within the institution and in use for the past six years, leads to the conclusion that the present system has become indispensable in electronic communication, both regarding the instructional process and student welfare. Thus, the Faculty of Transport and Traffic Sciences has become one of the leading faculties in Croatia in using the assistance of new technologies to aid students in the acquiring and evaluation of knowledge.

An overview of the statistical data concerning the use of the e-Learning system of the Faculty of Transport and Traffic Sciences indicates that there is great interest in system, and, what is more significant, the interest continues to grow. Both students and users have demonstrated a ready acceptance of the functionalities which have been developed so far, and they want and expect further expansion as well as the development of new system options. The greatest value of the system lies in the fact that it has been developed in its entirety at the Faculty of Transport and Traffic Sciences, in coordination with its teaching staff and students. Thus, the students may advance their knowledge and skills using new information and communication technologies and program tools on projects developing information systems, in a teamwork environment.

By means of conducting continuous surveys of user satisfaction with the current e-Learning system, as well as taking into consideration new requirements and rectifying possible errors, the new e-learning system of the Faculty will continue to develop.

Since a completely new system is about to be developed, it will be designed taking into consideration the goals mentioned above, but it will also aim to support and be integrated with Web 2.0 services. The new system will upgrade the processes of information exchange and publication, which will be adapted to the needs of each student by means of e-mail and text messages, as well as RSS feeds and integration with popular social networks such as Facebook and Twitter. Finally, people with disabilities will be able to use this LMS 2.0 model simply and efficiently. The LMS 2.0 will be designed fully in accordance with the model proposed in this paper, which is centered around the proposed use of up-to-date Web 2.0 technologies, compliance with European Commision Guidelines, adaptation to the Bologna education model, the ISVU system and the "Outcomes of Studying at the University of Zagreb" program.

6. References

COMMISSION OF THE EUROPEAN COMMUNITIES. (24 May 2000). COMMUNICATION FROM THE COMMISSION. *e-Learning – Designing tomorrow's education* .

EUROPEAN COMMISSION. (November 2003). Education and Training 2010. *Information and Communication Technologies in Education and Training* .

Grosseck, G. (2009). To use or not to use web 2.0 in higher education? *Procedia - Social and Behavioral Sciences* , *1* (1), 478-482.

Hage, H., & Aimeur, E. (2010). E-learning for the new generations, a Web 2.0 approach. In M. Buzzi, & M. Buzzi (Ed.), *E-learning* (pgs. 1-18). Vukovar, Croatia: In-teh.

Hargadon, S. (04. March 2009). *Web 2.0 Is the Future of Education.* Retrieved on 15 January 2010 from Steve Hargadon's blog: http://www.stevehargadon.com/2008/03/web-20-is-future-of-education.html

Hijon-Neira, R., & Velazquez-Iturbide, A. (2010). From the discovery of students access patterns in e-learning including web 2.0 resources to the prediction and enhacements of students outcome. In R. Hijon-Neira, & R. Hijon-Neira (Ed.), *E-learning, experiences and future* (pgs. 275-294). Vukovar, Croatia: In-Teh.

Huddlestone, J., & Pike, J. (2008). Seven key decision factors for selecting e-learning. *Cognition, Technology & Work* , *10* (3), 237-247.

Mason, R., & Rennie, F. (2007). Using Web 2.0 for learning in the community. *Internet and Higher Education* , *10*, 196-203.

Meyer, K. A. (2010). A comparison of Web 2.0 tools in doctoral course. *Internet and Higher Education* , xxx-xxx.

Molina, A. I., Giraldo, W. J., Jurado, F., Redondo, M. A., & Ortega, M. (2010). Evolution of Colaborative Learning Enviroment based on Desktop Computer to Mobile Computing: A Model-Based Approach. In S. Soomro, & S. Soomro (Ed.), *E-learning, experiences and future* (pgs. 261-274). Vukovar, Croatia: In-Teh.

Peraković, D., & Remenar, V. (2007). Security Audit and Mechanism of Protecting e-Learning System an the Faculty of Transport and Traffic Sciences. *Proceedings of the 10th International Conference on Information and Intelligent Systems* (pgs. 311-317). Varaždin: Faculty of Organization and Informatics.

Peraković, D., & Remenar, V. (2007). Security Audit and Mechanism of Protecting e-Learning System at the Faculty of Transport and Traffic Sciences. *Proceedings of the 10th International Conference on Information and Intelligent Systems* (pgs. 311-317). Varaždin: Faculty of Organization and Informatics.

Peraković, D., Jovović, I., & Forenbacher, I. (2009). Analysis of the Possibilities and Effects of Implementing Interactive Application for Mobile Terminal Devices in m-Learning System at the Faculty of Transport and Traffic Sciences. *Posters Abstracts of the ITI 2009 31st International Conference on Information Technology Interfaces* (pgs. 27-28). Cavtat / Dubrovnik: SRCE, University Computing Centre.

Peraković, D., Kavran, Z., & Remenar, V. (2006). The Impact of Introducing e-Learning System in the Teaching Process at the Faculty of Traffic and Transport Sciences. In B. Katalinić (Ed.), *Proceeding of 17th International DAAAM Symposium* (pgs. 299-300). Vienna: DAAAM International.

Peraković, D., Remenar, V., & Grgurević, I. (2008). Possibility of Applying Fuzzy Logic in the e-Learning System. *Proceedings of the 19th Central European Conference on Information and Intelligent Systems* (pgs. 89-94). Varaždin: Faculty of Organization and Informatics.

Peraković, D., Remenar, V., & Jovović, I. (2010). Conceptual Model of Developing a New LMS System of the Faculty of Transport and Traffic Sciences. *Poster Abstracts of The ITI 2010 32th International Conference on Information Technology Interfaces.* Cavtat / Dubrovnik: SRCE, University Computing Centre.

Peraković, D., Remenar, V., & Kavran, Z. (2008). Drawbacks of Implementing e-Learning System in the Teaching Process at the Faculty of Transport and Traffic Sciences. *Poster Abstracts of The ITI 2008 30th International Conference on Information Technology Interfaces* (pgs. 225-26). Cavtat / Dubrovnik: SRCE University Computing Centre.

Peraković, D., Remenar, V., & Šašek, Z. (2007). Analysis of Operation and Possibilities of Improving e-Learning System in Traffic Engineering. *Promet - Traffic&Transportation* , *19* (3), 167-172.

RSS. (7 June 2010). Retrieved on 24 June 2010 from the Wikipedia: http://en.wikipedia.org/wiki/RSS

Saeed, N., & Sinnappan, S. (2010). Effects of Media Richness on User Acceptance of Web 2.0 Technologies in Higher Education. In R. Hijón-Neira, & R. Hijón-Neira (Ed.), *Advanced Learning (str. 233-244). Vukovar, Croatia: In-Teh.*

SveZaIU. (2009). Ishodi učenja na Sveučilištu u Zagrebu. Zagreb: Sveučilište u Zagrebu.

Usluel, Y. K., & Mazman, S. G. (2009). Adoption of Web 2.0 tools in distance education. *Procedia - Social and Behavioral Sciences* , *1* (1), 818-823.

Wal, T. V. (2 February 2007). *vanderwal.net*. Retrieved on 25 January 2010 from Folksonomy: http://vanderwal.net/folksonomy.html

11

BEPU Approach in Licensing Framework, including 3D NK Applications

F. D'Auria, N. Muellner, C. Parisi and A. Petruzzi
Nuclear Research Group San Piero a Gardo University of Pisa, Pisa, Italy

1. Introduction

During the recent years, a world-wide renewed interest in the exploitation of nuclear energy for electricity production is seen among both the Western and the new industrializing Countries (e.g., China and India). As a result, 61 reactors are now under construction and more than 100 units are planned for the incoming decade. Such impressive development is totally based on Light and Heavy Water Reactor (LWR & HWR) technologies [1], on designs that are an evolution of the robust and reliable Nuclear Power Plants (NPP) designed and built during the seventies-eighties of the last century. At that time, the need to guarantee an high safety level on one side and on the other the limited computational capabilities and the scarce knowledge of some phenomena, drove the main nuclear safety authorities to establish extremely conservative rules. Nowadays, after that tremendous progress has been made into the computational power availability, models accuracy and the knowledge of relevant phenomena, there is the need to go toward more realistic safety analyses and to relax some levels of conservativeness without compromising the always elevated safety level of the nuclear industry.
The aim of this Chapter is to give an overview of the current trends in the licensing frameworks for a NPP. International best-practices are presented and discussed and sample applications derived from works of the San Piero a Grado Nuclear Research Group of the University of Pisa (GRNSPG/UNIPI) on existing industrial facilities are also reported.

2. The licensing framework and the best international practices

Three internationally recognized fundamental safety objectives [2] constitute the basis from which the requirements for minimizing the risks associated to NPPs shall be derived. A general nuclear safety objective is stated as "to protect individuals, society and the environment from harm by establishing and maintaining in nuclear installations effective defenses against radiological hazards". Two complementary safety objectives deal, respectively, with radiation protection and technical aspects.
The established terms for the technical safety objective are: "to take all reasonably measures to prevent accidents in nuclear installations and to mitigate their consequences should they occur; to ensure with a high level of confidence that, for all possible accidents taken into account in the design of the installation, including those of very low probability, any

radiological consequences would be minor and below prescribed limits, and to ensure that likelihood of accidents with serious radiological consequences is extremely low".
In a recent publication of the International Atomic Energy Agency (IAEA) [3] the safety objectives have been rephrased into one safety objective and ten safety principles.
To demonstrate that all applicable safety requirements are fulfilled by the design and the operation of a nuclear power plant, a systematic evaluation must be conducted throughout the lifetime of the installation. According to well established safety practices [4], this systematic assessment should follow two complementary paths (or twofold strategy): a comprehensive safety analysis of the plant, and a thorough evaluation of engineering factors embedded in the design and operation of the installation.
The comprehensive safety analysis shall address the dynamic response of the plant to a sufficiently broad spectrum of faults and accidents scenarios to demonstrate that states that could result in high radiation doses or radioactive releases are of very low probability of occurrence, and plant states with significant probability of occurrence have only minor or no potential radiological consequences. For performing plant safety analysis, methods of both deterministic and probabilistic analysis shall be applied [4].
The twofold safety assessment strategy shall be consolidated and documented in the so called Safety Analysis Report (SAR) which, according to recognized safety standard [5], must support the regulatory decision making process within the plant licensing framework. The requirements on SAR format and contents are dependent on country's regulatory regime, although some consolidated practices have been widely followed [6], [7].
The achievement of a high level of safety should be demonstrated primarily in a deterministic way. The deterministic approach typically considers a limited number of events for which conservative rules for system availability and parameter values are often applied.
Recently, with the development of code capabilities supported by experimental data basis, best estimate (BE) methods have also been applied within the design basis spectrum of events. In such situation, however, for licensing applications there is a need to address uncertainties in the calculations.
For the deterministic BE analysis this kind of uncertainty (epistemic) results among others from imperfect knowledge of the physical phenomena or of values of code model parameters. Instead, aleatory uncertainty, resulting from inherent randomness or stochastic variability, is by its very nature the subject of Probabilistic Safety Analysis (PSA) type of analysis.
Any BE analysis considering uncertainties (in short Best Estimate Plus Uncertainty, BEPU) which is applied in the field of licensing should be consistent with [8] and [9]. More details on the international framework on BEPU analysis in licensing please could be found in [10] and [11].

3. The BEPU approach

GRNSPG/UNIPI developed a procedure for the consistent application of BEPU in deterministic safety analysis (Fig. 1 shows a simplified flowchart), which has been applied in the licensing process of Atucha 2 NPP. The approach adheres to common practices of categorizing postulated initiating events (PIE) according to their frequency, and establishing more stringent acceptance criteria for more likely events. The following subsection provides an explanation on the terms, while the subsection after introduces a procedure for BEPU application.
The procedure follows well accepted design philosophy for NPPs, which recognizes the principle that plant states which could result in high radiation doses or radioactive releases

are of very low probability of occurrence, and plant states with significant probability of occurrence have only minor or no radiological consequences.

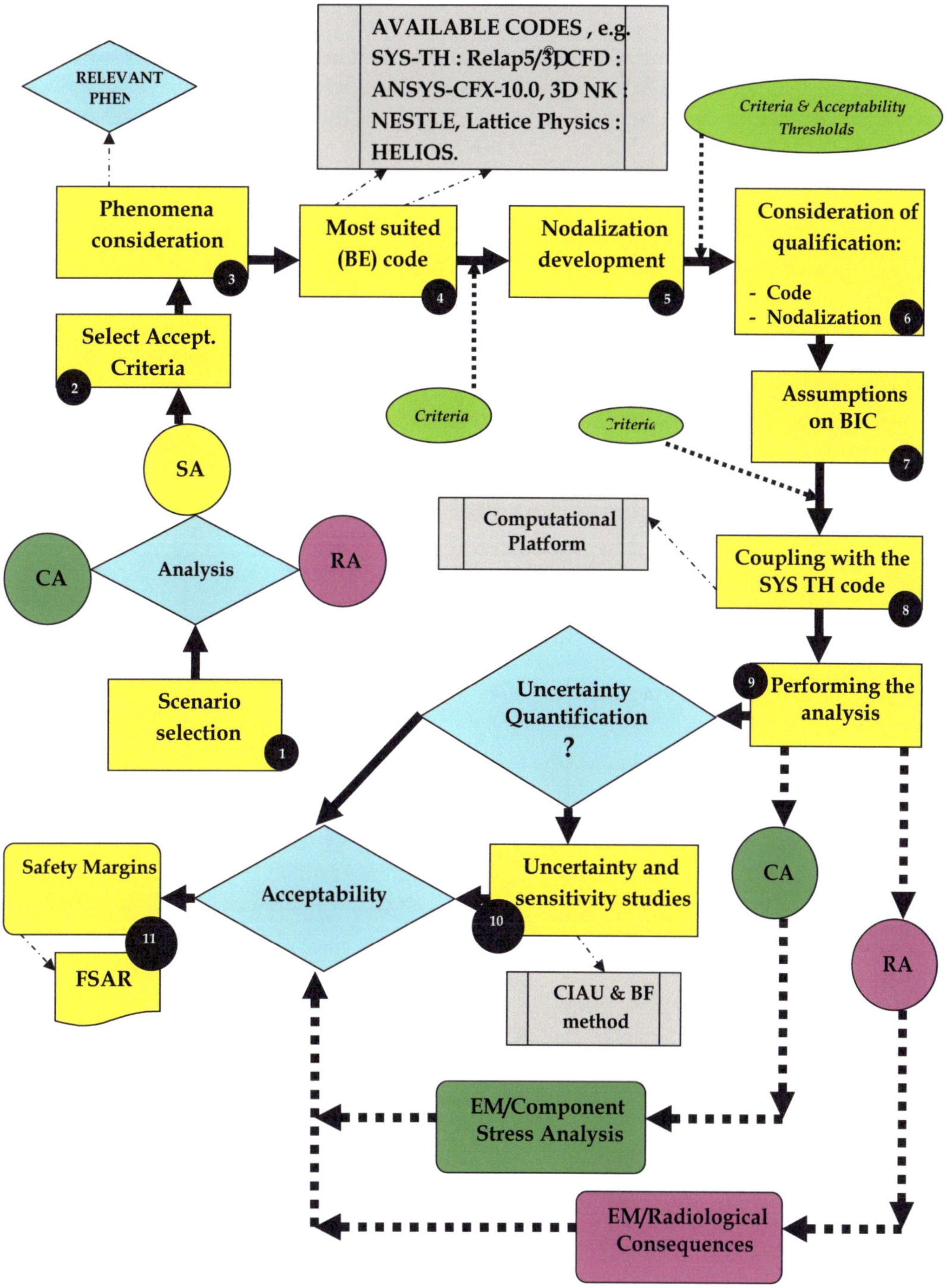

Fig. 1. General scheme of applying BEPU in accident analysis

3.1 The preliminary categorizing of events and acceptance criteria

Before the scenario selection for the analysis, a preliminary categorization of the events leading to such scenario and the acceptance criteria for the evaluation of the safety margins have to be fixed.

The event sequences postulated in the design of the plant are analyzed to demonstrate that in operational states, in and following a design basis accident and, to the extent practicable, on the occurrence of some selected accident conditions that are beyond the design basis accidents, the following three fundamental safety functions are performed:

- Safe shutdown and long term subcriticality
- Residual heat removal
- Limitation of radioactive releases.

PIE are grouped according to their anticipated probability of occurrence in anticipated operational occurrences (AOO), design basis accidents (DBA), beyond design basis accidents (BDBA) and severe accident (SA) (e.g., see Tab. 1). In the specific case of the Atucha-2 Final Safety Analysis Report (FSAR), an intermediate category or the selected beyond design basis accident (SBDBA) was introduced to address specific scenarios beyond design basis, including Double-Ended Guillotine Break (DEGB) Loss-of-Coolant-Accident (LOCA) and Anticipated Transient Without Scram (ATWS). Also in the Atucha-2 FSAR, the proposed BEPU analysis scheme was applied only to AOO, DBA and SBDA, while the remaining ones (BDBA and SA) were treated through the probabilistic safety analysis.

Range of frequency (1/reactor-year)	Characteristic	Category	Common Terminology	Safety Consequences
10^{-2} to 1 (Expected in life of the plant)	Expected	AOO	Anticipated transients, transients, frequent faults, incidents of moderated frequency	No additional fuel damage
10^{-4} to 10^{-2} (Chance greater than 1% over the life of the plant)	Possible	DBA	Infrequent incidents, infrequent faults, limiting faults, emergency conditions	No radiological impact at all or no radiological impact outside exclusion area
10^{-6} to 10^{-4} (Chance less than 1% over the life of the plant)	Unlikely	BDBA	Faulted conditions	Radiological consequences outside the exclusion area but within limits
$<10^{-6}$ (Very unlikely to occur)	Remote	SA	Faulted conditions	Emergency response needed

Table 1. Category of event based on its expected frequency of occurrence

Acceptance criteria are applied in the deterministic safety analysis, following some rules and methods which have been developed to introduce conservatisms in plant safety evaluations.

Acceptance criteria are directly or indirectly related to the barriers against releases of radioactive material. Current adopted values and rules for using acceptance criteria have been developed considering some decoupling techniques that cover the range from plant processes to environmental impact. The decoupling should ensure that if, for example, the fuel safety criteria are fulfilled during the accident, then the radiological releases are limited and acceptable provided that the criteria for the two other barriers are also fulfilled.

Safety criteria are mostly derived from the radiological reference values by applying several decoupling actions where, for some cases a phenomenon is substituted to the primitive one (decoupling phenomena), for some others, more restrictive values are imposed in order to be sure that the original requirement is satisfied (decoupling parameter). At each step, conservatisms are introduced that can be considered as margins for safety. Frequent events should have minor consequences and events that may result in severe consequences should be of very low probability. In this sense, the risk across the spectrum of AOO and DBA should be approximately constant. Acceptance criteria are derived for each category of event based on its expected frequency of occurrence, as shown in the Tab. 1 above.

3.2 Grouping the events

Generally, all selected scenarios are grouped in a classical families of events (around ten different families) where each family covers events with similar phenomena.

For the FSAR Chapter 15 analyses, and for each category of events, the results of the analyses are assessed in terms of the fulfillment of safety functions which are graded according to the expected frequencies of occurrences for the correspondent PIE.

To keep a consistently flat risk profile over the entire spectrum of AOO and DBA, the more frequent the event is, the less tolerable its consequences are. In this sense, acceptance criteria are selected for different event categories, for safety parameters as fuel and cladding temperatures, departure from nucleate boiling ratio (DNBR), primary circuit pressure, containment pressures, and total effective dose equivalent (TEDE).

3.3 Selection of the evaluation models & phenomena consideration

The BEPU approach takes credit of the concept of evaluation model (EM, see below), and comprising three separate possible modules depending on the application purposes:

- for the performance of safety system countermeasures (EM/SA);
- for the evaluation of radiological consequences (EM/RA);
- for the review of components structural design loadings (EM/CA).

A fundamental step in performing safety analysis is the selection of the EM. With EM is generally intended the calculation framework for evaluating the behavior of the reactor system during Chapter 15 analyzed events. EM could include one or more codes, analytical models or also calculation procedure and all other information for use in the target application.

To start analyzing typical events scenarios for the chapter 15 of an FSAR, EMs rely mostly on system thermal-hydraulic (SYS TH) codes (as for EM/SA) to solve the transport of fluid mass, momentum and energy throughout the reactor coolant systems. The extent and complexity of the physical modes needed to simulated plant behavior are strongly dependent of the reactor design and of the transient itself.

For some scenarios, or regarding some analysis purposes, the SYS TH code may, for example, be complemented by (or coupled with) a three-dimensional neutron kinetics code

or the reference model may need an expansion to include a detailed simulation of controls and limitation systems which play a relevant role for determining the plant response.
For the scope of the proposed approach for accident analyses, the complexity of the evaluation model may range from a simplified qualitative evaluation (EM/QA) to a complete combination of the three possible modules (EM/SA + EM/RA + EM/CA).
The two main aspects which have been considered for developing the evaluation model with the ability of adequately predict plant response to postulated initiating events are intrinsic plant features and event-related phenomena characteristics.
For the two modules EM/SA and EM/CA, the first set of requirements for the evaluation model is imposed by the design characteristics of the nuclear power plant, its systems and components. Requirements on the capability of simulating automatic systems are of particularly importance for AOO, in which control and limitation systems play a key role on the dynamic response of the plant. For evaluation of radiological consequences, the EM/RA module has demanded additional appropriate site-related features to be built in.
The third set of requirements is derived from the expected evolution of the main plant process variables and the associated physical phenomena. For the proposed approach, this is performed through the process of identifying the Phenomenological Windows (PhW) and the Relevant Thermal-hydraulic Aspects (RTA). The relevant timeframe for the event is divided into well defined intervals when the behavior of relevant safety parameters is representative of the physical phenomena.

3.4 Selection of boundary and initial conditions

Additionally to the computers codes and the selection of modeling options, the established procedures for treating the input and output information are also recognized as comprising key parts of the evaluation model. The adopted procedures to select initial and boundary conditions (BIC), which follows the original design safety philosophy, are of particular importance for supporting the regulatory acceptability of the results provided by the EM. As the foreseen use of an EM is for licensing purposes, it is necessary to evaluate the suitability of conservative assumptions or to adopt BE approaches with the quantification of uncertainties.
Suitability of conservatism should be understood as addressing the issue of "how conservative is conservative enough". Alternatively, when a BE approach is adopted, then realistic assumptions will be input to BE models, conducting to realistic estimates for plant behavior. In these cases, licensing applications demand the quantification of uncertainties in the calculated results to ensure that safety margins are still available. For the scenarios were the conservative assumptions may provide enough safety margins, it can be derived by the analysts a criterion to determine the need for uncertainty calculations.
Typically, SBDBA and some DBA can involve quantification of uncertainties.

3.5 Selection of qualified tools

For the adequate simulation of the identified phenomena (step "3" of Fig. 1), computational tools have to be selected from those which have previous qualification using an appropriate experimental data base. Satisfactory qualification targets provide basis for acceptability of the postulated application (see later). As referenced in Fig. 1 (step "4"), "computational tools" expression comprises:

- BE computer codes

- qualified detailed nodalizations for the adopted codes
- established computational methods for uncertainty quantification
- computational platform for coupling and interfacing inputs and outputs of the selected codes

In the GRNSPG/UNIPI approach, with the full scope of application of BEPU quantification, a pre-requisite is the availability or the support of the most advanced-qualified computational tools available on the market.

Generally, for most event scenarios, the single purpose evaluation model EM/SA may be necessary and sufficient to be developed. In this sense, the availability and the application of qualified system thermal-hydraulic code and reliable uncertainty methodology (UM) is the minimum requirement.

Additionally, depending on the specific event scenario and on the purpose of the analysis, it is necessary the availability of calculation methods that are not embedded in the SYS TH code, as, e.g. for burst temperature, burst strain and flow blockage calculations. This may imply an evaluation model EM/CA composed by a fuel rod thermal-mechanical computer code. Another example is in transients where strong asymmetric neutron flux changes happen; this could require the adoption of a three-dimensional (3D) neutron kinetics (NK) code (see later).

3.6 Nodalization development & qualification

Nodalizations (i.e., codes input decks and plant schematization) should be developed according to predefined qualitative and quantitative acceptance criteria. Different methods and procedures are available, depending on the analyst choices and code types. Nevertheless, in a licensing process, it is fundamental to demonstrate to the Safety Authority the efficiency and the reliability of such qualification procedures. At GRNSPG/UNIPI, a suited set of criteria is proposed for SYS TH codes qualifications according to references [13] to [16].

A major issue in the use of mathematical models is constituted by the model capability to reproduce the plant or facility behaviour under steady-state and transient conditions. These aspects constitute two main checks for which acceptability criteria have to be defined and satisfied during the nodalization-qualification process. The first of them is related to the geometrical fidelity of the nodalization of the reference plant; the second one is related to the capability of the code nodalization to reproduce the expected transient scenario. A simplified scheme of a procedure for the qualification of a TH nodalization is depicted in Fig. 2. In the following, it has been assumed that the code has fulfilled the validation and qualification process and a "frozen" version of the code has been made available to the final user.

3.7 Couplings, including the Neutronics

As anticipated in Chapter 1.3.5, SYS TH codes have a leading role in the licensing analyses because of their capabilities to simulate with sufficient level of details the thermal-hydraulic phenomena of primary and secondary circuits of a LWR/HWR. In some specific transient analyses (that could belong to every category of the Tab. 1), it could be requested with a high level of detail the simulation of non-TH phenomena. In such cases, other BE codes have to be employed and coupled with robust and qualified procedures with the SYS-TH code 18]. A typical example is when a detailed core power reconstruction is needed, both during

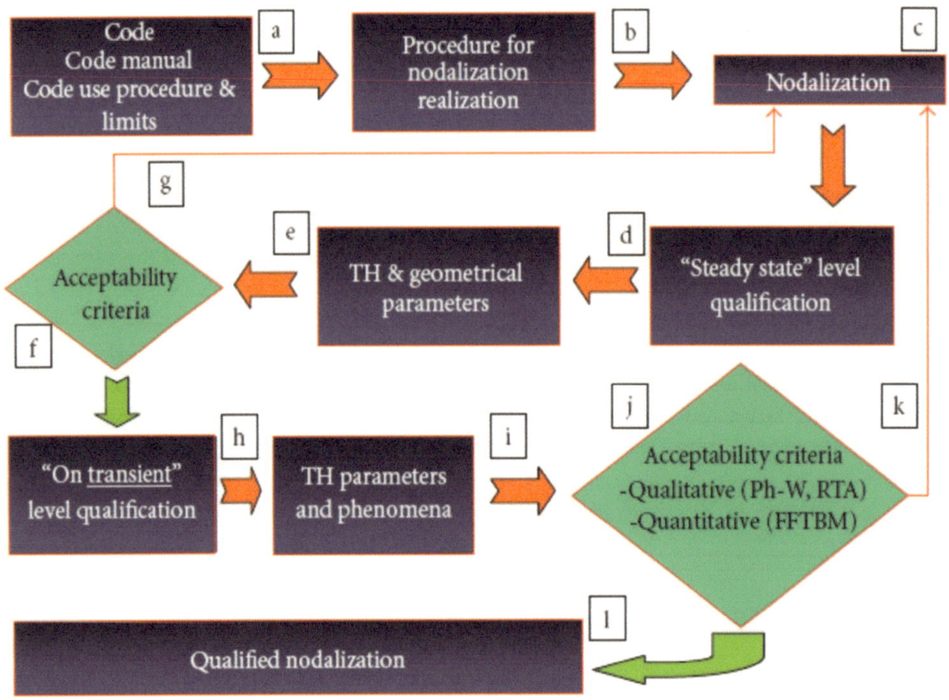

Fig. 2. Flow sheet of nodalization qualification procedure [17]

steady-state and transients analyses. A chain of Neutronic codes has to be employed, [starting from the Evaluated Nuclear Data Files processing, passing through the neutron transport simulations (both in deterministic and stochastic ways), the few group homogenized cross section libraries calculations and ending in the 3D core-wide NK simulations (eventually coupled with a SYS TH code). An example of this sophisticated chain of BE codes is given in Fig. 3, and it was applied by the GRNSPG/UNIPI team for the licensing calculation of the Atucha-2 HWR, Argentina. In this case, needs of detailed neutronic analyses were due to the evaluation of the safety margins during a SBDBA (the DEGB LBLOCA). Because the positive void reactivity coefficient of the reactor, that transient was, at the same time, also a Reactivity Initiated Accident (RIA). The peculiar characteristics of the reactor (e.g., oblique Control Rods and a second emergency scram system injecting boron solution in the full pressure moderator tank) requested the use of advanced Monte Carlo neutron transport simulations too.

Needs for coupling different codes with SYS TH codes, could be necessary also in other technological fields, e.g. for the Pressurized Thermal Shock analyses (coupling Structural Mechanics, Computational Fluid Dynamics and TH codes), the Fuel Pin Mechanics analyses (coupling of Fuel Pin Mechanics, Neutronics and SYS TH codes) or the Containment analyses (coupling of Containment and SYS TH codes).

Quality assurance of the coupling process as well as procedures for codes qualification has to be declared and demonstrated to the National Safety Authority. Example of procedure and validation campaigns of codes of different technological areas can be found in [19], [20], [21].

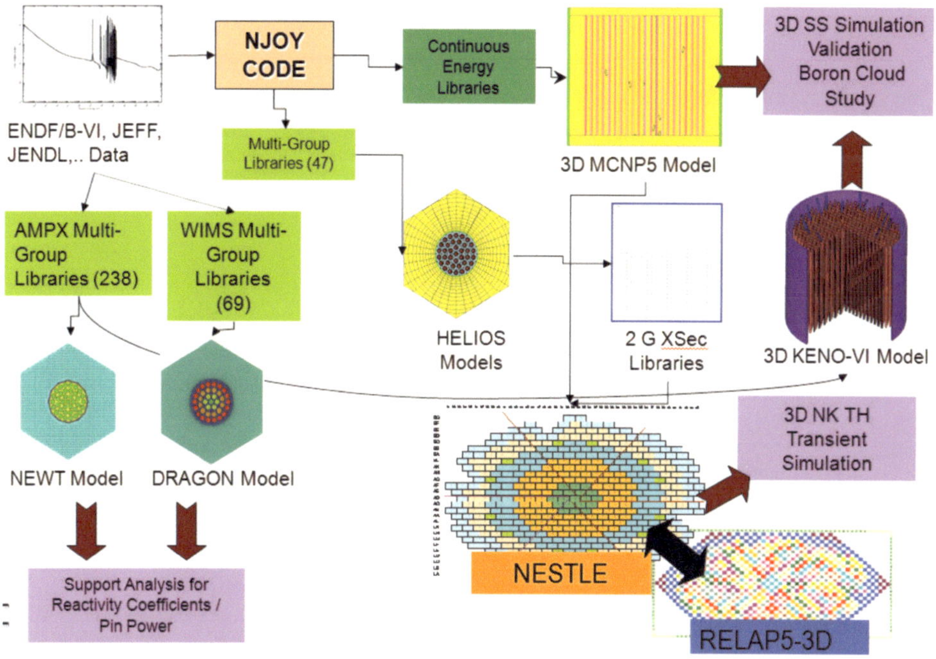

Fig. 3. Example of BE Neutronic Chain of Code use for a set of Licensing Calculations

3.8 Criteria for the application of uncertainty analyses

In many cases, BE calculations on plant behaviour demonstrate a performance with no significant challenges to the applicable safety limits to such extent that, even adding the maximum expected uncertainty, acceptance criteria are fulfilled. For this reason, the GRNSPG/UNIPI proposed BEPU approach derived a non-safety related criterion to decide upon the need for performing uncertainty calculation. Whenever the safety parameter, as calculated by the EM, comes within an established range or distance from the limit value, the uncertainty in the calculated results is quantified (BEPU application, see step "10" of Fig. 1). The development of the non-safety related criterion, implies to establish the "range or distance from the limit value" for the plant safety parameters. The GRNSPG/UNIPI general adopted formula (except for DNBR) can be represented by:

$$Par_{CALC}^{MAX} * \left(1 + U_{PAR}^{MAX}\Big|_{PS}\right) \geq Cr_{Cat}^{Par}$$

where:

- Par_{CALC}^{MAX} is the maximum value of the calculated parameter for the transient-event under consideration;
- $U_{PAR}^{MAX}\Big|_{PS}$ is the maximum uncertainty value affecting the best estimate prediction of the parameter. The maximum uncertainty value is derived from the CIAU database considering the phase-space (PS) characterizing each event-category (AOO, DBA non-LOCA, etc...). Therefore, the "range or distance from the limit value" is given by $Par_{CALC}^{MAX} * U_{PAR}^{MAX}\Big|_{PS}$. In some case, just the maximum value of the uncertainty for the whole phase space is considered (U_{PAR}^{MAX});

- Cr_{Cat}^{Par} is the limit value for the selected parameter and event-category (AOO, DBA non-LOCA, etc...) below of which no uncertainty analysis shall be performed. In some case, no distinction between event category is done (Cr^{Par}). Further information for each specific six non-safety related criteria can be found in [11].

4. The uncertainty quantifications

For licensing applications, evaluation of uncertainty constitutes a necessary complement of BE calculations, which are performed to understand accident scenarios in water-cooled nuclear reactors. The needs come from the imperfection of computational tools, on the one side, and from the interest of using such a tool to get more precise evaluation of safety margins. Several uncertainties methods were developed since the development of the code scaling, applicability and uncertainty (CSAU) evaluation methodology [22] by the U.S. Nuclear Regulatory Commission, e.g. see [23]. Hereafter, a brief description of the key features of the GRNSPG/UNIPI CIAU methodology for the uncertainty is given, together with some sample applications to different LWR.

4.1 A method for the uncertainty quantification: the CIAU method

The UMAE (Uncertainty Method based on the Accuracy Extrapolation) is the prototype method for the consideration of "the propagation of code output errors" approach for uncertainty evaluation. As described in section 3, the method focuses not on the evaluation of individual parameter uncertainties but on the propagation of errors from a suitable database calculating the final uncertainty by extrapolating the accuracy from relevant integral experiments to full scale NPP.

Considering integral test facilities of reference water cooled reactor, and qualified computer codes based on advanced models, the method relies on code capability, qualified by application to facilities of increasing scale. Direct data extrapolation from small scale experiments to reactor scale is difficult due to the imperfect scaling criteria adopted in the design of each scaled down facility. So, only the accuracy (i.e. the difference between measured and calculated quantities) is extrapolated. Experimental and calculated data in differently scaled facilities are used to demonstrate that physical phenomena and code predictive capabilities of important phenomena do not change when increasing the dimensions of the facilities.

Other basic assumptions are that phenomena and transient scenarios in larger scale facilities are close enough to plant conditions. The influence of user and nodalization upon the output uncertainty is minimized in the methodology. However, user and nodalization inadequacies affect the comparison between measured and calculated trends; the error due to this is considered in the extrapolation process and gives a contribution to the overall uncertainty.

The method utilizes a database from similar tests and counterpart tests performed in integral test facilities that are representative of plant conditions. The quantification of code accuracy is carried out by using a procedure based on the Fast Fourier Transform characterizing the discrepancies between code calculations and experimental data in the frequency domain, and defining figures of merit for the accuracy of each calculation. Different requirements have to be fulfilled in order to extrapolate the accuracy [24].

Calculations of both Integral Test Facility experiments and NPP transients are used to attain uncertainty from accuracy. Nodalizations are set up and qualified against experimental data by an iterative procedure, requiring that a reasonable level of accuracy is satisfied. Similar criteria are adopted in developing plant nodalization and in performing plant transient calculations. The demonstration of the similarity of the phenomena exhibited in test facilities and in plant calculations, accounting for scaling laws considerations, leads to the Analytical Simulation Model (ASM) or Evaluation Model, with a qualified nodalization of the NPP that The flow diagram of UMAE is given in Fig. 4. The bases of the methods and the conditions to be fulfilled for its application, including the use of the FFTBM can be found in references [25] and [26 to 29].

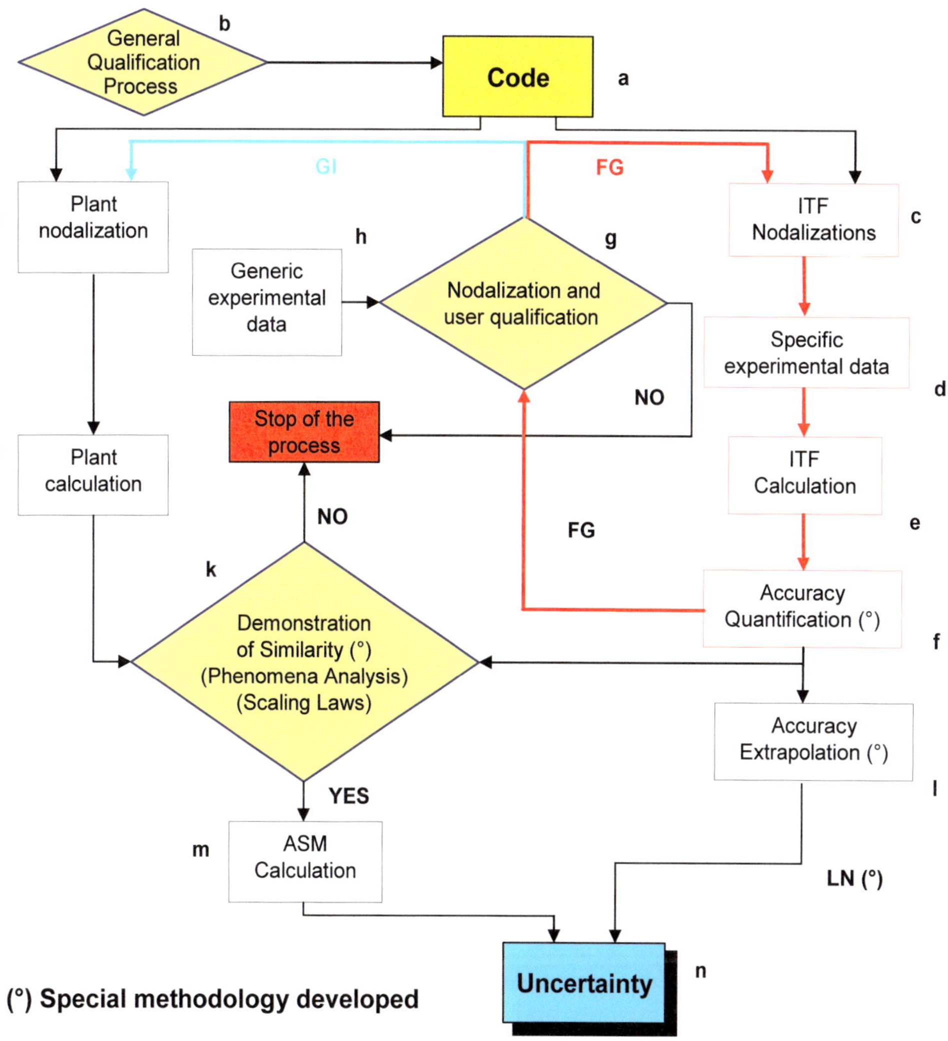

Fig. 4.UMAE flow diagram (also adopted within the process of application of CIAU).

All of the uncertainty evaluation methods are affected by two main limitations:

- The resources needed for their application may be very demanding, ranging to up to several man-years;
- The achieved results may be strongly method/user dependent.

The last item should be considered together with the code-user effect, widely studied in the past, e.g. reference [28], and may threaten the usefulness or the practical applicability of the results achieved by an uncertainty method. Therefore, the Internal Assessment of Uncertainty (IAU) was requested as the follow-up of an international conference jointly organized by OECD and U.S. NRC and held in Annapolis in 1996 [31]. The CIAU method [32] has been developed with the objective of reducing the above limitations.

The basic idea of the CIAU can be summarized in two parts, as per Fig. 5:

- Consideration of plant status: each status is characterized by the value of six relevant quantities (i.e. a hypercube) and by the value of the time since the transient start.
- Association of an uncertainty to each plant status.

In the case of a PWR the six quantities are: 1) the upper plenum pressure, 2) the primary loop mass inventory, 3) the steam generator pressure, 4) the cladding surface temperature at 2/3 of core active length, 5) the core power, and 6) the steam generator down-comer collapsed liquid level.

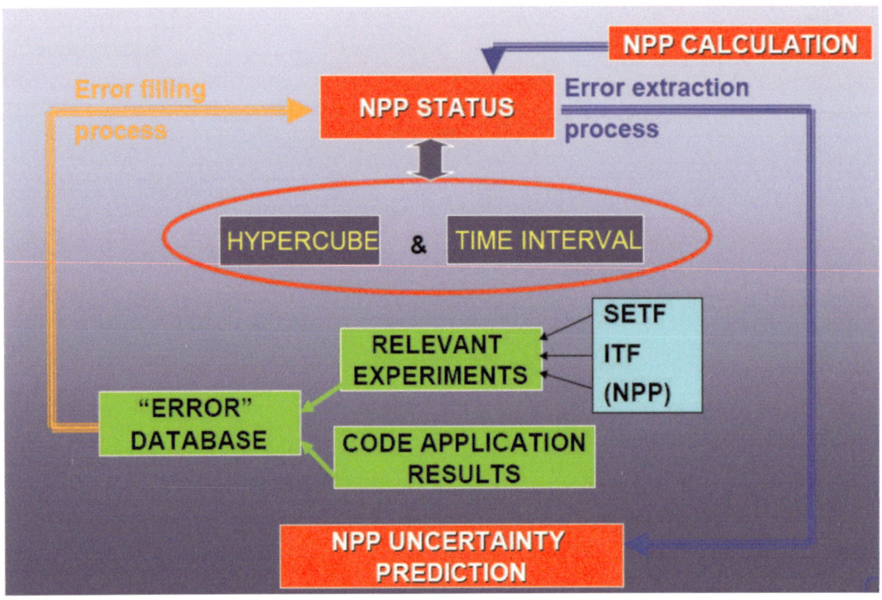

Fig. 5. Outline of the basic idea of the CIAU method.

A hypercube and a time interval characterize a unique plant status to the aim of uncertainty evaluation. All plant statuses are characterized by a matrix of hypercubes and by a vector of time intervals. Let us define Y as a generic thermal-hydraulic code output plotted versus time. Each point of the curve is affected by a quantity uncertainty (Uq) and by a time uncertainty (Ut). Owing to the uncertainty, each point may take any value within the rectangle identified by the quantity and the time uncertainty. The value of uncertainty, corresponding to each edge of the rectangle, can be defined in probabilistic terms. This satisfies the requirement of a 95% probability level to be acceptable to the NRC staff [33] for comparison of best estimate predictions of postulated transients to the licensing limits in 10 CFR (Code of Federal Regulation) Part 50.

The idea at the basis of CIAU can be made more specific as follows: the uncertainty in code prediction is the same for each plant status. A Quantity Uncertainty Matrix (QUM) and a

Time Uncertainty Vector (TUV) can be set up including values of Uq and Ut derived by an uncertainty methodology, Fig. 6.
At the moment the UMAE constitutes the 'engine' for the rotation of the CIAU shaft. The QAM and TAV, respectively Quantity Accuracy Matrix and Time Accuracy Vector in Fig. 6 are derived from an UMAE like process and are the precursor of QUM and TUV. However, within the CIAU framework, any uncertainty method can be used to derive directly QUM and TUV.

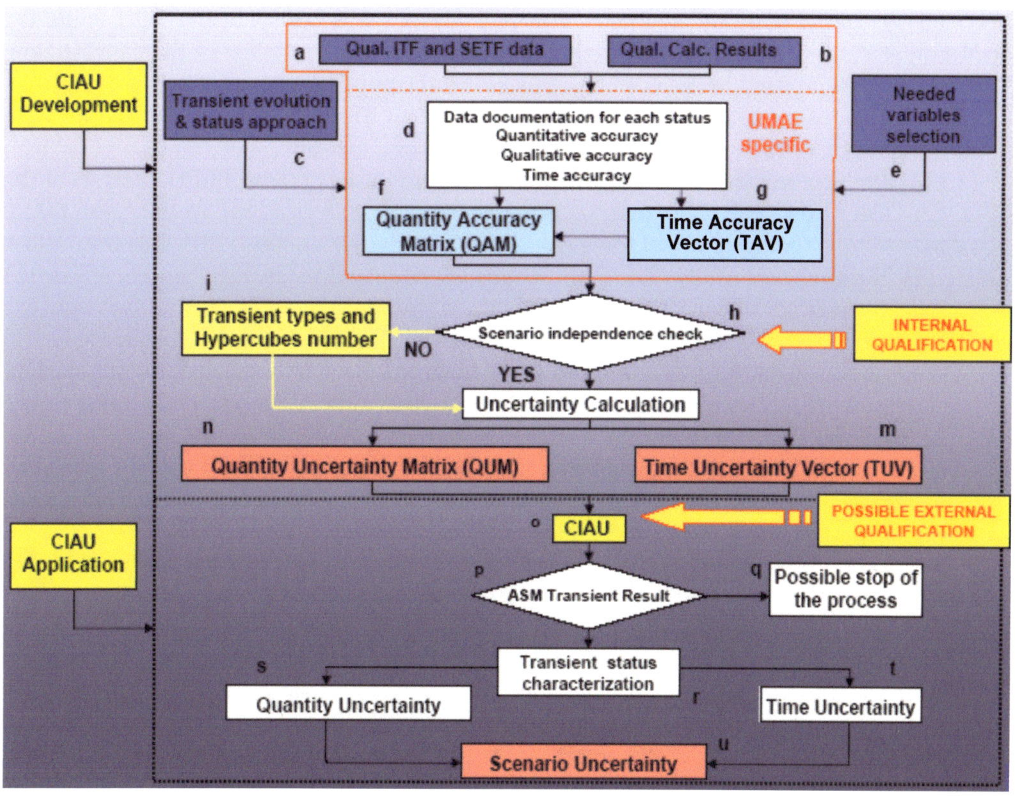

Fig. 6. Flow diagram at the basis of the CIAU methodology

4.2 The CIAU qualification

The UMAE and the CIAU have been used within the OECD/NEA/CSNI international projects UMS [23] and BEMUSE [34, 35 and 36]. The application of CIAU within the licensing process of Angra-2 PWR (a regulatory audit calculation, see later) [37 to 39] and the evaluation of safety of the Kozloduy-3 VVER-440 [40] can be recalled. Sensitivity studies and comparison with results from different uncertainty methods confirmed the qualification level of CIAU in those frameworks. Finally, different internationally available publications have been issued in relation to what is called external qualification process of CIAU, e.g. reference [41].

4.3 Sample applications of CIAU

Application of CIAU method to licensing analyses was performed by GRNSPG/UNIPI in various occasions, and last to the under-construction Atucha-2 NPP. Results of this work could not be included in this Chapter because at the time of writing the review process from the Safety Authority was still ongoing. Hereafter, there are shortly reported hereafter some other examples of CIAU applications produced in the last years.

4.3.1 Uncertainty Analysis of the LBLOCA-DBA of the Angra-2 PWR NPP

Angra-2 is a 4 loop 3765 MWth PWR designed by Siemens KWU. The NPP is owned and operated by the ETN utility in Brazil. The NPP design was ready in the '80s, while the operation start occurred in the year 2000 following about ten-year stop of the construction. The innovation proposed to the licensing process by the applicant consists in the use of a BE tool and methodology to demonstrate the compliance of the NPP safety performance with applicable acceptance criteria set forth in the Brazilian nuclear rule.

The CIAU application was employed for performing an 'independent' BEPU analysis of the LBLOCA-DBA of the NPP. The analysis was classified as 'independent' in the sense that it was carried out by computational tools (code and uncertainty method) different from those utilized by the applicant utility. The main results are summarized in Fig. 7 and 8, where the PCT and the related uncertainty bands obtained through the CIAU and through the computational tools adopted by the applicant, are given. The following comments apply:

- Continuous uncertainty bands have been obtained by CIAU related to rod surface temperature (Fig. 7), pressure and mass inventory in primary system. Only point values for PCT are considered in Fig. 8;
- The CIAU (and the applicant) analysis has been carried out as BE analysis: however, current rules for such analysis might not be free of undue conservatism and the use of peak factors for linear power is the most visible example;
- The conservatism included in the reference input deck constitutes the main reason for getting the 'PCT licensing' from the CIAU application above the acceptability limit of 1200 °C;
- The amplitude of the uncertainty bands is quite similar between the CIAU and the applicant. Discrepancies in the evaluation of the 'PCT licensing' outcome from the way of considering the 'center' of the uncertainty bands. In the case of CIAU, the 'center' of the uncertainty bands is represented by the phenomenological result for PCT obtained by the reference calculation (1100 °C in Fig. 7). In the case of applicant the 'center' of the uncertainty bands is a statistical value obtained from a process where the reference calculation has no role (796 °C in Fig. 8);
- The reference BE PCT calculated by the applicant (result on the left of Fig. 8) plus the calculated uncertainty is lower than the allowed licensing limit of 1473 K;
- The reference BE PCT calculated by CIAU (result on the right of Fig. 8) is higher than the PCT 'proposed' by the applicant and the upper limit for the rod surface temperature even overpasses the allowed licensing limit of 1473 K thus triggering licensing issues;
- Based on the results at the previous point, new evidences from experimental data have been made available by the applicant. This allowed to repeat the BE reference calculation (both for the CIAU and the applicant). The new reference BE PCT calculated by CIAU is lower than the previous (about 200 °C) and close to the new reference PCT calculated by the applicant ('base case' in Fig. 7) and it is shown that the new CIAU upper limit for the rod surface temperature is lower than the allowed licensing limit of 1473 K.

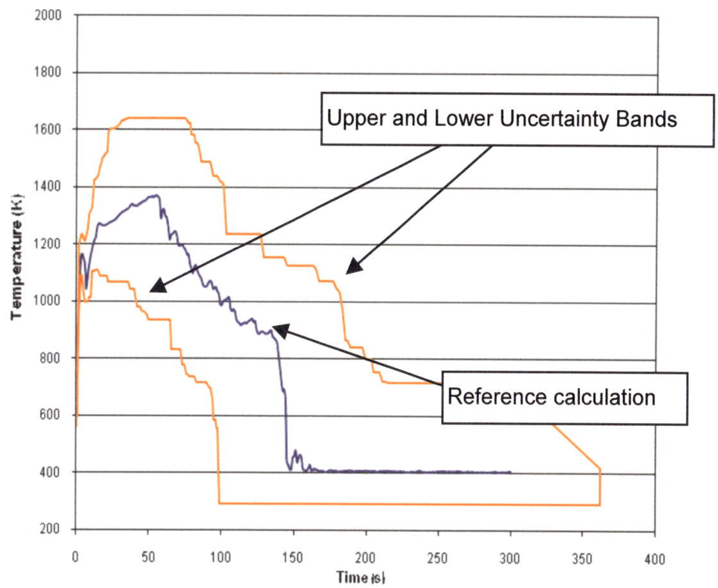

Fig. 7. Uncertainty bands for realistic hot rod

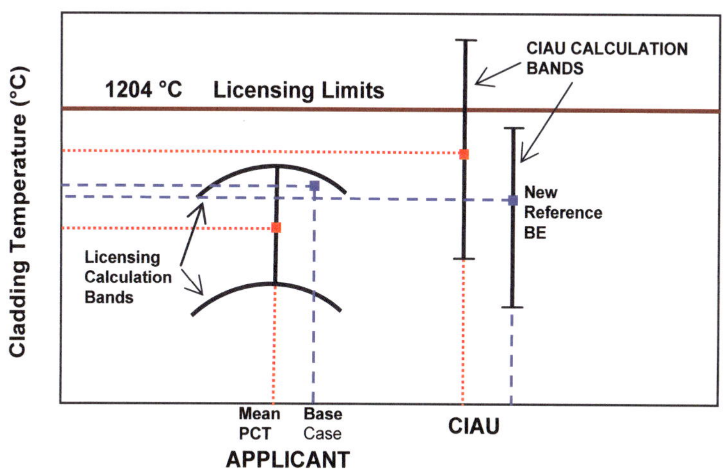

Fig. 8. Final result from the CIAU study and comparison with results of the applicant.

4.3.2 BE and uncertainty evaluation of LBLOCA 500 mm for Kozloduy-3

The analysis of the 'LBLOCA 500 mm' (DEGB in CL) transient [40] was carried out by adopting the Relap5 code. The specific purposes of the analysis included the assessment of the results and the execution of an independent safety analysis supported by uncertainty evaluation. A BE transient prediction of the 'LBLOCA 500 mm' was performed. Evaluation of the uncertainty was performed by CIAU for the RPV upper plenum pressure, the mass inventory in primary system and the hot rod cladding temperature. Only the last parameter is shown in Fig. 9 together with the uncertainty bands. The most relevant result is the demonstration that the PCT in the concerned hot rod is below the licensing limit. In the same Fig. 9, bounding results (PCT and time of quenching) from two conservative calculations (i.e. obtained by a BE code utilizing conservative input assumptions) are given: one is the conservative calculation ('driven' conservatism in Fig. 9) performed by the applicant, the other is the conservative calculation performed by GRNSPG/UNIPI ('rigorous' conservatism in Fig. 9). The following can be noted:

- The 'driven' conservative calculation has been performed by the applicant using a set of values for the selected conservative input parameters different respect to the values adopted in a previous analysis and accepted by the regulatory body;
- The 'driven' conservative calculation is not "conservative" and does not bind entirely the BEPU upper bound. This implies that code uncertainties are not properly accounted for by the adopted conservative input parameter values;
- The 'rigorous' conservative calculation performed by GRNSPG/UNIPI is correctly conservative (i.e. it use the same set of values for the selected conservative input parameters previously licensed), but its conservatism is such to cause PCT above the licensing limit;
- The comparison between the conservative PCT obtained by GRNSPG/UNIPI and the CIAU upper band of the BEPU calculation shows the importance of using a full BE approach with a suitable evaluation of the uncertainty.

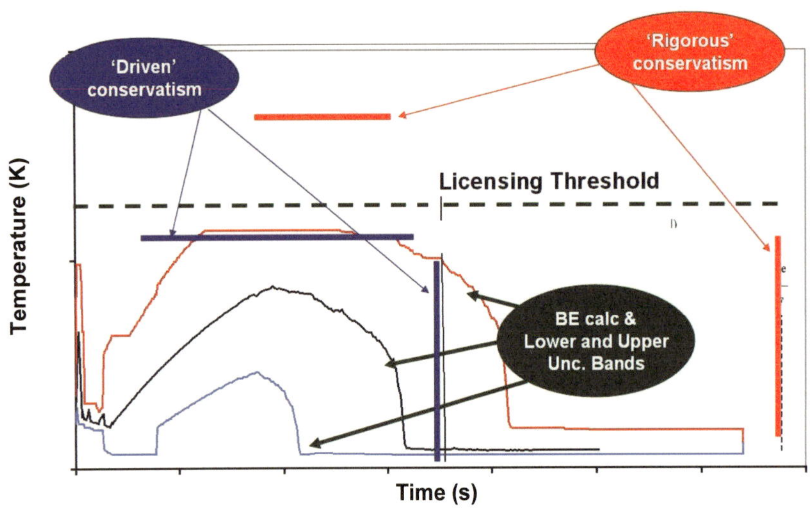

Fig. 9. Surface temperature at the PCT location in hot rod and uncertainty bands derived by CIAU application

4.3.2 The BEMUSE project

The last selected CIAU application constitutes a qualification study that at the same time allows a comparison with results of different uncertainty methods. Within the OECD (Organization for Economic Cooperation and Development) framework, two main activities related with the uncertainty evaluation have been performed (actually the second one is still in progress): the UMS [23] and the BEMUSE, [35], respectively. The objective of the BEMUSE project was to predict the LBLOCA performance of the LOFT experimental nuclear reactor

(i.e. test L2-5). The process included two steps: the derivation of a reference calculation, involving a detailed comparison between experimental and calculated data, and the derivation of uncertainty bands enveloping the reference calculation. The success of the application consisted in demonstrating that the uncertainty bands envelope the experimental data. Ten international groups participated to the activity. A sample result from the BEMUSE project is outlined in Fig. 10.
The application of the CIAU was performed by the GRNSPG/UNIPI (dotted vertical line in Fig. 10) while all other participants used uncertainty methods based on the propagation of the input errors supplemented by the use of the Wilks formula. The consistency between the CIAU results and the experimental data can be observed as well as the spread of results obtained by the other uncertainty methods based on the propagation of the input errors.

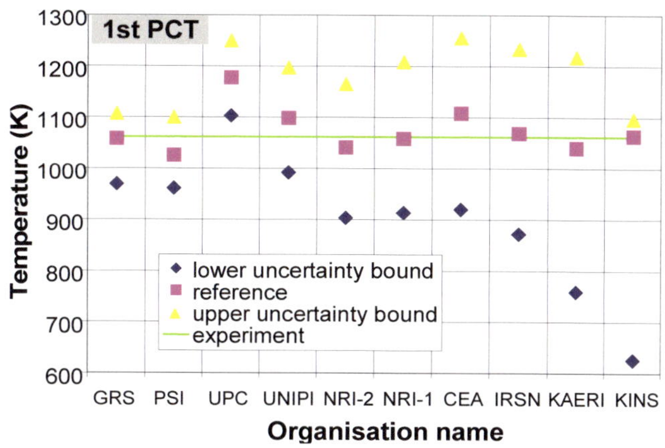

Fig. 10. Uncertainty bounds from each participant ranked by increasing band width from left to right related to the 1st PCT' of the LOFT experiment L2-5.

5. Conclusions

Massive and, at the same time, safe and reliable exploitation of nuclear energy constitutes a great challenge for the today world. Tremendous progresses were obtained during the last 30 years in the field of nuclear reactor simulation thanks to an extended knowledge of several physical phenomena, to the code validation & assessment campaigns and to the impetuous advance of (cheap) computer technology.
This chapter briefly presented the state-of-the-art in the licensing process for nuclear reactors, focusing on the potentialities and the advantages of the so-called BEPU approach. References to works and specific examples of the GRNSPG/UNIPI applications of such methodology to being-built or in operation nuclear power plants were also reported for giving a clearer idea of this sophisticated industrial process.

6. References

[1] INTERNATIONAL ATOMIC ENERGY AGENCY (IAEA), "Nuclear Power Reactors in the World - 2010 Edition", Reference Data Series No. 2, Vienna, 2010
[2] INTERNATIONAL ATOMIC ENERGY AGENCY (IAEA), "Safety of Nuclear Power Plants: Design -Requirements", Safety Standards Series No.NS-R-1, IAEA, Vienna, 2000

[3] INTERNATIONAL ATOMIC ENERGY AGENCY (IAEA), "Safety Standards, Fundamental Safety Principles" - IAEA SF-1, Vienna (A), 2006
[4] INTERNATIONAL ATOMIC ENERGY AGENCY (IAEA), "Safety Assessment and Verification for Nuclear Power Plants" - IAEA Safety Guide No. NS-G-1.2, Vienna, 2001
[5] USNRC NUREG-0800, "Standard Review Plan for the Review of Safety Analysis Reports for Nuclear Power Plants, LWR Edition" March 2007
[6] USNRC Regulatory Guide 1.70: "Standard Format and Content of Safety Analysis Reports for Nuclear Power Plants LWR Edition", Revision 3, November 1978.
[7] INTERNATIONAL ATOMIC ENERGY AGENCY (IAEA), "Format and Content of the Safety Analysis Report for Nuclear Power Plants" - IAEA Safety Guide No. GS-G-4.1, Vienna, (2004).
[8] INTERNATIONAL ATOMIC ENERGY AGENCY (IAEA), "Deterministic Safety Analysis for Nuclear Power Plants", Specific Safety Guide No SSG-2, IAEA 2010
[9] USNRC "Transient and Accident Analysis Methods", Regulatory Guide 1.203, 2005
[10] C. Camargo, R. Galetti, F. D'Auria, O. Mazzantini, "Exploitation of BEPU Approach for Licensing Purposes", ENS, TOPSAFE Dubrovnik, Croatia, 30.09 - 3.10.2008
[11] GRNSPG/UNIPI, "A Proposal For Performing The Atucha II Accident Analyses For Licensing Purposes, The BEPU Report", Internal Report, 2008
[12] INTERNATIONAL ATOMIC ENERGY AGENCY (IAEA), "Format and Content of the Safety Analysis Report for Nuclear Power Plants" - IAEA Safety Guide No. GS-G-4.1, Vienna, (2004)
[13] M. Bonuccelli, F. D'Auria, N. Debrecin. G.M. Galassi, "A Methodology for the qualification of thermal-hydraulic codes nodalizations", NURETH-6 Int. Conf. Grenoble (F) October 5-8 1993
[14] F. D'Auria, M. Leonardi, R. Pochard, "Methodology for the evaluation of Thermal-hydraulic Codes Accuracy", Int. Conf. On New Trends in Nuclear System Thermal-hydraulics - Pisa (I), May 30-June 2, 1994
[15] D'Auria F., Galassi G.M., "Code Validation and Uncertainties in System Thermal-hydraulics", J. Progress in Nuclear Energy, Vol. 33 No 1/2, pp 175-216, 1998
[16] A. Petruzzi, F. D'Auria, and W. Giannotti, "Description of the procedure to qualify the nodalization and to analyze the code results," DIMNP NT 557(05), May, 2005
[17] Petruzzi A., D'Auria F., "Thermal-Hydraulic System Codes in Nuclear Reactor Safety and Qualification Procedures", Hindawi Publishing Corporation Science and Technology of Nuclear Installations Volume 2008, Article ID 460795
[18] OECD/NEA, "Neutronics/Thermal-hydraulic Coupling in LWR Technology - Vol.3: Achievement and Recommendation Report", ISBN92-64-02085-3, © OECD 2004
[19] B. Ivanov, K. Ivanov, P. Grudev, M. Pavlova, V. Hadjiev, "V1000CT-1: Main Coolant Pump (MCP) Switching On", NEA/NSC/DOC(2002)6, November 2004
[20] F. Moretti, D. Melideo, D. D'Auria, T. Hohne, S. Kliem, "CFX Simulations of ROCOM Slug Mixing Experiments"
[21] K. Ivanov, M. Avramova, S. Kamerow, I. Kodeli, and E. Sartori, "Overview and Discussion of Phase I of the OECD LWR UAM Benchmark Activity", Invited paper for M&C 2009, Saratoga Springs, New York, May 3-7, 2009
[22] USNRC NUREG/CR-5249, "Quantifying Reactor Safety Margins: Application of Code Scaling, Applicability, and Uncertainty Evaluation Methodology to a Large Break Loss of Coolant Accident," December 1989.

[23] T. Wickett (Editor), F. D'Auria, H. Glaeser, E. Chojnacki, C. Lage (Lead Authors), D. Sweet, A.Neil, G.M. Galassi, S. Belsito, M. Ingegneri, P. Gatta, T. Skorek, E. Hofer, M. Kloos, M. Ounsy, J.I. Sanchez - "Report of the Uncertainty Method Study for advanced best estimate thermal-hydraulic code applications", Vols. I and II OECD/CSNI Report NEA/CSNI R (97) 35, Paris (F), June 1998.

[24] A. Petruzzi and F. D'Auria, "Accuracy quantification: description of the fast fourier transform based method (FFTBM)," DIMNP NT 556(05), May, 2005

[25] F. D'Auria, N. Debrecin, and G.M. Galassi, "Outline of the uncertainty methodology based on accuracy extrapolation," *Nuclear Technology*, vol. 109, no. 1, pp. 21–38, 1994

[26] Ambrosini W., Bovalini R., D'Auria F. - "Evaluation of Accuracy of Thermalhydraulic Codes Calculations", J. Energia Nucleare, Vol. 7 N. 2, May 1990

[27] Mavko B., Prosek A., D'Auria F. - "Determination of code accuracy in predicting small break LOCA experiment", J. Nuclear Technology, Vol. 120, pages 1-18, October 1997

[28] Bovalini R. D'Auria F. - "Scaling of the accuracy of Relap5/mod2 Code", J. Nuclear Engineering & Design, vol. 139, No. 2, pages 187-204, 1993

[29] Bovalini R., D'Auria F., Galassi G.M. - "Scaling of Complex Phenomena in System Thermal-hydraulics", J. Nuclear Science and Engineering, Vol. 115, pages 89-111, October 1993

[30] S. N. Aksan, F. D'Auria, H. Staedtke, "User Effects on the Thermal-hydraulic Transient System Codes Calculations". *Nucl. Eng. Des.*, 145, 1&2, (1993), OECD/CSNI Report NEA/CSNI R (94) 35, Paris (F), January 1995

[31] USNRC and OECD/CSNI, "Proceedings of the OECD/CSNI Workshop on Transient Thermalhydraulic and Neutronic Codes Requirements held in Annapolis", US NRC NUREG/CP-0159 and OECD/CSNI Report NEA/CSNI R(97)4, Washington, DC, USA.

[32] D'Auria F., Giannotti W., "Development of a Code with the Capability of Internal Assessment of Uncertainty", Nuclear Technology, Vol. 131, pp 159-196, 2000.

[33] USNRC Regulatory Guide 1.157, "Best-Estimate Calculation of Emergency Core Cooling System Performance", May 1989

[34] Petruzzi A., D'Auria F., Micaelli J-C., De Crecy A., Royen J. - "The BEMUSE programme (Best-Estimate Methods - Uncertainty and Sensitivity Evaluation)" Int. Meet. on Best-Estimate Methods in Nuclear Installation Safety Analysis (BE-2004) IX, Washington D.C. (US), Nov. 14-18, 2004, ANS copyright © 2004.

[35] A. Petruzzi, F. D'Auria, A. De Crecy, et al., "BEMUSE Programme. Phase 2 report (Re-Analysis of the ISP-13 Exercise, post test analysis of the LOFT L2-5 experiment)," OECD/CSNI Report NEA/CSNI/R(2006)2 - Issued June, 2006, JT03210882, Paris, France, OECD 2006, pages 1–625.

[36] OECD NEA/CSNI - "BEMUSE Phase III Report - Uncertainty and Sensitivity Analysis of the LOFT L2-5 Test", NEA/CSNI/R(2007)4, May 2007.

[37] Galetti M.R., D'Auria F., Camargo C., "Questions arising from the application of Best-Estimate Methods to the Angra 2 NPP Licensing Process in Brazil," International Meeting on "Best Estimate" Methods in Nuclear Installation Safety Analysis, Washington, DC, November, 2000.

[38] Galetti M.R., Galassi, G.M., "Angra 2 NPP Licensing - Regulatory Independent Calculation - Description of the process to achieve the LBLOCA reference case with Relap5/Mod3.2.2 code", NT-CODRE-Nº06/01, May 2001, CNEN, RESTRICTED.
[39] D'Auria F., Galassi G.M., "Best-Estimate Analysis and Uncertainty Evaluation of the Angra 2 LBLOCA DBA", University of Pisa Report DIMNP NT-433(01)-rev.1, July 2001, RESTRICTED
[40] D'Auria F., Galassi G. M., Giannotti W., Stanev I. - "Kozloduy 3 NPP large break loss of coolant accident sensitivity study", Int. Meet. Nuclear Power in Eastern Europe: Safety, European Integration, Free Electricity Market, with the Tenth Annual Conf. of The Bulgarian Nuclear Society, Varna (Bg) June 17-20, 2001.
[41] Borges R.C., D'Auria F., Alvim A.C.M. - "Independent qualification of the CIAU tool based on the uncertainty estimate in the prediction of Angra 1 NPP inadvertent load rejection transient", ASME-JSME Int. Conf. on Nuclear Engineering (ICONE-10) - Arlington (VA, USA), April 14-18 2002 (ICONE10-22135), CD Proceedings Copyright © 2002 by ASME.

12

A Review on Oscillatory Problems in Francis Turbines

Hermod Brekke
Profesor Emeritus,Norwegian Universityof Science and Technology, Trondheim Norway

1. Introduction

There is a worldwide requirement of hydropower turbines for peak load operation at increasing heads and for low head turbines operating with large head variations.
The reason for this requirement is the increasing demand for clean peak power production caused by the increasing number of wind mills and to reduce the peak power production from thermal power plants that is increasing the polluting production of CO_2.
In addition to this the weight of the turbines versus power has decreased by changing from cast steel to welded structures of high tensile strength steel plates. Further the flow velocity in the turbines has increased and thus increasing the danger of pressure pulsations, vibrations and noise in the power house. Fatigue problems caused by material defects also require a thoroughly inspection and regular maintenance of the machines.
This chapter is aimed towards three main targets:

- To avoid not acceptable noise and fatigue problems from high frequency pressure pulsations caused by blade passing through the guide vane wakes and vortex shedding from the trailing edges of vanes.
- To improve the design and analysis of steady oscillatory problems of hydro turbines during operation at off best point load.
- Present a reliable analysis of surge problems that may cause power oscillations in the grid.

2. Historical development of turbines after World War II

2.1 More powerful turbines and weight reduction

The traditional way of producing high head turbines was using steel castings in the pressure carrying part like spiral casings, head cover and bottom cover. Then the turbines became very heavy requiring a labor consuming production in foundries and workshops. Because of relatively small units this design was used up to around 1960 for the pressure carrying parts.
The reason for the possibility of a weight reduction to around 1/3 for the spiral casings, as illustrated in fig. 1, was the improved welding technology and the introduction of high tensile stress plates with yield point of 460 MPa.
The turbines illustrated in fig. 2 gives a clear indication of the weight reduction of high head Francis turbines. The main reason for the weight reduction was saved production cost caused by less man hours to produce the turbines. In addition the transport costs were reduced.

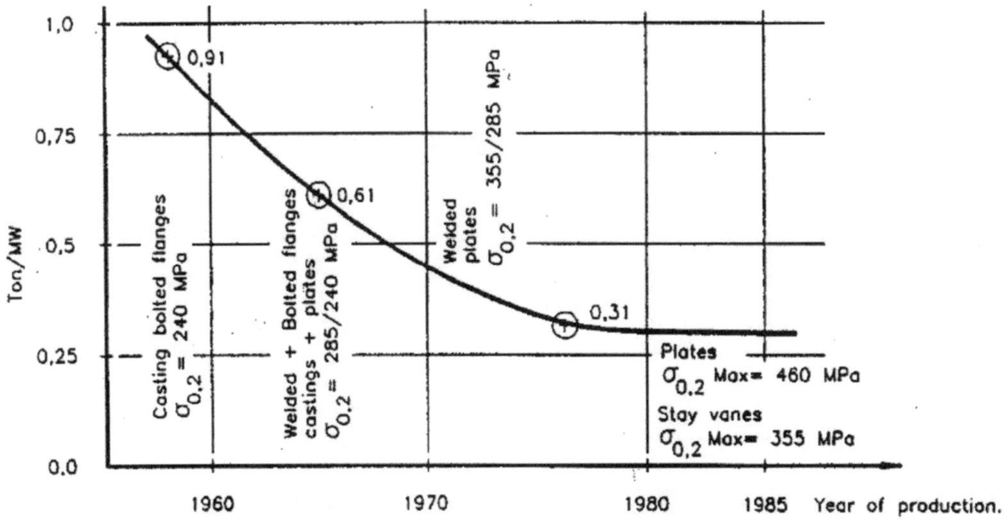

Fig. 1. Weight reduction of high head Francis turbines produced by KVAERNER from 1957 to 1985.

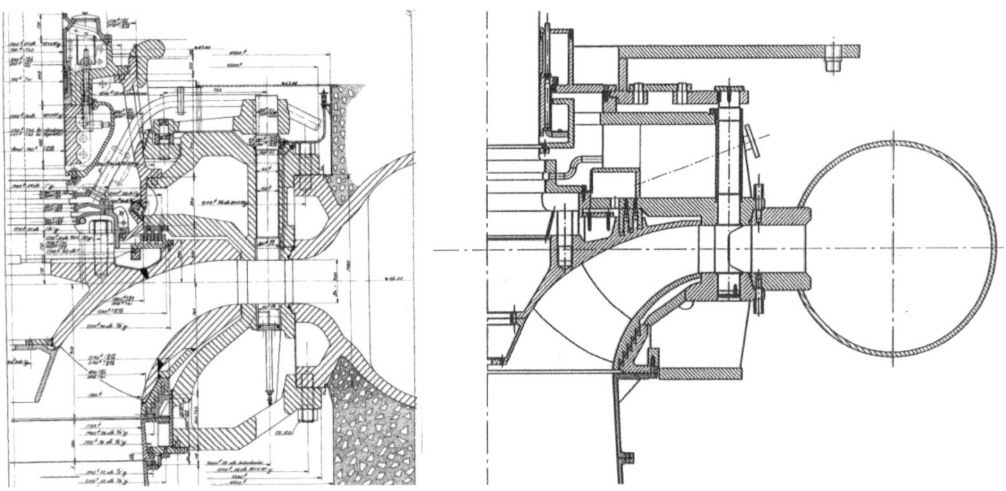

Fig. 2. A traditional high turbine for Tokke power plant made in 1959 (left) and a modern welded turbine of plate steel with yield point 460 MPa.

The draw back with the lighter turbines was that a lighter turbine gives less damping of pressure pulsations and the higher stresses required a more strict production control to reduce the size of material defects and geometrical inaccuracy. Also for Pelton turbines the design was changed from heavy casted bends and bifurcations to welded parts of rolled plates. In addition multi nozzle vertical units were introduced, mainly because the power for each unit increased from around 30 - 50 MW to 100- 315 MW in the period from 1957 to 1985. This development gave a weight reduction per kW similar to the high head Francis turbines.

However, this chapter is aiming at oscillatory problems in Francis turbines and a further description of the general trend in turbine design will not be included.

Another challenge for operation of the hydro turbines is the requirement for peak load operation from no load to overload in some cases. Further the growing demand for peak load support from the growing number of wind mills and the need for support to the coal fired power plants in order to run these power plants on planned more or less constant load setting in order to reduce the pollution from CO_2 production.
The next section in this chapter of the book will be aimed at dynamic high frequency problems in runners and adjacent components affected by high frequency pressure pulsations and draft tube surge problems at off best point operation.
Finally problems related to water hammer and pressure surges from draft tube voids acting as surge tanks will be described with examples of power oscillations with possible resonance from the electric system.

3. Dynamic problems of Francis turbines in the high frequency domain

3.1 The blade passing frequency

The dominating sound from a high head Francis turbine without cavitation problems is the blade passing frequency.
This sound is caused by the rotating pressure waves travelling on the outside of the runner in the water between head cover and bottom cover and through the guide vane cascade, stay vanes and spiral casing. These travelling pressure waves are caused by the passing of the runner blades through the wakes trailing from the guide vanes outlets. The frequency of these blade passings is the runner speed multiplied by the number of runner blades at the inlet i.e. including the splitter blades if the runner is furnished with splitter blades. However, inside the runner in the runner channel, the frequency of the pressure shocks will be the speed of the runner multiplied by the number of guide vanes.

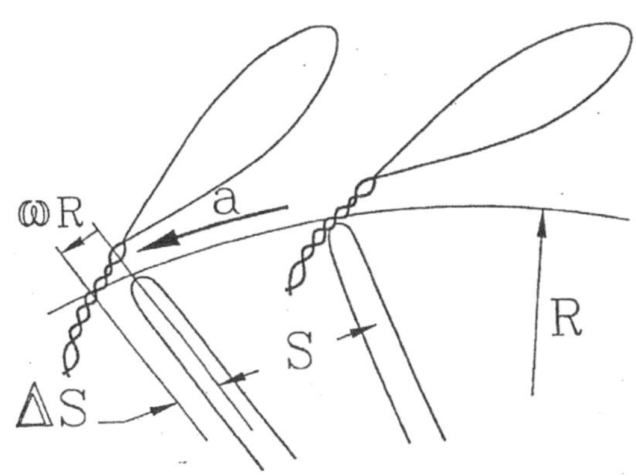

Fig. 1. Illustration of the blade passing of runner blades including splitter blades in a high head Francis turbine. (ω = rad/sec = angular speed)

The described "pressure shock" waves from the runner blades passing through the wakes from the guide vanes in a Francis turbine, may cause vibration and noise from the turbine above acceptable limits in special cases.
Several strain gauge measurements on runner blades in the field and pressure measurements on model turbine blades of high head Francis turbines have also proven the pressure pulsations in the blade channels.

In some cases these high frequency pressure pulsations have caused blade cracking after a short time in operation of a week or so depending on the speed of the turbine. The reason for this has normally been the shape and thickness of the blades and weld defects.
In special cases that will be described in the following, the reason has been very high pressure pulsations caused by a certain combination of runner blades and guide vanes numbers.
It should also be mentioned that resonance with the natural frequency or eigenvalues of the runner is not the reason for the problem of noise and or blade cracking of high head Francis turbines. This will be analyzed and explained later in this section. However, for low head Francis turbine there is normally not observed a similar dominating pressure pulsation of the blade passing frequency inside the runner.
The reason for this is a much larger distance from the guide vanes to the runner blade inlets at the crown compared to the small distance to the runner blades close to the band. The natural frequency of the low head runners may also in some cases be in resonance with the low frequency pressure surges in the draft tube with frequency around 1/3 of the runner speed.
However, by studying fig. 1, valid for high head runners we may find an interaction and amplification between the pressure shock wave from the regarded blade and the blade running in front of it, if the shock propagation speed in water reaches the blade in front of the regarded blade, simultaneously with the passing of this blade through the next guide vane wake.
Then an interaction and amplification of the high frequency pressure pulsations occurs as expressed by fulfillment of following equation and illustrated as in fig. 1:

$$\mathbf{R}\frac{\omega}{\Delta S} = a / s \tag{1}$$

In fig.1 and eq.(1), **S** = distance between the runner blades along the runner rim at runner inlet, **a** = wave propagation speed in water and **ωR** = rim speed of the runner blade inlets (**R** = radius of runner blades at the inlets and ω = angular speed).
In fig. 2 is shown the high head Francis turbine at Hemsil I in Norway. The shock propagating speed in the water was calculated to be around a = 900 m/sec taking into account the flexibility of the head and bottom cover. The technical specification for the turbine was Hn = 510 m, Pn = 35.7 MW and speed n = 750 rpm.
The number of guide vanes in the turbine was 28 and the first runner had 30 blades at the inlet. (15 splitter blades and 15 full length blades) and then an interaction of the pressure waves could be proven by a serious noise problem with 120 dB in the air surrounding of the turbine. Because of the high noise level and blade cracking the pressure pulsations were measures between the band at the runner inlet and the bottom cover as illustrated in fig.2.
The pressure pulsations had amplitudes peak to peak of 45 m with a frequency of number of runner blades of 30 multiplied by the speed per sec. = 375 Hz.
The runner with 30 blades was exchanged with a new runner with similar geometry, but with 16+16 = 32 blades and the noise level dropped to around 80-85 dB which is very good for a high head turbine operating at 510 m net head.
Later several measurements of stress amplitudes at the runner blade outlets of high head turbines have been made, mainly because of blade cracking caused by weld defects and residual stresses in the stainless composition of the weld deposits between blade and band or crown. Some blade cracking problems occurred around 1968 – 1970 because of unstable Austenite in the weld composite of the stainless 16% Cr 6% Ni alloy which was used. However, this problem was solved around 1970.

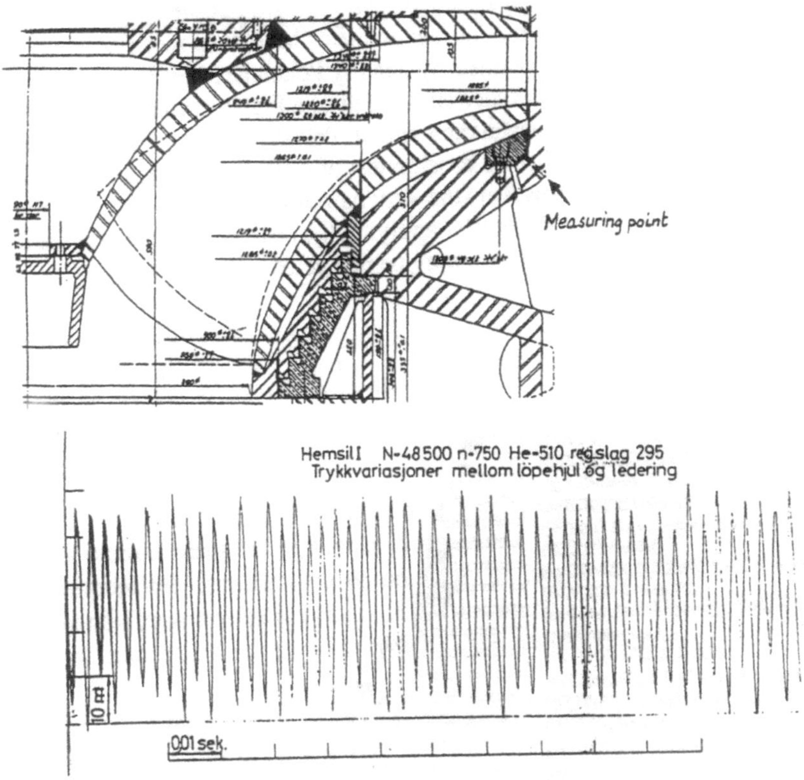

Fig. 2. Cross section of the runner at Hemsil I and the recorded pressure oscillations between runner and bottom cover..

However, the number of runner blades and guide vanes was carefully chosen to avoid interference as described for Hemsil I. An alternative to 32 runner blades and 28 guide vanes was 30 runner blades and 24 guide vanes which also gave a smooth running for the low specific speed Francis turbines operating at heads exceeding 300 m and sometimes exceeding 600 m.

In fig. 3 to the right the relative values of the measured static stress and stress amplitudes peak to peak at pressure side and suction side of the blade outlet of the runner at Tafjord is shown. In this turbine the number of guide vanes was 24 and number of runner blades was 30 and no interference with the runner blade passing the guide vane wakes, caused not acceptable noise. However, high residual stresses in the welds caused blade cracking also because of the pressure pulsations inside the runner in the blade channels even if these pulsations were moderate. The technical data for this turbine is: Pn = 70 MW, Hn = 420 m and speedn = 500 rpm.

To the right in fig. 3 the stress amplitudes from a similar measurement at one of the four turbines at Tonstad, is shown. The number of guide vanes and runner blades are the same as for Tafjord i.e. 24 guide vanes and 30 runner blades. The technical data for this turbine is: Pn = 165 MW, Hn = 430 m and speed n = 375 rpm. Note also the superimposed small stress amplitudes caused by the pressure and stress oscillations from the “wake passing” of the 5 blades in between the next guide vane passing of the regarded blade measured with strain gauges.

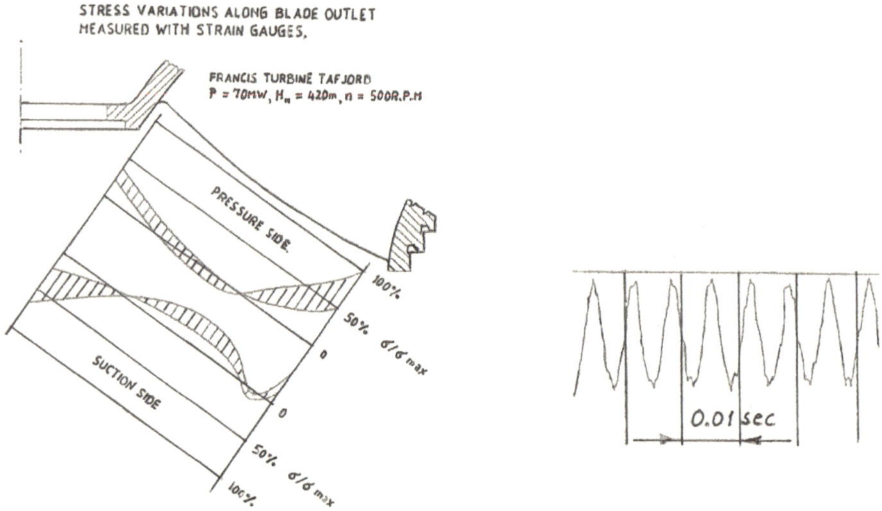

Fig. 3. Static stress and stress amplitudes measured along the outlet edge on pressure side and suction side of a runner blade of the turbine at Tafjord (left) and an illustration of the recorded stress amplitudes from a similar measurement at the blade outlet at one of the turbines at Tonstad. (Made in 1969.)

In fig. 4 is shown the result from a measurement of the stresses amplitudes at the blade outlet of a new runner installed in one of the turbines at Tonstad in 2004.

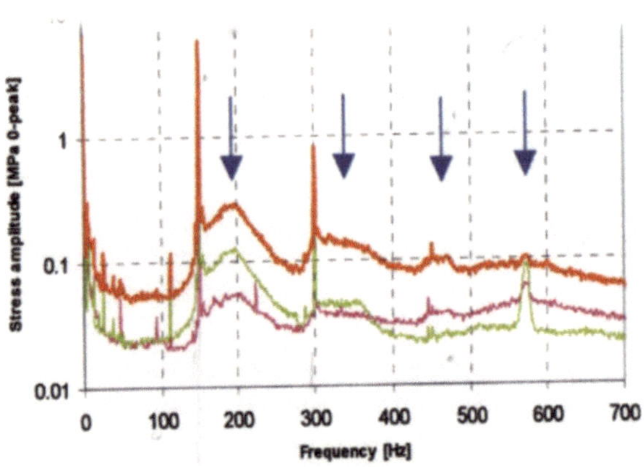

Fig. 4. Illustration of relative stress amplitudes measured at the blade outlets on a turbine at Tonstad compared with calculated natural frequencies marked with blue arrows. Number of guide vanes =24 and runner blades =30 (15 + 15) Speed =375 rpm.(REF. NORCONSULT, GE-KVAERNER)

The blade passing frequency of this turbine was 24*375/60=150Hz in the runner channels creating the main stress amplitudes as illustrated as in fig. 4.

The natural frequency of the runner was also calculated and shown with blue arrows in fig. 4. This measurement proves the statement that the stress amplitudes in the blades of low specific speed turbines will get the frequency of the blade passing and with a negligible influence from the natural frequency (eigenvalues) of the structure.

3.2 Vortex shedding and noise from the trailing edge of vanes

Vortex sheading from stay vanes and guide vanes in a Francis turbine may create vibrations and not acceptable noise which may cause problems for the operation of the turbines which will be discussed in this section.

The result from a basic study of the so called Von Kármán's vortex shedding, presented more than 50 years ago in Transaction of ASME in 1952 and Engineering of Power 1956 and 1960 **/Ref. 1/**, was used for reshaping the outlet of the fixed guide vanes in a small Francis turbine in Norway with noise problems shown in Fig. 6. The result of the research work presented in ASME **/Ref. 1/**, is illustrated in fig. 5.

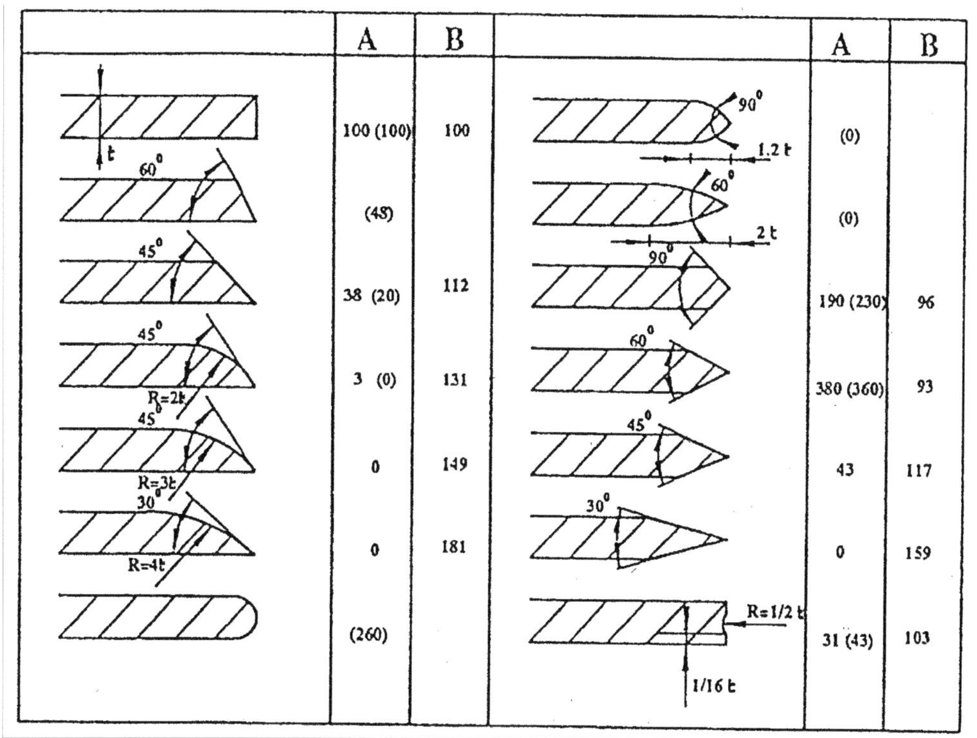

Fig. 5. Influence on frequency and amplitudes by the shape of the outlet edge of plates compared to a straight cut plate. A=relative amplitudes, B=relative frequency./Ref.1/

From the experiments presented in fig. 1 and described in **/Ref. 1/**, the frequency of the vortex shedding can be calculated by following equation:

$$f = 190 * \frac{B}{100} * \frac{C}{t + 0.56} \quad \text{(Hz)} \qquad (2)$$

The value = **B** in the equation can be taken from the table in **fig.5** and the relative amplitude = **A** can also be taken from **fig. 5**. Further: **C** = velocity of water (m/sec) at the outlet of the plate and **t** = thickness of the plate (mm). Following conclusion could be drawn by studying this work:

The outlet edge of all stay vane, guide vanes and runner blades in a turbine should have a skewed cut with an angle smaller or equal to 45 deg. measured relatively to the pressure side

of the vanes. However, a sharp corner should be avoided on the edge in contact with the suction side of the next guide vane in turbines with movable guide vanes, and a 1-3 mm wide skew cut should be made parallel to the suction side of the next guide vane side along the outlet edge for sealing in closed position. The width will depend on the size of the turbine.
The vortex shedding problem of the previously mentioned small turbine in Norway shown in fig 6, was clearly demonstrated by the noise that was finally cured by reshaping the outlet of the fixed guide vanes which also corrected the power of the turbine.

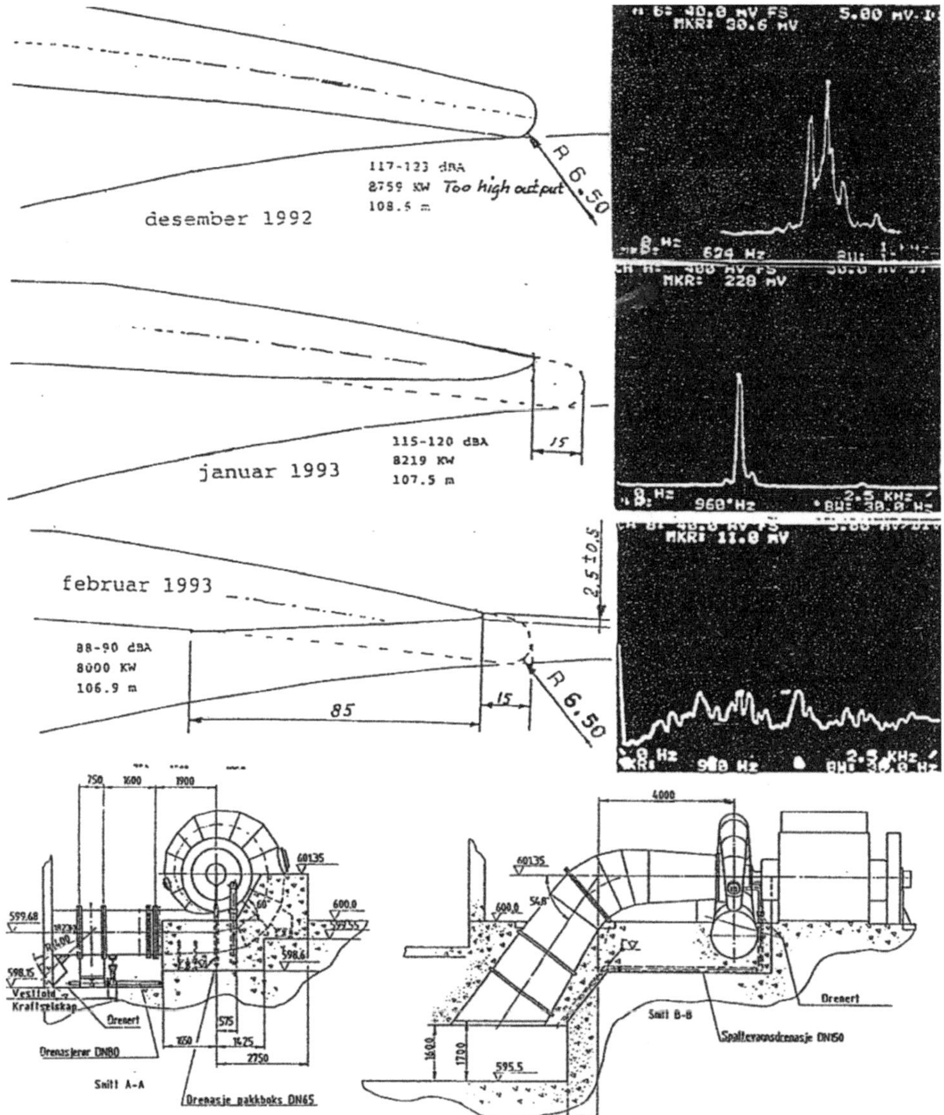

Fig. 6. Original shape and modified shape of stay vanes of a Francis turbine without movable guide vanes, but with stay vanes shaped for the design power with angle and outlet position similar to possible guide vanes. The reason for the illustrated modification was a very high noice and too high power of the turbine with the original vanes.

The original shape of the outlet edge of guide vanes was rounded which is the worst possible shape seen from the fluid machanical side. However, from the structural side the rounded outlet edge of these guide vanes which also were part of the pressure carrying structure of the turbine seemed to be a reasonable choice. (It should also be mention that a careful stress analysis had to be made for approval of the modificatione of the outlet edges.)
The history for the improvement of noice reduction in this turbine was as follows. (see Fig. 6).
The original shape with semicircle outlet profile:
Noise: 117-123 dBA, Frequency ω=624 Hz, Power: P=8759 kW, Net operational head: Hn=108.5 m.
After the first modification with skeewed cross section with a smaller flow angle towards a decreasing radius ending with a tangen in 90 degrees to the pressure side of the vane:
Noise :115-120 dBA, Frequency ω=960 Hz, Power: P=8219 kW, Net operationalhead: Hn=107.5 m.
After the third and final modification with a skewed profile in approximately 30 degree towards a slighly rounded profile towards a stright cut of 2.5 mm thickness following improvement was obtained: (See fig 6. bottom)
Noise: 88-90 dBA, Frequency ω=980 Hz, Power: P=8000 kW, Net operational test head: Hn=106.9
By this last modification the problem was solved also for the corrected power of 8000 kW.

4. Off best point operation of Francis turbines with low frequency draft tube pressure pulsations

4.1 Influence from the runner design on part load pressure surges.

The dynamic behaviour at part load has been a major problem for low head and medium head Francis turbines. The main reason for this is unstable swirl flow in the draft tube which is normally caused by the cross flow from crown towards the band on the pressure side of the runner blades. In fig. 7 the so called swirling rope at part load and the contra rotating centre void at full load are illustrated.

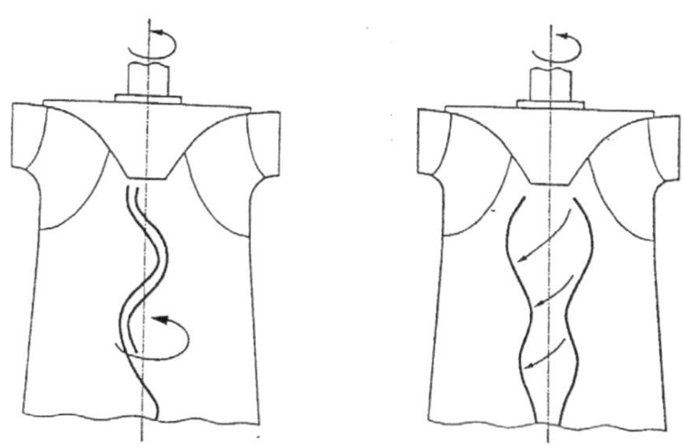

Fig. 7. Swirling part load draft tube rope that may cause unstable conditions and blade cracking (left) and rotating centre void at full load and overload (right).

A theoretical study of the flow in the runner blade channels regarding the cross flow from hub to band, has proven that it is possible to reduce this unfavourable cross flow by a negative blade lean at the inlet of the runner. Further the pressure distribution on the pressure side of the blades must be balanced all the way towards the blade outlet by adjusting the blade lean. This philosophy must be used during creation of a new runner.

The described method illustrated by the equations in fig. 8 bottom, is based on potential flow analysis which is valid for a runner with infinite number of blades. However, the

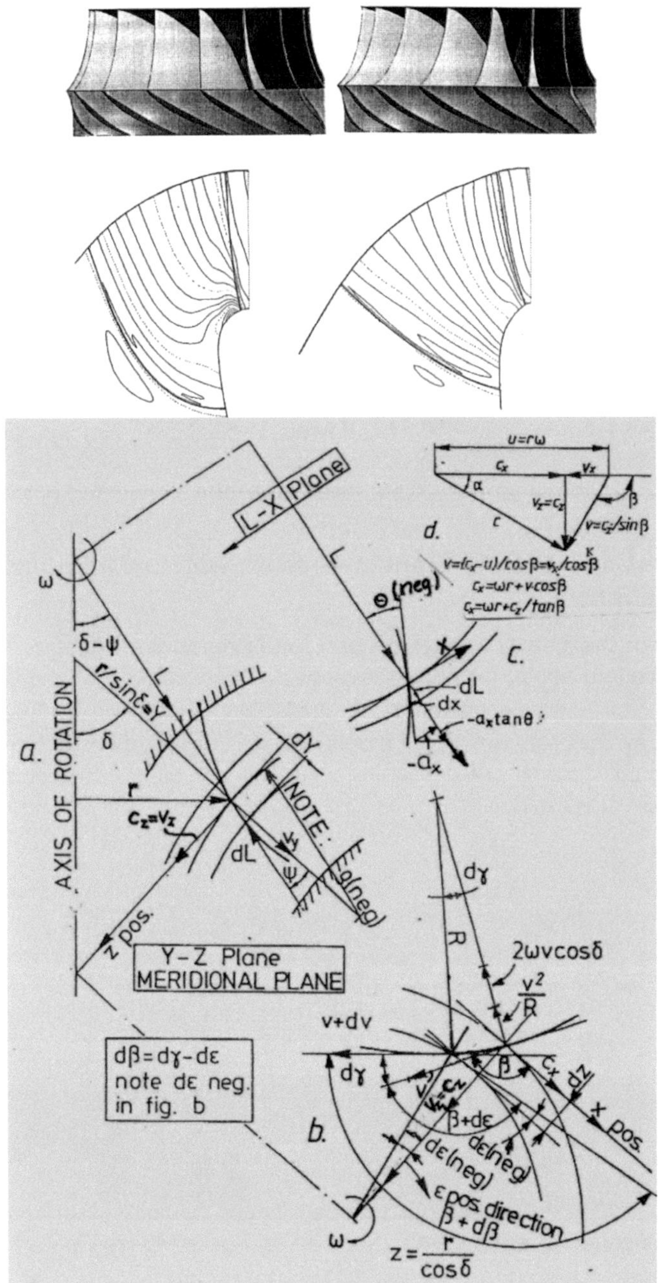

Fig. 8. Pressure distribution of a traditional low head runner (top left), an X-BLADE runner with the same specific speed (top right) and definition of geometry for a potential flow analysis for creation of the blade shape (bottom). Note the difference in pressure towards the band of the traditional runner compared with the X-Blade runner. (Top, left and right.)

following CFD analyses of runners designed by this method, has proven that the pre designed runner geometry gives an excellent starting point for creating a high efficiency runner with stable flow in the draft tube also at part load.
The most famous runners made by this method were the **X-Blade** runners for **Three Gorges Power Plant** in China. These runners operate from 61 m to 113 m net head with a non - restricted range of operation and no operating problems have occurred for the X-Blade runners according to reports. The blade lean angle at the inlet of a pressure balanced so called **X-Blade** runner, will be negative, but the blade must be twisted towards the outlet the other way to an angle similar to a traditional runner so the inlet edge and outlet edge forms an X when looking through the runner from inlet to outlet and this is the reason for the name X-Blade runner.
The blade lean angle for a runner blade is defined by the angle between the blade and a meridian plane perpendicular to the stream surface through the axis of rotation, denoted by Θ as illustrated in fig.8 (bottom) together with the other parameters used in the equations (3), (4) and (5). The dimensionless pressure gradient $d\underline{h}/dy = d(h/H_n)/dy$, perpendicular to the stream surfaces can be expressed by equation (3) as a function of the dimensionless meridian velocity component $\underline{C}_z = \underline{V}_z = C_z/(2gH_n)^{0.5}$ and the angular speed $\underline{\omega} = \omega/(2gH_n)^{0.5}$ (m^{-1}). The directions **x, y** and **z** are defined in fig.8 (bottom). The equation for the pressure gradient **($d\underline{h}/dy$)** is valid along the blade surface from the runner crown to band and is a function of $\underline{C}_z$ and the blade lean angle Θ as expressed in equation 3.

$$\begin{aligned} d\underline{h}/dy = {} & 2[[(1/R - \cos\delta\cos^3\beta/r)(\underline{C}_z{}^2/\sin^3\beta) - \underline{C}_z(\delta\underline{C}_z/\delta z)/\tan\beta + 2\underline{\omega}\cos\delta\underline{C}_z]\tan\Theta \\ & +(\sin\delta/(r\tan^2\beta) + 1/\rho)\underline{C}_z{}^2 + (2\underline{\omega}\sin\delta/\tan\beta)\underline{C}z + \underline{\omega}^2 r\sin\delta] \end{aligned} \tag{3}$$

For a pressure balanced blade **dh/dy = 0** when the blade lean angle Θ is adjusted correctly as illustrated by the X-blade runner in **fig. 8** to the right on top.
Equation (3) is as describer above based on the equilibrium of forces and is two dimensional, valid for a runner with infinite number of blades i.e. potential flow.
In addition to equation (3) **the ROTHALPY equation (4),** expressing the hydraulic pressure along a stream line, must be established together with the **equation of continuity (5).** The presented theory is as mentioned two dimensional (i.e. potential theory valid for an infinite number of blades). However, this theory is still useful in order to obtain a physical understanding of the quantitative influence of the blade lean during the basic design of a Francis runner. The final detailed shaping will be made by a full 3D viscous analysis. By making this simplified study, it quite clear that a negative blade lean on the inlet of the runner was necessary in order to reduce the cross flow on the pressure side of the blade and the **X-BLADE** runner was a result of this study in 1996 at turbine manufacturer KVAERNER where the author of this chapter was working part time as technical advisor**./Ref.2/.**

$$\underline{h} = \underline{\omega}^2 r^2 - \underline{C}_z{}^2/\sin^2\beta + (1 - 2\underline{U}_1\underline{C}_{u1}) - J \quad \text{(J is estimated loss along a stream line.)} \tag{4}$$

In addition to eq. (3) and eq.(4) the equation of continuity in the channel between two stream surfaces is used to find the meridian velocity C_z. **(J = 0.01-0.02)**

$$\underline{C}_z = (Q/N)/(2\pi\, r\, b\, \Phi) \tag{5}$$

(**b**= distance between the estimated and later adjusted stream surfaces, **N**= number of blades, and Φ = influence from blade thickness.)

4.2 Part load problems of Francis Turbines

In **Fig.9** to the right is shown the cross section of a traditionally designed runner installed in the turbine at the power plant Fröystul in Norway and to the left the relative values of the measured stresses on the outlet of the runner blades is shown. The magnitudes of the oscillating stresses did not allow continuous operation below 65 % load for this turbine due to danger of blade cracking. After synchronizing, the opening of the guide vanes had to be fast to reach the minimum allowable opening. The technical data for Fröystul is: P=37.7 MW, Hn=54, m, n=214 rpm.

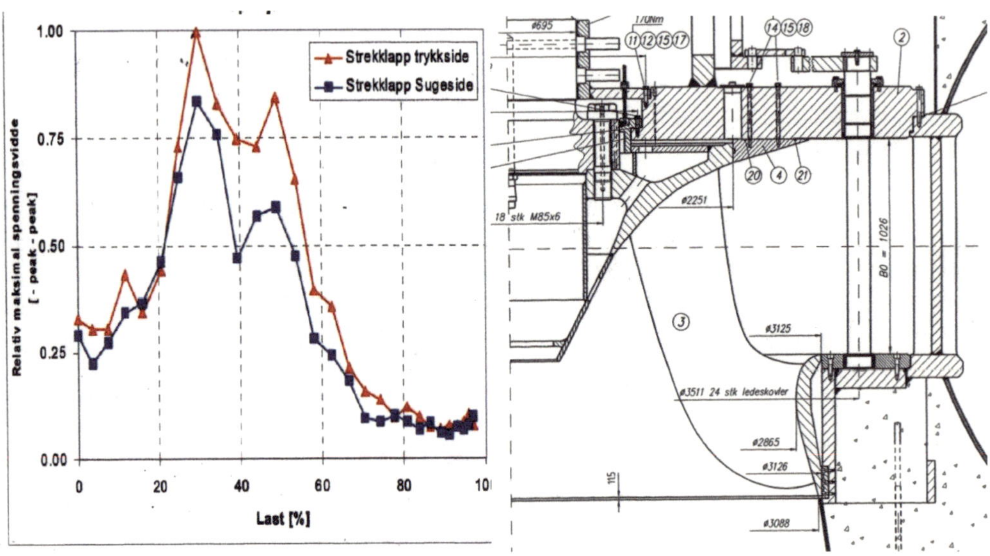

Fig. 9. Cross section of the low head turbine for Fröystul power plant in Norway (right) and the relative values of stress amplitudes on the runner blade outlets (left). (Courtesy: KVAERNER, NORCONSULT)

However, the first runner of the X-blade design was installed in the turbine at Bratsberg Power Plant in Norway proving a very stable operation over the whole range of operation. It should be noted that the highest value of stress amplitudes measured on this first X-BLADE was below 25% compared to the same scale as illustrated in the diagram to the left in fig. 9. From 30% load and up to full load the stresses were below 15% with the same relative scale.

For comparison of geometry the cross section of the runner for Bratsberg is shown to the right in fig. 10. Even if the turbine at Bratsberg has a lower specific speed than the turbine for Fröystul, a comparison is still valid.

The technical data for the turbine for Bratsberg are: P= 60 MW, Hn= 130 m and n=300 rpm.

The technical data for the turbine for Three Gorges proves a higher specific speed than Bratsberg and closer to Fröystul, but still this turbine is running very smoothly as illustrated in fig. 11. The technical data are: Pn = 710 MW at Hn = 80,6 m and n=75 rpm and P=852 MW at Hn = 92 **m /Ref. 4/.**

Finally the measured pressure oscillations in the model turbine for Three Gorges is shown in fig.11./**Ref.5**/. However, the results should be similar for X-blade runners for high and medium specific speed if they are well designed.

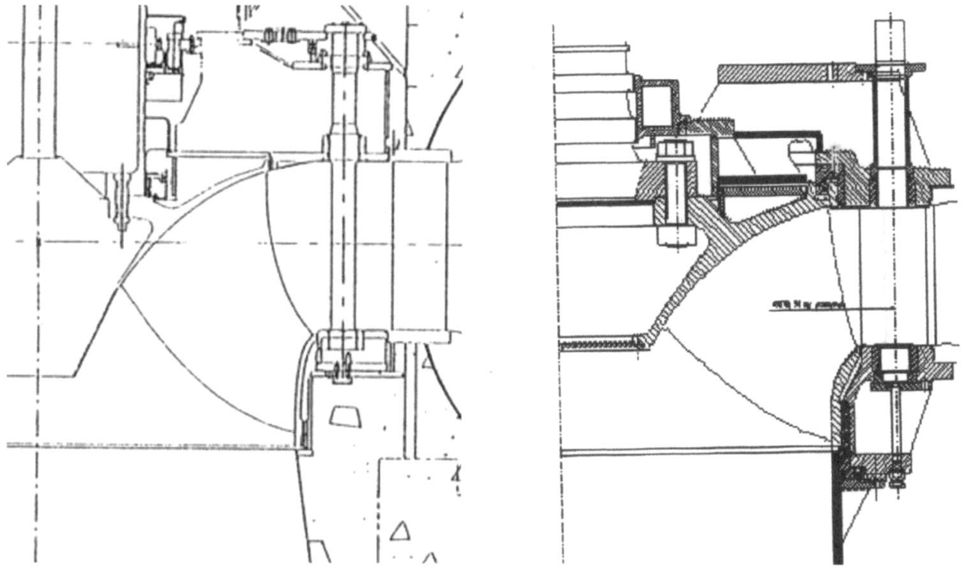

Fig. 10. Cross section of the Francis turbine for Bratsberg with the first X-BLADE runner in operation. (Right) Pn =60. MW, Hn=130 m, n=300 rpm and the runner for Three Gorges with a higher specific speed (Left) with Pn =710. MW at Hn =80.6 m and n =75 rpm.

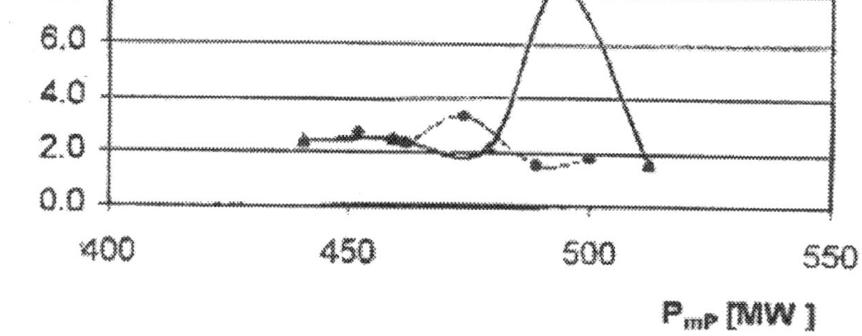

Fig. 11. Measured pressure oscillations peak to peak in % of nominal net head measured in the draft tube of the model turbine for Three Gorges before (full line) and after an optimized semi tapered centre piece was installed at the outlet of the hub (dotted line). Note the power refers to prototype. (Courtesy: GE Hydro (Kvaerner))/**Ref.5**/

During model testing of the X-blade runner for Three Gorges Power Plant a thoroughly research on pressure pulsation was made. Special attention was made on the influence of a tapered centre cone attached to the outlet cone of the runner.

An identical centre piece i.e. a semi tapered cone (not shown in fig.10) has been installed in the prototype. However, the influence from different shapes of centre pieces is well known among the turbine manufacturers.
It should also be mentioned that no centre piece was installed in the first X-BLADE runner that was installed in the turbine at Bratsberg and still the operation was smooth. However, for Three Gorges the operation range of head was very large from 61 m to 113 m so special care was taken to ensure a smooth running over the whole range of operation./ **Ref. 5/**

4.3 Pressure surges at full load that may create power oscillations

Part load pressure surges in the draft tube is recognized as the so called Reihnganz pressure surges created by the rotating vapour filled cork screw shaped rope, swirling in the same direction as the runner with a frequency of approximately 1/3 of the runner speed. The surging frequency and amplitudes of the pressure oscillations will depend on the runner design as discussed in the previous section.
However at full load a vapour- and air- filled void is normally built up in the centre below the runner cone rotating in opposite direction of the runner as illustrated in fig. 7 in this chapter.
The described centre void at full load underneath the runner, will act as a surge chamber or air accumulator connected to the draft tube ending at the nearest free water level forming a mass oscillation system as illustrated in fig. 12.

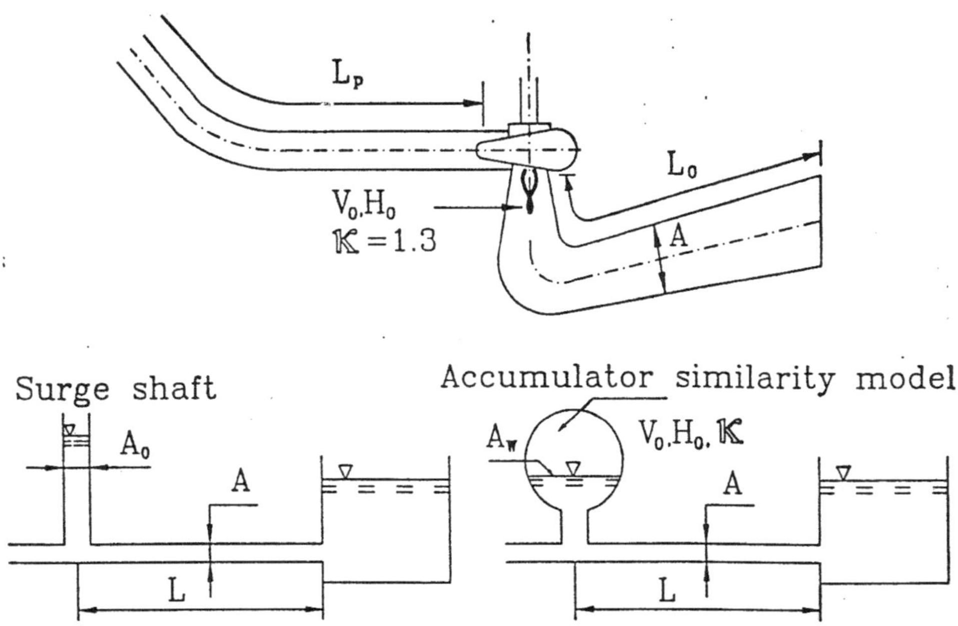

Fig. 12. Illustration of the rotating air and vapour filled centre void formed at full load and overload below the runner centre of a Francis turbine with illustration of the mass oscillation system.

The equation for natural frequency of the void underneath the runner and the mass of water in the draft tube to the first free water surface downstream of the turbine yields:

$$f = \left(\frac{1}{2\pi}\right)\sqrt{\frac{gA}{LAeq}} \tag{6}$$

In equation (6), **g** = the gravity constant (=9.81), **A**= middle cross section of draft tube, **L** = length of draft tube and **Aeq** = the equivalent cross section of a fictitious area of a water level in a surge shaft for a tunnel expressed by the air volume and the polytropic exponent **κ** of the saturated air in the void underneath the centre of the runner as expressed in equation (7), based on experience from air cushioned surge chambers in Norwegian power plants. (Note also that the area of the water level A_W in fig. 12 has been neglected due to smallness in the accumulator system illustrated.)

$$Aeq = Vo / (\kappa * Ho) \tag{7}$$

In equation 7, **Vo** = volume of the vapour and air filled void underneath the runner in the draft tube at full load, **Ho** = water pressure in the void underneath the runner, **κ = 1.3** is the polytropic exponent for water saturated air in accumulators **/Ref. 3/** and **Ho** = water pressure in draft tube underneath the runner.
In the system illustrated in fig. 12 attention should also be paid at the natural frequency of the penstock with length = L_P from the turbine to the water level in the surge shaft at the top of the penstock. For the two turbines with the shortest penstocks at the large high head power plant at Kvilldal in Norway large pressure pulsations and power oscillations occurred at full load partly caused by resonance with the penstock. The equation for the first, the third, and fifth harmonic of the reflected water hammer pressure waves yields:

$$f_p = a / 4L_p,\ 3a / 4L_p,\ 5a / 4L_p,\ ---- \tag{8}$$

In addition to these frequencies we know that the natural frequency of the speed of the generator rotor in the magnetic field of the stator will be between 1 and 2 Hz depending on the connection to the machines in the power house.
In the power house of Kvilldal 4 turbines of P_n = 315 MW, H_n = 520 m and n = 333.3 rpm each was installed connected to separate penstocks leading to a common air filled surge chamber.
By running the two turbines connected to the shortest penstocks, power oscillations and pressure surgs occurred at full load with the frequency between 1.4 and 1.5 Hz. The power oscillations were recorded to be 60 MW even with locked guide vanes in fixed position. However, by switching off the voltage governor during testing the power oscillations stopped. Further, by again switching on the voltage governor the oscillations came back.
The possible natural frequencies of the penstock were according to eq. 8 also between 1.4 and 1.5 Hz.
By using data for the draft tube and assuming we had an air and vapour volume = **Vo** = 0.5 m^3 underneath the runner based on observations from model tests and the absolute pressure **Ho** = 5.0 m WC and the average cross section of the draft tube **A** = 5.0 m^2 with known length of **L** = 30 m and setting the polythropic exponent for vapour saturated air **κ** = 1.3, we finally got the natural frequency of the draft tube system to be:

f = 1.46 Hz which was very close to frequency of the measured oscillations.

Further the natural frequency of the generator rotor was reported to be between 1.4 and 1.5 Hz and the first harmonic of the water hammer frequency of the penstock was observed during shut down tests and calculated to be **fp= a/(4L_P) = 1.4** Hz.

The reason for possible power oscillations of the two high head Francis turbines with the shortst penstocksat Kvilldal may than be explained as follows, based on the fact that the turbine characteristic diagram for a high head Francis turbine has strongly sloping characteristic curves where any increase in speed will reduce the flow and create pressure rices. /Ref. 4/

Following conclusion based on the full load surges at Kvilldal yields:

A rising pressure will increase the speed and then the flow will decrease because of the sloping turbine characteristic of the high head Francis turbines. We then get a positive feed back in the system because the reduced flow caused by the increased spud will in turn give a further increase in the pressure rise especially because the flow oscillations in the penstock and the pressure surges in the draft tube had the same natural frequency, as the generator rotor.

To solve the problem at Kvilldal a damping device was installed in the voltage governor which damped the rotor oscillations and then the connection to the turbine characteristic was broken and the no power oscillations occured.

5. Concluding remarks

Besides the research on oscillatory problems to avoid not acceptable pressure pulsations and fatigue problems, the goal for the research of turbine design must be aimed at following important factors:

EFFICIENCY, SAFETY, MAINTENANCE AND LIFETIME.

The reason for this is the growing need for peak power and because hydro power is the most important non-polluting source for peak power and then safety, reliability and life time will be most important.

6. References

[1] Transaction of the ASME: 1952 p.733, Engineering for power, April 1960 p.103 and 1956 p.1141 (in brackets).

[2] H.Brekke: "State of the art in turbine design" Proceedings of the XXIX IAHR Congress 21 Century: New Era for Hydraulic Research and Its Application, Beijing China. 2001.

[3] H.Brekke: "A STABILITY STUDY ON HYDROPOWER PLANT GOVERNING INCLUDING THE INFLUENCE FROM A QUASI NONLINEAR DAMPING OF OSCILLATORY FLOW AND FROM THE TURBINE CHARACTERISTICS" Dr. Tehn. Thesis NTH 1884. (NTNU Norway)

[4] Adrian P. Boldy, Duan Chang Guo "THE CURRENT STATE OF TECHNOLOGY IN HYDRAULIC MACHINERY" Gover Publishing Company (ISBN0566090249)

[5] J.T. Billdal, B.G. Holt: " Three Gorges Project: Review of GE Energy Norwy´s hydraulic design". Symposium on Hydraulic Macinery an Systemems, 20th IAHR Symposium. Charlotte, North Carolina, August 6 - 9, 2000.

13

Capabilities and Performances of the Selective Laser Melting Process

Sabina L. Campanelli, Nicola Contuzzi,
Andrea Angelastro and Antonio D. Ludovico
*Polytechnic of Bari, Department of Management and Mechanical Engineering,
Viale Japigia, 182
Italy*

1. Introduction

The current market is in a phase of accelerated process of change, that leads companies to innovate in new techniques or technologies to respond as quickly as possible to the ever-changing aspects of the global environment. The economy of a country is heavily dependent on new and innovative products with very short development time.
Companies, currently, have considerable success, only if they develop the ability to respond quickly to changing of customer needs and to use new innovative technologies. In this context, the companies that can offer a greater variety of new products with higher performance resulting in advantage over the other.
At the heart of this environment there is a new generation of customers, who forced organizations to research new technologies and techniques to improve business processes and accelerate product development cycle. As a direct result, factories are forced to apply a new philosophy of engineering as the Rapid Response to Manufacturing (RRM). The concept of the RRM uses products previously designed to support the development of new products.
The RRM environment was developed by integrating the various technologies, such as CAD-based modelling, the knowledge-based engineering for integrated product and process management and the direct production concepts. Direct production uses prototyping, tooling and rapid manufacturing technologies to quickly test the design and build the part (Cherng et al., 1998).
Among RRM technologies, Rapid (RP) and Virtual (VP) Prototyping are revolutionizing the way in which artefacts are designed.
Rapid Prototyping (RP) technologies embraces a wide range of processes for producing parts directly from CAD models, with little need for human intervention; so, designers can produce real prototypes, even very complex, in a simple and efficient way, allowing them to check the assembly and functionality of the design, minimizing errors, product development costs and lead times (Waterman & Dickens, 1994).
The SLS technology was developed, like other RP technologies, to provide a prototyping technique to decrease the time and cost of the product cycle design. It consists of building a three dimensional object layer by layer selectively sintering or partial melting a powder bed by laser radiation.

The SLS objects appear as rigid and porous, in fact, the density of the material depends on the temperature evolution. The molten layer, when the temperature decreases, solidifies and binds to the underlying layer, but the presence of the liquid phase, although happens for a few time, leads to a shrinkage of the powder–liquid mixture. Thus, the porosity cannot be completely eliminated, but the density of the manufactured part is usually higher than the powder density. Moreover, due to temperature variation, mechanical stresses are induced. As a consequence, the final state of the part (dimensions, density, residual stress levels, etc.) strongly depends on the process evolution (Kolossov et al., 2004).
The success of SLS as a rapid prototyping and rapid manufacturing technology results mainly to the possibility to process almost any type of materials (polymers, glass-filled nylon, metal and composites) to accommodate multiple applications throughout the manufacturing process, but high density is desired for the production of functional metallic parts.
The Selective laser sintering process can be indirect or direct:

1. The indirect SLS uses a polymer coating of about 5 μm in thickness for metal powders and ceramics. The metal powder particles are coated with the polymer and the action of the laser melts the polymer, bonding the metal particles together to produce a green part. It is necessary, therefore, a post-treatment in the oven at high temperature, so remove the polymer and sintering particles by creating a metal-metal link.
2. In Direct SLS (DMLS) a low melting point component is melted and employed as a matrix in which the higher melting point components sit. In this process are used or a single powder with two different grain sizes (a slight and a coarse grains) or binary systems. Typical binary phase systems investigated include Ni-Cu, Fe-Cu and Cu-Pb/Sn (Kruth et al. 2008; Lu et al., 2001). The disadvantage of the above processing routes is that the components produced exhibit the mechanical properties and characteristics of their weakest composite phase, thus lacking the full mechanical functionality required for heavy-duty tasks (Dewidar et al., 2008).

To obtain high density, for functional metallic prototypes, parts or tools, different powder binding mechanisms is necessary, so, Selective Laser Melting (SLM) was developed.
The SLM represents a variation from the classical SLS. The substantial differences compared to the latter are:

- using an integral powder metal without adding low melting point elements;
- the need to provide a much higher energy density, to bring fusion the powder.

In SLM near full density parts can be produced without the need for post-processing steps, while the same materials can be used as in serial production.
The advantage is to get an element a massively dense, close to 100%, with metallurgical characteristics similar to the objects achievable with conventional production processes and without need of post-treatments.
The laser must have a greater power than the SLS and at the end of the process the manufactured objects is quite similar to series production, in fact, it does not require special surface finishes and may be subjected quietly at conventional machining.
In addition, with Selective Laser Melting process there is the theoretical possibility of produce any complex geometry. This result is attributable to Layer Manufacturing, each individual point of the component can be reached by the laser beam at any time during the process. In contrast, the process isn't competitive for large lots or for slight complex parts, because the technique is rather slow and, therefore, very expensive.

In order to reach a high density, the metallic powder particles are fully molten, laser melting process is accompanied by the development of residual stresses, that derive from high thermal gradients in the material. These stresses can cause distortion of the part, cracks or delaminations (Fischer et al., 2003; Pohl et al., 2001; Nickel et al., 2001).
Another undesirable phenomenon is the vaporization, that occurs when the bed of powder is irradiated with high energy intensity. During the laser melting process, the temperature of the powder particles exposed to the laser beam exceed the melting temperature of the material. A further increase in temperature (about twice the material melting temperature) causes the evaporation of the powder, so, there are a fast-moving expansion of evaporated particles, which generate a overpressure on the melted zone and the material is ejected from its bed (Hauser et al., 2003).
Another problem that may occur during the SLM process is the "balling" or spheroidization phenomenon, i.e. the formation of isolated spheres with a diameter equal to the laser beam focus, which inhibits deposition and decreases the density of produced part. It occurs when the molten material is unable to fully wet the substrate because of surface tension.
The phenomenon is caused by an excessive amount of energy, which gives to the melted powder a too low value in viscosity (Nickel et al., 2001; Kannatey-Asibu, 2009; Niu & Chang, 1999).
The aim of this chapter will be to describe capabilities and performances of the SLM process. First, the capacity of the system in realizing micro-components with both high dimensional accuracy and maximum density, with mechanical properties similar to those assured by traditional technologies, will be investigated. Later on it, the chapter will describe some applications of the SLM process, such as the manufacture of moulds with conformal channels, of customised jigs for welding and the design and fabrication of reticular structures, to be used in automotive and aerospace field or in the medical one for custom prosthesis fabrication.

2. The process

Selective Laser Melting Equipment technology provide a laser system, a set of optical laser beam focusing, a powder feeding system (loader and roller or coater) and a control center (Fig. 1).
The piece to be achieved should be drawn using a three dimensional solid or surface modeler, then the mathematical model is developed in a CAD format compatible with the management software of the RP machine.
The currently accepted graphic standard by all manufacturers is the STL (solid to layer), introduced by 3D Systems, which provides the mesh of internal and external surfaces of the workpiece through triangular elements.
The STL file is then processed for the orientation and slicing phases. The first step allows to select the best growth direction of the piece, which greatly influences the dimensional accuracy and production time.
The slicing step, instead, is a critical because it determines the accuracy of the product; in fact, it provides a breakdown of the object with orthogonal planes to the direction of growth, obtaining the contour of each section, which describes the path of scanning beam laser, and the thickness of the layers.
The process takes place inside an enclosed chamber, filled with nitrogen gas, to minimize oxidation and degradation of the powdered material.

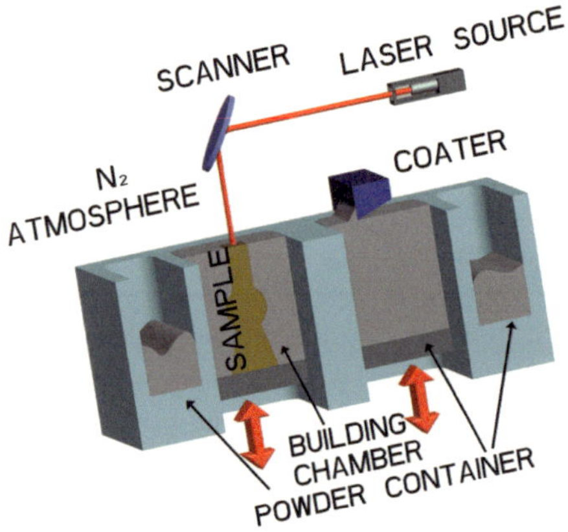

Fig. 1. SLM equipment

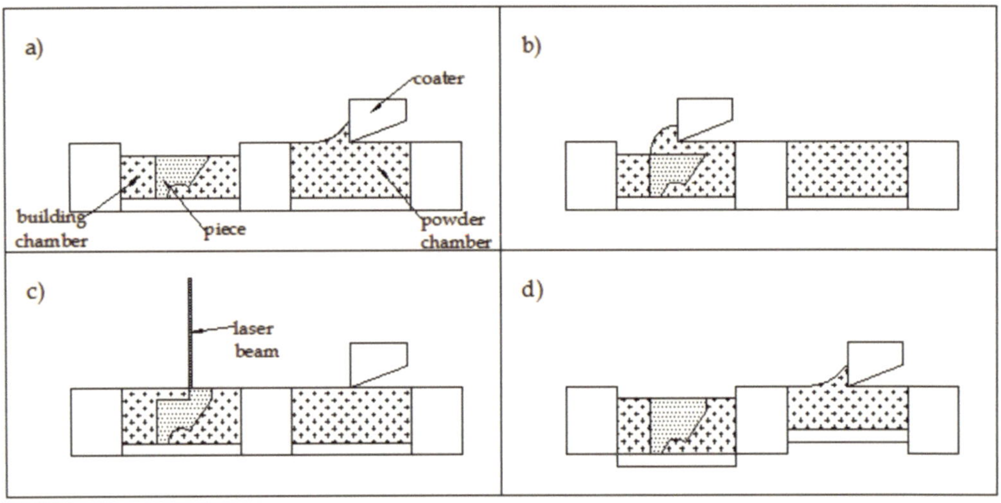

Fig. 2. Phases of the SLM process

In some technological solutions, to minimize distortions and residual stresses, the powder in the build platform is maintained at an elevated temperature (below the melting point of the powdered material); in other technological solution the preheating phase is used to minimize porosity on the built piece (Slocombe et al., 2001)

The most common solutions include:

- infrared heaters, that are placed about the building platform to maintain an elevated temperature around the part that being formed;
- feed cartridges, to pre-heat the powder prior to spreading over the build area.
- resistive heaters, around the building platform.

This powder pre-heating and the next maintenance of an elevated and uniform temperature, within the build platform, is necessary, also, to minimize the power laser requirements of the process, to improve absorption of the laser beam, to prevent warping of the part during the build due to non-uniform thermal expansion and contraction (curling), to reduce the temperature gradients and improve the wetting properties.

Preheating of the powder is used to improve absorption of the laser to reduce the temperature gradients and improve the wetting properties.
The Control Centre commands the RP machine power system: the powder room is raised by an amount such that the deposited powder by roller or coater, on the surface of the building chamber, deposits a layer of a thickness equal to that given during slicing, while the working chamber decreases the thickness of a layer (Fig.2a).
The coater deposits, in uniform way, the powder on the building chamber (Fig.2b).
The laser beam selectively scans the surface of the powder, that is, then, sintered (Fig.2c).
When the first two-dimensional layer is created, the process is repeated until the object is built, with a lowering, equal to the thickness of the layer, of the building chamber (Fig. 2.d) and the subsequent distribution of the powder with the coater and the selective scanning by the laser beam. Surrounding powder remains loose and serves as support for subsequent layers, thus eliminating the need for the secondary supports which are necessary for photopolymer vat processes.
For pre-heated solution, a cooling period is required. If the high temperature parts and powders are prematurely exposed to external atmosphere, they may degrade for the presence of oxygen and the powder cannot be recovered and used for further processing; also, the built objects may warp due to uneven thermal contraction. So, to allow the parts to uniformly come to a low-enough temperature that they can be handled and exposed to ambient temperature and atmosphere.
Finally, the parts are removed from the powder bed, loose powder is cleaned off the parts, and further finishing operations, if necessary, are performed.

3. Materials

In order to develop a process that can produce functional high strength parts, in a first time, successful results, with DMLS, have been obtained with high-strength powder mixtures, that containing only two metal powders, such as Fe-Cu, WC-Co, TiC-Ni/Co/Mo, TiCN-Ni, TiB_2-Ni, ZrB_2-Cu and Fe_3C-Fe, melting only the powder having the lower melting point (Kruth et al., 1997; Kruth et al., 1996; Laoui et al., 1998).
However, a substantial amount of work has been carried out in the field of laser sintering of metal parts, some of which are as follows. Niu and Chang (Niu and Chang, 1998; Niu and Chang, 1999; Niu and Chang, 1999; Niu and Chang, 2000) have studied the SLS process on HSS powder with a carbon di-oxide laser of 25 W, while Fischer (Fischer et al., 2003) has studied the behaviour of commercial titanium. Song (Song, 1997) have performed experiments on direct sintering of pre-alloyed bronze as a low melting metallic powder on a laboratory test facility. Abe (Abe et al., 2001) has studied the behavior of nickel base alloys and has reported that the process does not exhibit balling phenomenon but only part deflections and cracking.
At this day, few studies have been performed on SLM process. Badrossamay (Badrossamay & Childs, 2007) has performed a further studies on M2 tool steel, 316L and 314S-HC stainless steel, while Abe (Abe et al., 2001) has investigated the effects of a single scanning test on several kinds of materials: aluminium, copper, iron, stainless steels (SUS 316L), chromium, titanium and nickel-based alloy. Kruth (Kruth et al., 2004) has studied a mixture of different types of particles (Fe, Ni, Cu and Fe_3P) specially developed for SLM.

4. Experimental tests and analysis of results

4.1 Experimental plan

Experimental tests have been performed in this work to understand the capabilities of the SLM process in terms of density, roughness, dimensional accuracy and mechanical properties.
Several process parameters can be modified in order to obtain optimum quality of laser sintered samples. Some of these parameters are related to the process (single layer thickness, scan velocity, hatch spacing, scan strategy), others depend on the laser (laser power, spot diameter, wavelengths, the energy of the pulse laser), the type of material and other characteristics of the used powder (such as particle size, distribution, shape, material type, the percentage composition of materials), others by the final component which is to be realized (shape, size, etc...).
The energy density E_d of a single track can be calculated by the relation between laser power (P), scan speed (v) and spot diameter (d) (Eq. 1) (Lu, 2009):

$$E_d = \frac{P}{v \cdot d}\left[\frac{J}{mm^2}\right] \quad (1)$$

where:

- P is the laser power used to scan a part;
- v is the scan speed or the velocity by which the laser beam moves over the powder surface.
- d is the spot diameter, equal to 0.2 [mm].

A full factorial plan (3^2 plan) was used to project experiments. A contemporary variation of the two parameters, scanning speed (measured in mm/s) of the laser power (measured in Watt) was considered, in order to find out the parameters combination leading to the maximum mechanical properties. The combination of these two parameters is responsible for the melting mechanism and influence the quality of built parts.
Nine samples were built, with three levels of scanning speed (180, 200, 220) and three levels of laser power (57, 86, 100). Moreover, five replications for each combination of the parameters were realized.

4.2 Machine set up

The machine used to perform experiments was equipped with a Rofin Nd:YAG laser source characterized by a wavelength of 1.064 µm, a spot diameter of 200 µm and a maximum output power of 100 W. The laser beam was moved over the powder surface by means of scanning mirrors in order to draw selectively every layer of the powder.
The powder deposition system consisted in a building chamber, a powder chamber and a coater. Powder layers were deposited in one direction using a coater and the layer thickness was set to 30 µm. Moreover, the chamber was filled, in slight overpressure, with nitrogen to prevent oxidation of the parts and to reduce the initial oxygen level at 0.8%.

4.3 Material

A material with the typical composition of maraging steels, reinforced with cobalt, was used for the present study. Specifically, the composition was very close to the 18 Ni Marage 300 steel and the powder had spherical particles with average size below 40µm. The chemical composition of the used powder, determined by Energy Dispersive X-ray (EDX) analysis, consisted in 4.2 wt.% Mo, 0.88 wt.% Ti, 65.9 wt.% Fe, 10.2 wt.% Co, 18.8 wt.% Ni and 0.02 wt.% C.

Maraging steels are a special class of low-carbon ultra-high-strength steels, which derive their strength not from carbon, but from precipitation of inter-metallic compounds; they were developed for high performance applications, especially those in which high strength and good toughness are required.

18 Ni Marage 300 steel presents excellent mechanical properties, high value of yield and tensile strength, toughness, ductility and impact strength, high fatigue limit, high compressive strength, hardness and wear resistance suitable for many machining tools.

This type of steel is conventionally used for produce tools suitable for complex machining, to achieve high-performance industrial parts (aerospace applications), for manufacturing of dies for hot injection moulding, for inserts for moulds of all standard thermoplastic materials, for casting and for the direct realization of objects for engineering applications.

4.4 Mechanical characterisation of built samples

Small square samples of 15 x 15 x 10 mm were built to measure density, hardness and to study microstructure.

First, density was measured with the 'Archimedes-method' by weighting the samples in air and subsequently in ethanol after coating them with a lacquer. The coating prevented absorption of ethanol by the specimen during measurements.

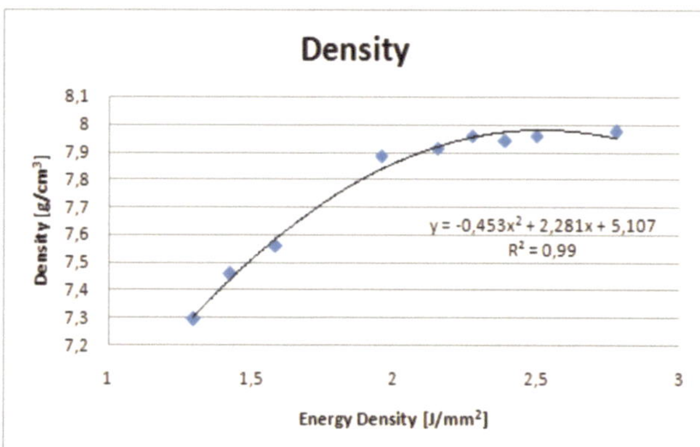

Fig. 3. Density versus Energy Density

Fig. 3 shows the tendency of Density versus the Energy Density E_d. Values of Ed were calculated by means of Eq. 1.

It is evident that density increases with the increase of E_d and it is maximal for the maximum E_d, which means scan speed and laser power set respectively to the minimum (180 mm/s) and the maximum value (100W) of the considered range. A maximum value of 8.0 g/m^3 was found for density; thus the porosity was approximately 0.01%. This means that it was possible to produce nearly full dense parts with properties comparable to those of the bulk material.

Hardness and tensile tests were performed on specimens built with the maximum density, setting the scan speed to 180 mm/s and laser power to 100W.

Rockwell C test was used for determining hardness. Experimental results gave an average value of 34 HRC.

Tensile tests were performed using an Instron 4467 machine equipped by an extensometer (with a 12.5 mm gage length and maximum elongation of 5 mm). Results of tensile test brought to a value of 985 MPa for Yield Strenght, 1145 MPa for Tensile Strength and 7.6 % for Elongation to break.

Fig.4(a) and Fig.4(b), respectively, show micrographs at magnifications 200X and 1000X obtained with an optical microscope. It is evident that the metal powder is completely fused and constituted by molten/re-solidified zones with curved edges (approximately parabolic). The laser tracks overlap in order to produce a non-porous part. This means that each part is welded onto the layers surrounding it. The presence of pores is very limited as it is possible to see from black spots in the pictures.

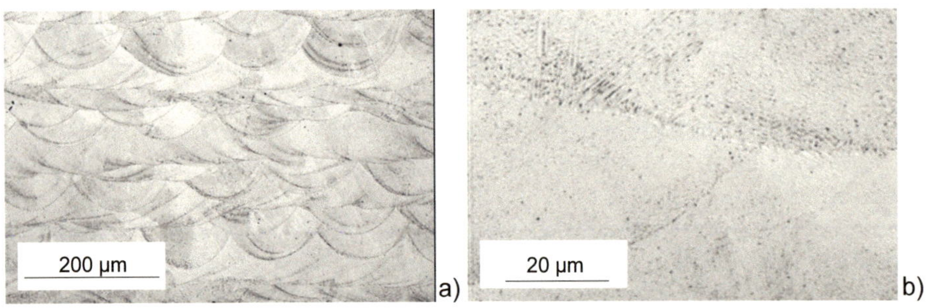

Fig. 4. 200x magnification (a) and 1000x magnification (b) of the sample obtained with P=100W and v=180mm/s

4.5 Dimensional accuracy

A specific part test was developed in order to test the dimensional accuracy of SLM parts with the maximum density achieved. The proposed geometry with maximum dimensions 70 x 70 x 25 mm is shown in (Fig. 5). It was thought to determine capabilities of the process in terms of:

- minimum feasible feature size;
- dimensional achievable accuracy.

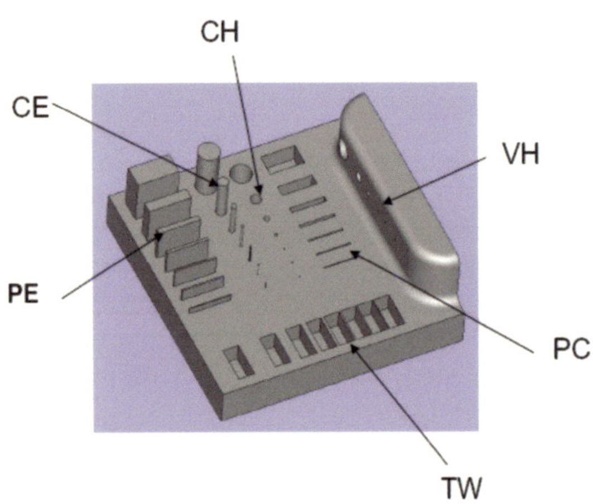

Fig. 5. Benchmark geometry

The feasible precision and resolution of the process were tested by:

- cylindrical holes (CH), ranging from 0.2 to 6 mm diameter (CH1. CH2, CH3, CH4, CH5, CH6, CH7);
- cylindrical extrusions (CE), ranging from 0.2 to 6 mm diameter (CE1, CE2, CE3, CE4, CE5, CE6, CE7);
- parallelepiped extrusions (PE), ranging from 0.2 to 6 mm thickness (PE1, PE2, PE3, PE4, PE5, PE6, PE7);
- parallelepiped cavities (PC), ranging from 0.2 to 6 mm width (PC1, PC2, PC3, PC4, PC5, PC6, PC7);
- vertical holes (VH) ranging from 0.2 to 6 mm diameter (VH1, VH2, VH3, VH4, VH5, VH6, VH7);
- thin walls (TW) ranging from 0.2 to 6 mm thickness (TW1, TW2, TW3, TW4, TW5, TW6, TW7);

The built benchmark (Fig. 6) was measured using a coordinate measuring machine (DEMeet 400) together with an optical microscope.

Fig. 6 shows a picture of the built benchmark. It is evident how all designed features could be realized, except for cylindrical extrusions CE1, CE2, CE3. They were probably built, but they were removed by the coater during the deposition of the powder.

The dimensional error E% was introduced in order to perform an analytical study of the accuracy (Eq. 2). In this equation N_d and M_d represent respectively the nominal and the measured dimension.

$$E\% = \frac{N_d - M_d}{N_d} \cdot 100 \qquad (2)$$

Results of measurements (Table 3) showed that it was possible to build:

- thin walls with a maximum error of 15% for a thickness of 0.2 mm and with a minimum error of 1.17% for a thickness of 6 mm;
- parallelepiped cavities with a maximum error of 20% for a width of 0.2 mm and with a minimum error of 4% for a width of 6 mm;
- parallelepiped extrusions with a maximum error of 20% for a thickness of 0.2 mm and with a minimum error of 1.33% for a thickness of 6 mm;

Fig. 6. Built benchmark

Element	Measurable size	Nominal dimension [mm]	Measured Dimension [mm]	E%
CE1	diameter	0.2	-	-
CE2	diameter	0.3	-	-
CE3	diameter	0.4	-	-
CE4	diameter	0.8	0.87	-8.75
CE5	diameter	1.5	1.6	-6.67
CE6	diameter	3	2.99	0.33
CE7	diameter	6	5.99	0.17
CH1	diameter	0.2	0.23	-15.00
CH2	diameter	0.3	0.34	-13.33
CH3	diameter	0.4	0.41	-2.50
CH4	diameter	0.8	0.82	-2.50
CH5	diameter	1.5	1.53	-2.00
CH6	diameter	3	2.91	1.67
CH7	diameter	6	5.94	1.00
PE1	thickness	0.2	0.16	20.00
PE2	thickness	0.3	0.32	-6.67
PE3	thickness	0.4	0.36	10.00
PE4	thickness	0.8	0.88	-10.00
PE5	thickness	1.5	1.64	-9.33
PE6	thickness	3	3.24	-8.00
PE7	thickness	6	6.08	-1.33
PC1	width	0.2	0.24	-20.00
PC2	width	0.3	0.33	-10.00
PC3	width	0.4	0.44	-10.00
PC4	width	0.8	0.84	-5.00
PC5	width	1.5	1.37	8.67
PC6	width	3	2.84	5.33
PC7	width	6	5.43	9.50
VH1	diameter	0.2	0.25	-25.00
VH2	diameter	0.3	0.36	-20.00
VH3	diameter	0.4	0.43	-7.50
VH4	diameter	0.8	0.85	-6.25
VH5	diameter	1.5	1.55	-3.33
VH6	diameter	3	2.73	9.00
VH7	diameter	6	5.46	9.00
TW1	thickness	0.2	0.17	15.00
TW2	thickness	0.3	0.33	-10.00
TW3	thickness	0.4	0.37	7.50
TW4	thickness	0.8	0.75	6.25
TW5	thickness	1.5	1.44	4.00
TW6	thickness	3	3.06	-2.00
TW7	thickness	6	6.07	-1.17

Table 3. Results of measurements

- cylindrical extrusions with a maximum error of 8.75% for a diameter of 0.8 mm and with a minimum error of 0.17% for a diameter of 6 mm;
- cylindrical holes with a maximum error of 15% for a diameter of 0.2 mm and with a minimum error of 1% for a diameter of 6 mm;
- vertical holes with a maximum error of 25% for a thickness of 0.2 mm and with a minimum error of 3.33% for a thickness of 1.5 mm.

The following considerations can be drawn:
The process has a good accuracy for nominal dimensions over 0.4 mm; E% increases for values lower than 0.4 mm because the nominal dimension is gradually approaching to the laser spot diameter.

5. SLM Applications

5.1 Cellular structures

The cellular metal structures have been used in various industrial applications such as heat exchangers, in reconstructive surgery, in chemistry, in automotive and aerospace industries. They possess valuable characteristics as low density, high strength, good energy absorption, good thermal and acoustic properties (Dotcheva et al., 2008).
Periodic cellular structures are highly porous structures, with only 20% or less of their internal volume occupied by solid material (Evans et al., 2001). These advantageous characteristics make them very desirable, but difficulties in their production limits their application. Layer manufacturing technologies such as Direct Metal Additive Manufacturing, the Electron Beam Melting, the Direct Metal Laser Sintering and Selective Laser Melting (SLM), allow the manufacture of solid parts of any geometry, using laser technology and layers of metal powder.
The SLM technology offers the possibility to produce parts with complex engineering materials (stainless steel, tool steel, titanium alloy and cobalt-chromium alloys) relatively quickly, directly from a 3D CAD data.
This gives designers the freedom to use the cellular materials where there is the need to create a better functionality of a product without sacrificing its mechanical properties.
The SLM technology, used as a process for the production of cellular structures, can be beneficial as it has:

- The complete freedom in defining the geometry of the part (Kruf et al., 2006);
- The reduction of the production cycle (Santos et al., 2006).

In this field, few works have been performed.
Rehme (Rehme,2010) have been investigated the manufacturability and scaling laws for mechanical properties of periodic lattice structures built with SLM technology. In his work has built eight different unit cell types, that possess low relative density and response well to compressive load. Dotcheva et al. (Dotcheva et al., 2008) have been used cellular truss structures as a core material for injection moulding tools. They investigated the capabilities of SLM technology for fast building of 3D complex geometry and to improve the thermal management of the injection moulding process.
At DIMeG of the Polytechnic of Bari, has been tested the ability of the SLM to build lattice structures (Contuzzi, 2010).
To test the process, it was chosen, for its versatility, a lattice structure reinforced vertically and with 45° truss columns (Fig. 7).

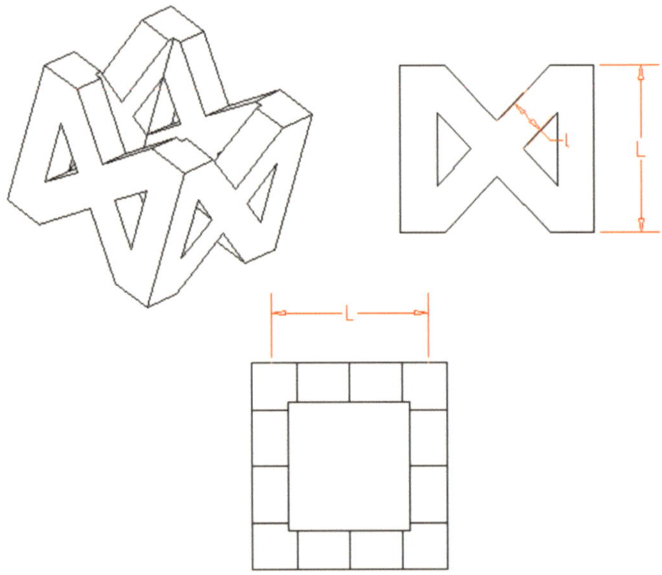

Fig. 7. Single cell and its dimension

The studied configurations are:

A. cell of 2 mm (L) with side of the truss of 500 µm (l) (Fig. 8);

B. cell of 3 mm (L) with side of the truss of 700 µm (l) (Fig. 9).

Overall, the configuration A (2 mm cells, 500 µm side truss) has the dimensions of 16x16x16 mm, while the B configuration (3 mm cells, 700 µm side truss) has the dimensions of 21x21x15 mm.

Between multi-functional applications of periodic cellular structures one of the most interesting is one that uses they as heat sinks (resistors).

Fig. 8. CAD model of sample A

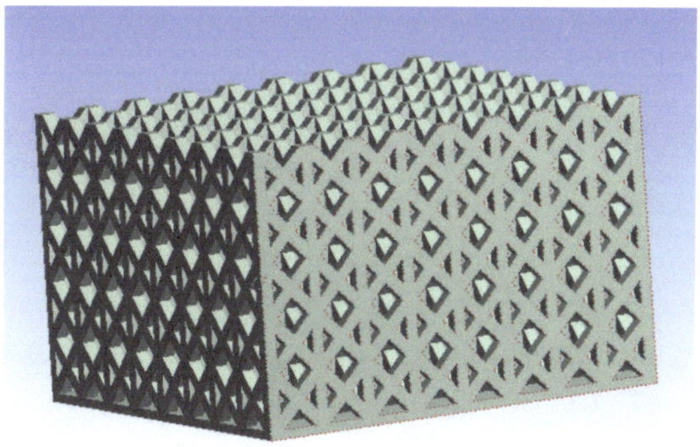

Fig. 9. CAD model of sample B

Its function is to extend the physical dimensions of the device, allowing, thus, a better heat dissipation, because it deprives heat, for Joule effect, and transfers it to the surroundings ambient, preventing the excess of the temperature limit. The presence of the heat sink leads to a reduction in overall thermal resistance, allowing, thus, the reduction of the temperature reached of the device, or, wanting exploit the maximum operating temperature, have a higher power dissipated.
The studied lattice structure have a good behavior as resistors, allowing a disposal, in both configurations, well over 100 watts (Fig. 10) at room temperature of 0°C and over 60 W at room temperature of 220°C.
The structures show, also, good mechanical properties. A compressive test was performed (Fig. 11)
The sintered AISI MARAGE 300 has an elastic behavior almost to the breaking point, this means that, according to Timoshenko & Gere (Gere & Timoshenko ,1984), the lattice structure, at a certain point, begins to deform plastically, but continues to support the stress because of the hardening of the trusses, the beams themselves then collapsing for plastic instability.
For this reason the tests were not performed until the complete collapse of structures, but were stopped when the load is stable for a specified time interval. In this interval the trusses begin to destabilized, but the load is supported by the intact trusses.

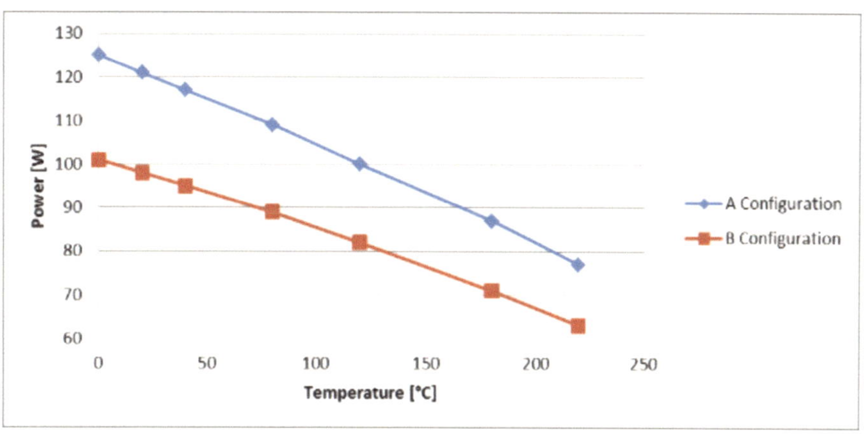

Fig. 10. Maximum power dissipated

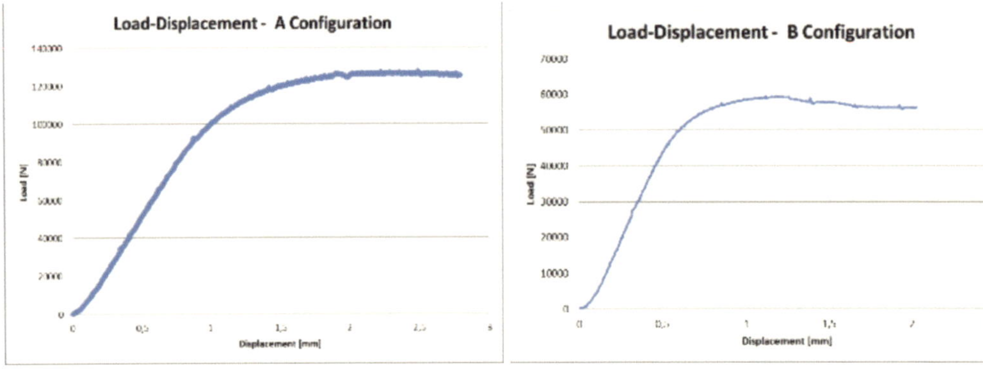

Fig. 11. Load-Displacement curve

5.2 Tools with three-dimensional conformal channels

In recent years, SLM have found widespread use in the field of rapid tooling, especially in the injection mould. The basic principle of injection moulding is that a solid polymer is molten and injected into a cavity inside a mould, which is then cooled and the part ejected from the machine. Therefore, the main phases in an injection moulding process involve filling, cooling and ejection. The cost-efficiency of the process is dependent on the time spent in the moulding cycle. Correspondingly, since the cooling phase is the most lenghty step among the three, it determines the rate at which the parts are produced. A reduction in the time spent on cooling the part before it is ejected would drastically increase the production rate, hence reduce costs (Dimla et al., 2005). Historically, this result has been achieved by creating several channels inside the mould and forcing a cooler liquid to circulate and conduct the excess heat away. The channels are constituted by holes as close as possible to the actual moulding area of the die. Up to now, the methods used for producing these holes rely on the conventional machining process such as drilling or boring. However, this simple technology can only create straight holes which, besides having to intersect each other and requiring the closure of entry points of the tool with a plug, cannot remain at a constant distance from the moulding area to cool it uniformly. Uneven cooling will result in an increase of the mould defects, like warpages or residual stresses, and of the cooling time. A more acceptable cooling method, the conformal cooling, is performed by the coolant flows in a pattern that closely matches the geometry of the part being moulded (Au & Yu, 2007). The term conformal means that the geometry of the cooling channel follows the mould surface geometry. The aim is to maintain a steady and uniform cooling performance for the moulding part.

Ring et al. investigated the effectiveness of conformal channels by through the construction of three different moulds with and without conformal cooling. They showed that the conformal cooling channels technique led to significant improvements and a general reduction of the cycle time while increasing heat transfer (Ring et al., 2002).

A comparison between conformal channels and drilled cooling channels has also been conducted by Sachs et al. Their analysis shows that the conformal channel mould reaches

operational temperature faster than the conventional one, attaining a more uniform temperature distribution with efficient heat transfer capacity (Sachs et al., 2000).
While the use of conformal cooling channels on the one hand greatly enhances the injection molding process, on the other hand it is much more complicated fabricate them on respect of the straight ones.
The advancement of Solid Freeform Fabrication gives rise to the production of injection mould with intricate cooling channel geometry. Rapid tooling based on SFF technology includes SLM. Much research has focused on improving the geometric design of the cooling channel via SFF technologies. In 2001, Xu studied injection mould with complex cooling channels based on SFF processes. He described the conformal cooling layout that can be realized with substantial improvements in part quality and productivity (Xu et al., 2001).
At DIMeG of the Polytechnic of Bari, in collaboration with Elfim srl company of Gravina in Puglia (Ba), SLM was used to create a jig for welding of constituent parts of a titanium alloy intramedullary nail. The realization of this jig had two basic problems:

- cooling of the jig;
- conveyance of shielding gas on the welding area.

The cooling circuit of each component of the jig was designed with conformal channels in order to follow the welding areas and the nail seat, as the conformal cooling channels follow the geometry of the cavity in the dies for injection moulding. However, in this case, the situation was more complicated because of the need to provide even the conveyance of shielding gas.
The use of the shielding gas was necessary due to the high chemical reactivity of titanium, at high temperature, in presence of the oxygen. So, for a good quality joints and to prevent the oxides formation, the shielding of an inert gas was indispensable both during welding and during the cooling of the seam. Therefore, with the same concept usually adopted for the conformal cooling, was created a system of cylindrical channels, whit a diameter down to 1 mm, in order to direct the shielding gas on weld areas.
Fig.12a and 13a show respectively the 3D CAD geometry of the upper and the lower component of the jig; Fig 12b and Fig 13b show the cooling (blue) and shielding gas (yellow) channels that were designed for the jig. Fig. 14 and Fig. 15 show respectively the fabricated upper and lower part of the jig, while Fig. 16 shows the coupling of the jig components.

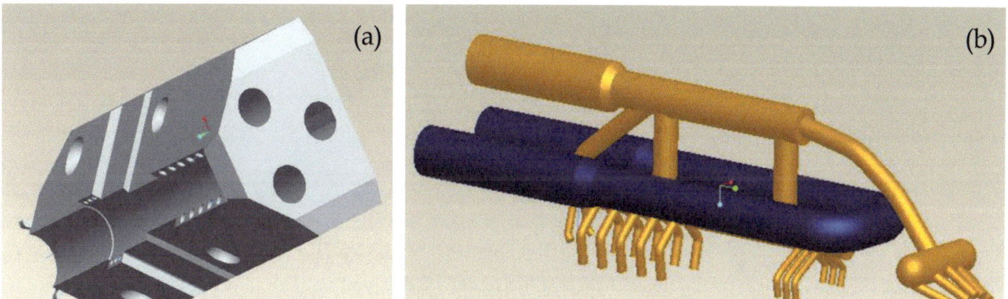

Fig. 12. 3D-CAD model of (a) the upper component of the jig; (b) cooling (blue) and shielding gas (yellow) channels in it

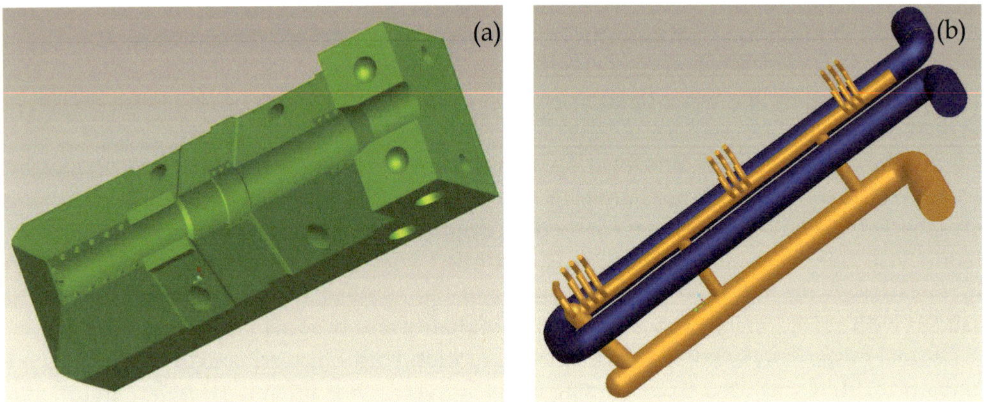

Fig. 13. 3D-CAD model of (a) the lower component of the jig; (b) cooling (blue) and shielding gas (yellow) channels in it

Fig. 14. Laser sintered upper component of the jig

Fig. 15. Laser sintered lower component of the jig

Fig. 16. Coupling of the jig components

6. Conclusions

The aim of this chapter was to describe capabilities and performances of the SLM process. First, experimental tests were performed in order to investigate characteristics of SLM parts. It was found that SLM parts could be produced with almost full density and with mechanical properties similar to those assured by traditional technologies. Later on it, a benchmark was built to study the accuracy of the process. It was found that the process has a good accuracy for nominal dimensions over 0.4 mm; the dimensional error increases for values lower than 0.4 mm because the nominal size is gradually approaching to the laser spot diameter.

Finally, the capacity of the process in producing customized jigs for welding and reticular structures was illustrated.

7. References

Abe F., Osakada K., Shiomi M., Uematsu K. & Matsumoto M. (2001). The manufacturing of hard tools from metallic powders by selective laser sintering. *Journal of Materials Processing Technology*, Vol. 111, No. 1-3, April 2001, 210–213, ISSN 0924-0136

Au K.M. & Yu K.M. (2007). A scaffolding architecture for conformal cooling design in rapid plastic injection moulding, *International Journal of Advanced Manufacturing Technology*, Vol. 34, No. 5-6, September 2007, 496-515, ISSN: 0268-3768

Badrossamay, M. & Childs, T.H.C. (2007). Further studies in selective laser melting of stainless and tool steel powders. *International Journal of Machine Tools & Manufacture*, Vol. 47, No. 5, April 2007, 779–784, ISSN: 0890-6955

Campanelli, S.L; Contuzzi, N. & Ludovico, A.D (2009). Selective Laser Melting: Evaluation of the performances of 18 AISI Marage 300 steel parts. *Proceedings of 9th A.I.Te.M. Conference*, ISBN 8895057074, pp. 109-110, September 2009, Torino, Italy

Campanelli, S.L; Contuzzi, N. & Ludovico, A.D (2010). Manufacturing of 18 Ni Marage 300 steel samples by selective laser melting. *Advanced Materials Research*, Vol. 83-86, February 2010, 850-857, ISBN 0-87849-297-6

Cherng, J. G.; Shao, X.; Chen Y. & Sferro P. R. (1998). Feature-Based Part Modeling And Process Planning For Rapid Response Manufacturing. *Computers and Industrial Engineering*, Vol. 34, No. 2, April 1998, 515-530, ISSN 0360-8352

Contuzzi, N. (2010). Analisi numerica e sperimentale di strutture innovative create mediante Fusione Laser Selettiva. *Ph.D. Thesis in Mechanical and Biomechanical Design*

Dewidar, M. M.; Lim J. K. & Dalgarno, K.W. (2008). A Comparison between Direct and Indirect Laser Sintering of Metals. *Journal of Materials Sciences and Technology*, Vol. 24, No.2, 227-232, ISSN 1005-0302

Dimla, D.E.; Camilotto, M. & Miani, F. (2005), Design and optimisation of conformal cooling channels in injection moulding tools. *Journal of Materials Processing Technology*, Vol. 164–165, May 2005, 1294–1300, ISSN 0924-0136

Dotcheva, M.; Millward, H. & Thomas, D. (2008), Investigation of Rapid Manufactured Cellular Structures for Injection Moulding Tools. *Proceedings of 6th CIRP Int. Conf. on Intelligent Computation in Manufacturing Engineering- CIRP ICME'08*, ISBN 978-88-900948-7-3, pp.369-374, July 2008, Naples, Italy

Evans, A. G.; Hutchinson, J. W.; Fleck, N. A.; Ashby, M. F. & Wadley, H. N. G. (2001). The topological design of multifunctional cellular metals. *Progress in Materials Science*, Vol. 46, No. 3-4, April 2001, 309–327, ISSN 0079-6425

Fischer, P.; Romano, V.; Weber, H.P.; Karapatis, N.P.; Boillat, E. & Glardon, R. (2003). Sintering of commercially pure titanium powder with a Nd:YAG laser source, *Acta Materialia*, Vol. 51, No. 6, April 2003, 1651-1662, ISSN 1359-6454

Fischer, P.; Romano, V.;. Weber, H.P.; Karapatis, N.P.; Boillat, E. & Glardon, R. (2003). Sintering of commercially pure titanium powder with a Nd:YAG laser source, *Acta Materialia*, Vol. 51, No. 6, April 2003, 1651-1662, ISSN 1359-6454

Gere, J. M. & Timoshenko, S. P. (1984). Mechanics of materials, Stanley Thornes Publishers, ISBN 074873998X, Boston, USA

Hauser, C.; Childs, T.H.C. ; Taylor, C.M. & Badrossamay, M. (2003). Direct selective laser sintering of tool steel powders to high density. Part A. Effect of laser beam width and scan strategy, *Proceedings of the Solid Freeform Fabrication Symposium*, pp. 644–655, ISBN 1845440803, Austin, TX, August 2003, Emerald, Bingley, UK

Kannatey-Asibu, E. Jr. (2009). *Principles Of Laser Materials Processing*, John Wiley & Sons Inc., ISBN 978-0-470-17798-3, Hoboken, New Jersey,

Kolossov, S.; Boillat, E.; Glardon, R.; Fischer P. & Locher M. (2004). 3D FE simulation for temperature evolution in the selective laser sintering process. *International Journal of Machine Tools & Manufacture*, Vol. 44, No. 2-3, February 2004, 117–123, ISSN 0890-6955

Kruf, W.; Van de Vorst, B.; Maalderink, H. & Kamperman, N. (2006). Design for Rapid Manufacturing Functional SLS Parts, *Proceedings of 5th CIRP Int. Conf. on Intelligent Computation in Manufacturing Engineering- CIRP ICME'06*, ISBN 978-88-95028-01-9, pp. 609-613, July 2006, Ischia, Italy

Kruth, J.P.; Van der Schueren, B.; Bonse, J.E. & Morren, B. (1996). Basic powder metallurgical aspects in selective metal powder sintering. *CIRP Annals - Manufacturing Technology*, Vol. 45, No. 1, 1996, 183-186, ISSN 0007-8506

Kruth, J.P.; Froyen, L.; Morren, B. & Bonse, J.E. (1997). Selective laser sintering of WC-Co 'hard metal' parts. *Proceedings of 8th Int. Conf. on Production Engineering (ICPE-8), Rapid Product Development*, pp.149-56, ISBN 0-412-81160X, Hokkaido University, Sapporo, Japan, August 1997, Chapman and Hall, London

Kruth, J.P.; Froyen, L.; Van Vaerenbergh, L.; Mercelis, P.; Rombouts M. & Lauwers, B. (2004). Selective laser melting of iron-based powder. *Journal of Materials Processing Technology*, Vol. 149, No. 1-3, June 2004, 616-622, ISSN 0924-0136

Laoui, T.; Froyen, L. & Kruth, J.P.(1998). Selective laser sintering of hard metal powders. *Proceedings of Rapid Prototyping and Manufacturing Conference*, pp. 435–467, ISSN : 1018-7375, Dearborn, Michigan, USA, April 1998

Lu, L.; Fuh, J.Y.H. & Wong, Y.S. (2001). *Laser-Induced Materials and Processes for Rapid Prototyping*, Kluwer Academic Publishers, ISBN 0792374002, Norwell, USA

Lu, T. J. (1998). Heat transfer efficiency of metal honeycombs. *International Journal of Heat and Mass Transfer*, Vol. 42, No 11, June 1998 , 2031-2040, ISSN 0017-9310

Nickel, A.H.;. Barnett, D.M & Prinz, F.B. (2001). Thermal stresses and deposition patterns in layered manufacturing. *Materials Science and Engineering A*, Vol. 317, No. 1-2, October 2001, 59-64, ISSN 0921-5093

Niu, H.J. & Chang I.T.H. (1998). Liquid phase sintering of M3/2 high speed steel by selective laser sintering. *Scripta Materialia*, Vol. 39, No.1, June 1998, 67-72, ISSN 1359-6462

Niu, H.J. & Chang I.T.H. (1999). Selective laser sintering of gas and water atomized high speed steel powders. *Scripta Materialia*, Vol. 41, No. 1, June 1999, 25–30, ISSN 1359-6462

Niu, H.J. & Chang I.T.H. (2000). Selective laser sintering of gas atomized M2 high speed steel powder. *Journal of Materials Science*, Vol.35, No. 1, January 2000, 31-38, ISSN 0022-2461

Niu, H.J. & Chang, T.H. (1999). Instability of scan tracks of selective laser sintering of high speed steel powder. *Scripta Materialia*, Vol. 41, No. 11, November 1999, 1229–1234, ISSN: 1359-6462

Pohl, H.; Simchi, A.; Issa, M. & Dias, H.C. (2001). Thermal stresses in direct metal laser sintering, *Proceedings of the Solid Freeform Fabrication Symposium*, pp. 366–372, ISBN 1845440803, Austin, TX, August 2001, Emerald, Bingley, UK

Queheillalt, D. T.; Murtyb, Y. & Wadleya, H. N.G. (2008). Mechanical properties of an extruded pyramidal lattice truss sandwich structure, *Scripta Materialia*, Vol. 58, No. 1, January 2008, 76-79, ISSN 1359-6462

Rehme, O. (2010). *Cellular Design for Laser Freeform Fabrication*, Cuvillier Verlag, ISBN 3869552735, Hamburg, Germany

Ring, M.; Dimla, D.E. & Wimpenny, W.I. (2002). An Investigation of the effectiveness of Conformal Cooling channels and Selective Laser Sintering material in injection moulding tools, *RPD 2002 Advanced Solutions in Product Development*, Miranda Grande, Portugal, October 2002

Sachs, E.; Wylonis, E.; Allens, S.; Cima, M. & Guo, H. (2000). Production of Injection Moulding Tooling with Conformal Cooling Channels Using the Three Dimensional Printing Process, *Polymer Engineering and Science*, Vol. 40, No. 5, 2000, 1232-1247.

Santos, E. C.; Shiomi, M.; Osaka, K. & Laoui, T. (2006). Rapid manufacturing of metal components by laser forming, *International Journal of Machine Tools and Manufacture*, Vol. 46, No. 12-13, October 2006, pp. 1459-1468, ISSN 0890-6955

Slocombe, A. & Li, L. (2001). Selective laser sintering of TiC–Al2O3 composite with self-propagating high-temperature synthesis. *Journal of Materials Processing Technology*, Vol. 118, No. 1-3, December 2001, 173-178, ISSN 0924-0136

Song, Y. (1997). Experimental study of the basic process mechanism for direct selective laser sintering of low-melting metallic powder. *CIRP Annals - Manufacturing Technology*, Vol. 46, No. 1,1997, 127–130, ISSN 0007-8506

Van der Schueren, B. & Kruth, J.P. (1995). Laser Based Selective Metal Powder Sintering : A Feasibility Study. *Proceedings of the Laser Assisted Net Shape Engineering*, pp. 517-523, October 1994, Erlangen, Germany

Waterman, N.A.; Dickens P. (1994). Rapid product development in the USA, Europe and Japan. *World Class Design To Manufacture*, Vol. 1, No. 3, 27–36, ISSN 1352-3074

Xu, X.; Sachs, E. & Allen, S. (2001). The design of conformal cooling channels in injection molding tooling. *Polymer Engineering And Science*, Vol. 41, No. 7, July 2001, 1265–1279, ISSN 0032-3888

Yadroitsev, I.; Shishkovsky , I.; Bertrand P. & Smurov I. (2009). Manufacturing of fine-structured 3D porous filter elements by selective laser melting. *Applied Surface Science*, Vol. 255, No. 10, March 2009, 5523–5527, ISSN 0169-4332

Yadroitsev, I.; Thivillon, L.; Bertrand, P. & Smurov, I. (2007). Strategy of manufacturing components with designed internal structure by selective laser melting of metallic powder. *Applied Surface Science*, Vol. 254, No. 4, December 2007, 980-983, ISSN 0169-4332

Zhang, J. & Ashby, M. F. (1992). The out-of-plane properties of honeycombs, *International Journal of Mechanical Sciences*, Vol. 34, No. 6, June 1992, pp. 475–489, ISSN 0020-7403

14

Experimental Analysis of the Direct Laser Metal Deposition Process

Antonio D. Ludovico, Andrea Angelastro and Sabina L. Campanelli
Polytechnic of Bari, Department of Management and Mechanical Engineering
Italy

1. Introduction

In the current economic situation, with a liberalized and worldwide trade, the weight of time and costs reduction for developing new products increases. In particular, the tooling processes, which are already expensive and time consuming, recently became a critical trouble. The geometric complexity, the number of variants and the amount of requirements that characterize the products that are required by market increase this trouble. Reducing time to market and the international competitiveness are two of the biggest challenges for manufacturing companies of the twenty-first century. They not only have to produce component with high quality, low cost and better functionality than before, but also have to respond to customer requests in a more reactive way as possible. To do this they have to accelerate, of course, the tooling stage and to reduce processing times. Indeed, it is known that a delay of some weeks during the development and the commercialization of a new product may result in a loss of profits of 30%. These factors thrust the manufacturing companies to find solutions both in technology and in organization.
The Material Accretion Manufacturing (MAM), consisting of additive technologies, and the modern operational structures, such as those provided by the Concurrent Engineering, are effective tools that help to reduce time and costs. Probably, they are the best weapons that western industry can take to survive the fierce competition with those countries, mainly asiatic and eastern europe countries, where labor has a very low cost, but is nowadays able to achieve good quality products.
Among the various technologies MAM, thanks to which it is possible to apply the basic concepts of Rapid Prototyping to realize finished products, stands out, with its huge capacity, the Direct Laser Metal Deposition (DLMD) process of metal powders. But what are the key features of this process? What are the properties of the products it is able to accomplish? For what types of applications have already been used or are expected to do it? This chapter will give a comprehensive answer to these questions.
The DLMD pertain the group of technologies called Material Accretion Manufacturing (MAM), and, in particular, is based on the principles of rapid prototyping (RP) and laser cladding (Peng et al., 2005). The MAM technologies start by the 3D design to get the object in a single pass through an additive processing, that is overlapping each other layers with a small thickness. Other words, there is a conversion of the three-dimensional piece into N two-dimensional overlapping pieces, that constitute the layers.

The idea of making plastic, metal and ceramic functional pieces by overlapping layers of material led to the creation first of the rapid prototyping and then of the MAM technologies. In fact, the principles behind these technologies were already used in 1934, when a process which, by means of a blowtorch, allowed the fabrication of metal pieces by welding metal plates placed one above the other was invented. It followed the cladding technique made by means of the welding arc first and of the laser then, combined with the supply of material in powder or wire. However, the limitations in accuracy, in resolution and in mechanical properties of parts obtained by both processes still hampered the production of functional metal components. The development of the lasers and of the microprocessors used for an accurate position and speed control of the beam led, in the early '80s, to the birth of rapid prototyping, and in particular of stereolithography (Lewis & Schlienger, 2000).
The aim of the mechanical industry, however, has always been to use the basic principles of rapid prototyping to direct manufacture metal parts with mechanical properties as close as possible to those of parts obtained with conventional processes. This led to research and improvement of various technologies in that direction. With the development of computer aided design (CAD) and computer aided manufacturing (CAM) technologies, particularly with the advent of CAD/CAM and CNC systems, it was possible to combine the 3D CAD representation of the desired component and the deposition process that was able to achieve it (Zhong et al., 2001).
The accurate control of manufacturing processes so obtained allowed an increasing use of technologies like MAM for the production of metal parts with good mechanical properties. In conclusion, from prototyping, which creates objects with only an aesthetical, conceptual, technical or pre-series function, it has arrived to MAM technologies, which generates the finished and with good mechanical properties products (Angelastro, 2010; Santon et al., 2006; Keicher & Smugeresky, 1997).
The creation of a component with a MAM technology, usually, takes place with the succession of the following phases:

1. *Creating the CAD model of the piece* - The piece has to be CAD designed using a three dimensional or surface modeling.
2. *Conversion of the CAD model into the STL format (Solid To Layer)* - The STL format implies tessellation (or mesh) of the inside and outside surfaces of the piece through triangular elements, each identified by the spatial coordinates of three vertices and three direction cosines of the external normal.
3. *Workpiece layout and orientation and support generation* - The direction of optimal growth of the product greatly influences dimensional accuracy, surface finishing, production time and costs. In some MAM technologies support generation is essential to support possible overhangs.
4. *Slicing* - The STL model is dissected by a series of planes perpendicular to the growth direction. Slicing can be uniform, resulting in layers of uniform thickness, or adaptive, in which case the thickness will be determined by the surface curvature, to minimize the steps appearance of the surface (staircase effect).
5. *Sections fabrication* - The physical construction of the piece is performed overlapping layers each other.
6. *Supports remove, cleaning and finishing* – Eventual supports are taken away from the piece; so the piece is cleaned, by removing the excess material attached to the inner and outer surfaces, and then finished.

2. Process

The development of Direct Laser Metal Deposition has been pursued simultaneously by a lot of researchers for several years and finally gave birth, in 1995-96, three very similar processes. These processes are now acknowledged worldwide, under the names of Directed Light Fabrication (DLF), Light Engineered Net Shaping (LENS) and Direct Metal Deposition (DMD), developed respectively at the Los Alamos National Laboratory (Los Alamos, New Mexico, USA), at Sandia National Laboratory (Albuquerque, New Mexico, USA) and at the University of Michigan (Ann Arbor, Michigan, USA) (Lewis & Schlienger, 2000).
The DLMD process combines powder metallurgy, solidification metallurgy, laser, CAD-CAM, CNC, rapid prototyping and sensors technologies (Zhong et al., 2004). This process allows to produce metal components ready to use, in a single step, without the need for dies, molds or tools and using a wide variety of metals, including those very difficult to work with conventional techniques (Milewski et al., 1998; Choi & Chang, 2005).

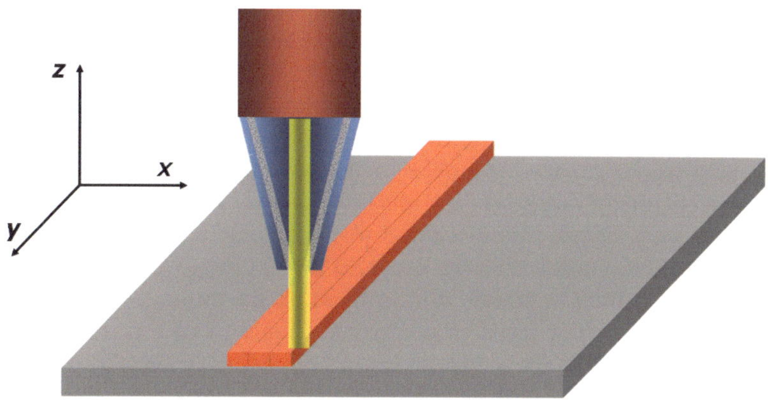

Fig. 1. Schematic representation of the process

The DLMD of metal powders is an additive process in which a metal powder is deposited on the base material, layer by layer, and melted with it. The connection between the metal powder and the below surface is similar to a weld with a high mechanical strength; subsequently the added metal can be subjected to any machining process. Fig. 1 shows a schematic drawing of the process.
The DLMD process starts from the three-dimensional CAD representation of the part and then fabricate it using the technique of layer by layer building (Dwivedi et al., 2007). Now two different processing solutions can be adopted, depending on the type of system used, which can be a 3-axis or a 5-axis CNC system. With the 3-axis systems the steps are identical to those described above for all MAM processes, and so it need for the conversion of CAD files in STL files. The model obtained is then equipped by supports, useful to sustain overhangs, and finally subjected to the stage of slicing. The slicing consists in a dissection of the model into a series of layers whose normal is parallel to the Z direction of construction. In systems with 5 or more axis, in addition to translations along X, Y, Z, the head of the machine can perform rotations around the same directions. For these systems, increasingly popular, the slicing operation is carried out directly on 3D-CAD model, with layers having its normal always parallel to construction direction, which may differ from the Z axis. In this case there is an advantage: it not needs the conversion of 3D-CAD model into the STL model

(Milewski et al., 1998). Furthermore, with 5-axis systems, components which require the deposition of layers with different orientation with respect to the previous ones can be fabricated by tilting the laser beam and nozzle that supplies powder or by tilting the piece under construction. Thus, the axis of the laser beam is always normal to the plane of deposition. This possibility provides another advantage: it is possible to produce protruding or overhanging parts without the need of supports or of a powder bed below them, like, for example, in SLS.

With the slicing, the geometry of each layer is determined and it is possible to define the path of the laser beam. Therefore, the movements of the machine head and of the worktable and the commands to control the laser and the powder feeder were defined. Because of the influence of high thermal stresses and of surface tension, the tool path for the DLMD is more complicated than that one of the traditional rapid prototyping processes. It is important to design an appropriate path, as it not only improves the quality of the final product, but also increases the efficiency of the manufacturing process. In the realization of the layers, DLMD technology incorporates laser cladding, a technique of metal surfaces coating performed by laser: so, someone calls it "three-dimensional laser cladding" (Zhang et al., 2003; Lu et al., 2001; Qian et al., 1997).

The laser beam, through its heat energy, creates a molten pool with dimensions, on the XY plane, varying from half to five times the diameter of the focal spot, according to the adopted power and scanning speed. A flow of metal powder is continuously injected in the molten pool, created on the substrate by the laser. Powder melts completely and then solidifies, making a very compact and welded to the substrate track. The beam scanning path causes the formation of a set of tracks, one next to the other, with a designated amount of overlap. Usually, each track is started and stopped on the outer edges of the piece until the entire first section, the first layer, is completed. The size of the molten pool is also influenced by the heat stored in the substrate or in the previously deposited material. If the heat loss occurs very quickly the width of the molten pool decreases, on the contrary, if the heat loss occurs very slowly the width of the molten pool increases: setting the laser power and the processing speed the width of molten pool can be determined. This is important because the size of the molten pool affects the properties of the final product (Lewis & Schlienger, 2000).

The first layer is built on a metal substrate that is the construction base of the piece. Usually, the substrate is made of the same material of the component has to be built. Sometimes a material best-able to dissipate the heat can be used as substrate material (Angelastro et al., 2007a).

After deposition of the first layer, laser head, that includes focusing lens and powder delivery nozzle, moves in the construction direction (Z direction in the case of 3 axis systems) and starts deposition of the second layer. Therefore, process recurs, line-by-line and layer-by-layer, until the entire component is built up (Atwood et al. 1998; Angelastro et al., 2007b). In some cases, the laser head is fixed and the worktable move.

3. Comparison with conventional manufacturing technologies

Typically, metal parts are produced by means of thermomechanical processes, including casting, rolling, forging, extrusion, machining and welding. Conventionally, to obtain the finished pieces multiple steps and heavy equipment, forms, dies and tools are required. However, the use of this equipment is justified and advantageous in mass and in large

batches production. But when the part is unusual in form, has complex internal cavities or has to be made in small batches costs and time to prepare the production rapidly grow and, therefore, conventional technologies are practically unenforceables (Zhang et al., 2003).
On the other hand, the DLMD process fabricates, in a single step, the required components already with the final material, with mechanical properties close to or, in some cases, higher than those obtained with conventional processes. It can make parts with extremely complex geometries, usually oversized of 0.025 mm, so after a quick cleaning, to obtain the tolerances and the surface finishes required, they are ready to use. For example, in the experimental tests of Lewis and Schlienger, the tensile properties of some deposited materials (AISI 316 stainless steel, Inconel 690 and Ti-6Al-4V) were measured and compared with those of the same conventional processed materials. These data show that it is possible to obtain equivalent and superior properties to those of wrought annealed materials. So, multiple thermomechanical treatments, required in conventional procedures, can be avoided (Lewis & Schlienger, 2000).
Using DLMD of metal powders chemical segregation is eliminated. This problem occurs in parts obtained by casting and to cure it homogenizing heat treatments and plastic deformations for the grain refinement are indispensable. Chemical homogenization is achieved through randomization of composition by using powders as input material and by limiting chemical diffusion in the liquid state to the boundaries of the small molten pool that is used to deposit the entire component. (Lewis & Schlienger, 2000). Moreover, the high solidification rate, resulting from the narrow HAZ (Heat-Affected Zone), and the high temperature gradients, which are generated around the molten pool, allows a fine grain microstructure (Zhang et al., 2003).
Another positive aspect of this process consists in the production speed increasing. Indeed, a study by the National Center of Manufacturing Science, Michigan, showed that the direct laser deposition of metal powders can reduce by 40% the time to produce a mold.
The convenience of this technology is also determined by the reduction of the labor required and of the equipment cost, since there is only one machine that runs the entire process automatically. Later, a machine tool or EDM is used to finish the piece. This results in a decrease of the floor area needed for the factory and for the warehouse.
Even compared to the use of CNC machine tools, in many cases, DLMD is suitable because of at least further two advantages: no waste and higher material purity of the final component. In fact, the processes performed by the CNC machine tools do not result only in the finished part, but also in metal chips, fluids, lubricants for cutting and worn tools that need to dispose of. The DLMD process results only in the desired component. There are no wastes.
Since this process can be achieved in a controlled atmosphere, with high purity inert gas, powder that was not melted by the laser can be sucked and used again. This feature makes DLMD particularly attractive when it has to use very expensive or hazardous materials, which require containment during processing. Then, the material purity of the final piece is kept high by eliminating contamination resulting from contact with cutting tools, surfaces of the molds or forms, lubricants and chemicals for cleaning (Milewski et al., 1998).

4. Process equipment

The system which allows DLMD process of metal powders consists of a set of subsystems that are interconnected. These subsystems, during the process, provide a full control of laser power, speed, position of the spot, powder flow rate and size of molten pool.

All subsystems currently used consist basically of:

- control workstation;
- laser source;
- powder feeding system;
- multi-axis CNC system.

In some cases there are also a feedback control system and a controlled atmosphere chamber.

The control workstation is a PC that receives the part data in the CNC instructions format, derived from 3D-CAD model by means of a CAD/CAM software. Also it communicates with all the subsystems and allows, therefore, the intelligent control of the entire deposition process.

The laser source used must have a high power, typically between 500 W and 6000 W, to be able to melt both the metal substrate and the metal powder. By means of mirrors or optical fibers, the laser beam is conveyed to a lens that focalises it to an optimal spot size.

The powder feeding system make available the powder in the work area and also can mix the several types of powders when it needs to use them simultaneously. It consists of some containers and related devices for the mass flow rate control, a mixer and a nozzle to deposit the material into the molten pool. In systems with several containers it is possible to generate mixtures of powders, filling each one with a different powder and combining them. The mixing of powders can be obtained directly in the molten pool, feeding them simultaneously, or within the mixer. This solution makes the process able to produce different types of alloys or to vary the composition of the material, within the piece under construction, depending on the desired properties. The mechanisms for the mass flow rate control can be based on different operating principles.

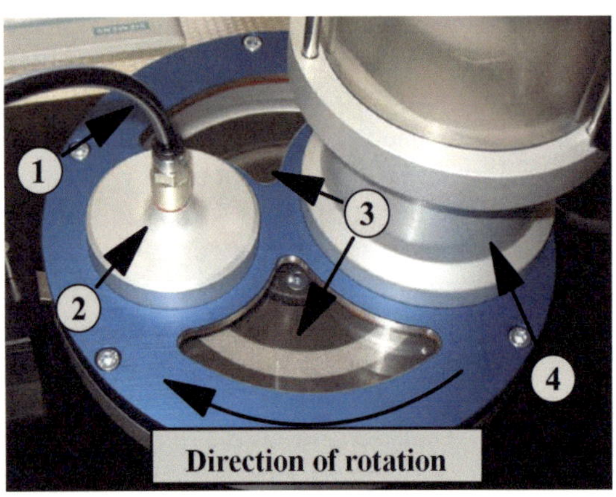

1. **Powder feeding pipe**
2. **Powder outlet**
3. **Slot on the disc**
4. **Container**

Fig. 2. Disc powder feeder

For example, a mechanism is constituted by a hopper, under each container, having a horizontal shaft with some caves on the lateral surface. The rotation speed of the shaft can be varied to control the mass flow rate. Another type of powder feeding system consists of a container from which the powder flows by gravity into a slot created on a rotating disc. The powder is transported to a suction unit, which through a carrier gas flow, sent it to a nozzle. The powder mass flow rate depends on the slot size and the rotating speed of the disc. The last type of powder feeding system consists of a pneumatic screw feeder; in this case the mass flow rate of powder depend on the rotating speed and on the size of the screw. Then the powder has to be transferred to the work area: this can be done by means of a carrier gas (argon, helium, nitrogen) or simply by means of gravity.

The powder feeding system ends with a nozzle that has to inject the powder into the molten pool (Lin, 1999b). So that the capture efficiency of powder by the molten pool can be high, the nozzle has to be placed very close to the pool itself (Lin, 1999a). This implies that the nozzle is subjected to high thermal loads and, therefore, is usually made of brass. So it also is able to reflect the laser radiation which is back-reflected from the molten pool (Lin, 1999c). The nozzle can have different configurations. There are two basic ones: with a coaxial supply and a lateral supply of powder (Weerasinghe & Steen, 1983; Zekovic et al., 2007). In the first one, there is the advantages of the independence from the moving direction of the head, of the controlled heating of the powder before it goes into the molten pool and of the high capture efficiency of powder by the molten pool. In this case, the powder moves coaxially to the laser beam. The nozzle, cone-shaped, includes two internal annular interstices. The carrier gas, or primary gas, is used to transport powder outer the first interstice. A secondary gas, usually argon or helium, is injected into the inner interstice to perform various functions: to avoid fraying of powder flow, ensuring the confinement along the nozzle axis, and to protect the molten pool from oxidation. Not all the components, however, may be accessible by a coaxial powder nozzle. On the other hand, the lateral supply of powder allows the treatment of all types of geometries, designing dedicated nozzles. Basically, the nozzles are lateral pipe with tailored length, shape and diameter (Schneider, 1998).

Fig. 3. Coaxial nozzle

The multi-axis CNC system monitors the position, the speed and the motion path of the laser spot. The motion path is determined by slicing the STL model or directly from 3D-CAD model, if the system has 3 or more axis (usually 5) respectively.

In some cases it is also envisaged the use of a feedback control system. Indeed, actually, the deposited powder mass results to be irregular because of the oscillation of the laser power and of the difficulty of controlling the powder mass flow rate (Fearon & Watkins, 2004; Mazumder et al., 2000). This factors result in a defective deposition, characterized by uneven surfaces and lack of adhesion between the layers. To control the height of the deposition, that is corresponding to the layer thickness, a feed back control system can be used. This subsystem makes the control of some variables, such as the energy density and powder flow rate (Choi & Chang, 2005; Krantz & Nasla, 2000). It generally consists of a detection height unit (Height Sensing Unit), a processing unit of the feedback signal (Signal Processing Unit) and an optical sensors apparatus consisting of CCD cameras. The detection height unit elaborates the images from the optical sensors apparatus, which generally consists of at least two cameras in order to neutralize the effect of the laser head movement (Mazumder et al., 1999). Trough the processing unit it is possible to remedy an anomalous deposition. During the trial, a deposition lower than desired may be covered by repeating the path on the same layer before moving to the next. Instead, a deposition greater than desired can be resolved reducing the laser power and the powder flow rate as quickly as possible or (Bi et al., 2006). When the use of shielding gas ejected from the head is insufficient to have a good deposition, the process can be performed in a sealed chamber in which it is created a controlled atmosphere consisting of inert gases, such as argon, helium or nitrogen. Without this condition, the fabricated component may have excessive porosity and oxidation, with insufficient adhesion between layers and, consequently, with poor mechanical properties (Erzincanh & Ermurat, 2005). When a controlled atmosphere chamber is used, DLMD becomes a free waste process. Indeed, the powders not fused by the laser during the process can be retrieved through a suction system, filtered and reused. This characteristic makes it particularly attractive when it has to process very expensive or hazardous materials, that require containment during processing.

5. Process materials

Lots of materials can be used with this process, including those very hard to work with conventional technologies. Some of tested materials are (Alemohammad et al., 2007; Simchi, 2006; Wu et al., 2002; Yellup, 1995):

- stainless steels (316, 304L, 309, 420);
- maraging steels (M300, M2);
- tool steels (H13, P20, P21, S7, D2);
- nickel-based alloys (Inconel 600, 625, 690, 718);
- titanium alloys (Ti-6Al-4V);
- aluminium;
- copper and its alloys;
- stellite;
- tungsten carbide.

Although the material can be supplied both in powder and in wire form, powder is generally used for several reasons: the process is more flexible because, at any time, the changes in the layer thickness and in the composition of material are possible; many more elements and alloys are available in powder; there is no direct contact with the molten pool; the laser beam can pass through the powder flow, instead of being intercepted by the wire.

The flexibility that acquires the process through the use of powders of different metals is extraordinary (Liu & DuPont, 2003). In fact, the properties of the product can be controlled almost point by point by combining different powders in various amounts during the process. The powders of different metals can be premixed and then conveyed by a single powder feeder to the work area. But in this case, due to different densities, sizes and shapes of particles, vibrations in the feeding system can cause segregation of the powders, which, in turn, may change uncontrollably composition and properties of deposited material. This problem is completely eliminated conveing the different powders toward the work area using separate powder feeder and injecting them simultaneously, with desired proportion, directly into the molten pool (Schwendner et al., 2001). Thus, the desired functional composition gradients can be obtained by increasing or decreasing the feeding speed of the various metals. Also, any type of composition, with the types of powders available, can be easily used, eliminating all the necessary steps for the production of the powders of the desired alloy (Lewis & Schlienger, 2000).

6. Process parameters

The quality of the parts that can be achieved with DLMD is strongly influenced by numerous process parameters. The problem of determining what are the parameters that influence the final properties and in what measure is complicated by the significant interactions between them (Choi & Chang, 2005).

A careful control of process parameters becomes crucial to obtain products having a satisfactory dimensional accuracy and good mechanical properties. The main process parameters are:

- powder feed rate;
- laser power;
- scanning speed;
- beam diameter;
- layer thickness;
- overlap percentage;
- energy density.

The powder feed rate is the amount of powder per time unit which exits from the nozzle. It affects especially the layer thickness: high flow rates result in very thick layers. In turn, the thickness of the deposited layer determines not only the dimensional and geometrical accuracies of the product, but its mechanical properties also. In fact, with the same laser power, it becomes more difficult to obtain good adhesion between the layers when layer thickness increases.

With the same material to be deposited, the laser power depends on the type of laser used. Currently, the laser sources most widely used for DLMD process are CO_2 and Nd:YAG sources. The power of lasers which are usually used for the process stands in the range from 1 to 18 kW. The main difference between the CO_2 and Nd:YAG lasers is the wavelength. The Nd:YAG laser has a wavelength of 1.06 μm while that of the CO_2 laser is 10.6 μm. The absorption of the laser energy for most metals increases when the wavelength decreases. Various studies showed a greater depth of molten pool through the use of a Nd:YAG laser due to increased energy absorption. However, most commercial machines are equipped with a CO_2 laser because of its high efficiency, lower cost and easier maintenance than Nd:YAG laser.

The scanning speed is the speed with which the laser beam executes the tool path about the section that has to be built. If the scanning speed increases, the time required to complete sections and, thus, to realize the component decreases. However, in this case there are some disadvantages that affect the quality of the final component. For example, because of the high speeds, the layers may not achieve the desired thickness or a difference in height between the edges and the central part of the part may originate: in fact, to reverse the scanning direction the laser head slows down until to stop in correspondence of the layer edge and then accelerates to the set speed: if the speed increases the time needed to slow down and accelerate increases accordingly, leading, with a constant powder feed rate, to the deposition of a greater amount of material at the edge than the center of each line.

The beam diameter is a key parameter for the laser processing. It determines the spot diameter, which is equal to the beam diameter at the focus, and, consequently, the size of the track. In turn, the size of the track affects the surface roughness and the minimum size of achievable features. Finally, the beam diameter affects directly the energy density, which is after defined.

The thickness of the layer, as seen for all MAM processes, is defined by the slicing of the CAD model. One of the greatest difficulties in the DLMD is just to maintain the desired layer thickness. Therefore, it is preferable to use a feedback control system.

The overlap percentage is defined as the ratio between the area of overlap between two adjacent tracks and the total area of the track itself, measured on the work plane (Schneider, 1998):

$$\%O = \frac{A_{overlapped}}{A_{tot}} \tag{1}$$

The overlap percentage affects the porosity and, consequently, the density and mechanical properties of the deposited material.

Several studies have shown that the quality and mechanical properties of the DLMD components are strongly influenced by the size of the molten pool and by the residual stresses, which can be directly controlled from the laser energy input (Peng et al., 2005). The laser energy input can be described by the specific energy or energy density E [J/mm^2], defined by the following equation:

$$E = \frac{P}{hv} \tag{2}$$

where P is the laser power [W], h is the hatch spacing [mm], defined as the distance between the vectors drawn from the center of the laser spot in two parallel and consecutive tracks, v is the scanning speed [mm/s]. In the case of overlapping zero, h becomes equal to the spot diameter.

7. Main experimental results

In this paragraph, the main results on experimental tests conducted by the authors on the Nickel alloy (Colmonoy 227-F) will be illustrated. This material, usually used to manufacturing moulds, is characterized by an elevated hardness (22-27 Rockwell C), a density of 8.53 g/cm^3 and a melting point of 915 °C; it is able to tolerate extreme job conditions because of its elevated resistance to abrasion, to corrosion, to stresses and to high temperatures. Colmonoy 227-F is furnished in powder form with granules of spherical shape which have a maximum size of 106 μm.

The aim of experiments was to obtain deposited layers, welded to the substrate and between them, in order to realize good quality 3D parts by means of the DMLD process. A six axis machine, equipped with a welding head and with a CO_2 laser with a maximum power of 3kW, was used for experiments.

7.1 Gravity distribution system

First, a gravity distribution powder apparatus was designed and built. In this system the powder was deposited by gravity over an AISI 304 steel substrate with a thickness of 10 mm. The main components of this system are illustrated in Fig. 4 and consisted in:

- a powder container;
- a device for ON/OFF regulation of the powder feed rate controlled electrically by means of signals coming from the laser machine;
- a nozzle for the coaxial deposition of the powder on the melt pool created from the laser, working by gravity or by a carrier fluid.

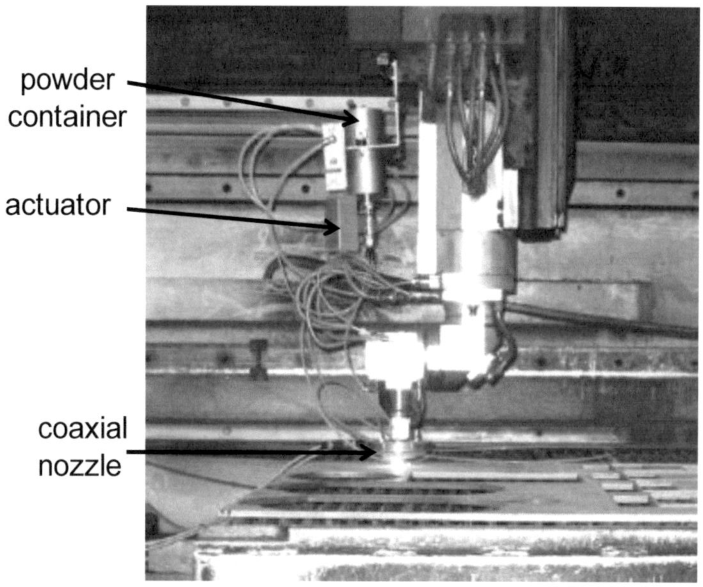

Fig. 4. Gravity powder feeding system

One layer specimens were produced with this system changing the two factors laser power and scanning speed.
Table 1 shows the experimental plane that was used in this step.

Test Number	Laser Power [kW]	Scanning Speed [m/min]	Energy Density [J/mm^2]
A1	0.5	1.2	83.3
B1	0.5	3.0	33.3
C1	0.5	4.8	20.8
D1	1	1.2	166.7
E1	1	3.0	66.7
F1	1	4.8	41.7
G1	1.5	1.2	250
H1	1.5	3.0	100
I1	1.5	4.8	62.5

Table 1. Experimental plane

The Energy Density was calculated by means of equation 2 setting overlapping to zero and the value of hatch distance equal to the spot diameter (0.3 mm in this case). The built samples were analyzed in terms of adhesion to the substrate, porosity and cracks.
Fig. 5 shows transversal sections of the nine samples. The analysis of results showed that good result could be obtained setting laser power to 1.5 kW and the scanning speed to 3.0 m/min. Moreover, if the powder feed rate could be regulated, better structure would be achieved with specific energies lower than 100 J/mm^2, using both lower values of the laser power and of the scanning speed.
Sections orthogonal to laser scanning direction (Fig. 5) showed that almost all samples had a wide cavity between deposited metal and substrate. In these specimens only few initial tracks were stuck to the substrate, whereas all the others stayed above it. Only samples H1 and I1 showed a full adhesion between the deposited metal and the substrate.

Fig. 5. Transversal cross sections of specimens

The sample I1 realized with a specific energy of 62.5 J/mm^2, even if was almost free of porosity, was afflicted with a very large number of vertical cracks (Fig. 6), probably due to the very high thermal gradient and the excessive cooling rate caused by the application of a very high laser power (1500 W) at a high scanning speed (4.8 m/min). Therefore, it was possible that cracks were generated from residual stresses melted and re-solidified metals of the deposited layer and of the substrate, having different expansion coefficients, couldn't be able to contrast stresses with an adequate strain.
At last, considering such factors as adhesion on the substrate, porosity and presence of cracks it can be concluded that, the best specimen was H1. This sample (Fig. 6) was obtained

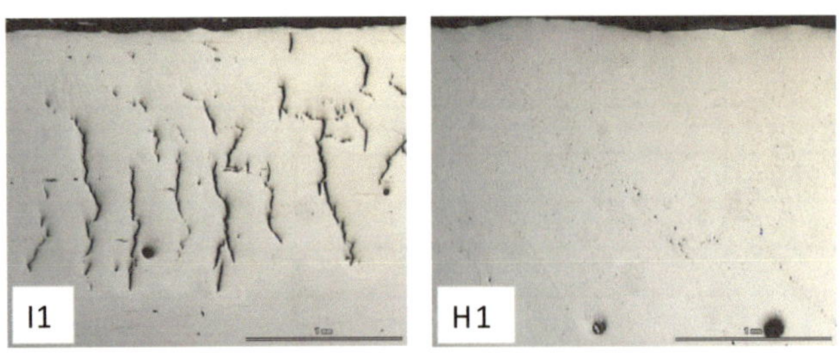

Fig. 6. 50X magnification of transversal sections for specimens H1 and I1

with a laser power of 1 kW and a scanning speed of 3.0 m/min, resulting in a specific energy of 100 J/mm². Nevertheless, it showed some spherical cavities, although limited in number, probably due to the gas among the powder grains before the process or generated itself from decomposition of metal during the process.

7.2 Pneumatic distribution system

Later on it, a pneumatic system for the distribution of the powder was designed and built; it was able to make a discreet control of the powder feed rate. This system was installed again on the 6 axes laser machine, which was used before together with the gravity system (Fig. 7). This distribution system allowed a control of the powder feed rate in a range between 20.1 g/min and 29.2 g/min, acting on pressure and flow rate of carrier gas (Argon), which was taken directly from the laser machine.

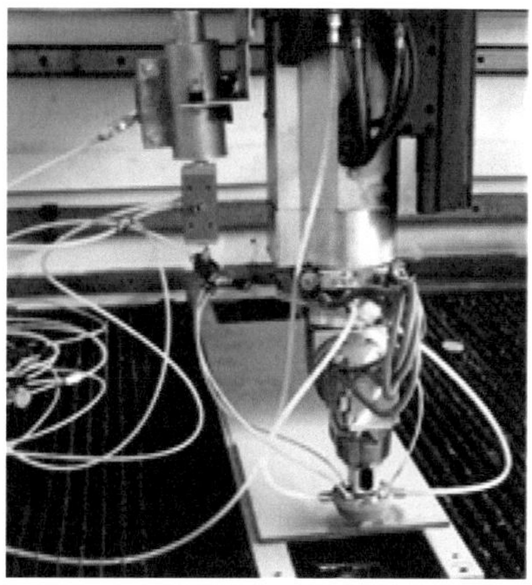

Fig. 7. Pneumatic distribution system

A reduced orthogonal experimental plan was chosen to plan experiments. Three factors were considered: powder flow rate (q), laser power (P) and scanning speed (v). They were changed on three levels, as shown in Table 2, resulting in 9 combinations of the parameters.

Test Code	Powder Flow Rate [g/min]	Laser Power [W]	Scanning Speed [m/min]	Energy Density [J/mm²]
A2	20.1	500	1.2	83.3
B2	20.1	750	3.0	50.0
C2	20.1	1000	4.8	41.7
D2	23.8	500	3.0	33.3
E2	23.8	750	4.8	31.3
F2	23.8	1000	1.2	166.7
G2	29.2	500	4.8	20.8
H2	29.2	750	1.2	125.0
I2	29.2	1000	3.0	66.7

Table 2. Experimental Plane used for the pneumatic distribution system

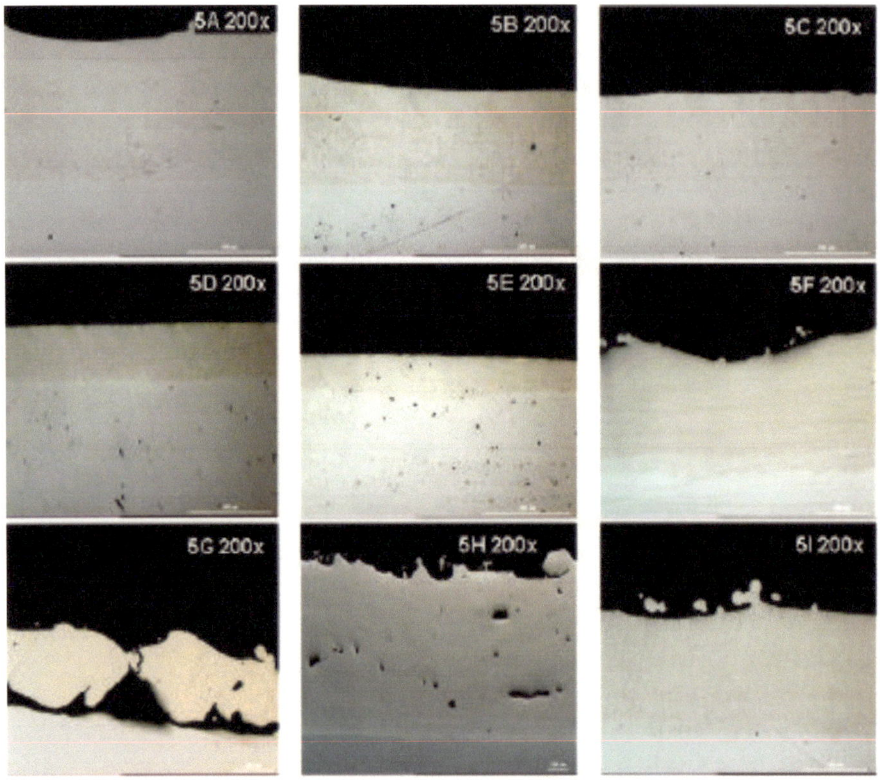

Fig. 8. 200X magnification of cross sections of five layers samples.

Values of the Energy Density E are also reported in Table 2.

One layer samples were obtained with the pneumatic powder feeding system. The quality of the parts, in terms of porosity and adhesion to the substrate, was evaluated. First of all, the 9 one layer samples were observed at the optical microscope in order to evaluate the adhesion between the layer and the substrate. The analysis of samples sections revealed that the material deposited has a good adhesion, excluding samples G2, H2 and I2, all three achieved with greatest powder flow rate (29.2 g/min).

Following a qualitative evaluation, it was possible to say that in the samples made with low and medium powder flow rate levels the porosity of deposited material was very low, so as to be comparable with that of the substrate. Porosity of samples G2, H2 and I2, made with greatest powder flow rate, amounting to 29.2 g/min, seemed to be worse than the others. Subsequently, the same experimental plan were used to realize 5 layers samples.

The 5 layers samples were analysed with the optical microscope, assessing, in this case too, adhesion between the first layer and substrate and between a layer and the contiguous one. Fig. 8 shows transversal sections for five layer samples. Also in this case, it was found that low (20.1 g/min) and medium (23.8 g/min) powder flow rate generate better results than high powder flow rate. In fact, specimens 5A, 5B, 5C, 5D, 5E, 5F, shows a good adhesion between a layer and the contiguous one; 5G, 5H and 5I were characterized by the worse adhesion between a layer and the contiguous one.

Roughness tests were also performed on built samples. Results for roughness showed that it increases with powder flow rate and laser power but it decreases with scanning speed.

It means that good surface samples can be obtained setting the powder flow rate and laser power to the minimum value and the scanning speed to the maximum value; on the other hand, considering factors as adhesion and presence of pores and cracks, the best results can be obtained in correspondence of minimum values for the three analysed process parameters. Thus, 3D components can be realised by means of the DLMD process using low values for powder flow rate, laser power and scanning speed (20.1 g/min, 500 W, 1.2 m/min, in this case).

8. DLMD applications

The above mentioned capacities of DLMD process allow to build, modify and repair any metal piece reaching results unimaginable at a previous time. The application field of the process is very broad and its expansion rate does not mention decrease. Some of today's applications cover the following areas:

- fabrication of metal products and functional prototypes without tools;
- cooling of molds and inserts used in casting;
- tools reconfiguration;
- repair and restoring of tools;
- cladding;
- smart components (sensors embedded in the component);
- functionally graded materials (materials with a gradual variation of composition);
- fabrication and repair of components for aviation and aerospace industries;
- fabrication and repair of military equipment;
- rapid production of molds and dies from metal sheets;
- creation of custom implants and surgical instruments.

Thanks to the know-how acquired over several years of experimentation on the DLMD process, performed in collaboration with the ELFIM company, it was possible to achieve the claddings and restorings of worn parts, some of them having very complex geometries.

Nowadays, it is widely known that the application of coatings by means of conventional processes, such as CVD (Chemical Vapour Deposition), PVD (Physical Vapour Deposition), PTA (Plasma Transferred-Arc) and other thermal spray methods, can significantly improve wear and corrosion resistance of a surface. However, these processes can generate only coatings very thin (1-10 μm) and somewhat porous, even causing heat affected zones of considerable size. On the other hand, DLMD is able to create a strong metallurgical bond between very different materials and allows to obtain coatings from 10 to 100 times thicker than those conventionally produced. Currently, this process is used to deposit a wide variety of stellites and nickel based alloys on the tool surfaces subject to very hard working conditions, increasing their life of about 5 times compared to traditional ones.

This is the case of the impeller of a centrifugal pump used for the extraction of natural bitumen (Fig. 9). Due to the passage of this highly viscous and containing sulphur liquid, the impeller suffers wear and corrosion that damage and put it out of use very quickly. Therefore, a manufacturer of these pumps has required the cladding with tungsten carbide, about 10 mm thick, upon specific surfaces of the impeller. Because of the enormous diversity of the expansion coefficients of the two materials, it was not possible to deposit tungsten carbide directly on steel of the impeller. So, it was used a nickel alloy as buffer material. Therefore, first it was deposited the pure nickel alloy and then the gradually variable composition layers were fabricated. Exactly, in their composition while the percentage of the nickel alloy reduces that of the tungsten carbide increases, to obtain a semi-finished whole surface of tungsten carbide.

Another application of the DLMD process was performed on a cochlea for the vegetable oil industry. This stainless steel screw, used to knead the olive paste, is subject to high wear, especially on the surfaces far from the rotation axis, where, of course, the tangential velocity is highest. Therefore, as shown in the Figure 10, a cladding of nickel alloy, which is more wear-resistant than stainless steel, was made.

The replacement of a broken or worn tool, generally, is not a convenient choice because of high costs and delays to support in production. However, the conventional repair processes, almost always, result in tools with properties below of the quality standards and in long waiting time due to pre- and post-heating and final finishing. Indeed, for example, to prevent cracking and excessive deformation, the steel tools, before welding, are heated to about 550 °C, after that all the parts which can not withstand heat up to this point are removed, held at this temperature for several hours and then post-heated after the repair. Moreover, the part repaired by welding is not as strong as the original tool and tends to be again damaged during the use. Thus, many companies do not allow the use of welding on tools because of the risk of damaging their base material.

Fig. 9. Coated pump impeller

Fig. 10. Coated cochlea

(a) (b)

Fig. 11. a) Restored mold component; b) Restored component of the crushing machine

The DLMD makes it easy to repair or restore damaged or worn out tools so that they become like new. In fact, this process allows to deposit, without creating porosity, the metal having the same composition and properties of the base material, creating a high resistance bond with it. The almost total absence of heat affected zone allows to preserve the properties of the tool, eliminates the pre-and post-heating, avoids the risk of causing deformation and minimizes the time required for final finishing.

This application has been used to rebuild the worn out parts of a mold for the production of glass bottles and a tool of a crushing machine, which is able to crushing stones. In Fig. 11a and Fig 11b the components after deposition and before the finishing are shown.

9. Conclusions

This chapter has been organized to pursue the following objectives: a) to describe the DLMD process and its performances, in terms of capacities and applications; b) to seek, through experimental analysis, the functional relationship between certain attributes of quality of built parts and the parameters used to control the process.

First, a description of the DLMD systems in terms of equipments and process parameters was furnished. Later on it, some experimental tests were presented in order to obtain 3D good quality parts.

Finally, some industrial applications were illustrated. It was demonstrated how it is possible to achieve the claddings and the restoring of worn parts, some of them having very complex geometries.

10. Acknowledgments

The authors wish to acknowledge Elfim srl Company for the support given to the experimentation.

11. References

Alemohammad, H.; Toyserkani, E. & Paul, C.P. (2007). Fabrication of smart cutting tools with embedded optical fiber sensors using combined laser solid freeform fabrication and moulding techniques. *Optics and Lasers in Engineering*, Vol. 45, No. 10, April 2007, pp. 1010-1017, ISSN 0143-8166

Angelastro, A.; Campanelli, S.L.; Casalino, G. & Ludovico. A.D. (2007a). Analysis of a tool steel sample obtained by Direct Laser Deposition, *Annals of DAAAM for 2007 and Proceedings of the 18th International DAAAM Symposium*, pp. 23-24, ISBN 3-901509-58-5, Zadar, Croatia, October 2007, Published by DAAAM International, Wien, Austria

Angelastro, A.; Campanelli, S.L.; Casalino, G. & Ludovico, A.D. (2007b). Dimensional and metallurgical characterization of free-formed Colmonoy 227-F samples obtained by laser radiation. *ICALEO 2007 Congress Proceedings, Laser Materials Processing Conference*, pp. 206-215, ISBN: 978-0-912035-88-8, Orlando, FL, USA, October-November 2007, Published by Laser Institute of America

Angelastro, A.; Campanelli, S.L.; Ludovico, A.D. & D'alonzo, M. (2009). Design and development of an experimental equipment for the Direct Laser Metal Deposition process - Proceedings of 9th A.I.Te.M. Conference 2009, pp. 113-115, ISBN 8895057074, Torino, Italia, September 2009

Angelastro, A.; Campanelli, S.L. & Ludovico, A.D. (2010). Characterization of Colmonoy 227-F samples obtained by direct laser metal deposition. *Advanced Materials Research*, Edited by Hashmi, M.S.J.; Yilbas, B.S. & Naher, S.,Vols 83-86, pp. 842-849, Trans Tech Publications Ltd., ISBN 0878492976

Angelastro, A. (2010). Studio e sviluppo di un sistema sperimentale per la deposizione laser diretta di polveri metalliche. *Ph.D. Thesis in Advanced Manufacturing Systems.* March 2010

Atwood, C.; Griffith, M.; Harwell, L.; Schlinger, E.; Ensz, M.; Smugeresky, J.; Romero, T.; Greene, D. & Reckaway, D. (1998). Laser Engineered Net Shaping (LENS TM): a tool for direct fabrication of metal parts. *Proceedings of the laser materials processing conference ICALEO'98*, ISBN 10 0912035587, Orlando, FL, USA, November 1998, Published by Laser Institute of America, Orlando.

Bi, G.; Gasser, A.; Wissenbach, K.; Drenker, A. & Poprawe R. (2006). Characterization of the process control for the direct laser metallic powder deposition. *Surface and Coatings Technology*, Vol. 201, No. 6, December 2006, pp. 2676-2683, ISSN 0257-8972

Choi, J. & Chang, Y. (2005). Characteristics of laser aided direct metal/material deposition process for tool steel. *International Journal of Machine Tools and Manufacture*, Vol. 45, No. 4-5, April 2005, pp. 597-607, ISSN 0890-6955

Dwivedi, R.; Zekovic, S. & R. Kovacevic. (2007) A novel approach to fabricate unidirectional and branching slender structures using laser-based direct metal deposition. *International Journal of Machine Tools and Manufacture*, Vol. 47, No. 7-8, June 2007, pp. 1246-1256, ISSN 0890-6955

Erzincanh, F. & Ermurat, M. (2005). Comparison of the direct metal laser fabrication technologies. *Technical report, Gebze Institute of Technology, Dept. Design and Manufacturing Engineering*, Gebze, Turkey, April 2005

Fearon, E. & Watkins, K.G. (2004) Optimisation of layer height control in direct laser deposition, *Proceedings of the laser materials processing conference ICALEO'04*, Vol. 97, San Francisco, CA, USA, October 2004, Published by Laser Institute of America

Keicher, D.M. & Smugeresky, J.E. (1997). The laser forming of metallic components using particulate materials. *Journal of the Minerals, Metals and Materials*, Vol. 49, No. 5, May 1997, pp. 51-55, ISSN 1047-4838

Krantz, D. & Nasla, S. (2000). Intelligent process control for laser direct metal deposition, *Proceedings of the laser materials processing conference ICALEO'00*, Vol. 89, pp. D1-D10, ISBN-10: 0912035625, Dearborn, MI, USA, October 2000, Published by Laser Institute of America

Lewis, G.K. & Schlienger, E. (2000). Practical considerations and capabilities for laser assisted direct metal deposition. *Materials and Design*, Vol. 21, No. 4, August 2000, pp. 417-423, ISSN: 0261-3069

Lin, J. (1999a) A simple model of powder catchment in coaxial laser cladding. *Optics and Laser Technology*, Vol. 31, No. 3, April 1999, pp. 233-238, ISSN 0030-3992

Lin, J. (1999b) Concentration mode of the powder stream in coaxial laser cladding. *Optics and Laser Technology*, Vol. 31, No. 3, April 1999, pp. 251-257, ISSN 0030-3992

Lin, J. (1999c) Temperature analysis of the powder streams in coaxial laser cladding. *Optics and Laser Technology*, Vol. 31, No. 8, November 1999, pp. 565-570, ISSN 0030-3992

Liu, W. & DuPont, J.N. (2003). Fabrication of functionally graded TiC/Ti composites by Laser Engineered Net Shaping. *Scripta Materialia*, Vol. 48, No. 9, May 2003, pp. 1337-1342, ISSN 1359-6462

Lu, L.; Fuh, J.Y.H. & Wong, Y.S. (2001). *Laser-Induced materials and processes for rapid Prototyping*, Kluwer Academic Publishers, ISBN 0-7923-7400-2, Norwell, Massachusetts, USA

Mazumder, J.; Schifferer, A. & Choi, J. (1999). Direct materials deposition: designed macro and microstructure. *Material Research Innovations*, Vol. 3, No. 3, October 1999, pp. 118-131, ISSN 1432-8917

Mazumder, J.; Dutta, D.; Kikuchi, N. & Ghosh, A. (2000). Closed loop direct metal deposition: art to part. *Optics and Lasers Engineering*, Vol. 34, No. 4-6, October 2000, pp. 397-414, ISSN 0143-8166

Milewski, J.O.; Lewis, G.K.; Thoma, D.J.; Keel, G.I.; Nemec, R.B. & Reinert, R.A. (1998). Directed light fabrication of a solid metal hemisphere using 5-axis powder deposition. *Journal of Materials Processing Technology*, Vol. 75, No. 1-3, March 1998, pp. 165-172, ISSN 0924-0136

Peng, L.; Taiping, Y.; Sheng, L.; Dongsheng, L.; Qianwu, H.; Weihao, X. & Xiaoyan, Z. (2005). Direct laser fabrication of nickel alloy samples. *International Journal of Machine Tools and Manufacture*, Vol. 45, No. 11, September 2005, pp. 1288-1294, ISSN 0890-6955

Qian, M.; Lim, L.C.; Chen, Z.D. & Chen, W.L. (1997). Parametric studies of laser cladding processes. *Journal of Materials Processing Technology*, Vol. 63, No. 1-3, January 1997, pp. 590-593, ISSN 0924-0136

Santosa, E.C.; Shiomi, M.; Osakada, K. & Laoui, T. (2006). Rapid manufacturing of metal components by laser forming. *International Journal of Machine Tools and Manufacture*, Vol. 46, No. 12-13, October 2006, pp. 1459-1468, ISSN 0890-6955

Schneider, M.F. (1998). Laser cladding with powder, effect of some machining parameters on clad properties. *Ph.D. Thesis*. Enschede, March 1998

Schwendner, K.I.; Banerjee, R.; Collins, P.C.; Brice, C.A. & Fraser, H.L. (2001). Direct laser deposition of alloys from elemental powder blends. *Scripta Materialia*, Vol. 45, No. 10, November 2001, pp. 1123-1129

Simchi, A. (2006). Direct laser sintering of metal powders: mechanism, kinetics and microstructural features. *Materials Science and Engineering: A*, Vol. 428, No. 1-2, July 2006, pp. 148-158, ISSN 0921-5093

Weerasinghe, V.M. & Steen, W.M. (1983). Laser cladding with pneumatic powder delivery. *Proceedings of 4th International Conference on Lasers in Materials Processing*, pp. 166-175, Los Angeles, CA, USA, January 1983

Wu, X.; Sharman, R.; Mei, J. & Voice, W. (2002). Direct laser fabrication and microstructure of a burn-resistant Ti alloy. *Materials and Design*, Vol. 23, No. 3, May 2002, pp. 239-247, ISSN 0261-3069

Yellup, J. (1995). Laser cladding using the powder blowing technique. *Surface and Coatings Technology*, Vol. 71, No. 2, March 1995, pp. 121-128, ISSN 0257-8972

Zekovic, S.; Dwivedi, R. & Kovacevic, R. (2007). Numerical simulation and experimental investigation of gas-powder flow from radially symmetrical nozzles in laser-based direct metal deposition. *International Journal of Machine Tools and Manufacture*, Vol. 47, No. 1, January 2007, pp. 112-123, ISSN 0890-6955

Zhang, Y.; Xi, M.; Gao, S. & Shi, L. (2003). Characterization of laser direct deposited metallic parts. *Journal of Materials Processing Technology*, Vol. 142 No. 2, November 2003, pp. 582-585, ISSN 0924-0136

Zhong, M.; Ning, G.; Liu, W. & Yang, L. (2001). Fundamental aspects on laser rapid and flexible manufacturing of metallic components. *Chinese Journal of Lasers B*, Vol. 10, p. 1130, ISSN 1004-2822

Zhong, M.; Liu, W.; Ning, G.; Yang, L. & Chen, Y. (2004). Laser direct manufacturing of tungsten nickel collimation component. *Journal of Materials Processing Technology*, Vol. 147, No. 2, April 2004, pp. 167-173, ISSN 0924-0136

Low-Shock Manipulation and Testing of Micro Electro-Mechanical Systems (MEMS)

Carlo Ferraresi[1], Daniela Maffiodo[1], Francesco Pescarmona[1], Omar Bounous[1], Luciano Bonaria[2] and Maurizio Straiotto[2]
[1]*Department of Mechanics, Politecnico di Torino*
[2]*SPEA S.p.A*
[1,2]*Italy*

1. Introduction

A MEMS (Micro Electro-Mechanical System) is a block of different miniaturized components (mechanical, electric and electronic) integrated on a silicon layer. This kind of semiconductor system is able to combine the computational skills of microelectronics with the perception and control capabilities of micro sensors and micro actuators, with the main advantage of making the system compact. After the sensors gather information, the electronics processes the data and issues commands for a desired outcome or purpose.
Since MEMSs can perform optical, chemical, thermal, electronic, mechanical and biological functions, they can operate as inkjet heads, accelerometers, gyroscopes, pressure sensors, microphones and micro-fluidic systems, with the possibility of combining more than one function in the same MEMS. They are employed in several fields, such as automotive, electronics, consumer goods, high technology. As Fig. 1 shows, the MEMS market is increasing, with a change in the types produced. In fact, even if the currently most produced ones are inkjet heads for printers, market forecasts highlight that microphones for telephone market and accelerometers for multimedia consoles are going to be the main business in the next few years.
MEMSs require a high level of fabrication and design knowledge, in order to create compact and functional components. In this direction, a dedicated research is being carried on with the aim of improving fabrication and testing processes. Unlike electronic components made with integrated circuits, mechanical ones are fabricated selectively etching away or adding parts of the silicon substrate (Lindroos et al, 2010). Almost all kinds of MEMSs are built on wafer (thin films) of silicon like ICs, but there are other several different processes. Besides silicon and gallium arsenide commonly used in semiconductor industry, other materials as quartz, piezoelectric materials, Pyrex, polymers, plastics and ceramics are utilized (Gad-el-Hak, 2001).
Essentially there are 3 building steps: deposition of thin film of material on a substrate; application of a patterned mask on top of the films by photolithographic imaging, and etching of the films selectively to the mask (Hsu, 2008). Normally the deposition methods are classified in 2 big groups: using chemical reactions or physical reactions. The lithographic process typically consists in the transfer of a pattern to a photosensitive material by selective exposure to a radiation source such as light. Finally, in order to form a

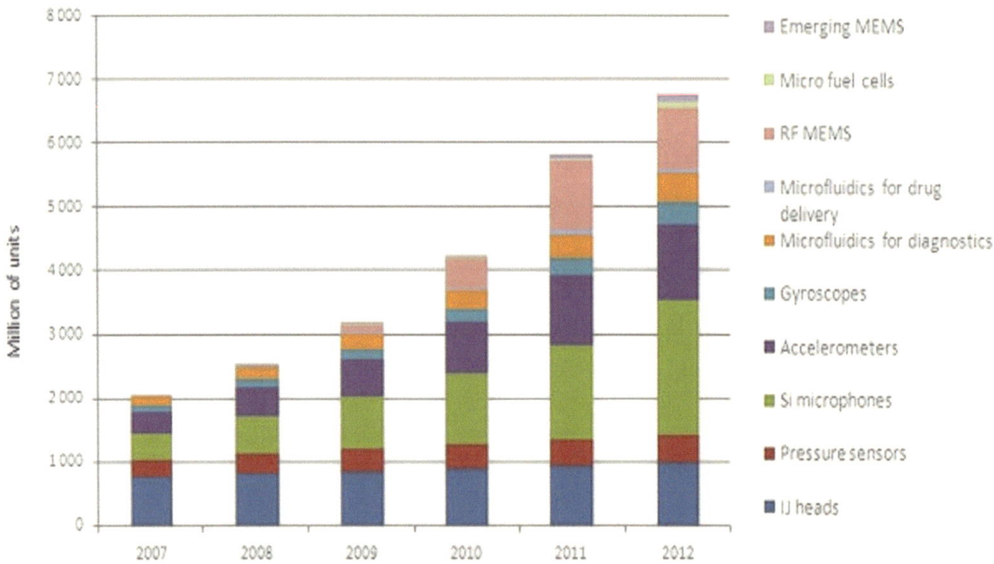

Fig. 1. 2007-2012 MEMS market forecast (millions of units) (Yole Développement, 2009)

functional MEMS structure on a substrate, it is necessary to etch thin films previously deposited. These processes consist in removal of material from desired areas by means of physical or chemical techniques and they are classified as wet or dry. However, each manufacturer uses their own technology, and consequently also the building process is quite particular. Moreover, MEMSs can differ for shape and types of contact (Fig. 2).

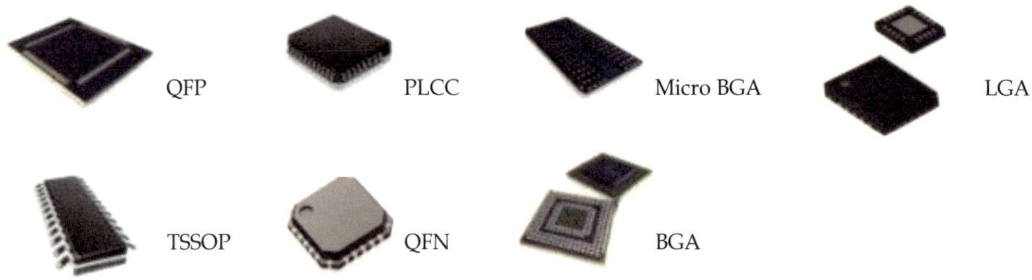

Fig. 2. Different types of MEMSs tested with SPEA Test Cells

Since MEMSs are manufactured using batch fabrication techniques like other electronic devices, fabrication and material costs must be maintained relatively low. For this reason, in order to keep functional test costs as low as possible, those tests are done by specific MEMS Test Cells (MTC).

Currently, there are two ways of testing MEMSs:

- On strips (components physically joint together as they come from the manufacturing process)
- As singulated components (already diced and finished)

The former approach is normally applied at the end of the manufacturing process just before dicing, while the latter is done on single MEMSs. Obviously, both ways of testing have pros and cons: diced MEMSs imply more issues for manipulation; in the case of strip test, MEMSs damaged during the subsequent dicing cannot be detected. Some MEMS manufacturers opt for testing single components, after which they can directly pack the products for customers. This strategy requires a testing and packing process that does not damage the MEMS.

During handling and testing, MEMSs undergo different stresses (inertial forces, collisions, negative pressure). Some typologies, e.g. low-*g* accelerometers and gyroscopes, are particularly sensible to shock and acceleration. Datasheets declare maximum acceptable acceleration (e.g. 10000 *g*[1] for 0.1 ms) to avoid permanent mechanical damage, but practical experience shows that MEMSs could be troubled by lower stresses lasting longer, for instance losing calibration. For this reason, the handling process needs to be designed to guarantee low shock manipulation, or in other words low mechanical impacts and accelerations. There are few studies and data on these phenomena, therefore the aim of this work was also to evaluate forces and accelerations on the MEMS during the whole test cycle (from its entrance in the MEMS Test Cell to its exit) and to highlight the most critical phases.

Automatic Test Cells for single MEMS have different architectures and every kind of MEMS needs different tests, even if some phases (feeding, handling, etc.) are common. Without loss of generality - as much as possible - the study has been done referring to MEMS Test Cells produced by SPEA S.p.A. for low-*g* accelerometers and gyroscopes.

2. Types and structure of a MEMS Test Cell (MTC)

The MEMS Test Cell for testing single components is based on several modules with different functions. Mainly, the entire test cell is composed of five modules (Fig. 3):

- Feeding equipment
- Handler unit
- MEMS Physical Stimulus
- Tester
- Unloading equipment

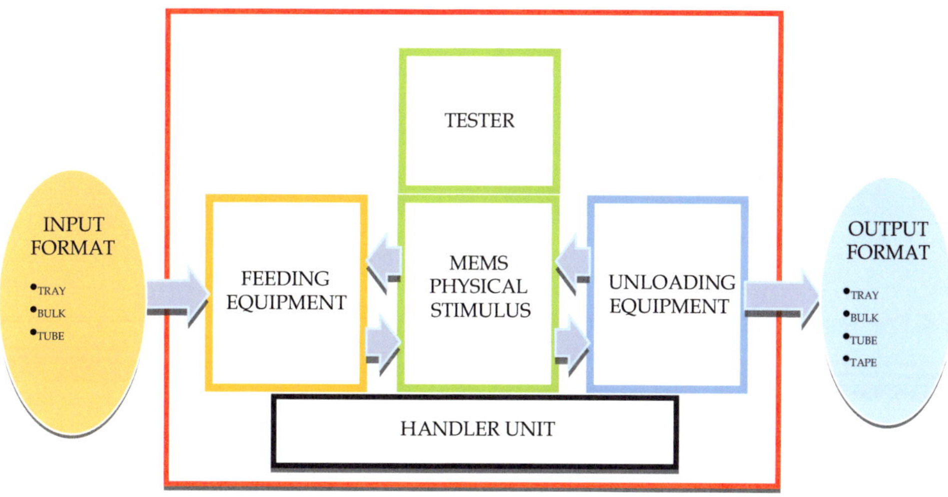

Fig. 3. Block scheme of MTC for singulated MEMS

[1]It is common use to indicate accelerations with reference to the acceleration of gravity g. This Chapter uses the same notation wherever applicable.

Before and after the test, MEMSs can be stocked and carried in different ways. The most common one is the use of a tray: a specific plate in plastic material with proper pockets. It is specific to every kind of MEMS, but it is standardized following the semiconductor laws. Hereinafter the five modules are presented.

2.1 Feeding equipment

The main function of the feeding equipment is to prepare the MEMSs to be picked by the handler unit. MEMSs are carried to the final test by tray, bulk or tube. Obviously these ways influence the feeding equipment since they need to be customized using quite different technologies. Since during the test MEMSs have to be contacted univocally, their orientation is an issue to solve. Tray and tube contain orientated MEMSs since they are prepared in advance, while in bulk they come unoriented.
SPEA's feeding equipment for tray is called TSL (Tray Stack Loader) and consists in a module with the function of tray storage. An opportune mechanical device permits preparing a tray filled with MEMS, in order to be ready for the picking phase (Fig. 4a). This approach guarantees a high throughput, since a high number of MEMSs can be quickly picked up by the handler unit. The bigger constraint is the predetermined pitch, influencing all the equipments downstream.
If MEMSs are carried in bulk, a bowl feeder is used to load them in the MTC (Fig. 4b). Since MEMSs come disorderly, this module is able to orientate and to convey them to a station representing the input for the next module. This equipment must be customized according to the test cycle.
Finally, tubes containing a defined number of MEMSs are also used to carry them. Its feeding equipment is based on gravity to extract the MEMSs from the tube, implying some issues due to impact shocks.

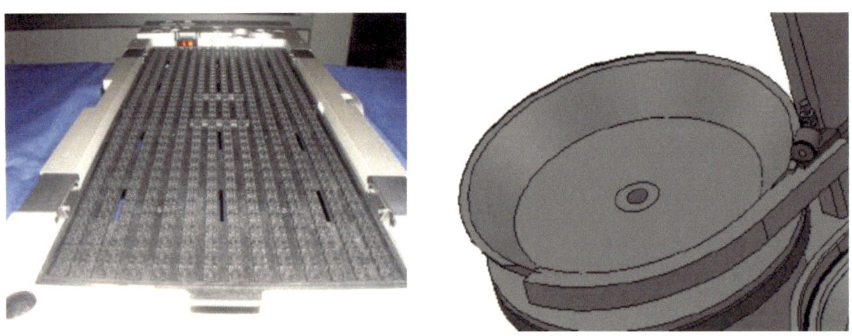

a) b)

Fig. 4. Feeding equipments: a) TSL, b) Bowl Feeder

2.2 Handler unit

This unit is fundamental for the testing of single MEMSs. In fact, it permits to pick the MEMSs prepared on the feeding equipment and to place them in the test cell, ready to be tested. Sometimes, the same handler unit permits to pick MEMSs in the test cell and to take them to the unloading equipment. This module can be made using different technologies. SPEA uses a head (Fig. 5), moving along an horizontal work plane (XY), with picking tools (pick-up) moving along the Z axis. Such devices, thanks to a movement in the Z-direction done by pneumatic actuators, achieve the grasp of MEMSs.

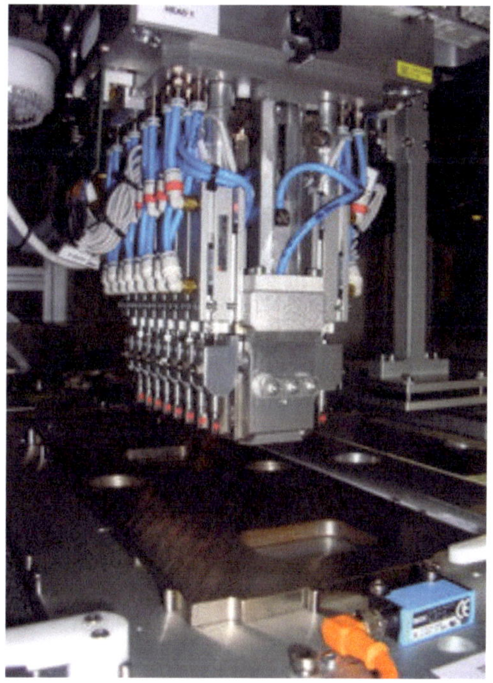

Fig. 5. SPEA's head (handler unit)

2.3 Physical Stimulus Unit

The test cell is based on a stimulus unit aimed at testing the MEMSs with an adequate stimulation, depending on their typology and contacting technology. In order to be tested, MEMSs need to be inserted in a specific contacting unit with electric pins. These pins permit to acquire the electric signals from the MEMSs under test and to transmit them to the tester, stating if the test result is pass or fail. As an example, the stimulus unit used for low-*g* accelerometers and gyroscopes consists of a board, with a defined number of pockets where the MEMSs are lodged, mounted on a system of orthogonal axes. This system permits 2 angular degrees of freedom (roll and pitch), or 3 (roll, pitch and yaw), in order to turn the board as needed.

2.4 Tester

This unit acquires signals from the MEMSs under test and, after an elaboration, it declares the results of the test as pass or fail.

2.5 Unloading equipment

Finally, the unloading equipment permits to sort the MEMSs in proper packages, ready to be sold to customers. Normally, this equipment allows to separate defective MEMSs from valid ones, according to the different failures that may have occurred. Some MEMS manufacturers stock the semiconductor components in tape on reel (Fig. 6). The tape presents specific pockets where MEMSs are individually stored and closed by a thin film. This way to carry MEMSs is not commonly used during the manufacturing process, since it permits a limited throughput. On the contrary it is quite common after the final test.

Fig. 6. Tape on reel

3. Manipulation techniques

As presented in section 2, the handler unit of the MTC has the main role of carrying MEMSs from a module to another. This phase could be done by using a pick-and-place device which manipulates MEMSs. Even if MEMSs are of millimetric size, their very low mass permits to assimilate their manipulation to micro manipulation.

Micro manipulation is based on physical laws and principles different from those of large scale mechanics. In fact, in this field, some forces which are negligible in the large scale (electrostatic, Van der Waals, surface tension) become dominant; on the contrary, gravitational and inertial forces are not so significant. For this reason, the manipulation assumes an irreversible nature and the unloading of the component becomes the most critical phase in the whole process (Tichem et al., 2004). In fact, the bigger issue is the detachment of the micro component from the actuator. This section presents the state of the art of current technology for micro component handling, and a brief classification of physical principles useful in this field.

Micro manipulation can be classified in two main categories: with or without contact. The former category is the most used and well known. It includes the use of forces of mechanical, fluidic or molecular nature.

The most common grasping technique is based on suction. A suction pad or a sucker touches the object, and then the suction is created between the two parts. This technique is widely used for pick-and-place phases, because it allows short grasping and releasing times. In Fig. 5 SPEA's handler unit is shown. It has eight picking tools using suction and operating in parallel.

A further method is the use of grippers (Ansel et al., 2002). Fig. 7 shows the two basic principles of functioning.

The opening and the closure of the gripper, corresponding to grasping and releasing the object, can be done through different actuations: shape memory alloys (Zhong & Yeong, 2006), MEMS (Mayyas et al., 2007; Skidmore et al., 2003), piezoelectric (Agnus et al., 2003) and chemical (Randhawa et al., 2008) actuators. As an example Fig. 8 shows a MEMS micro gripper.

Another solution is to use a passive gripper, characterized by the absence of any type of direct actuation for the manipulation of the object (Tsui et al., 2004).

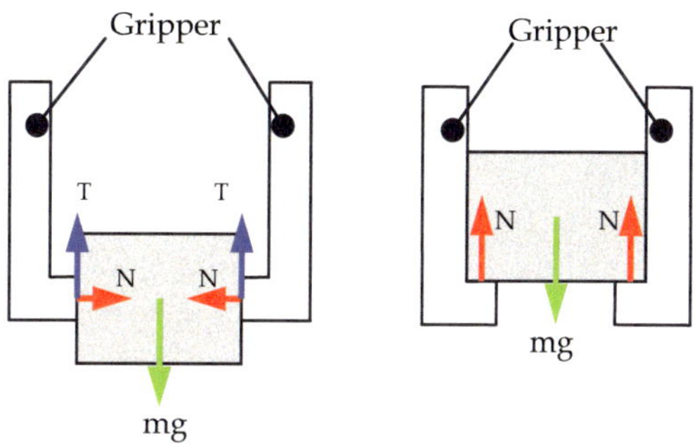

Fig. 7. Gripping by friction (left) and form closure (right).

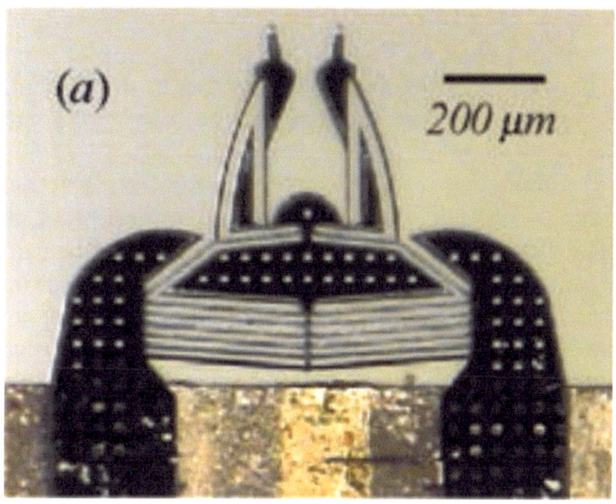

Fig. 8. MEMS micro gripper.

Other interesting manipulation techniques use forces of fluidic nature.
Some possibilities are sketched on Fig. 9. Fig. 9a shows a surface tension method. This method utilizes a low viscosity fluid that evaporates without leaving residuals. For small objects with limited weight, the capillarity force and the surface tension of a liquid between the grasping equipment and the object are sufficient to guarantee the grasp without any damage to the component (Bark et al., 1998). Moreover, using an appropriate shape of the gripping tool, the surface force is able to orient the component.
Fig. 9b shows the principle of micro holes heating. This method is based on the pressure variation inside micro holes on the actuator surface due to a temperature change (Arai & Fukuda, 1997). In detail, before the contact between the picking tool, the temperature is increased. After the tool touches the component, temperature is decreased again, so the pressure inside the hole decreases. In this way, vacuum is created inside the holes, allowing the picking of the object. Finally, the release is done by heating again. The actuator heating can be realised in different ways: a laser beam, a micro heater on the actuator, conduction.

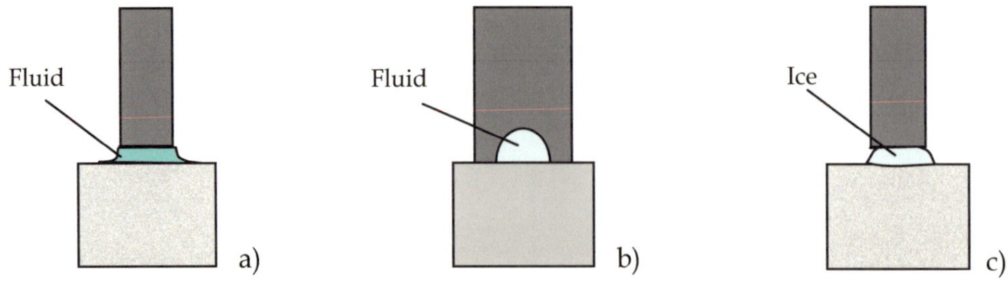

Fig. 9. Manipulation with fluidic forces: a) surface tension, b) micro holes heating, c) cryogenic.

Fig. 9c shows the cryogenic method. The grasp using the adhesion property of ice consists on freezing a small quantity of liquid between the picking tool and the object (Liu et al., 2004, Ru et al., 2007). The release of the object is done by breaking or melting the frozen material.

In addition to cryogenic grippers working in air, some submerged freeze microgrippers working in water have been developed in order to avoid issues due to capillarity force (Walle et al., 2007).

Forces of molecular nature such as Van der Waals force can be used for manipulation (Feddema et al., 2001). The main issue of this method is the difficulty to control it, in particular during the release phase that must be done by skewing the actuator tip in order to decrease the adhesion force.

The second category, without contact between the tool and the manipulated object, includes the use of magnetism, electrostatic, ultrasonic pressure, optical pressure, fluidic principles (Vandaele et al., 2005). They present various advantages such as the fact that surface forces can be completely neglected, friction is reduced, it is possible to manage breakable and delicate objects, surface contamination is completely avoided. They are less known and currently not reliable, but they could possibly represent the future of micro manipulation.

4. Experimental detection of shocks during the test cycle

As above described, the automatic test cycle is generally composed of the following phases: MEMSs are loaded on trays (or bowl feeder), then a pick-and-place robotic head carries the MEMSs to the test area where it undergoes the test. After the test, another (or the same) pick-and-place unit carries the MEMSs to different destinations, according to the test results (pass or fail). Usually a code reader is placed between the loading area and the test area.

Each phase shows specific circumstances that are potentially critical from the shock point of view.

The careful analysis of the whole test cycle leads to the identification of three critical situations, requiring in-depth examination:

1. during the pick phases, there could be an impact between pick-up and MEMS; in particular the pick-up of the robot head could collide with the MEMS during its downward motion, depending on the shape and the calibration of the device;
2. when the MEMS is already picked, it undergoes possibly dangerous accelerations during the head movement;
3. before the test phase, MEMS is clamped to the socket board; generally the clamp is controlled by means of a single-acting pneumatic cylinder, in which the spring causes the clamp impact on MEMS.

4.1 Pick-up force measurement

The contact force during the pick phase is strongly dependent on the pick-up typology.

In a first case (Fig. 10a) the pick-up end is provided with a seat shaped to allocate the MEMS, at the end of the descent stroke the pick-up hits the tray surface, so avoiding a direct contact with the component; the grasping is operated by the subsequent suction, sparing any shock.

A second solution is shown in Fig. 10b. In this case the pick-up directly hits the MEMS surface; this solution is less expensive and therefore very diffused but, since the contact force may be critical, an accurate analysis of the operating condition is usually necessary. In the following such analysis, both experimental and analytical, is described.

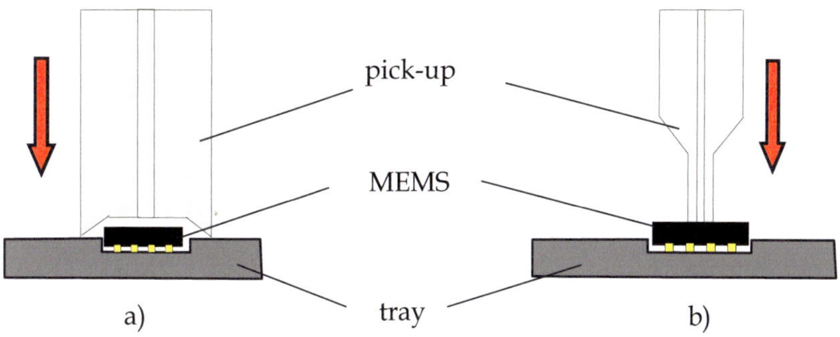

Fig. 10. Pick-up typologies a) without impact on MEMS, b) with impact on MEMS

A load cell (Brüel & Kjær Load Sensor, 0- 5000 N, resolution: 0.001 N) replaced the MEMS to be hit by the pick-up, as the MEMS is during the pick-and-place operation. The main difference between the real situation and the experimental layout was the replacement of the tray with an adequate aluminium support for the load cell.

The load cell support was positioned on the test machine and a setup was made to repeat the pick-up spring compression of 1.5 mm that was measured with a speed cam on real test cycle. The test was repeated using all 8 pick-up's from the same robot head and using two springs with different stiffness inside the pick-up. Fig. 11 shows the experimental test setup.

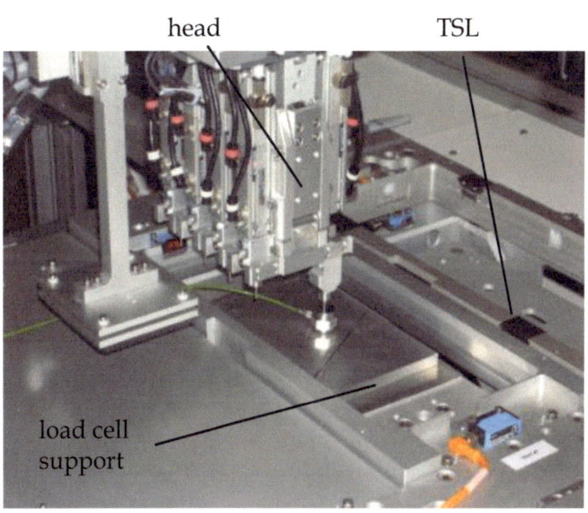

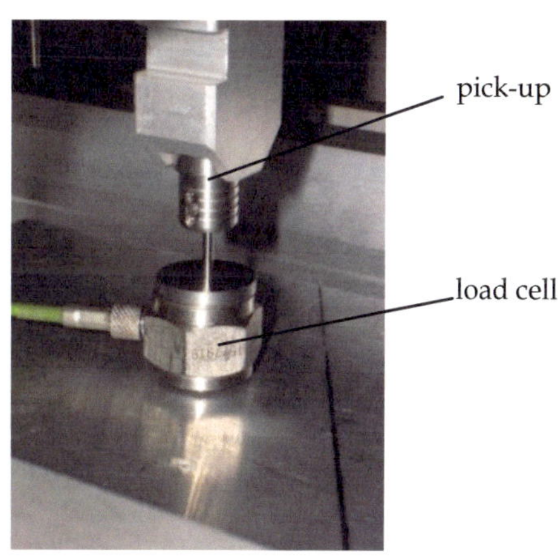

Fig. 11. The experimental test setup

Fig. 12 shows the results of the test performed using the 8 different pick-up's from the same head. The impact force peak is about 5 N. The results of tests performed using springs with different stiffness do not show a significant dependence of the force peak value on spring type. Performing the test again in the same conditions with the same pick-up demonstrated a good repeatability.

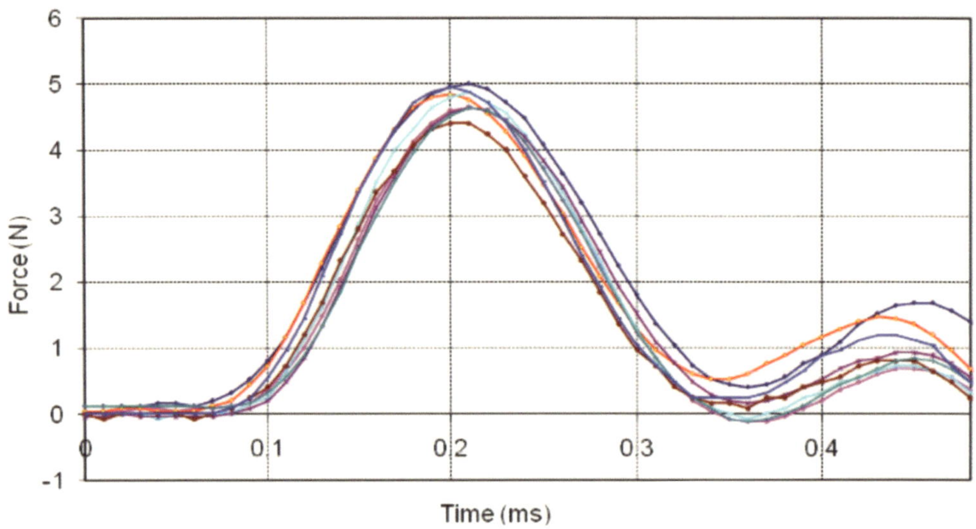

Fig. 12. Impact force in pick phase, using the 8 different pick-up's from the same head

4.2 Pick-up acceleration measurement

Measurements of pick-up acceleration during its descent and back were performed in order to investigate on the possibly dangerous accelerations during the head movement. The X and Y axis movement are operated by electro-mechanical drives on pneumostatic slides so that their acceleration is well known and controllable, whereas the Z-axis movement (vertical) is operated by a pneumatic cylinder. Its acceleration is unknown and not controlled. Therefore the first step is to investigate on these values in order to evaluate its possible effects and to consider the need to limit or control the Z-axis movement.

The pick-up was replaced with an accelerometer (Silicon Design 2210, ± 100 *g*) and the pick-up movement along Z-axis was reproduced, recording the results. Fig. 13 shows the experimental setup.

The mass difference between the pick-up and the sensor is about one gram, but considering that there are also other masses on movement like the piston and the piston rod, this variation is not significant. The test was repeated using 3 cylinders on the same head.

Fig. 14 shows an example of the test results. In Fig. 14a, the stroke time is about 110 ms and the acceleration peak is around 100 *g*. Therefore the machine design and set up needs a direct attention to the required compromise between the operation velocity and the possible component shock. If the MEMS needs a smoother manipulation it is possible to adjust the pneumatic circuit in order to obtain lower velocity and, consequently, lower acceleration. As an example, Fig. 14b shows a test result with the same pick-up and head conditions, with different pneumatic resistances of the circuit obtained by properly varying the diameter of intake and exhaust throats. In this case, the acceleration peak decreases to about 40 *g*, with a stroke time increasing to 140 ms.

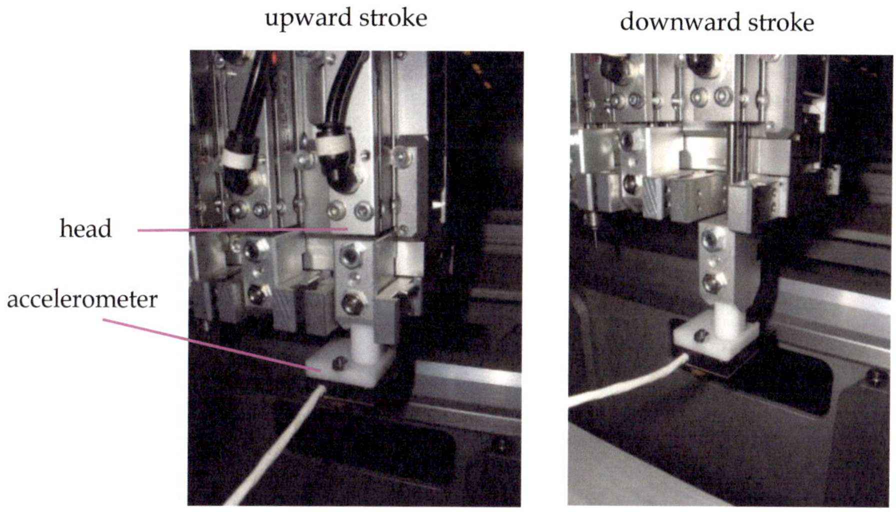

Fig. 13. Experimental setup for the pick-up acceleration measurements

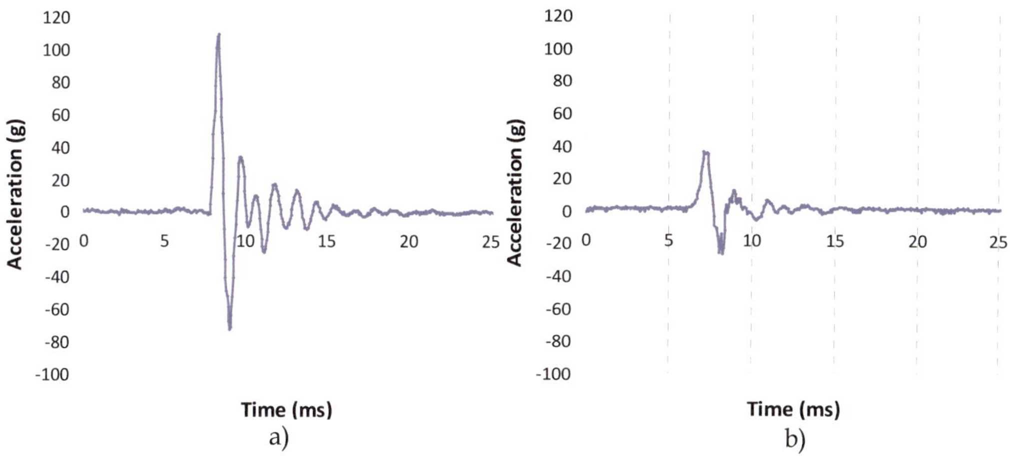

Fig. 14. Examples of the pick-up acceleration

4.3 Clamping force measurement

There are different socket types, but generally each of them has a clamp device aiming at constraining the MEMS during the test. A commonly used device is the turret socket, which includes clamps operated by pneumatic cylinders. Fig. 15 shows a functional scheme: the opening of the clamp is done by the air supply, the clamp closure is passively carried out by springs exerting force on the upper face of the cylinder when the compressed air is exhausted. The clamp moves downwards and hits the MEMS, which is positioned on an elastic support, thus generating a potentially critical phase in the manipulation cycle.

The force exerted by the clamp could be critical for the MEMS, therefore an investigation has been carried out in order to quantify the impact force.

Since there are many difficulties in equipping the Test Cell with sensors without altering the system, a separated test setup was designed and realised.

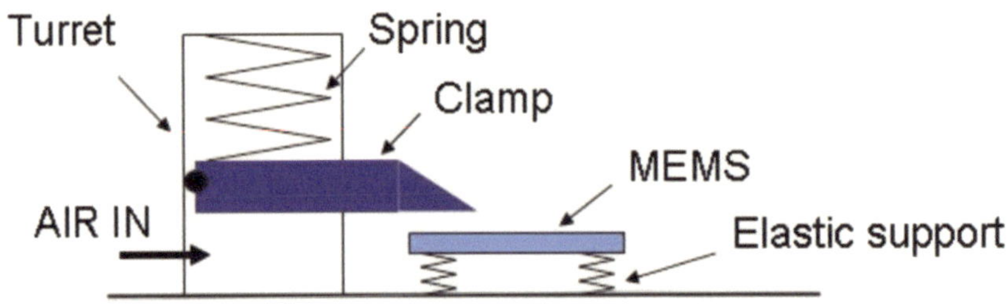

Fig. 15. Functional scheme of the clamping

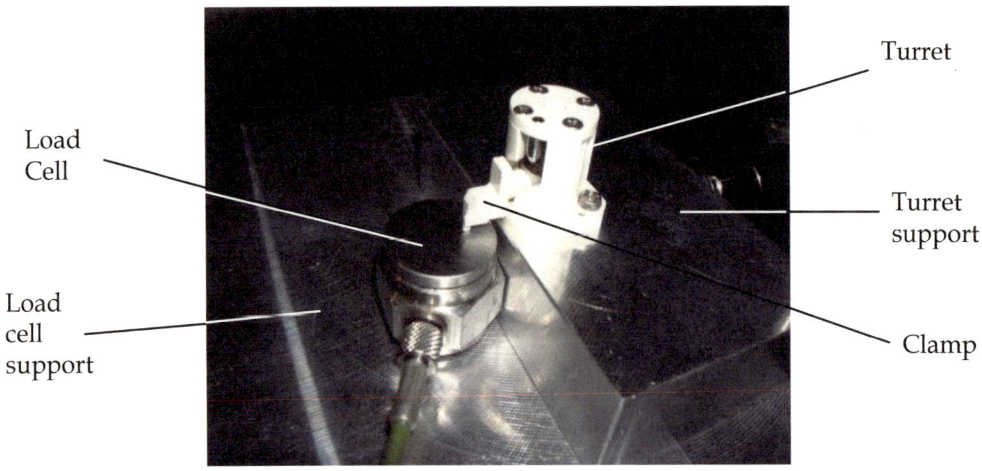

Fig. 16. Experimental setup for clamping force measurement

The experimental test setup is shown in Fig. 16: the MEMS is replaced with a load cell with the aim of evaluating the impact force during the clamp closure dynamics, with a rigid load cell support.

The turret is from a SPEA MEMS Test Cell. The load cell is a Brüel & Kjær (Range: 0-5000 N; Resolution: 0.001 N).

Similarly to the real situation, the clamp hits the load cell 0.3 mm before its total stroke. On the other hand, there are some important differences between the actual situation and the test setup. In the real application the MEMS support has a significant compliance and the air supply to the 8 turrets is given by proper channels calibrated by adequate pneumatic throats; on the contrary, in the test setup the load cell support is practically rigid and the turret is supplied by a short and direct duct.

Tests are repeated using three different supply pressures, with or without an outlet silencer and inserting different throats on the line.

Fig. 17 shows experimental test results for different supply pressures. Each curve is the average of at least three tests. There was a good repeatability and the supply pressure influence on peak value is almost negligible. The impact force peak is between 20 and 27 N.

Fig. 18 shows a comparison between a clamping performed by the experimental turret with no throat in the pneumatic line and one performed with a throat of 0.8 mm diameter.

These tests highlight that the impact force may be controlled by a proper selection of the pneumatic circuit parameters.

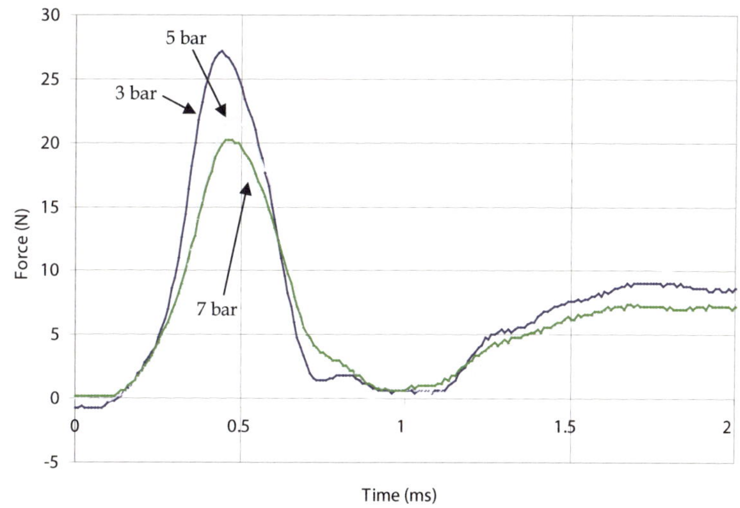

Fig. 17. Impact force of the clamp on the load cell at different supply pressures

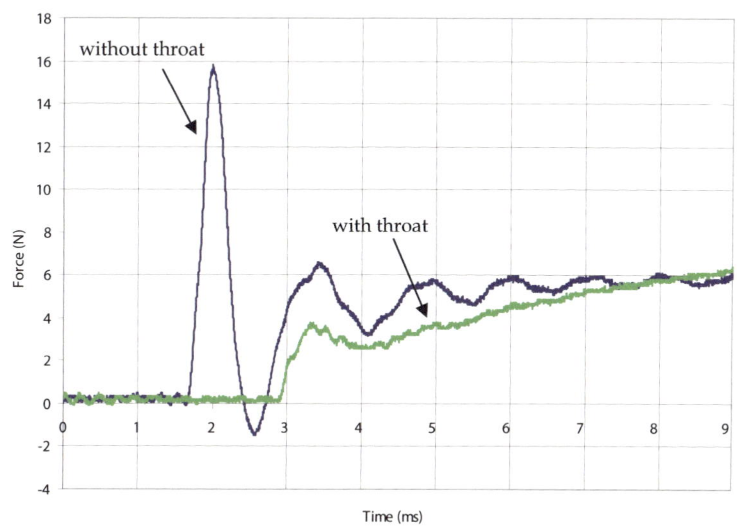

Fig. 18. Impact of the clamp on the load cell with and without throat in pneumatic line

On the other hand, because of the differences between the actual Test Cell and the experimental setup, these results cannot be directly used to evaluate the impact forces exerted on the MEMS during the actual test cycle.

For this reason, a numerical model was realised and validated against the experimental results. Such a model can therefore be used both to verify the MEMS condition during the actual test cycle, and to design future Test Cells.

5. Numerical model of clamping

An AMESim model has been realised to simulate the interaction between clamp and MEMS. Its aim is to analyse the process and find out the parameters which most influence the

strength of the impact and the operation speed. Once the model is complete with all the physical parameters involved and is validated against experimental data, it is possible to simulate a change in one or more parameters in order to mitigate the impact or to increase the operation speed.
This spares the necessity to set up a test bench for that, whilst remaining confident in the validity of the results for application to the real world. The scheme of the model is shown in Fig. 19.

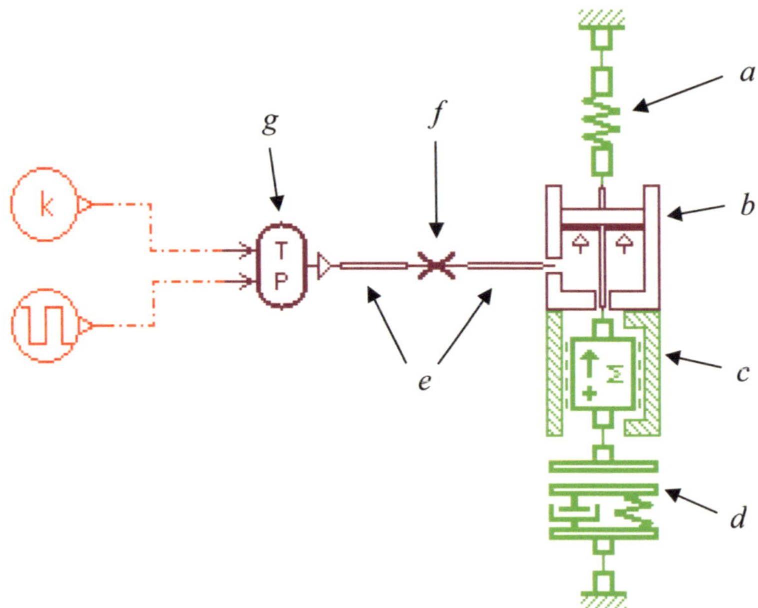

Fig. 19. AMESim model of the clamp-MEMS contact

Each block of the model represents a physical feature of the real device. In particular, the clamp is modelled as follows: the spring *a* works against the piston *b*. The mass *c* represents the total moving mass of the clamp; the corresponding block includes friction and end stops. The block *g* imposes a constant temperature and a user defined pressure law upstream the pipe *e* and the included pneumatic resistance *f*.
The contact between the clamp and the MEMS is modelled by the block *d*, which includes the internal rigidity and damping of the MEMS and its support as a whole.
The numeric parameters of each block have been obtained through specific tests on the real components.
By comparing the results from the model and those from experimental tests, the model itself can be validated. In particular, the model has been set up to duplicate the conditions of the experimental tests shown in Fig. 17. The comparison in the case of no pneumatic throat is shown in Fig. 20, while Fig. 21 shows a zoom-in of the force peak in the case of the pneumatic throat with a diameter of 0.8 mm. Note the X axis scale in the latter figure: it represents a time frame of about 4 ms.
The model presents a very good validity against experimental data, even in different set up conditions. In particular, the impact peaks from the model are nearly the same as the real ones, and also the pulsation frequencies correspond. The tiny differences between the curve pairs

depend on the very small time scale of the physical phenomenon and on the uncertainty in the determination of a few parameters such as friction, stiction and internal damping.
Therefore, the model can be used to test virtually the possible changes in the system configuration, with an excellent degree of confidence. Altering the parameters of the model, for instance the spring rigidity, the turret piping volume or the pneumatic resistance, the performance of the system can be adapted to different needs, such as higher clamping speed or lower impact or static force, performing virtual tests with good reliability until the desired conditions are found.

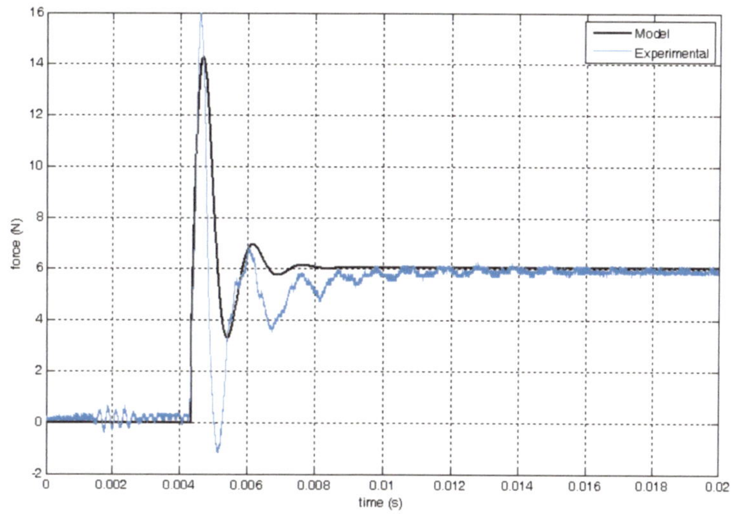

Fig. 20. Comparison between model and experimental data in the case of no pneumatic throat

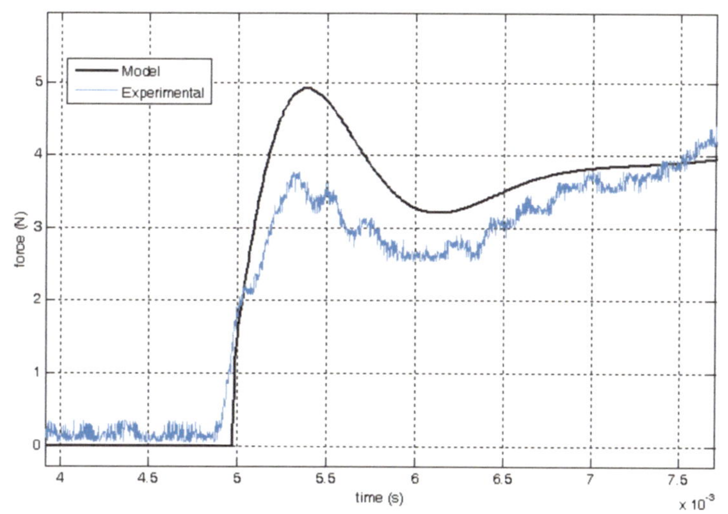

Fig. 21. Comparison between model and experimental data in the case of a 0.8 mm pneumatic throat.

As an example, the model can simulate the behaviour of the system when different pneumatic throats are used. This result is shown in Fig. 22, where the corresponding diameters of the throat inserted along the pneumatic line are: A=0.5 mm, B=0.8 mm, C=1.1 mm, D=1.4 mm, E=2.5 mm.

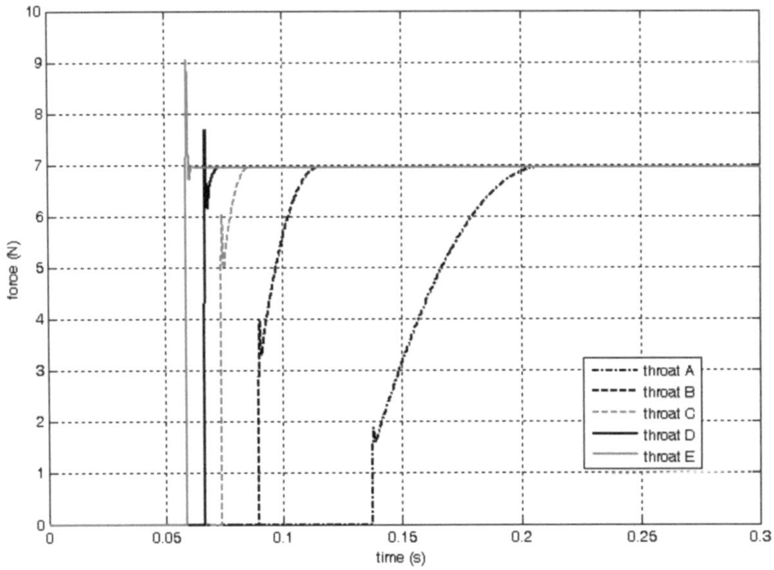

Fig. 22. Simulation of the clamp behaviour with different pneumatic throats

6. Conclusion

During the last few years, MEMSs have become more and more diffused in a large variety of applications requiring a high level of reliability. At the same time they have become more and more complicated, being able to allow different operations.
Since the reliability of MEMSs must be tested for every single component by special automatic test equipments, we decided to study the functioning of a specific MEMS Test Cell (MTC), in order to verify the existence of potentially critical situations for the tested components.
In particular we have focused on three cycle phases in which some kinds of MEMS, e.g. gyroscopes and low-*g* accelerometers, may suffer damage because of shocks that the component can undergo during the test cycle, namely acceleration peaks and impulsive contact forces.
To allow the producer to limit and control these shocks during the test cycle it is necessary to evaluate each phase and to design adequate tools. For this reason we have performed experimental tests on critical phases highlighting the parameters that may cause shock on MEMS.
As concerns acceleration peaks, they may occur mainly during manipulation operated by pneumatic drives, which present difficulties in motion control. Usually a direct measurement of the acceleration values is possible and that allows to individuate proper remedies, which consist mainly in paying particular attention in design and selection of the various components of the pneumatic circuit.

On the other hand, impulsive contact forces take place in pick and clamping phases. A direct measurement was carried out in the pick phase, thus obtaining definite indication about possible occurrence of critical values. On the contrary, contact forces could not be measured in real conditions during the clamping phase, since the experimental apparatus influenced the measure itself. So it was necessary to realise a numerical model of the clamping system, validated by comparison with experimental measures in altered condition. In this way the model represents a possible useful tool capable of simulating the MEMS real behaviour in the Test Cell, allowing a correct analysis of actual condition and the design of future test equipments.

7. Acknowledgement

This work has been supported by Piedmont Region in the frame of Axis 1 Project "Innovation and Productive Transition" of POR-FESR 2007/2013 in the target "Regional Competitivity and Occupation".

8. References

Agnus, J.; De Lit, P.; Clévy, C. & Chaillet, N. (2003). Description and Performances of a Four-Degrees-of-Freedom Piezoelectric Gripper, *Proceedings of the IEEE International Symposium on Assembly and Task Planning,* pp. 66 - 71, Besançon, France

Ansel, Y.; Schmitz, F.; Kunz, S.; Gruber, H. P. & Popovic, G. (2002). Development of tools for handling and assembling microcomponents. *Journal of Micromechanics and Microengineering,* Vol. 12, No. 4, pp. 430 - 437

Arai, F. & Fukuda, T. (1997). A new pick up & release method by heating for micromanipulation, *Proceedings of the IEEE 10th Annual International Workshop on Micro Electro Mechanical Systems,* pp. 383 – 388, Nagoya, Japan

Bark, C.; Binnenböse, T.; Vögele, G.; Weisener, T. & Widmann, M. (1998). *Proceedings of the IEEE 11th Annual International Workshop on Micro Electro Mechanical Systems,* pp. 301 - 305, Heidelberg, Germany

Feddema, J. T.; Xavier, P. & Brown, R. (2001). Micro-assembly planning with van der Waals force. *Journal of Micromechatronics,* Vol. 1, No. 2, pp. 139-153

Gad-el-Hak, M. (2001). *The MEMS handbook,* CRC Press, ISBN 0849300770, Notre Dame, Indiana

Hsu, T.R. (2008). *MEMS and Microsystems: design, manufacture, and nanoscale engineering,* Wiley, ISBN 9780470083017, Hoboken, New Jersey

Lindroos, V.; Tilli, M.; Lehto, A. & Motooka, T. (2010). *Handbook of silicon based MEMS materials and Technologies,* Elsevier, ISBN 9780815515944, Oxford, UK

Liu, J.; Zhou, Y. X. & Yu, T. H. (2004). Freeze tweezer to manipulate mini/micro objects. *Journal of Micromechanics and Microengineering,* Vol. 14, No. 2, pp. 269-276

Mayyas, M.; Zhang, P.; Lee, W. H.; Shiakolas, P. & Popa, D. (2007). Design Tradeoffs for Electrothermal Microgrippers, *Proceedings of the IEEE International Conference on Robotics and Automation,* pp. 907-912, Roma, Italy

Randhawa, J. S.; Leong, T. G.; Bassik, N.; Benson, B. R.; Jochmans, M. T. & Gracias, D. H. (2008). Pick-and-Place Using Chemically Actuated Microgrippers. *Journal of the American Chemical Society,* Vol. 130, No. 51, (December 2008), pp. 17238-17239

Ru, C.; Wan, X.; Ye, X. & Guo, S. (2007). A new ice gripper based on thermoelectric effect for manipulating micro objects. *Proceedings of the 7th IEEE International Conference on Nanotechnology,* pp. 438-441, Hong Kong

Skidmore, G.; Ellis, M.; Geisberger, A.; Tsui, K.; Saini, R.; Huang, T. & Randall, J. (2003). Parallel assembly of microsystems using Si micro electro mechanical systems. *Microelectronic Engineering,* Vol. 67-68, pp. 445-452

Tichem, M.; Lang, D. & Karpuschewski, B. (2004). A classification scheme for quantitative analysis of micro-grip principles. *Assembly Automation,* Vol. 24, No. 1, pp. 88-93

Tsui, K.; Geisberger, A. A.; Ellis, M. & Skidmore, G. D. (2004). Micromachined End-effector and Techiniques for Directed MEMS Assembly. *Journal of Micromechanics and Microengineering,* Vol. 14, No. 4, (April 2004), pp. 542-549

Vandaele, V.; Lambert, P. & Delchambre, A. (2005). Non-contact Handling in Microassembly: Acoustical Levitation. *Precision Engineering,* Vol. 29, No. 4, (October 2008), pp. 491-505

Walle, B. L.; Gauthier, M. & Chaillet, N. (2007). A Submerged Freeze Microgripper for Micromanipulations. *Proceedings of the IEEE International Conference on Robotics and Automation,* pp. 826-831, Roma, Italy

Yole Développement (2009). *www.yole.fr*

Zhong, Z. W. & Yeong, C. K. (2006). Development of a gripper using SMA wire. *Sensors and Actuators A: Physical,* Vol. 126, No. 2, pp. 375-381

16

Magnetically Nonlinear Dynamic Models of Synchronous Machines: Their Derivation, Parameters and Applications

Gorazd Štumberger[1], Bojan Štumberger[1], Tine Marčič[2]
Miralem Hadžiselimović[1] and Drago Dolinar[1]
[1]*University of Maribor, Faculty of Electrical Engineering and Computer Science*
[2]*TECES- Research and Development Centre of Electrical Machines*
[1,2]*Slovenia*

1. Introduction

This chapter deals with the magnetically nonlinear dynamic models of synchronous machines. More precisely, the chapter focuses on the dynamic models of the permanent magnet synchronous machines and reluctance synchronous machines. A general procedure, which can be applied to derive such magnetically nonlinear dynamic models, is presented. The model is of no use until its parameters are determined. Therefore, some available experimental methods, that are appropriate for determining parameters of the discussed models, are presented. The examples given at the end of the chapter show, how the magnetically nonlinear dynamic models of discussed synchronous machines can be applied.
Generally, in all synchronous machines the resultant magneto-motive force and the rotor move with the same speed. This condition is fulfilled completely only in the case of steady-state operation. However, during the transient operation, the relative speed between the resultant magneto-motive force and the rotor can change. In the case of permanent magnet synchronous machines, the force or torque that causes motion appears due to the interaction between the magnetic fields caused by the permanent magnets and the magnetic excitation caused by the stator currents. On the contrary, in the reluctance synchronous machines the origin of motion is the force or torque caused by the differences in reluctance. Most of the modern permanent magnet synchronous machines utilize both phenomena for the thrust or torque production.
A concise historic overview of the development in the field of synchronous machine modelling, related mostly to the machines used for power generation, is given in (Owen, 1999). When it comes to the modern modelling of electric machines, extremely important, but often neglected, work of Gabriel Kron must be mentioned. In the years from 1935 to 1938, he published in General Electric Review series of papers entitled ``The application of tensors to the analysis of rotating electrical machinery.'' With these publications as well as with (Kron, 1951, 1959, 1965), Kron joined all, at that time available and up to date knowledge, in the fields of physics, mathematics and electric machinery. In such way, he set a solid theoretical background for modern modelling of electric machines. Unfortunately, the generality of Kron's approach faded over time. In modern books related to the modelling and control of

electric machines, such as (Fitzgerald & Kingslley, 1961), (Krause et al., 2002), (Vas, 1992, 2001), (Boldea & Tutelea, 2010), (Jadrić & Frančić, 1997), and (Dolinar & Štumberger, 2006), the tensors proposed by Kron are replaced with matrices. The electric machines are treated as magnetically linear systems, whereas the magnetically nonlinear behaviour of the magnetic cores, which are substantial parts of electric machines, is neglected.

The agreement between measured responses and the ones calculated with the dynamic model of an electric machine cannot be satisfying if magnetically nonlinear behaviour of the magnetic core is neglected. Moreover, if such a model is applied in the control design, the performances of the closed-loop controlled system cannot be superb. In order to improve the agreement between the measured and calculated responses of electric machines and to improve performances of the closed-loop controlled systems, individual authors make an attempt to include the magnetically nonlinear behaviour of magnetic cores into dynamic models of electric machines.

In the research field of synchronous machines, the effects of saturation are considered by the constant saliency ratio and variable inductance (Iglesias, 1992), (Levi & Levi, 2000), (Vas, 1986). In (Tahan, & Kamwa, 1995), the authors apply two parameters to describe the magnetically nonlinear behaviour of the magnetic core of a synchronous machine. The authors in (Štumberger B. et al., 2003) clearly show that only one parameter is insufficient to properly describe magnetically nonlinear behaviour of a permanent magnet synchronous machine. A magnetically nonlinear and anisotropic magnetic core model of a synchronous reluctance machine, based on current dependent flux linkages, is presented in (Štumberger G. et al., 2003). The constraints that must be fulfilled in the conservative or loss-less magnetic core model of any electric machine are given in (Melkebeek & Willems, 1990) and (Sauer, 1992). A new, energy functions based, magnetically nonlinear model of a synchronous reluctance machine is proposed in (Vagati et al., 2000), while the application of this model in the sensorless control is shown in (Capecchi et al., 2001) and (Guglielmi et al., 2006). One of the first laboratory realizations of the nonlinear input-output linearizing control, considering magnetic saturation in the induction machine, is presented in (Dolinar et al., 2003). Similar approach is applied in the case of linear synchronous reluctance machine in (Dolinar et al., 2005). The design of this machine is presented in (Hamler et al., 1998), whereas the magnetically nonlinear model of the magnetic core, given in the form of current and position dependent flux linkages, is presented in (Štumberger et al., 2004a). Further development of the magnetically nonlinear dynamic model of synchronous reluctance machine, based on the current and position dependent flux linkages expressed with Fourier series, is presented in (Štumberger et al., 2006). An extension to the approach presented in (Štumberger et al., 2004a, 2004b, 2006) towards the magnetically nonlinear dynamic models of permanent magnet synchronous machines is made in (Hadžiselimović, 2007a), (Hadžiselimović et al., 2007b, 2008), where the impact of permanent magnets is considered as well.

The experimental methods appropriate for determining parameters of the magnetically nonlinear dynamic model of a linear synchronous reluctance motor are presented in (Štumberger et al., 2004b). They are based on the stepwise changing voltages generated by a voltage source inverter controlled in the *d-q* reference frame. Quite different method for determining parameters of the synchronous machines, based on the supply with sinusoidal voltages, is proposed in (Rahman & Hiti, 2005). A set of experimental methods, appropriate for determining the magnetically nonlinear characteristics of electromagnetic devices and electric machines with magnetic cores, is presented in (Štumberger et al., 2005). However, the methods presented in (Štumberger et al., 2005) could fail when damping windings are wound around the magnetic core, which is solved by the methods proposed in (Štumberger et al., 2008b). In the cases, when the test required in the experimental methods cannot be performed on the electric machine, the optimization based methods, like those presented in (Štumberger et al., 2008a) and (Marčič et al., 2008), can be applied for determining parameters required in the magnetically nonlinear dynamic models.

Many authors use magnetically nonlinear dynamic models of synchronous machines. However, it is often not explained how the models are derived and how their parameters are determined. In the section 2, the three-phase model of a permanent magnet synchronous machine with reluctance magnetic core is written in a general form. The effects of saturation, cross-saturation as well as the interactions between the permanent magnets and slots are considered by the current and position dependent characteristics of flux linkages. Since only those dynamic models with independent state variables are appropriate for the control design, the two-axis magnetically nonlinear dynamic model written in the d-q reference frame is derived. The derived model of the permanent magnet synchronous machine is simplified to the model of a synchronous reluctance machine, which is further modified to the model of a linear synchronous reluctance machine. The experimental methods appropriate for determining parameters, that are required in the magnetically nonlinear dynamic models, are presented in the section 3. The performance of the magnetically nonlinear dynamic model as well as applications of the models is shown in the section 4. The chapter ends with the conclusion given in the section 5 and references given in the section 6.

2. Derivation of the magnetically nonlinear dynamic model

The magnetically nonlinear properties of the ferromagnetic material are normally described in the form of $B(H)$ characteristics, defined with the magnetic flux density B and the magnetic field strength H (Boll, 1990). These characteristics can be determined by different tests and measurements performed on a material specimen (Abdallh, 2009). In the case of an isotropic material, the properties changing along the hysteresis loop can be described by the differential permeability defined with the partial derivative $\partial B/\partial H$. However, in the case of an anisotropic material, the local relations between the flux density vector $\mathbf{B}$ and the magnetic field strength vector $\mathbf{H}$ are given in the form of the differential permeability tensor, defined with the partial derivative $\partial \mathbf{B}/\partial \mathbf{H}$.

An electric machine or an electromagnetic device consists at least from a magnetic core and windings wound around the core, which means that the materials with different properties are combined. Thus, the magnetically nonlinear behaviour of the entire electric machine or electromagnetic device is influenced by different materials. This influence is present also in the variables measured on the terminals of the machine. The time behaviour of the variables measured on the terminals can be used to describe the magnetically nonlinear behaviour of the machine. In the case of an electric machine with only one winding, the magnetically nonlinear behaviour of the machine can be described in the form of the $\psi(i)$ characteristic, where the flux linkage ψ and the current i can be determined with the measurement of currents and voltages on the terminals of the machine. Similarly as in the case of $B(H)$ characteristic, the local behaviour along the $\psi(i)$ characteristic can be described with the partial derivative $\partial\psi/\partial i$. However, when more than one winding is wound around the common magnetic core, the flux linkages and currents of individual windings can be used to compose the flux linkage vector $\boldsymbol{\psi}$ and the current vector $\mathbf{i}$. Again, the partial derivative $\partial\boldsymbol{\psi}/\partial\mathbf{i}$ can be used to describe the locally nonlinear behaviour of the magnetic core. It is wise to compose the flux linkage vector and the current vector from the linearly independent variables.

2.1 Permanent magnet synchronous machine

The derivation performed in this subsection is given for the two-pole, three-phase permanent magnet synchronous machine (Hadžiselimović, 2007a), (Hadžiselimović et al., 2007b, 2008). It is schematically shown in Fig. 1. The axes *a, b* and *c* are defined with the magnetic axes of the phase *a, b* and *c* windings.

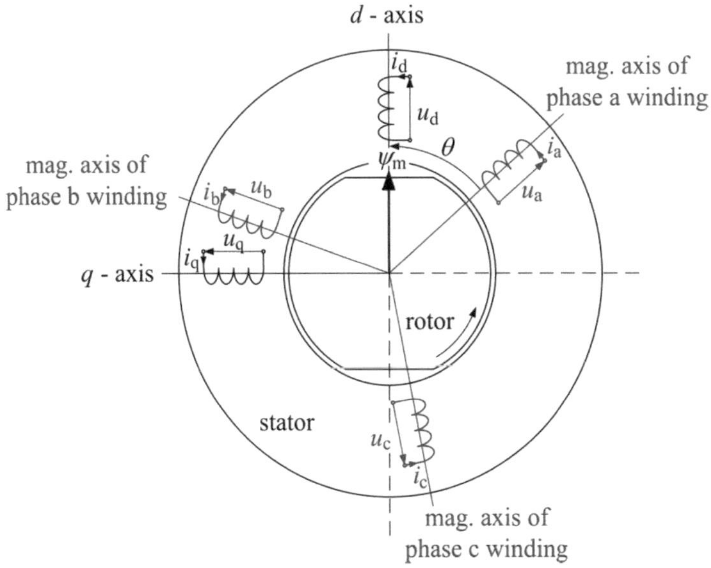

Fig. 1. Schematic presentation of a two-pole, three-phase permanent magnet synchronous machine

The model windings shown in Fig. 1 are aligned with the magnetic axes of the actual phase windings, whereas the effect of the permanent magnets is considered through the flux linkage vector $\boldsymbol{\psi}_m$ with the length ψ_m. The *d*-axis is aligned with the flux linkage vector and, the *q*-axis is displaced by an electric angle of $\pi/2$. Since the permanent magnets are located on the rotor, the angle θ represents the rotor position or the displacement of the *d*-axis with respect to the phase *a* magnetic axis. The magnetic axes of the phase *a, b* and *c* windings are treated as the reference frame *abc*. The voltage balances in the individual phase windings of the machine shown in Fig. 1 are described by (1) and (2):

$$\mathbf{u}_{abc} = \mathbf{R}\mathbf{i}_{abc} + \frac{d}{dt}\boldsymbol{\psi}_{abc} + \frac{d}{dt}\boldsymbol{\psi}_{mabc} \tag{1}$$

$$\mathbf{u}_{abc} = \begin{bmatrix} u_a \\ u_b \\ u_c \end{bmatrix};\ \mathbf{i}_{abc} = \begin{bmatrix} i_a \\ i_b \\ i_c \end{bmatrix};\ \boldsymbol{\psi}_{abc} = \begin{bmatrix} \psi_a(i_a, i_b, i_c, \theta) \\ \psi_b(i_a, i_b, i_c, \theta) \\ \psi_c(i_a, i_b, i_c, \theta) \end{bmatrix};\ \boldsymbol{\psi}_{mabc} = \begin{bmatrix} \psi_{ma}(\theta) \\ \psi_{mb}(\theta) \\ \psi_{mc}(\theta) \end{bmatrix};\ \mathbf{R} = \begin{bmatrix} R_a & & \\ & R_b & \\ & & R_c \end{bmatrix} \tag{2}$$

where $k \in \{a,b,c\}$ denotes the three phases, u_k, i_k, and R_k are the phase *k* voltage, current, and resistance, respectively. ψ_k is the phase *k* flux linkages due to the stator current excitation, while ψ_{mk} is the phase k flux linkages due to the permanent magnets. The flux linkages caused by the stator current excitation are treated as position and current dependent $\psi_k(i_a,i_b,i_c,\theta)$, whereas the flux linkages caused by the permanent magnets are treated as position dependent $\psi_{mk}(\theta)$. In such way, the impact of the stator currents on the flux linkages, caused by the permanent magnets, is neglected. The equations (1) and (2) are completed by (3) describing motion:

$$J\frac{d^2\theta}{dt^2} = t_e(i_a, i_b, i_c, \psi_m, \theta) - t_l - b\frac{d\theta}{dt} \tag{3}$$

where J is the moment of inertia, t_e is the electric torque, t_l is the load torque, b is the coefficient of the viscous friction, while $\omega = d\theta/dt$ is the angular speed. The effects of slotting, saturation and cross-saturation as well as the effects, caused by the interactions between the permanent magnets and slots, are included in the model with the current and position dependent characteristics of flux linkages and electric torque. They can be determined either experimentally or by applying some of the numerical methods, like finite element method. When the resistances and characteristics of the flux linkages and electric torque are known, the model can be applied for calculation of different transient and steady-state conditions. Unfortunately, the electric machines used in the controlled drives are mostly wye-connected, which means that the sum of all three currents equals zero. Thus, the state variables of the proposed model, given by (1) to (3), are linearly dependent, which causes algebraic loops and problems with convergence during simulations. Moreover, only the dynamic models with independent state variables can be used for the control synthesis. It seems that all the aforementioned effects are considered in the model given by (1) to (3), however, the model itself is useless. The question is: how the model should be modified to become useful?

One of the possibilities is the reduction of dependent state and input variables. In this case, the model should be described using only two independent currents and only two independent line to line voltages. Another possibility is the transformation of the model into the d-q reference frame, which is a common approach in the case of magnetically linear models. According to (Fitzgerald et al., 1961) and (Krause et al., 2002), this transformation can be performed exclusively for the magnetically linear models, which is true. However, it is also true, that not the model, but its variables, can be transformed into any arbitrary reference frame, if the inverse transformation exists whereas the variables can be written in a unique way in the new reference frame. According to Fig. 1, the relations between the axes a, b and c, which define the reference frame abc, and the d-axis, q-axis and 0-axis, which define the reference frame $dq0$, are described by the transformation matrix $\mathbf{T}$ (4):

$$\mathbf{T} = \sqrt{\frac{2}{3}}\begin{bmatrix} \cos(\theta) & -\sin(\theta) & 1/\sqrt{2} \\ \cos\left(\theta + 4/3\,\pi\right) & -\sin\left(\theta + 4/3\,\pi\right) & 1/\sqrt{2} \\ \cos\left(\theta + 2/3\,\pi\right) & -\sin\left(\theta + 2/3\,\pi\right) & 1/\sqrt{2} \end{bmatrix} \tag{4}$$

where the 0-axis, not shown in Fig. 1, is orthogonal to the d-axis and q-axis. The transformation matrix $\mathbf{T}$ is orthogonal, which means that the inverse matrix equals the transposed original matrix $\mathbf{T}^{-1}=\mathbf{T}^{T}$. Moreover, the orthogonal transformation matrix assures the invariance of power.

Let us express the voltages, currents and flux linkages, written in the *abc* reference frame in (2), with the transformation matrix and voltages, currents and flux linkages written in the *dq0* reference frame (5), (6).

$$\mathbf{u}_{abc} = \mathbf{T}\mathbf{u}_{dq0};\ \mathbf{i}_{abc} = \mathbf{T}\mathbf{i}_{dq0};\ \boldsymbol{\psi}_{mabc} = \mathbf{T}\boldsymbol{\psi}_{mdq0};\ \boldsymbol{\psi}_{abc} = \mathbf{T}\boldsymbol{\psi}_{dq0}; \tag{5}$$

$$\mathbf{u}_{dq0} = \begin{bmatrix} u_d \\ u_q \\ u_0 \end{bmatrix};\ \mathbf{i}_{dq0} = \begin{bmatrix} i_d \\ i_q \\ i_0 \end{bmatrix};\ \boldsymbol{\psi}_{dq0} = \begin{bmatrix} \psi_d(i_d,i_q,i_0,\theta) \\ \psi_q(i_d,i_q,i_0,\theta) \\ \psi_0(i_d,i_q,i_0,\theta) \end{bmatrix};\ \boldsymbol{\psi}_{mdq0} = \begin{bmatrix} \psi_{md}(\theta) \\ \psi_{mq}(\theta) \\ \psi_{m0}(\theta) \end{bmatrix} \tag{6}$$

By inserting (5) into (1), equations (7) to (10) are obtained.

$$\mathbf{T}\mathbf{u}_{dq0} = \mathbf{R}\mathbf{T}\mathbf{i}_{dq0} + \frac{d}{dt}\left\{\mathbf{T}\boldsymbol{\psi}_{dq0}\right\} + \frac{d}{dt}\left\{\mathbf{T}\boldsymbol{\psi}_{mdq0}\right\} \tag{7}$$

$$\mathbf{u}_{dq0} = \mathbf{T}^{-1}\mathbf{R}\mathbf{T}\mathbf{i}_{dq0} + \mathbf{T}^{-1}\frac{d}{dt}\left\{\mathbf{T}\boldsymbol{\psi}_{dq0}\right\} + \mathbf{T}^{-1}\frac{d}{dt}\left\{\mathbf{T}\boldsymbol{\psi}_{mdq0}\right\} \tag{8}$$

$$\mathbf{u}_{dq0} = \mathbf{T}^{-1}\mathbf{R}\mathbf{T}\mathbf{i}_{dq0} + \mathbf{T}^{-1}\frac{d}{dt}\left\{\mathbf{T}\right\}\boldsymbol{\psi}_{dq0} + \mathbf{T}^{-1}\mathbf{T}\frac{d}{dt}\left\{\boldsymbol{\psi}_{dq0}\right\} + \mathbf{T}^{-1}\frac{d}{dt}\left\{\mathbf{T}\right\}\boldsymbol{\psi}_{mdq0} + \mathbf{T}^{-1}\mathbf{T}\frac{d}{dt}\left\{\boldsymbol{\psi}_{mdq0}\right\} \tag{9}$$

$$\mathbf{u}_{dq0} = \mathbf{T}^{-1}\mathbf{R}\mathbf{T}\mathbf{i}_{dq0} + \mathbf{T}^{-1}\frac{d}{dt}\left\{\mathbf{T}\right\}\boldsymbol{\psi}_{dq0} + \frac{d}{dt}\left\{\boldsymbol{\psi}_{dq0}\right\} + \mathbf{T}^{-1}\frac{d}{dt}\left\{\mathbf{T}\right\}\boldsymbol{\psi}_{mdq0} + \frac{d}{dt}\left\{\boldsymbol{\psi}_{mdq0}\right\} \tag{10}$$

After considering (4) and (6) in (10), the voltage equation (11) is obtained:

$$\begin{bmatrix} u_d \\ u_q \\ u_0 \end{bmatrix} = R\begin{bmatrix} i_d \\ i_q \\ i_0 \end{bmatrix} + \begin{bmatrix} \frac{\partial\psi_d}{\partial i_d} & \frac{\partial\psi_d}{\partial i_q} & \frac{\partial\psi_d}{\partial i_0} \\ \frac{\partial\psi_q}{\partial i_d} & \frac{\partial\psi_q}{\partial i_q} & \frac{\partial\psi_q}{\partial i_0} \\ \frac{\partial\psi_0}{\partial i_d} & \frac{\partial\psi_0}{\partial i_q} & \frac{\partial\psi_0}{\partial i_0} \end{bmatrix}\frac{d}{dt}\begin{bmatrix} i_d \\ i_q \\ i_0 \end{bmatrix} + \frac{d\theta}{dt}\left\{\begin{bmatrix} \frac{\partial\psi_d}{\partial\theta} \\ \frac{\partial\psi_q}{\partial\theta} \\ \frac{\partial\psi_0}{\partial\theta} \end{bmatrix} + \begin{bmatrix} -\psi_q \\ \psi_d \\ 0 \end{bmatrix} + \begin{bmatrix} \frac{\partial\psi_{md}}{\partial\theta} \\ \frac{\partial\psi_{mq}}{\partial\theta} \\ \frac{\partial\psi_{m0}}{\partial\theta} \end{bmatrix} + \begin{bmatrix} -\psi_{mq} \\ \psi_{md} \\ 0 \end{bmatrix}\right\} \tag{11}$$

where $R=R_a=R_b=R_c$ denotes the stator resistance. It must be explained that after performing similar derivation for the magnetically liner dynamic model in (Fitzgerald et al., 1961) and (Krause et al., 2002), the model written in the *dq0* reference frame is obtained together with the model parameters. However, in the case of magnetically nonlinear model, treated in this chapter, only the form of the matrix voltage equation is obtained. The model is of no use until the characteristics of flux linkages in the *dq0* reference frame are determined. To carry out this, a voltage source inverter controlled in the *dq0* reference frame can be applied.

Due to the wye connected winding, the sum of currents i_a, i_b and i_c equals zero, whereas considering the transformation matrix **T** (4), (5) and (6) yields (12).

$$i_a + i_b + i_c = 0 \;\Rightarrow\; i_0 = \sqrt{\frac{2}{3}}\frac{1}{\sqrt{2}}\left(i_a + i_b + i_c\right) = 0 \tag{12}$$

Thus, the current in the neutral conductor i_0 equals 0. The neutral point voltage u_0 appears due to the changing flux linkages ψ_0 and ψ_{m0} which are caused by the changing level of the saturation in individual parts of the machine. The flux linkage ψ_0 is caused by the current excitation and the flux linkage ψ_{m0} appears due to the permanent magnets. When the discussed machine is closed-loop current controlled, the current controllers implicitly compensate u_0. Thus, the neutral conductor current i_0 and the neutral point voltage u_0 equal zero. They are caused by the changing flux linkages ψ_0 and ψ_{m0}, which influence only on u_0 and i_0, and are not coupled with the d-axis or q-axis components. Therefore, in the case of closed-loop current controlled machine, the zero component voltage, current and flux linkages in (11) can be neglected, which leads to the voltage equation (13).

$$\begin{bmatrix} u_d \\ u_q \end{bmatrix} = R \begin{bmatrix} i_d \\ i_q \end{bmatrix} + \begin{bmatrix} \frac{\partial \psi_d}{\partial i_d} & \frac{\partial \psi_d}{\partial i_q} \\ \frac{\partial \psi_q}{\partial i_d} & \frac{\partial \psi_q}{\partial i_q} \end{bmatrix} \frac{d}{dt} \begin{bmatrix} i_d \\ i_q \end{bmatrix} + \frac{d\theta}{dt} \left\{ \begin{bmatrix} \frac{\partial \psi_d}{\partial \theta} \\ \frac{\partial \psi_q}{\partial \theta} \end{bmatrix} + \begin{bmatrix} -\psi_q \\ \psi_d \end{bmatrix} + \begin{bmatrix} \frac{\partial \psi_{md}}{\partial \theta} \\ \frac{\partial \psi_{mq}}{\partial \theta} \end{bmatrix} + \begin{bmatrix} -\psi_{mq} \\ \psi_{md} \end{bmatrix} \right\} \quad (13)$$

Thus, the $dq0$ reference frame is reduced to the dq reference frame. In order to complete the magnetically nonlinear dynamic model of the permanent magnet synchronous machine, the torque expression in (3) must be written in the dq reference frame too. According to (Krause et al., 2002), in the cases of magnetically nonlinear magnetic core, the co-energy approach can be used to determine the torque expression. Since the co-energy approach is rather demanding, another approach, which is often used in the machine design, is applied in this section. It gives very good results in the case of surface permanent magnet machines whereas in the other cases, it normally represents an acceptable approximation. If the leakage flux linkages are neglected and the energy in the magnetic field due to the permanent magnets is considered as constant, the back electromotive forces multiplied by the corresponding currents represent the mechanical power at the shaft of the machine $d\theta/dt\ t_e$ (14).

$$\frac{d\theta}{dt} t_e = e_d i_d + e_q i_q = \frac{d\theta}{dt} \left(\frac{\partial \psi_d}{\partial \theta} + \frac{\partial \psi_{md}}{\partial \theta} - \psi_q - \psi_{mq} \right) i_d + \frac{d\theta}{dt} \left(\frac{\partial \psi_q}{\partial \theta} + \frac{\partial \psi_{mq}}{\partial \theta} + \psi_d + \psi_{md} \right) i_q \quad (14)$$

With the comparison of the left hand side and the right hand side of (14), the electric torque can be expressed by (15):

$$t_e = \left(\frac{\partial \psi_d}{\partial \theta} + \frac{\partial \psi_{md}}{\partial \theta} - \psi_q - \psi_{mq} \right) i_d + \left(\frac{\partial \psi_q}{\partial \theta} + \frac{\partial \psi_{mq}}{\partial \theta} + \psi_d + \psi_{md} \right) i_q \quad (15)$$

which represents the torque equation of a two-pole synchronous permanent magnet machine. By neglecting the partial derivatives in (15), the effects of slotting and the effects of interactions between the slots and permanent magnets are neglected. By neglecting additionally the terms ψ_d, ψ_q, and ψ_{mq}, (15) reduces to the well known torque expression of a permanent magnet synchronous machine (Krause et al., 2002).

The magnetically nonlinear two-axis dynamic model of a two-pole permanent magnet synchronous machine, written in the dq reference frame, is given in its final form by the voltage equation (13), the torque equation (15) and the equation describing motion (16).

$$J\frac{d^2\theta}{dt^2} = t_e - t_l - b\frac{d\theta}{dt} \tag{16}$$

The current and position dependent characteristics of flux linkages $\psi_d(i_d, i_q, \theta)$ and $\psi_q(i_d, i_q, \theta)$, caused by the current excitation, as well as the position dependent characteristics of flux linkages $\psi_{md}(\theta)$ and $\psi_{mq}(\theta)$, caused by the permanent magnets, are applied in the model. They are used to account for the effects of saturation and cross-saturation, the effects of slotting and the effects caused by the interactions between the permanent magnets and slots.

2.2 Reluctance synchronous machine

The magnetically nonlinear dynamic model of a synchronous reluctance machine (Štumberger G. et al., 2003, 2004a, 2004b) can be derived from the model of the permanent magnet synchronous machine, by neglecting all the terms related with the permanent magnets. Thus, all the terms containing ψ_{md} and ψ_{mq} in (13) and (15) are dropped out. Only the flux linkages caused by the current excitation appear in the equations. In the case of the reluctance motor, the flux linkage vector due to the permanent magnets is missing. Therefore, the d-axis and the q-axis are defined with the directions of the minimum and the maximum reluctance, respectively. Both axes are displaced by the electric angle of $\pi/2$. In such way, the two-axis magnetically nonlinear dynamic model of the two-pole, three-phase synchronous reluctance machine is obtained. It is given by the voltage equation (17), the torque equation (18) and the equation describing motion (19).

$$\begin{bmatrix} u_d \\ u_q \end{bmatrix} = R\begin{bmatrix} i_d \\ i_q \end{bmatrix} + \begin{bmatrix} \frac{\partial \psi_d}{\partial i_d} & \frac{\partial \psi_d}{\partial i_q} \\ \frac{\partial \psi_q}{\partial i_d} & \frac{\partial \psi_q}{\partial i_q} \end{bmatrix} \frac{d}{dt}\begin{bmatrix} i_d \\ i_q \end{bmatrix} + \frac{d\theta}{dt}\left\{\begin{bmatrix} \frac{\partial \psi_d}{\partial \theta} \\ \frac{\partial \psi_q}{\partial \theta} \end{bmatrix} + \begin{bmatrix} -\psi_q \\ \psi_d \end{bmatrix}\right\} \tag{17}$$

$$t_e = \left(\frac{\partial \psi_d}{\partial \theta} - \psi_q\right) i_d + \left(\frac{\partial \psi_q}{\partial \theta} + \psi_d\right) i_q \tag{18}$$

$$J\frac{d^2\theta}{dt^2} = t_e - t_l - b\frac{d\theta}{dt} \tag{19}$$

The effects of slotting, saturation, and cross-saturation are considered in the model by the current and position dependent characteristics of flux linkages.

2.3 Linear reluctance synchronous machine

The dynamic model of the rotary synchronous reluctance machine, given by (17) to (19) must be modified, to be suitable for appropriate description of the dynamic conditions in the linear synchronous reluctance machine (LSRM) (Štumberger G. et al., 2003, 2004a, 2004b). The design of this machine is described in (Hamler et al. 1998), whereas similar machines are described in (Daldaban & Ustkoyuncu, 2006, 2007, 2010). It has a short and moving primary called mover and a long reluctance secondary. It is supplied from the primary where windings are placed. The LSRM, schematically presented in Fig. 2, performs translational motion.

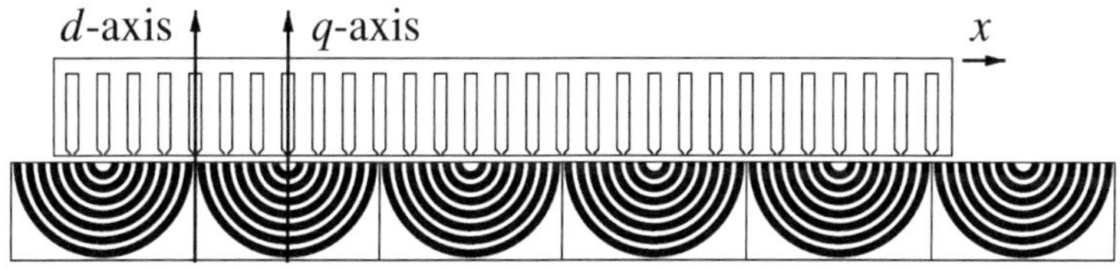

Fig. 2. Schematic presentation of a linear synchronous reluctance machine

In (17) to (19), the angular position θ is expressed with the position x and the pole pitch τ_p (20).

$$\theta = \frac{\pi}{\tau_p} x \tag{20}$$

This leads to changes in (19), where the moment of inertia J is replaced with the mass of the mover m, while the electric torque t_e and the load torque t_l are replaced with the thrust F_e and the load force F_l. The obtained magnetically nonlinear dynamic model of the linear synchronous reluctance machine is given by (21) to (23), where F_f is the friction force.

$$\begin{bmatrix} u_d \\ u_q \end{bmatrix} = R \begin{bmatrix} i_d \\ i_q \end{bmatrix} + \begin{bmatrix} \frac{\partial \psi_d}{\partial i_d} & \frac{\partial \psi_d}{\partial i_q} \\ \frac{\partial \psi_q}{\partial i_d} & \frac{\partial \psi_q}{\partial i_q} \end{bmatrix} \frac{d}{dt} \begin{bmatrix} i_d \\ i_q \end{bmatrix} + \frac{\pi}{\tau_p} \frac{dx}{dt} \left\{ \begin{bmatrix} \frac{\partial \psi_d}{\partial x} \\ \frac{\partial \psi_q}{\partial x} \end{bmatrix} + \begin{bmatrix} -\psi_q \\ \psi_d \end{bmatrix} \right\} \tag{21}$$

$$t_e = \frac{\pi}{\tau_p} \left(\psi_d i_q - \psi_q i_d \right) + \frac{\partial \psi_d}{\partial x} i_d + \frac{\partial \psi_q}{\partial x} i_q \tag{22}$$

$$m \frac{d^2 x}{dt^2} = F_e - F_l - F_f \tag{23}$$

In the case of linear synchronous reluctance motor, the current and position dependent characteristics of flux linkages are used to consider not only the effects of slotting, saturation and cross-saturation, but also the end effects. The end effects appear only in the linear machines due to the finite lengths of the primary and secondary. In linear machines, the end-poles and the windings in these poles cannot be symmetrical. On the contrary, in the rotational machines the stator and rotor are cylindrical and closed in themselves. The path along the circumference of the stator or rotor apparently never ends, whereas all the poles and phase windings are symmetrical.

For the given constant values of the currents i_d and i_q, the position dependent flux linkages ψ_d and ψ_q can be expressed in the form of Fourier series (24) and (25):

$$\psi_d = \psi_{d0} + \sum_{h=1}^{N} \left(\psi_{dch} \cos(\frac{\pi}{\tau_p} hx) + \psi_{dsh} \sin(\frac{\pi}{\tau_p} hx) \right) \tag{24}$$

$$\psi_q = \psi_{q0} + \sum_{h=1}^{N} \left(\psi_{qch} \cos(\frac{\pi}{\tau_p} hx) + \psi_{qsh} \sin(\frac{\pi}{\tau_p} hx) \right) \tag{25}$$

where N is the number of considered higher order harmonics, h denotes the harmonic order while ψ_{d0}, ψ_{dch}, ψ_{dsh} and ψ_{q0}, ψ_{qch}, ψ_{qsh} are the Fourier coefficients. After inserting (24) and (25) into (22), the position dependent thrust $F_e(x)$ can be expressed by (26):

$$F_e(x) = F_0 + \sum_{h=1}^{N} \left(F_{ch} \cos(\frac{\pi}{\tau_p} hx) + F_{sh} \sin(\frac{\pi}{\tau_p} hx) \right) \tag{26}$$

where F_0, F_{ch} and F_{sh} are the Fourier coefficients. The mean value of the thrust is given by F_0 (27), whereas the position dependent thrust pulsation is described by F_{ch} (28) and F_{sh} (29).

$$F_0 = \frac{\pi}{\tau_p}\left(\psi_{d0} i_q - \psi_{q0} i_d\right) \tag{27}$$

$$F_{ch} = \frac{\pi}{\tau_p}\left(\psi_{dch} i_q - \psi_{qch} i_d + h\left(\psi_{dsh} i_d + \psi_{qsh} i_q\right)\right) \tag{28}$$

$$F_{sh} = \frac{\pi}{\tau_p}\left(\psi_{dsh} i_q - \psi_{qsh} i_d - h\left(\psi_{dch} i_d + \psi_{qch} i_q\right)\right) \tag{29}$$

The mean value of the thrust, given by F_0 (27), is well known thrust equation of linear synchronous reluctance machines.
All the dynamic models presented in this section can be easily realized in the program packages like Matlab/Simulink. However, they are of no use without model parameters, given in the form of current and position dependent characteristics of flux linkages. They can be easily implemented in the models using the multi-dimensional lookup tables.

3. Determining dynamic model parameters

The magnetically nonlinear dynamic models of synchronous machines presented in the section 2 are of no use until the model parameters are determined in the form of current and position dependent characteristics of flux linkages. In the subsection 3.1, the experimental method appropriate for determining the position dependent characteristics of flux linkages due to the permanent magnets is presented (Hadžiselimović et al., 2007b). Similarly, the subsection 3.2 describes the experimental method appropriate for determining the current and position dependent characteristics of flux linkages caused by the current excitation (Štumberger G. et al., 2004b). The experimental methods which can be used to determine the thrust and the friction force characteristics of a linear synchronous reluctance motor are presented in the subsection 3.3 (Štumberger G. et al., 2004b). In order to determine the aforementioned characteristics, a special experimental setup is applied. The tested machine is supplied from a voltage source inverter controlled in the dq reference frame. The closed-loop current control in the one axis is combined with the open-loop voltage control in the other axis.

3.1 Position dependent flux linkages caused by the permanent magnets

The angular position θ dependent characteristcis of flux linkages $\psi_{md}(\theta)$ and $\psi_{mq}(\theta)$, required in (13) and (15), can be determined from the waveforms of the three-phase back electromotive forces e_a, e_b and e_c. They can be measured on the open terminals of the tested permanent magnet synchronous machine, driven by another motor at the constant speed ω

$=d\theta/dt$. Using (4) in (5), the e_a, e_b and e_c are transformed into dq reference frame. In such way, the waveforms of e_d and e_q are determined. Since the machine terminals are open, the currents i_d, i_q and consequently the flux linkages ψ_d, ψ_q, caused by these currents, equal zero. Considering this, (13) is reduced to (30).

$$\frac{e_d}{\omega} = -\psi_{mq} + \frac{\partial \psi_{md}}{\partial \theta}, \qquad \frac{e_q}{\omega} = \psi_{md} + \frac{\partial \psi_{mq}}{\partial \theta} \tag{30}$$

The second partial derivatives of expressions in (30) give (31), whereas after considering (30) in (31), (32) is obtained.

$$\frac{\partial}{\partial \theta}\left(\frac{e_d}{\omega}\right) = -\frac{\partial \psi_{mq}}{\partial \theta} + \frac{\partial^2 \psi_{md}}{\partial \theta^2}, \qquad \frac{\partial}{\partial \theta}\left(\frac{e_q}{\omega}\right) = \frac{\partial \psi_{md}}{\partial \theta} + \frac{\partial^2 \psi_{mq}}{\partial \theta^2} \tag{31}$$

$$\frac{\partial^2 \psi_{md}}{\partial \theta^2} + \psi_{md} = \frac{\partial}{\partial \theta}\left(\frac{e_d}{\omega}\right) + \frac{e_q}{\omega}, \qquad \frac{\partial^2 \psi_{mq}}{\partial \theta^2} + \psi_{mq} = \frac{\partial}{\partial \theta}\left(\frac{e_q}{\omega}\right) - \frac{e_d}{\omega} \tag{32}$$

The known waveforms of e_d and e_q, as well as the unknown flux linkage $\psi_{md}(\theta)$ and $\psi_{mq}(\theta)$, can be represented in the form of Fourier series (33) to (36):

$$e_d = e_{d0} + \sum_{h=1}^{N} \left(e_{dch}\cos(h\theta) + e_{dsh}\sin(h\theta)\right) \tag{33}$$

$$e_q = e_{q0} + \sum_{h=1}^{N} \left(e_{qch}\cos(h\theta) + e_{qsh}\sin(h\theta)\right) \tag{34}$$

$$\psi_{md} = \psi_{md0} + \sum_{h=1}^{N} \left(\psi_{mdch}\cos(h\theta) + \psi_{mdsh}\sin(h\theta)\right) \tag{35}$$

$$\psi_{mq} = \psi_{mq0} + \sum_{h=1}^{N} \left(\psi_{mqch}\cos(h\theta) + \psi_{mqsh}\sin(h\theta)\right) \tag{36}$$

where N is the number of considered higher order harmonics, e_{d0}, e_{dch}, e_{dsh} and e_{q0}, e_{qch}, e_{qsh} are the Fourier coefficients of back electromotive forces, while ψ_{md0}, ψ_{mdch}, ψ_{mdsh} and ψ_{mq0}, ψ_{mqch}, ψ_{mqsh} are the Fourier coefficients of flux linkages. The expressions (33) to (36) are applied to calculate the partial derivatives required in (32). The results are given in (37) and (38) for the harmonic order h. With the comparison of the terms on the left hand side and on the right hand side of equations (37) and (38), the expressions for calculation of ψ_{mdch}, ψ_{mdsh}, ψ_{mqch}, and ψ_{mqsh} are obtained. They are given in (39).

$$\left(1-h^2\right)\left(\psi_{mdch}\cos(h\theta) + \psi_{mdsh}\sin(h\theta)\right) = \frac{he_{dsh} + e_{qch}}{\omega}\cos(h\theta) + \frac{-he_{dch} + e_{qsh}}{\omega}\sin(h\theta) \tag{37}$$

$$\left(1-h^2\right)\left(\psi_{mqch}\cos(h\theta) + \psi_{mqsh}\sin(h\theta)\right) = \frac{he_{qsh} - e_{dch}}{\omega}\cos(h\theta) + \frac{-he_{qch} - e_{dsh}}{\omega}\sin(h\theta) \tag{38}$$

$$\psi_{mdch} = \frac{he_{dsh} + e_{qch}}{\left(1-h^2\right)\omega},\quad \psi_{mdsh} = \frac{-he_{dch} + e_{qsh}}{\left(1-h^2\right)\omega},\quad \psi_{mqch} = \frac{he_{qsh} - e_{dch}}{\left(1-h^2\right)\omega},\quad \psi_{mqsh} = \frac{-he_{qch} - e_{dsh}}{\left(1-h^2\right)\omega} \tag{39}$$

More details related to this method can be found in (Hadžiselimović, 2007b), (Hadžiselimović et al., 2007b).

3.2 Current and position dependent flux linkages caused by the current excitation

The characteristics of the current and position dependent flux linkages are determined using the experimental setup schematically presented in Fig. 3.

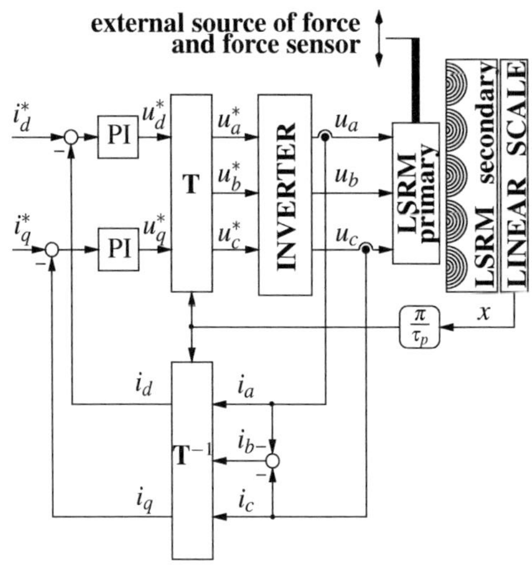

Fig. 3. Schematic presentation of the applied experimental setup

It consists of a tested linear synchronous reluctance machine (LSRM), controlled voltage source inverter, external driving motor used as an external source of force, measurement chains, and control system with digital signal processor (DSP). The DSP is applied to realize the closed-loop current control and the open-loop voltage control in the *dq* reference frame. The PI controllers are used for the closed-loop control, whereas the required transformations are performed using the transformation matrix **T** (4). The reference values in Fig. 3 are marked with *. The inverter, measurement chains, and transformations make possible treatment of the tested three-phase LSRM through its two-axis dynamic model, given by (21) to (23). The current and position measurement chains are used to close the control loops in the applied control algorithm. The force measurements chain is used off-line.

In order to determine the current and position dependent characteristics of flux linkages $\psi_d(i_d,i_q,x)$ and $\psi_q(i_d,i_q,x)$, the mover of the LSRM is locked at the chosen position. The current in one axis is closed-loop controlled in order to maintain the preset constant value, while the voltage in the orthogonal axis is changed in a stepwise manner. Let us suppose that the mover is locked, while the current i_q is closed loop controlled and can be treated as constant. In such conditions, the u_d in (21) can be expressed by (40), which leads to (41).

$$u_d = Ri_d + \frac{\partial \psi_d}{\partial i_d}\frac{di_d}{dt} = Ri_d + \frac{d\psi_d}{dt} \tag{40}$$

$$\psi_d(t) = \psi_d(0) + \int_0^t \left(u_d(\tau) - Ri_d(\tau)\right) d\tau \tag{41}$$

If the resistance R and the time behaviours of u_d and i_d are known, the time behaviour of ψ_d can be determined by (41), where $\psi_d(0)$ is the flux linkage due to the remanent flux. Fig. 4 shows the applied stepwise changing voltage $u_d(t)$ and the responding current $i_d(t)$ measured during the test.

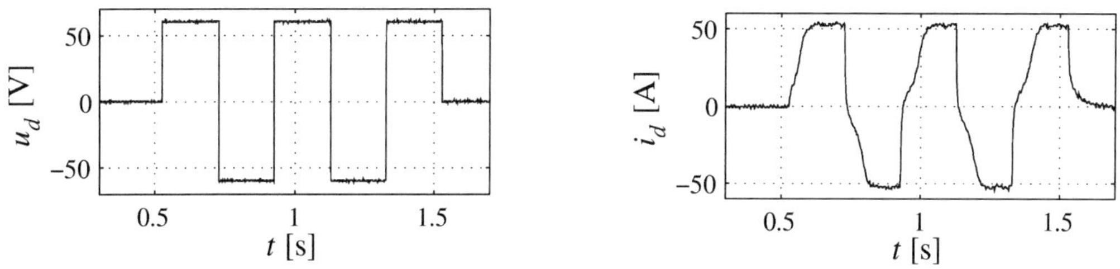

Fig. 4. Applied stepwise changing voltage $u_d(t)$ and responding current $i_d(t)$

The calculated time behaviour of the flux linkage $\psi_d(t)$ is shown in Fig. 5a. Fig. 5b shows the flux linkage $\psi_d(t)$ presented as a function of $i_d(t)$ in the form of a hysteresis loop $\psi_d(i_d)$. By calculating the average value of the currents for each flux linkage value, the unique characteristic $\psi_d(i_d)$, shown in Fig. 5c, is determined. The unique characteristic $\psi_d(i_d)$ shown in Fig. 5d is obtained by mapping the part of the characteristic $\psi_d(i_d)$, shown in Fig. 5c, from the third quadrant into the first quadrant and by calculating the average value of the currents for each value of the flux linkages again. By repeating the described procedure for different and equidistant values of the closed-loop controlled currents in both axes, the characteristics $\psi_d(i_d,i_q)$ and $\psi_q(i_d,i_q)$ can be determined for the given position x. To determine

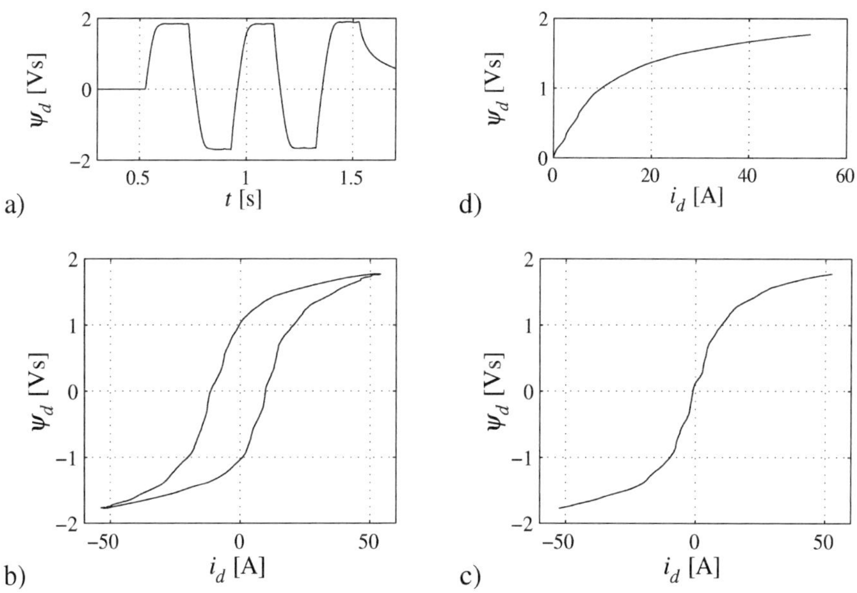

Fig. 5. Flux linkage time behaviour $\psi_d(t)$ a); hysteresis loop $\psi_d(i_d)$ b); unique characteristic $\psi_d(i_d)$ in the first and third quadrant c); unique characteristic $\psi_d(i_d)$ in the first quadrant d)

the current and position dependent characteristics $\psi_d(i_d,i_q,x)$ and $\psi_q(i_d,i_q,x)$, the tests must be repeated for different equidistant positions of the locked mover over at least one pole pitch. During the tests, the magnitude of the stepwise changing voltage should be high enough to assure that the responding current covers the entire range of operation. The frequency of the applied voltage must be low enough to allow the current to reach the steady-state before the next change. Since the expression (41) represents an open integrator without any feedback, the obtained results are extremely sensitive to the incorrect values of the resistance R. The steady-state values of the voltage and current can be used to calculate the resistance R after each voltage step change. The voltages generated by an inverter are normally pulse width modulated, whereas the measurement of such voltages could be a demanding task. The method for determining characteristics of flux linkages presented in this subsection gives acceptable results even when the voltage reference values are used instead of the measured ones. However, calculation of the resistance value after each voltage step change is substantial. Some of the methods that can be also applied for determining the magnetically nonlinear characteristics of magnetic cores inside electromagnetic devices and electric machines are presented in (Štumberger, 2005, 2008).

3.3 Thrust and friction force

The experimental setup shown in Fig. 3 is applied to determine the thrust F_e and the friction force F_f. The d-axis current and the q-axis current are closed-loop controlled in order to keep the preset constant values. The external source of force, in the form of a driving motor, is applied to move the mover of the LSRM at the constant speed of 0.02 m/s over two pole pitches from left to right and back again. The force causing the motion of the mover is measured by a force sensor. Since the friction force always opposes the force causing motion, one half of the difference between the forces measured for the moving left and right is the friction force, while the average value of both measured forces is the thrust. Fig. 6 shows the position dependent forces measured for the moving left and right together with the thrust and friction force. They are given over two pole pitches. To determine the thrust and friction force characteristics over the entire range of operation, the tests are repeated for different preset constant values of the closed-loop controlled currents in the axes d and q.

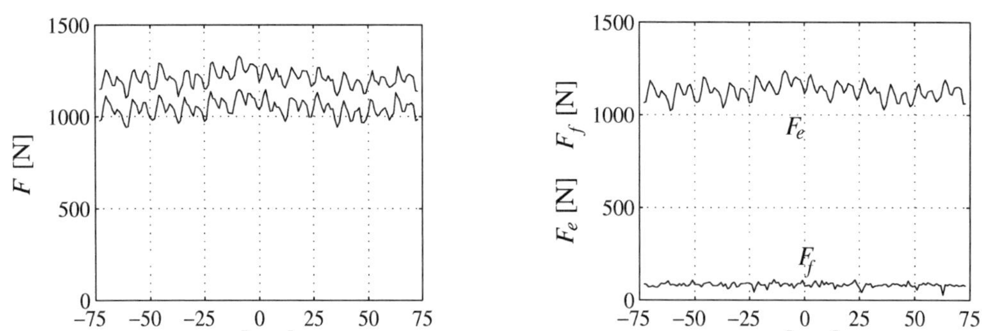

Fig. 6. Force F measured for moving left and right, thrust F_e and friction force F_f.

4. Results

The magnetically nonlinear dynamic models of permanent magnet and reluctance synchronous machines are presented in the section 2. These models are of no use without their parameters, given in the form of current and position dependent characteristics of flux linkages. Some of the methods, appropriate for determining these parameters are presented in the section 3. This section focuses on the linear synchronous reluctance motor. The characteristics of flux linkages, thrust and friction force, determined by the methods described in section 3, are presented. Applications of the proposed models are shown at the end of this section.

Figs. 7 and 8 show the characteristics of flux linkages $\psi_d(i_d,i_q,x)$ and $\psi_q(i_d,i_q,x)$. They are determined over the entire range of operation using the method presented in the subsection 3.2. Fig. 9 and 10 show the thrust characteristics determined by the method described in the subsection 3.3.

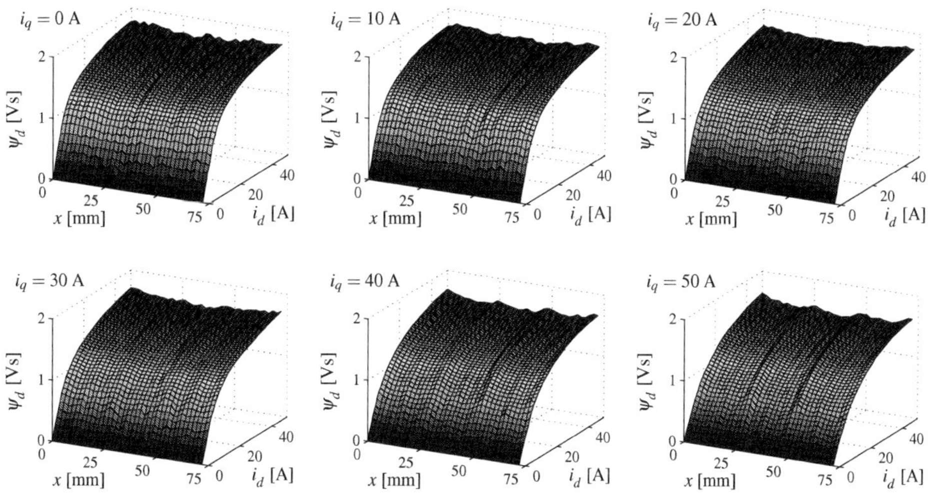

Fig. 7. Characteristics $\psi_d(i_d,i_q,x)$ given over one pole pitch for different constant values of current i_q

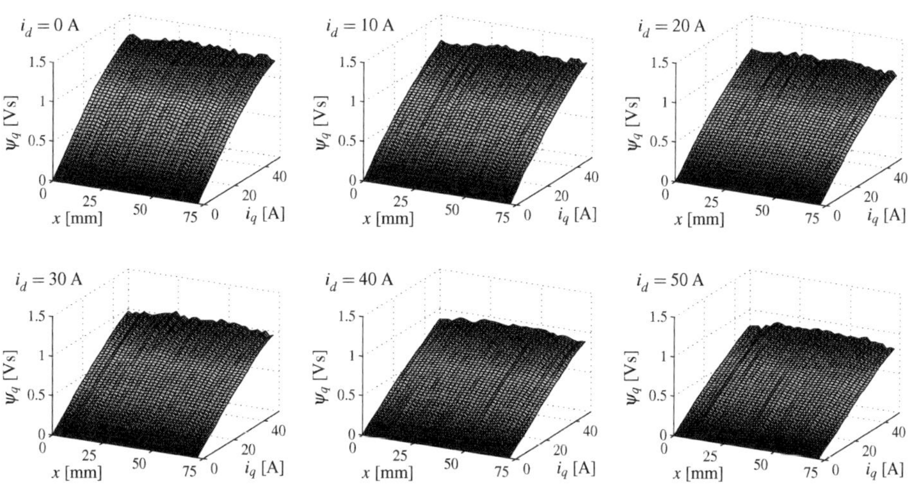

Fig. 8. Characteristics $\psi_q(i_d,i_q,x)$ given over one pole pitch for different constant values of current i_d

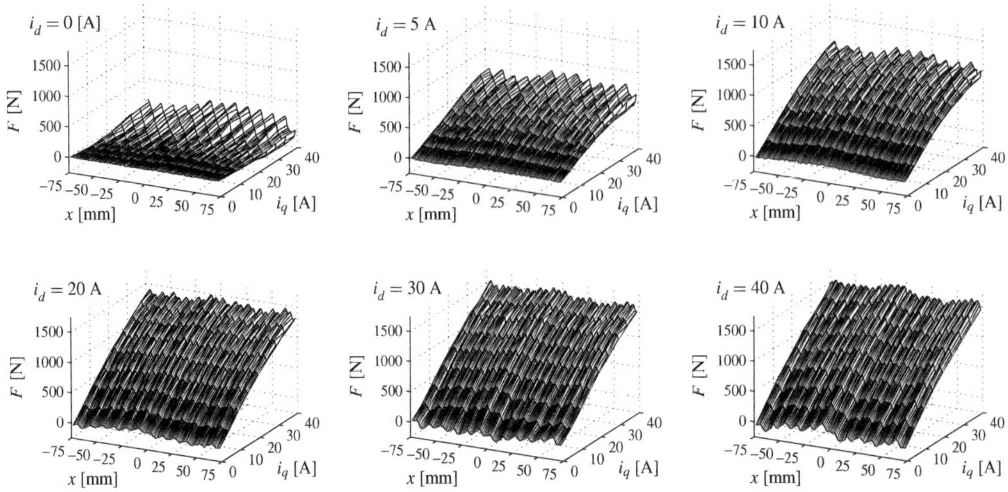

Fig. 9. Characteristics $F_e(i_d,i_q,x)$ given over two pole pitches for different constant values of current i_d

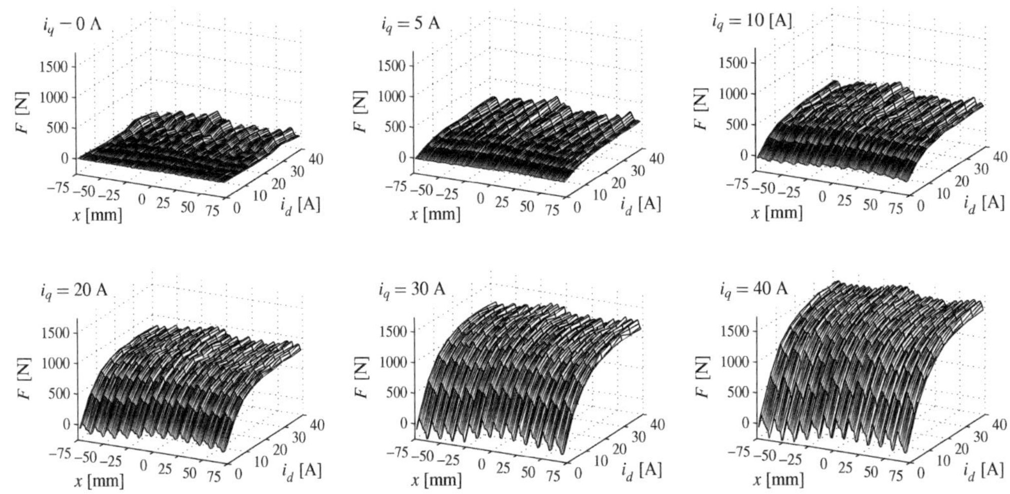

Fig. 10. Characteristics $F_e(i_d,i_q,x)$ given over two pole pitches for different constant values of current i_q

Figs. 11 to 14 show the trajectories of individual variables for the case of a kinematic control performed with experimental setup shown in Fig. 3. Figs. 11 and 12 give results for the kinematic control performed at higher speed, whereas the results presented in Figs. 13 and 14 are given for the kinematic control performed at much lower speed. Figs. 11 and 13 show the trajectories of the position x and current i_d measured during the experiment. These trajectories are used in the model to calculate the corresponding trajectories of the speed v and current i_q. The measured and calculated trajectories are shown in Figs. 12 and 14. The calculations are performed with the dynamic model of the LSRM given by equations (21) to (29), considering characteristics of flux linkages given in Figs. 7 and 8.

Fig. 11. Trajectories of position x and current i_d measured on the experimental system in the case of kinematic control

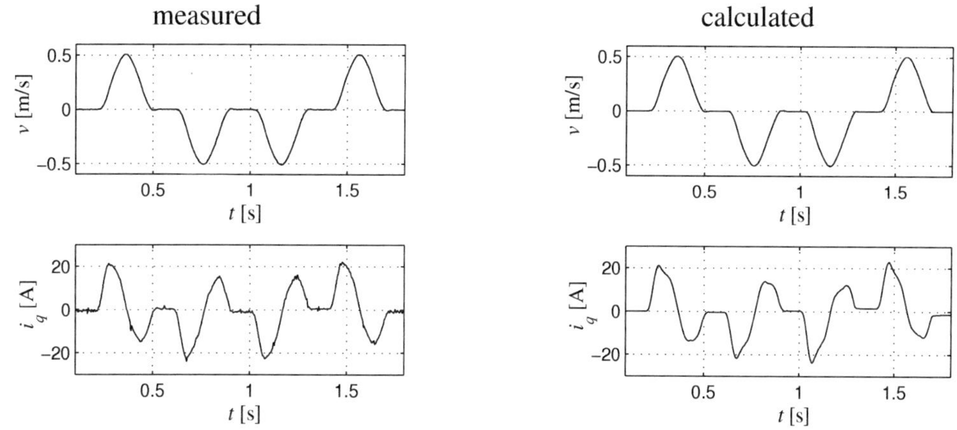

Fig. 12. Measured and calculated trajectories of speed $v=dx/dt$ and current i_q in the case of kinematic control with position trajectory given in Fig. 11

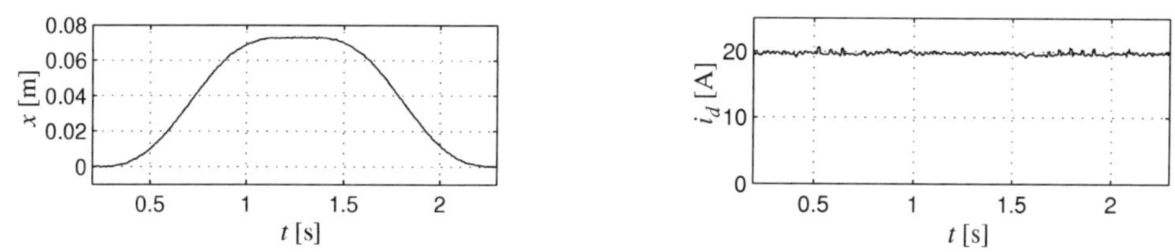

Fig. 13. Trajectories of position x and current i_d measured on the experimental system in the case of low speed kinematic control

The results presented in Fig. 14 clearly show that the measured as well as the calculated speed trajectories are deteriorated due to the effects of slotting. The mass of the mover filters these effects out at higher speeds, as shown in Fig. 12.

The results presented in Fig. 14 show that the magnetically nonlinear dynamic models of synchronous machines presented in this chapter contain the effects of slotting. In the next example, the models are involved in the design of nonlinear control with input-output decoupling and linearization described in (Dolinar, 2005). Fig. 15 shows the results of experiments performed on the experimental setup shown in Fig. 3. Compared are the results of low speed kinematic control obtained with two different control realizations. In the first realization, the control design is based on the magnetically linear model of tested linear synchronous reluctance machine. In the second one, the control design is based on the

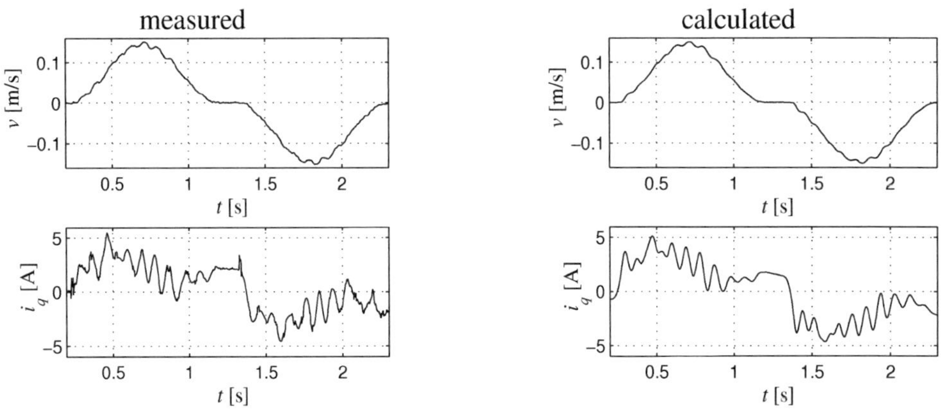

Fig. 14. Measured and calculated trajectories of speed $v=dx/dt$ and current i_q in the case of low speed kinematic control with position trajectory given in Fig. 13

magnetically nonlinear model presented in this work. Fig. 15 compares the trajectories of position reference, position, position error, speed reference, speed, speed error, *d*- and *q*-axis currents, and *d*- and *q*-axis voltages for both control realizations. The results presented in Fig. 15 clearly show a substantial improvement of the low speed kinematic control performance in the case when the magnetically nonlinear dynamic model is applied in the nonlinear control design. The position error is reduced for more than five times while the speed error is reduced for more than two times. However, as it is shown in Figs. 11 to 14, the magnetically nonlinear dynamic models presented in this chapter can substantially contribute to the position error reduction at very low speeds, whereas at higher speeds the mass of the mover filter these effects out.

5. Conclusion

This chapter deals with the magnetically nonlinear dynamic models of synchronous machines. The procedure that can be used to derive the magnetically nonlinear dynamic model of synchronous machines is presented in the case of rotational permanent magnet synchronous machine. The obtained model is then modified in order to be suitable for description of the rotary and linear synchronous reluctance machine. Since the model is useless without its parameters, the experimental methods suitable for determining model parameters are described. They are presented in the form of current and position dependent characteristics of flux linkages, given for the tested linear synchronous reluctance machine. The effects of slotting, saturation, cross-saturation, and end effects are accounted for in the models. The models can help to reduce the tracking errors when used in the control design.

6. References

Abdallh, A.A.E.; Sergeant, P.; Crevecoeur, G.; Vandenbossche, L.; Dupre, L. & Sablik, M. (2009). Magnetic Material Identification in Geometries with Non-Uniform Electromagnetic Fields Using Global and Local Magnetic Measurements. *IEEE Transactions on Magnetics,* Vol. 45, No. 10, November 2009 page numbers 4157-4160, ISSN: 0018-9464

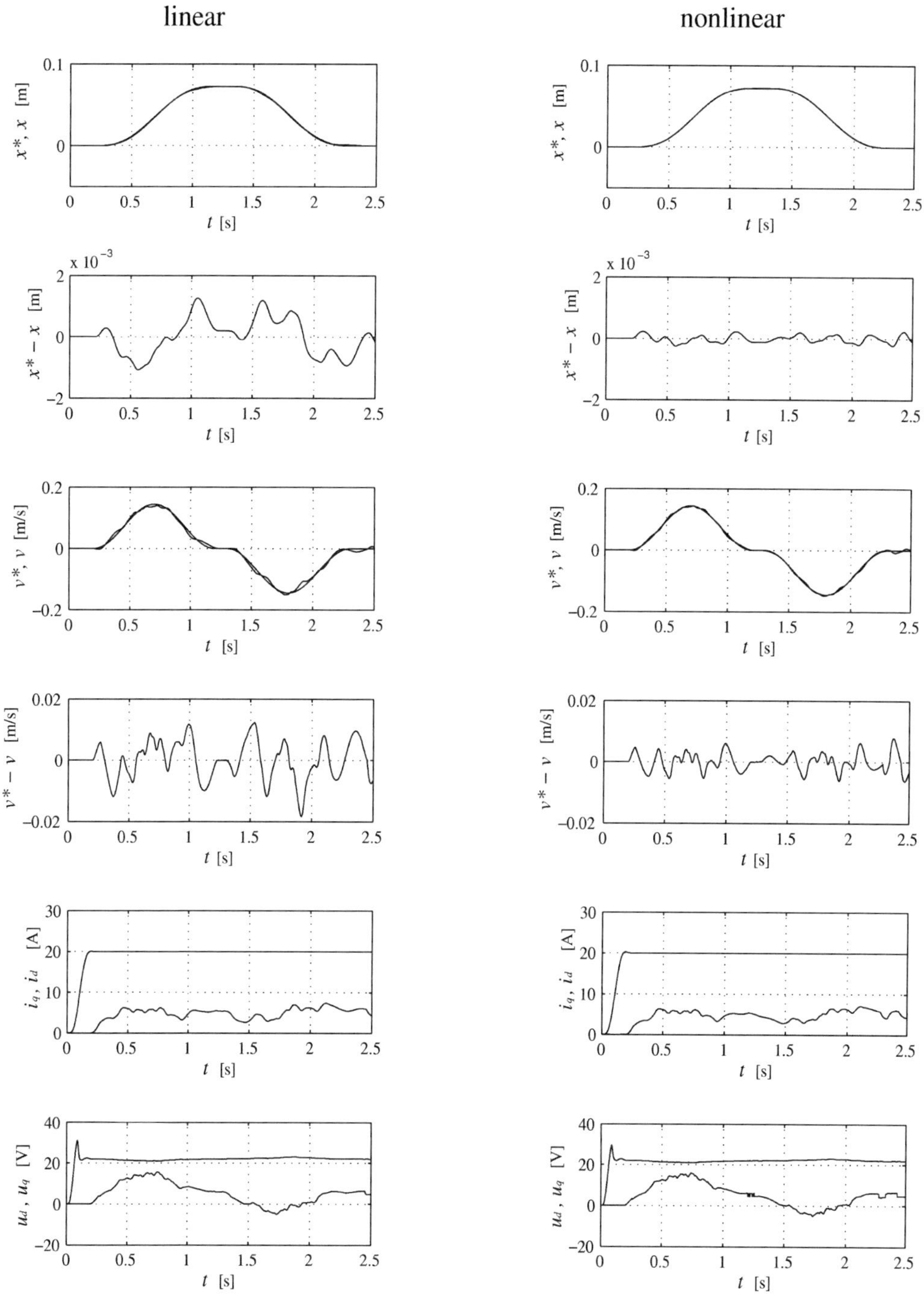

Fig. 15. Trajectories of position x and position reference x^*, position error x^*-x, speed v and speed reference v^*, speed error v^*-v, currents i_d and i_q, and voltages u_d and u_q measured on experimental setup for the cases when the linear and nonlinear LSRM dynamic models are applied in the control design

Boldea, I. & Tutelea, L. (2010). *Electric machines: steady states, transients, and design with Matlab,* CRC Press, ISBN: 978-1-4200-5572-6, New York

Boll, R. (1990). *Weichmagnetische Werkstoffe,* Vakuumschmelze GmbH, Hanau

Capecchi, E.; Guglielmi, P.; Pastorelli, M. & Vagati, A. (2001). Position-sensorless control of the transverse-laminated synchronous reluctance motor. *IEEE Transactions on Industry Applications,* Vol. 37, No. 6, November/December 2001 page numbers 1768-1776, ISSN: 0093-9994

Daldaban, F. & Ustkoyuncu, N. (2006). A new double sided linear switched reluctance motor with low cost. *Energy Conversion and Management,* Vol. 47, No. 18-19, November 2006 page numbers 2983-2990, ISSN: 0196-8904

Daldaban, F. & Ustkoyuncu, N. (2007). New disc type switched reluctance motor for high torque density. *Energy Conversion and Management,* Vol. 48, No. 8, August 2007 page numbers 2424-2431, ISSN: 0196-8904

Daldaban, F. & Ustkoyuncu, N. (2010). A novel linear switched reluctance motor for railway transportation systems. *Energy Conversion and Management,* Vol. 48, No. 8, March 2010 page numbers 465-469, ISSN: 0196-8904

Dolinar, D; Ljušev, P. & Štumberger, G. (2003). Input-output linearising tracking control of an induction motor including magnetic saturation effects. *IEE Proceedings - Electric Power Applications,* Vol. 150, Issue 6, November 2003 page numbers 703-711, ISSN: 1350-1352

Dolinar, D; Štumberger, G. & Milanovič, M. (2005). Tracking improvement of LSRM at low-speed operation. *European Transactions on Electrical Power,* Vol. 15, Issue 3, May/June 2005 page numbers 257-270, ISSN: 1430-144X

Dolinar, D. & Štumberger, G. (2006). *Modeliranje in vodenje elektromehanskih sistemov,* Založniška dejavnost FERI, ISBN: 86-435-0672-9, Maribor

Fitzgerald, A. E. & Kingslley., C. Jr. (1961). *Electric machinery,* McGraw-Hill Book Company, New York

Guglielmi, P.; Pastorelli, M. & Vagati, A. (2006). Impact of cross-saturation in sensorless control of transverse-laminated synchronous reluctance motors. *IEEE Transactions on Industrial Electronics,* Vol. 53, No. 2, April 2006 page numbers 429-439, ISSN: 0278-0046

Hadžiselimović, M. (2007a). *Magnetically nonlinear dynamic model of a permanent magnet synchronous motor,* Ph.D. Thesis, University of Maribor, Maribor

Hadžiselimović, M.; Štumberger, G.; Štumberger, B. & Zagradišnik, I. (2007b). Magnetically nonlinear dynamic model of synchronous motor with permanent magnets. *Journal of Magnetism and Magnetic Materials,* Vol. 316, Issue 2, 2007 page numbers e257-e260, ISSN: 0304-8853

Hadžiselimović, M.; Štumberger, B.; Virtič, P.; Marčič, T. & Štumberger, G. (2008). Determining parameters of a two-axis permanent magnet synchronous motor dynamic model by finite element method. *Przegłęad Elektrotechniczny,* Vol. 84, No. 1, 2008 page numbers 77-80, ISSN: 0033-2097

Hamler, A.; Trlep, M. & Hribernik, B. (1998). Optimal secondary segment shapes of linear reluctance motors using stochastic searching. *IEEE Transactions on Magnetics,* Vol. 34, No. 5, September 1998 page numbers 3519-3521, ISSN: 0018-9464

Iglesias, I.; Garcia-Tabares, L. & Tamarit, J. (1992). A d-q model for the self-commutated synchronous machine considering magnetic saturation. *IEEE Transactions on Energy Conversion,* Vol. 7, No. 4, December 1992 page numbers 768-776, ISSN: 0885-8969

Jadrić, M.; & Frančić, B. (1997). *Dinamika električnih strojeva,* Sveučilište u Splitu, ISBN: 953-96399-2-1, Split

Krause, P. C.; Wasynczuk, O. & Sudhoff, S. D. (2002). *Analysis of electric machinery,* IEEE Press, ISBN: 0-471-14326-X, New York

Kron, G. (1951). *Equivalent circuits of electric machinery,* J. Wiley & Sons, New York

Kron, G. (1959). *Tensor for circuits,* Dover, New York

Kron, G. (1965). *Tensor analysis of networks,* MacDonald, London

Levi, E. & Levi, V. A. (2000). Impact of dynamic cross-saturation on accuracy of saturated synchronous machine models. *IEEE Transactions on Energy Conversion,* Vol. 15, No. 2, June 2000 page numbers 224-230, ISSN: 0885-8969

Marčič, T.; Štumberger, G.; Štumberger, B.; Hadžiselimović, M. & Virtič, P. (2008). Determining parameters of a line-start interior permanent magnet synchronous motor model by the differential evolution. *IEEE Transactions on Magnetics,* Vol. 44, No. 11, November 2008 page numbers 4385-4388, ISSN: 0018-9464

Melkebeek, J. A. & Willems, J. L. (1990). Reciprocity relations for the mutual inductances between orthogonal axis windings in saturated salinet-pole machines. *IEEE Transactions on Industry Applications,* Vol. 26, No. 1, January/February 1990 page numbers 107-114, ISSN: 0093-9994

Owen, E. L. (1999). History: Charles Concodia 1999 IEEE Medal of Honor. *IEEE Industry Applications Magazine,* Vol. 5, No. 3, May/June 1999 page numbers 10-16, ISSN: 1077-2618

Rahman, K.A. & Hiti, S. (2005). Identification of machine parameters of a synchronous motor. *IEEE Transactions on Industry Applications,* Vol. 41, No. 2, March/April 2005 page numbers 557-567, ISSN: 0093-9994

Sauer, P. W. (1992). Constraints on saturation modeling in ac machines. *IEEE Transactions on Energy Conversion,* Vol. 7, No. 4, March 1992 page numbers 161-167, ISSN: 0885-8969

Štumberger, B.; Štumberger, G.; Dolinar, D.; Hamler, A. & Trlep, M. (2003). Evaluation of saturation and cross-magnetization effects in interior permanent-magnet synchronous motor. *IEEE Transactions on Industry Applications,* Vol. 39, No. 5, September/October 2003 page numbers 1265-1271, ISSN: 0093-9994

Štumberger, G.; Štumberger, B. & Dolinar, D. (2003). Magnetically nonlinear and anisotropic iron core model of synchronous reluctance motor. *Journal of Magnetism and Magnetic Materials,* Vols. 254/255, No. 1, January 2003 page numbers 618-620, ISSN: 0304-8853

Štumberger, G.; Štumberger, B.; Dolinar, D. & Težak, O. (2004a). Nonlinear model of linear synchronous reluctance motor for real time applications. *Compel,* Vol. 23, No. 1, 2004 page numbers 316-327, ISSN: 0332-1649

Štumberger, G.; Štumberger, B. & Dolinar, D. (2004b). Identification of linear synchronous reluctance motor parameters. *IEEE Transactions on Industry Applications,* Vol. 40, No. 5, September/October 2004 page numbers 1317-1324, ISSN: 0093-9994

Štumberger, G.; Polajžer, B.; Štumberger, B. & Dolinar, D. (2005). Evaluation of experimental methods for determining the magnetically nonlinear characteristics of electromagnetic devices. *IEEE Transactions on Magnetics,* Vol. 41, No. 10, October 2005 page numbers 4030-4032, ISSN: 0018-9464

Štumberger, G.; Štumberger, B. & Dolinar, D. (2006). Dynamic two-axis model of a linear synchronous reluctance motor based on current and position-dependent characteristics of flux linkages. *Journal of Magnetism and Magnetic Materials,* Vol. 304, No. 2, 2006 page numbers e832-e834, ISSN: 0304-8853

Štumberger, G.; Seme, S.; Štumberger, B.; Polajžer, B. & Dolinar, D. (2008a). Determining magnetically nonlinear characteristics of transformers and iron core inductors by differential evolution. *IEEE Transactions on Magnetics,* Vol. 44, No. 6, June 2008 page numbers 1570-1573, ISSN: 0018-9464

Štumberger, G.; Marčič, T.; Štumberger, B. & Dolinar, D. (2008b). Experimental method for determining magnetically nonlinear characteristics of electric machines with magnetically nonlinear and anisotropic iron core, damping windings, and permanent magnets. *IEEE Transactions on Magnetics,* Vol. 44, No. 11, November 2008 page numbers 4341-4344, ISSN: 0018-9464

Tahan, S. A. & Kamwa, I. (1995). Two-factor saturation model for synchronous machines with multiple rotor circuits. *IEEE Transactions on Energy Conversion,* Vol. 10, No. 4, December 1995 page numbers 609-616, ISSN: 0885-8969

Vas, P.; Hallenius, K. E. & Brown, J. E. (1986). Cross-saturation in smooth-air-gap electrical machines. *IEEE Transactions on Energy Conversion,* Vol. 1, No. 1, March 1986 page numbers 103-112, ISSN: 0885-8969

Vas, P. (1992). *Electrical machines and drives,* Clarendon Press, ISBN: 0-19-859378-3, Oxford

Vas, P. (2001). *Sensorless Vector and Direct Torque Control,* Oxford University Press, ISBN: 978-0-19-856465-1, New York

Vagati, A.; Pastorelli, M.; Scapino, F. & Franceschini, G. (2000). Impact of cross saturation in synchronous reluctance motors of the transverse-laminated type. *IEEE Transactions on Industry Applications,* Vol. 36, No. 4, July/August 2000 page numbers 1039-1046, ISSN: 0093-9994

17

Potential for Improving HD Diesel Truck Engine Fuel Consumption Using Exhaust Heat Recovery Techniques

D.T. Hountalas[1] and G.C. Mavropoulos[2]
[1]*Professor of Internal Combustion Engines, Dipl.Ing., Dr.Ing. (NTUA)*
[2]*Research Associate, Internal Combustion Engines Dipl.Ing., Dr.Ing. (NTUA), M. SAE, M. ACS*
Internal Combustion Engines Laboratory
Thermal Engineering Department
School of Mechanical Engineering
National Technical University of Athens (NTUA)
9 Iroon Polytechniou St., Zografou Campus, 15780 Athens, Greece

1. Introduction

Increased fuel costs and diminishing petroleum supplies are forcing both governments and industries to reduce engine fuel consumption. Up to now significant improvement has been achieved towards this direction using advanced engine techniques such as downsizing, VVT, advance fuel injection, advanced boosting etc. However the potential for further significant improvement is extremely limited.

Despite these advancements diesel engines still reject a considerable amount of fuel chemical energy to the ambience through the exhaust gas. This is approximately 30-40% of the energy supplied by the fuel depending on engine load. If part of this thermal energy is recovered and converted to power it could result to a significant reduction of engine bsfc. To achieve this various techniques are available some of which have been partially tested in the past. Mechanical turbocompounding is considered a base technology which combines the output of a diesel engine with that of an exhaust gas driven turbine located downstream of the turbocharger turbine. Electrical turbocompounding is a similar technique using an efficient turbocharger turbine and a high speed generator mounted on the turbocharger's shaft. The excess turbine power is converted to electricity. However both techniques can be significantly improved especially using advanced turbocharging components and relevant techniques to control and vary power turbine pressure ratio or exhaust manifold pressure (electric turbocompounding) with operating conditions. Another promising technology is the use of a Rankine Bottoming Cycle with steam or organic as working media. Its use has special difficulties for truck applications due to size limitations, packaging and the negative impact on the engine primary cooling system. Therefore special care is necessary when studying such an application.

In the present is made an effort to evaluate the potential bsfc improvement of HD diesel engines using the aforementioned technologies for exhaust gas heat recovery. For this reason results from the analysis on a heavy duty truck diesel engine at various speeds and loads are presented. Results are based on simulation models developed and used to simulate the aforementioned configurations. Special attention is given to the identification of the potential efficiency gain and mainly to the various problems that have to be resolved for a vehicle application. As shown there exists a strong potential for the application of these techniques since a reduction of fuel consumption from 3% up to 12-13% depending on the technology is achievable.
In 2006, the emissions from transportation sources in the USA reached an amount of 40% of the total Greenhouse gas emissions in this country. Medium and heavy-duty vehicles represent about 22% of the transportation emissions. The use of trucks is energy-intensive and accounted for 69% of freight energy use, consuming 2.35 million barrels of oil per day in 2008 and generating 363 million metric toned of carbon dioxide. Hence trucks are one of the top priorities when looking for energy savings and climate change mitigation in the transportation sector.

2. The NTUA engine simulation model

During the last years, the high interest in the examination of potential benefits from the use of exhaust gas heat recovery systems, has motivated a number of companies and research institutes to develop models appropriate for the simulation of such energy systems.
An extensive evaluation of potential benefits concerning engine performance and bsfc improvement for the available exhaust heat recuperation technologies was performed in the Internal Combustion Engine Laboratory of NTUA (National Technical University of Athens). The evaluation was based on a comprehensive thermodynamic simulation model of diesel engines developed in this Laboratory. The simulation has been used to estimate exhaust gas characteristics and available amount of exhaust energy. In addition it was used, after modifications, to simulate turbocompounding both mechanical and electrical considering for the interaction between the T/C, the power turbine and the engine. The basic elements of this model are briefly outlined in this section.

2.1 Brief outline of the NTUA engine simulation model

The simulation covers the entire engine cycle taking into account the gas exchange period using the filling and emptying technique. It considers in general the following engine subsystems: a) Engine Cylinder b) Fuel Injection System and c) Gas Exchange System. It is based on a three-dimensional multi-zone combustion model. The pressure inside the engine cylinder is considered to be uniform. The first law of thermodynamics and the conservation equations for mass and momentum are employed for the calculation of local conditions inside each zone [1,2].

2.1.1 Heat transfer

A turbulent kinetic energy viscous dissipation rate $k\sim\varepsilon_t$ model is used to determine the characteristic velocity for the heat transfer calculations. The heat transfer coefficient is estimated from the following correlation [3,4],

$$h_c = c\,Re^{0.8}\,Pr^{0.33}\,\frac{\lambda}{l_{car}} \tag{1}$$

The overall heat exchange rate is then distributed among the zones according to their mass, temperature and specific heat capacity.

2.1.2 Air swirl

The swirling motion of the intake air is described in a rather simple, but quite efficient way assuming a hybrid scheme consisting of a solid body core surrounded by a potential flow region [3,5]. The intake air swirl velocity is estimated from the angular momentum conservation equation using the angular momentum added to the engine cylinder during the intake process and considering for the part destroyed because of friction with the combustion chamber walls. Details concerning the analysis are provided in [1,2].

2.1.3 Spray model

After initiation of fuel injection, zones form and penetrate inside the combustion chamber. The zone velocity along the jet axis is obtained from the following relations depending on the time instant after injection [1,6].

$$u_p = c_d\sqrt{\frac{2\Delta P}{\rho_l}} \quad \text{for } t<t_b \tag{2a}$$

$$u_p = \frac{2.95}{2}\left(\frac{\Delta P}{\rho_a}\right)^{0.25} d_{inj}^{0.5}\,t^{-0.5} \quad \text{for } t_b<t<t_{hit} \tag{2b}$$

$$u_p = \frac{2.95}{2}\left(\frac{\Delta P}{\rho_a}\right)^{0.25} d_{inj}^{0.5}\,t^{-0.5} \quad \text{for } t>t_{hit} \tag{2c}$$

where t_b is the breakup time and t_{hit} is the time of impingement on the cylinder walls. The breakup time is obtained from Jung and Assanis [7].

$$t_b = 4.351\frac{\rho_l d_{inj}}{c_d\left(\rho_a \Delta P\right)^{0.5}} \tag{3}$$

The effect of air swirl upon the jet is considered for, using the local components of the air velocity in the radial and axial directions and using the momentum conservation equations in both axes.

2.1.4 Air entrainment into the zones

The air entrained into a zone from initiation of injection is calculated from the conservation of momentum using the following relation:

$$m_f \cdot u_{inj} = (m_a + m_f)\cdot u_p \rightarrow m_a = (m_f \cdot u_{inj} / u_p - m_f) \tag{4}$$

2.1.5 Droplet break up and evaporation

The injected fuel is distributed to the zones according to the instantaneous injection rate, while inside each zone the fuel is divided into packages (groups) using a chi-squared distribution [3], where the droplets have the same Sauter Mean Diameter. For evaporation the model of Borman and Johnson [8] is followed, as described in [1-3].

2.1.6 Combustion model

The amount of air entering a zone is mixed with evaporated fuel. The local reaction rate depends on the local mass concentration of fuel, oxygen and the local temperature. Ignition commences after an ignition delay period, which is estimated using the local conditions inside the zone as follows [1-3,6]:

$$S_{pr} = \int_0^1 \frac{1}{a_{del} P_g^{-2.5} \Phi_{eq}^{-1.04} \exp\left(5000/T_g\right)} dt = 1 \tag{5}$$

where "Φ_{eq}" is the local equivalence ratio of the fuel air mixture inside the zone and a_{del} is a constant. After ignition the combustion rate of fuel is obtained from the following relation [1-3]:

$$m_{fb} = K_b C_f^{af} C_o^{ao} \exp\left(-\frac{E_c}{R_m}\right) \frac{1}{6N} \tag{6}$$

where K_b is a constant, E_c the reduced activation energy and C_f, C_o the mass concentrations of fuel and oxygen respectively. During diffusion combustion, the combustion rate is practically controlled from the air entrainment rate and it's mixing with evaporated fuel.

2.1.7 Fuel injection

The fuel injection system is of great importance for the operation of the diesel engine [3,9,10]. In the present work, the injection rate has been obtained from the following relation:

$$\frac{dm_{inj}}{dt} = C_{dinj} \cdot A_{inj} \cdot \sqrt{2 \cdot \Delta P \cdot \rho_f} \tag{7}$$

where: $\Delta P = P_{cyl} - P_{inj}$ is the instantaneous pressure difference at the nozzle exit, C_{dinj} the discharge coefficient, A_{inj} the area of the nozzle holes and ρ_f the fuel density. Since experimental injection rates were not available a constant rail pressure has been used during injection, which was estimated to match the measured injection duration at each operating condition.

2.1.8 Gas exchange

For the simulation of the inlet and exhaust manifolds and the calculation of the mass exchange rate between them and the engine cylinder the filling and emptying method is used. One-dimensional, quasi-steady, compressible flow is assumed for the calculation of the mass flow rates through the inlet and exhaust valves during the gas exchange process.

2.1.9 Turbocharger simulation

For the present simulation model T/C maps were used, provided by the T/C manufacturer for both mechanical and electrical turbocompounding. In the case of electric turbocompounding the turbine effective flow area was calculated using an iterative procedure to provide the required exhaust pressure value before the turbine. Then the turbine excess power was estimated from the following relation:

$$P_{el,p} = P_{eT}\eta_{mT} - \frac{P_{eC}}{\eta_{mC}} \tag{8}$$

where P_{eC} is the compressor power, P_{eT} the power generated from the turbine and $P_{el,p}$ the excess power of the turbine which is then converted to electric power.
For the present application the mechanical efficiencies of the turbine and the compressor, n_{mT} and n_{mC} were taken equal to 0.95.

2.2 NTUA simulation model modification for turbocompounding

To consider for the existence of the power turbine, the NTUA engine simulation code has been undergone additional modifications. The power turbine has been mounted downstream of the existing T/C and its inlet conditions have been calculated using the first thermodynamic law and the power balance across the T/C. For this reason a new calculation procedure has been setup which estimates the pressure at the T/C turbine inlet to produce the required boost pressure that provides the necessary amount of air to the engine. For electrical turbocompounding, the main modification involved also the calculation procedure. A new methodology was developed al already mentioned to estimate the required turbine nozzle effective flow area to increase exhaust gas pressure before the turbocharger turbine to the required level [11]. Moreover, the simulation estimates the amount of turbine power required to drive the compressor and supply the necessary amount of air to the engine. Then the remaining amount of turbine power is the available one to be converted to electricity.
In both cases, the simulation estimates the total power produced from the system, which corresponds to the net engine power and the mechanical power produced from the power turbine or the electric generator in the case of electrical turbocompounding. The total power $P_{e,tot}$ produced is thus obtained from the following relation:

$$P_{e,tot} = P_{eE} + P_{eT,net}n_{GT} \tag{9}$$

where $P_{eT,net}$ is the power turbine power or the net power produced from the electric generator for electrical turbocompounding and P_{eE} is the net engine power. The term n_{GT} is the mechanical efficiency of the gear train for mechanical turbocompounding or the generator efficiency for electrical turbocompounding.

3. Mechanical turbocompounding

Mechanical turbocompounding refers to the addition of a power turbine after the T/C to extract mechanical power from the exhaust gas. Figure 1 depicts a representative schematic view of a diesel engine configuration equipped with an additional power turbine (Mechanical Turbocompounding). The power turbine is mounted downstream of the turbocharger (T/C) and is mechanically coupled to the engine crankshaft via a gear train. Caterpillar [11,12] used an axial power turbine on a 14.6-liter diesel and reported an average bsfc reduction of about 4.7% for a 50,000 miles extra-urban driving test in USA. Cummins [13] used a radial flow power turbine and achieved 6% improvement in bsfc at full load and

3% at part load. Finally, Scania [11] applied this technology on an 11-liter displacement 6-cylinder turbocharged diesel engine providing a 5% improvement of bsfc at full load.

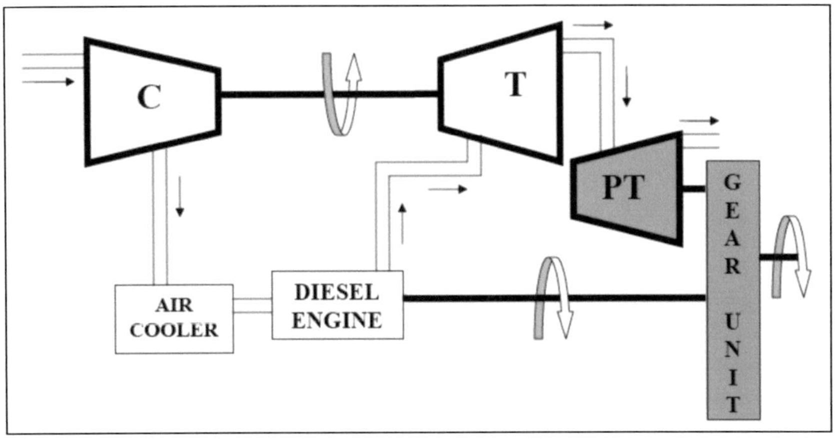

Fig. 1. Schematic view of Mechanical Turbocompounding.

Several theoretical studies have been conducted from various research groups to estimate the potential for exhaust heat recovery of a heavy-duty diesel engine when using mechanical turbocompounding. Using the model developed by the research group of NTUA-Internal Combustion Engines laboratory [1,2] a parametric investigation has been conducted using as a basis a diesel engine simulation model appropriately modified to consider for the effects of this technology [14]. The parametric investigation involved the effect of power turbine pressure ratio on engine performance and overall engine bsfc improvement. The power turbine pressure ratio has been varied in the range of 1.5-2.3. On the other hand, the power turbine efficiency has been kept constant at 80% The parametric analysis has been conducted for all engine-operating points shown in Table 1.

Speed (rpm)	**Load (%)**	**Fuel (kg/h/cyl)**	**Inlet Pressure (bar)**	**Injection Advance (deg ATDC)**
2100	25	3.00	1.36	-6
2100	50	4.87	1.78	-8
2100	75	6.90	2.06	-8
2100	100	8.77	2.20	-13
1700	25	3.45	1.50	-6
1700	50	6.10	2.23	-7
1700	75	8.90	2.65	-8
1700	100	11.80	3.08	-8
1300	25	3.25	1.61	-4
1300	50	5.87	2.36	-7
1300	75	8.83	3.01	-7
1300	100	12.00	3.73	-6

Table 1. Engine operating conditions considered for the investigation.

Before applying the simulation model for the examination of the various exhaust heat recovery techniques it has to be calibrated at a selected operating point. Its predictive ability concerning engine performance and emissions is also validated afterwards against available experimental data. This is a standard procedure when using a simulation model and is always performed before it is used in any theoretical investigation.

3.1 Effect on engine performance and potential bsfc improvement

The parametric analysis of mechanical turbocompounding is conducted considering a variation of power turbine pressure ratio from 1.5 to 2.3. The effect of this variation on exhaust manifold pressure is depicted in Fig. 2 for all engine loads at 1700 rpm engine speed. The maximum value of exhaust pressure is 5.76 bar at full engine load for the maximum power turbine pressure ratio. On the other hand the maximum pressure ratio is obviously lower and in the range of 2.6-2.7 which is quite acceptable.

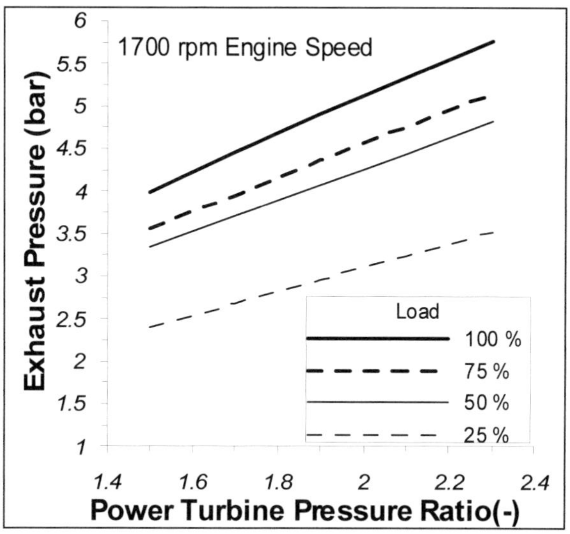

Fig. 2. Exhaust Pressure variation vs. power turbine pressure ratio at 1700 rpm for various engine loads.

During the parametric analysis, the fuelling rate for turbocompounding has been maintained constant. Therefore to estimate its effect on engine bsfc it has been considered the overall power output i.e. the power generated from the engine and the power turbine. In Fig. 3 is given the percentage change of overall engine bsfc vs. power turbine pressure ratio at 1700 rpm for loads ranging from 25% to 100%. As revealed, the introduction of mechanical turbocompounding results to a decrease of overall bsfc. In general the relative reduction of bsfc increases initially sharply with power turbine pressure ratio and then after a certain point the increase starts to decline. For part load operation after a certain pressure ratio bsfc improvement starts to decrease revealing that a maximum has been reached. Therefore, the optimum pressure ratio is shifted to higher values of power turbine pressure ratio with the increase of engine load. As depicted in Fig 3, the potential of bsfc decrease is extremely low at 25% load and increases to approximately 4.5% at full load. It is also noticeable that after a certain power turbine pressure ratio at medium and high loads the additional decrease of bsfc is rather limited while it has a negative effect on exhaust pressure as shown in Fig. 2. Therefore considering the overall results it appears that the overall optimum power turbine pressure ratio is in the range of 1.7-1.9.

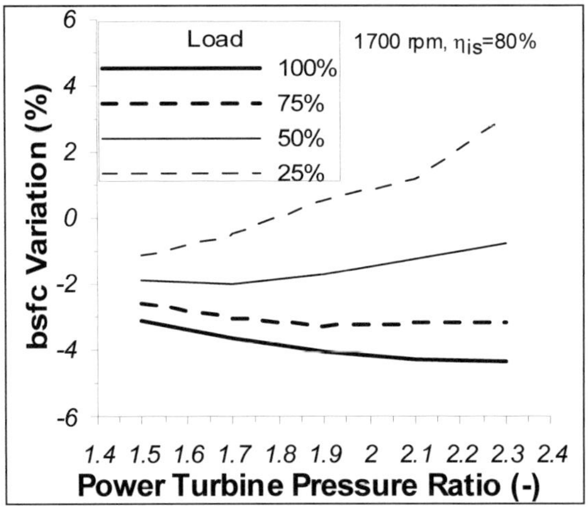

Fig. 3. bsfc variation vs. power turbine pressure ratio at 1700 rpm for various engine loads.

Another important issue is the effect of mechanical turbocompounding on engine power output. An increase of exhaust pressure before the T/C turbine (Fig. 1) is observed due to the addition of the power turbine. Due to this, gas exchange work is increased which is expected to have a negative impact on net engine power. This is verified by Fig. 4 providing the relative reduction of engine power output with power turbine pressure ratio for all loads examined. As observed, engine power decreases almost linearly with increasing power turbine pressure ratio the slope increasing with the reduction of engine load. The engine power reduction considering the previous value of the optimum pressure ratio (1.7~1.9) ranges from 8% at 100% load to 20% at 25% load. The absolute reduction of primary engine power is depicted in Fig. 5. As shown, net engine power decreases from 352 kW to 325 kW at full engine load linearly with power turbine pressure ratio the slope being the same for all engine loads.

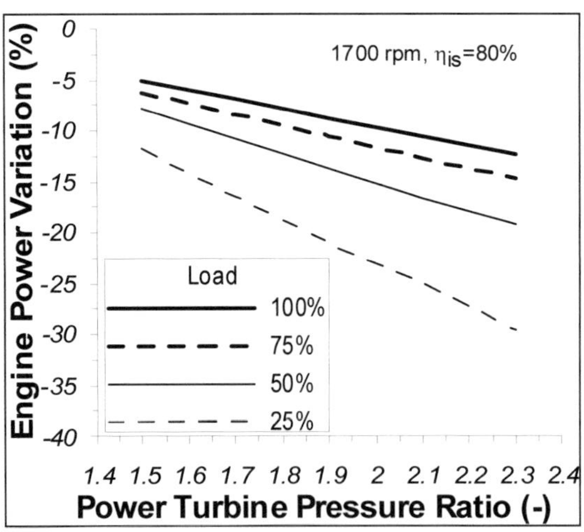

Fig. 4. Engine power variation vs. power turbine pressure ratio at 1700 rpm and various engine loads.

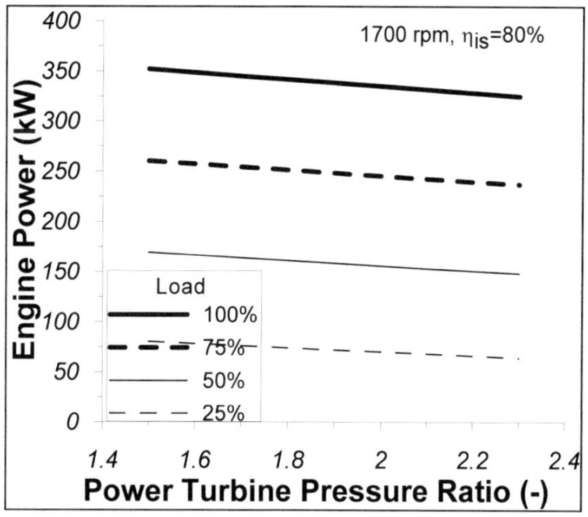

Fig. 5. Net engine power vs. power turbine pressure ratio at 1700 rpm and various engine loads.

On the other hand, in Fig 6 is given the variation of generated power at the power turbine vs. pressure ratio for all loads examined. It is observed an increase of generated power with increasing pressure ratio the slope decreasing with the increase of pressure ratio. The slope of the curve, as expected increases also with engine load and a maximum generated power value of approximately 35 kW is expected if the pressure ratio is maintained in the optimum range of 1.7~1.9.

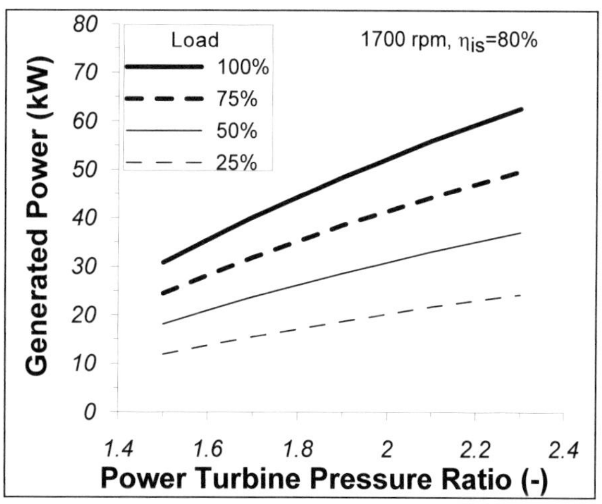

Fig. 6. Generated Power vs. power turbine pressure ratio at 1700 rpm and various engine loads.

4. Electrical turbocompounding

Electrical turbocompounding (Fig 7) recuperates part of the exhaust gas heat directly from the T/C using a high-speed generator. In this case the turbine produces more power compared to the one required to drive the compressor. This excess power is converted to electric power using a high speed generator incorporated into the TC casing. Caterpillar has considered this concept in a research program [15-16] providing indications for 5%

reduction of bsfc on a cycle basis and a maximum reduction of approximately 9-10% when using turbocharger components with high efficiency.

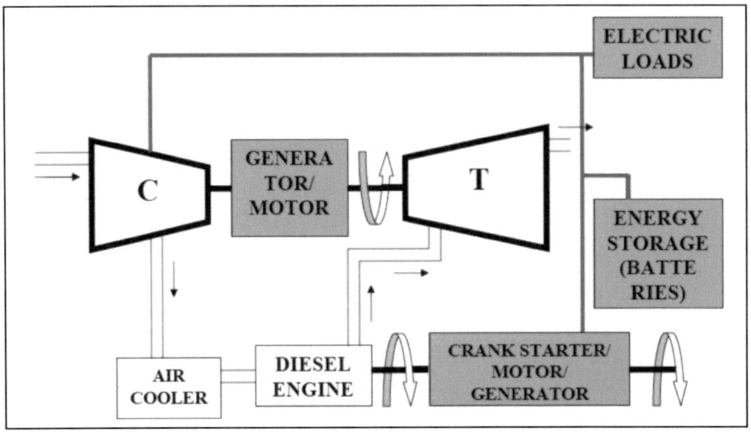

Fig. 7. Schematic view of Electrical Turbocompounding.

Using the NTUA engine simulation model described in section 2 a parametric investigation similar to the one conducted for mechanical turbocompounding has been conducted also for the case of electrical turbocompounding. The engine operating conditions examined are again the ones displayed in Table 1. However in the present investigation the parameters examined were turbocharger efficiency and exhaust pressure increase before the turbine. The investigation has been conducted for three values of overall T/C efficiency, i.e. the standard value corresponding to the provided T/C maps and two additional ones considering an increase of 10% and 20% on a percentage basis. For the standard T/C the maximum overall efficiency is approximately 49%, for the one with a 10% increase 54% and for the last with a 20% increase 60%. The last value i.e. 60% corresponds to an isentropic efficiency of both turbine and compressor in the range of 80% that can be considered realistic especially for future improved T/Cs.

In the case of electric turbocompounding, exhaust pressure has been increased approximately 1 bar above the value corresponding to the non-turbocompound case using the standard T/C efficiency. Obviously the potential for exhaust pressure increase is higher when using a T/C with increased efficiency. The value of 1 bar exhaust pressure increase has been selected to avoid excessive pressure ratios across the turbine of the turbocharger.

4.1 Effect on engine performance and potential bsfc improvement

During the investigation exhaust pressure has been increased approximately 1 bar above the one corresponding to the standard efficiency T/C. In Fig.8 are given the corresponding values of exhaust pressure for the standard efficiency T/C vs. exhaust pressure increase for 25% and 100% load. From this graph it appears that the maximum pressure ratio across the turbine is 4.0 which is high and explains the reason for increasing exhaust pressure only by 1 bar. Obviously the absolute exhaust pressure value compared to mechanical turbocompounding, Fig. 2 is significantly lower.

The effect of electrical turbocompounding on engine performance is given as function of turbine inlet pressure variation (compared to the standard value). To estimate bsfc improvement in this case, two different approaches are adopted, one by comparing bsfc

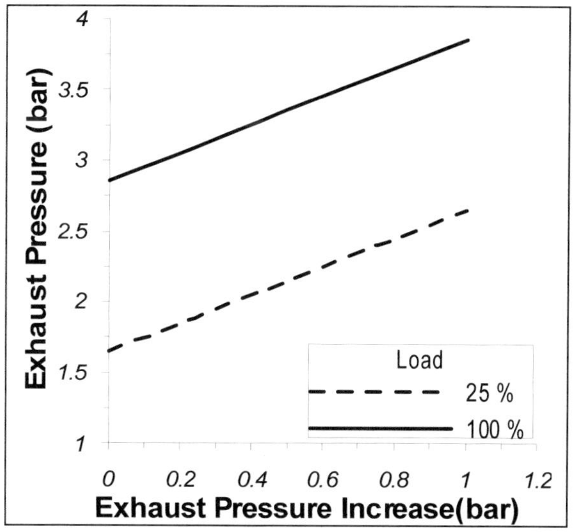

Fig. 8. Exhaust manifold pressure variation vs. exhaust pressure increase for 1700 rpm and 25% load.

variation to the non-turbocompound case for each T/C efficiency considered and a second by comparing the non-turbocompound case to the standard T/C efficiency (existing one). In the first case the actual benefit of electrical turbocompounding is revealed while in the second the result is the combined effect of both electrical turbocompounding and increased T/C overall efficiency.

In Figs 9a,b and 9c,d is given the overall bsfc change vs. exhaust pressure variation for 25% and 100% load respectively for the two approaches adopted. As observed, a different behavior is revealed at 25% and 100% load, Figs 9a,b and 9c,d respectively. At low engine load, the bsfc reduction is significantly lower compared to the one at full engine load.

In Fig 9a is depicted the bsfc variation compared to the one without turbocompounding and the current T/C efficiency at 25% load. As revealed, bsfc improvement starts to deteriorate after a certain value of exhaust pressure increase, which is shifted towards higher values as T/C efficiency increases. On the other hand in Fig.9b is given the bsfc variation compared to the non-turbocompound case for the standard T/C efficiency. Obviously in this case bsfc reduction is significantly higher. The observed maximum bsfc improvement attributed to electric turbocompounding alone is 1% for the highly efficient T/C efficiency while the total bsfc improvement due to both turbocompounding and increased T/C efficiency is 3.5%.

For 100% load, Figs 9c,d., bsfc reduction continues to improve with exhaust pressure increase the slope of the curve being reduced as exhaust pressure increases. The maximum bsfc improvement attributed to electric turbocompounding alone is 5% for the highly efficient T/C. On the other hand the total overall bsfc improvement due to both electric turbocompounding and increased T/C efficiency is 6.5%. Therefore a considerable bsfc improvement especially at low and medium load is experienced only using a high efficiency T/C. At 25% load the maximum bsfc improvement is attributed by 2/3 to T/C efficiency increase and by 1/3 to turbocompounding. At high load the situation is reversed and the overall improvement is attributed by 1/3 to T/C efficiency increase and 2/3 to electric turbocompounding. Therefore, the previous analysis reveals that electrical turbocompounding can be beneficial if we use a highly efficient T/C. Consequently the need for increasing T/C efficiency is clearly revealed.

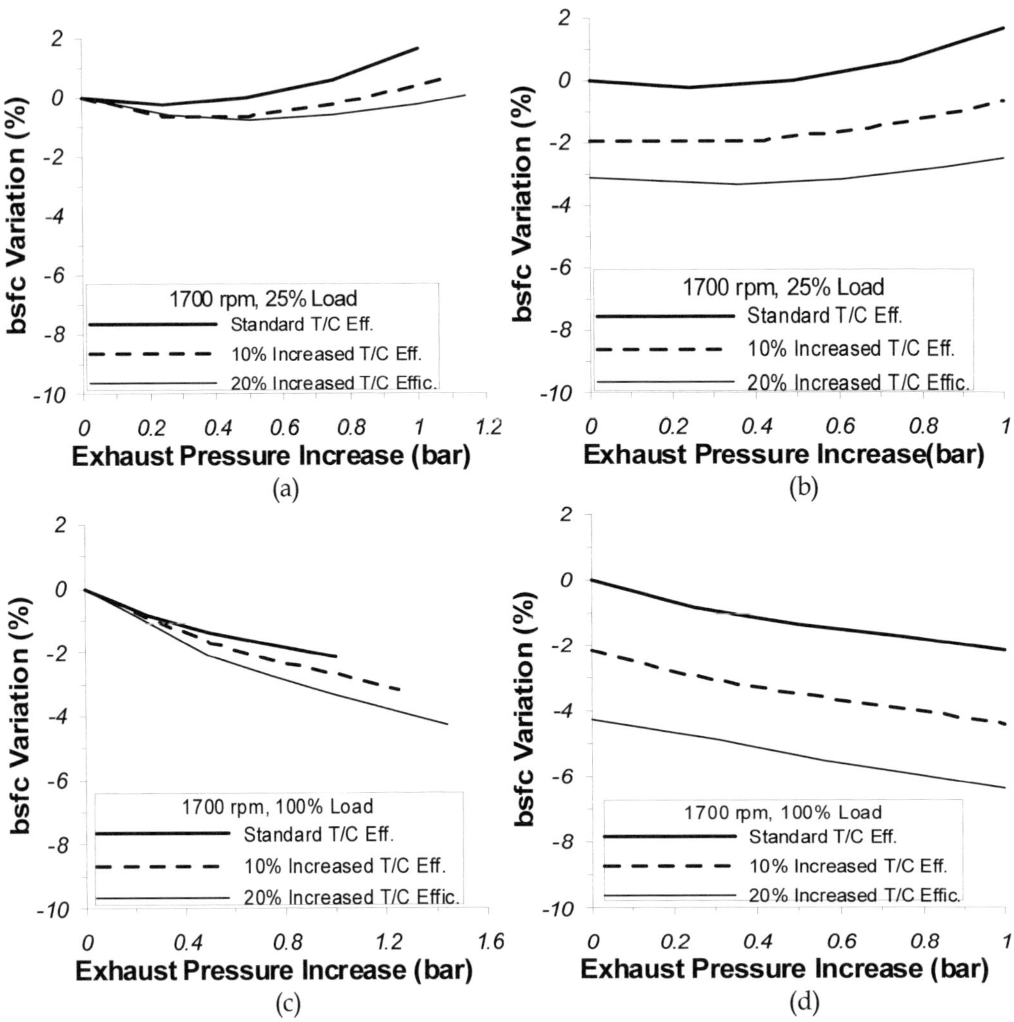

Fig. 9. (a) bsfc variation vs. exhaust pressure increase for 1700 rpm and 25% load. (b) bsfc variation compared to the standard T/C efficiency vs. exhaust pressure increase for various T/C efficiencies at 1700 rpm and 25% load. (c) bsfc variation vs. exhaust pressure increase for 1700 rpm and 25% load. (d) bsfc variation compared to the standard T/C efficiency vs. exhaust pressure increase for various T/C efficiencies at 1700 rpm and 100% load.

In the case of electric turbocompounding, generated electric power is significantly important because it defines the size of the required electric generator. As witnessed from Figs 10a-b corresponding to 25% and 100% load generated electric power increases with exhaust pressure at a decreasing slope. Generated electric power increases significantly with the increase of T/C efficiency. For the case examined the maximum generated electric power ranges from 15 kW at 25% engine load to 40 kW at 100% engine load for the highly efficient T/C.

Finally, in Figures 11a-b is given the relative decrease of net diesel engine power for 25% and 100% respectively as function of exhaust pressure increase. The relative decrease has a maximum value of 18% at low engine load and 6% at full engine load. The relative reduction of net engine power does not depend on T/C efficiency and varies linearly with exhaust pressure increase. Compared to mechanical turbocompounding the relative reduction is lower since a lower exhaust manifold pressure has been used as already shown.

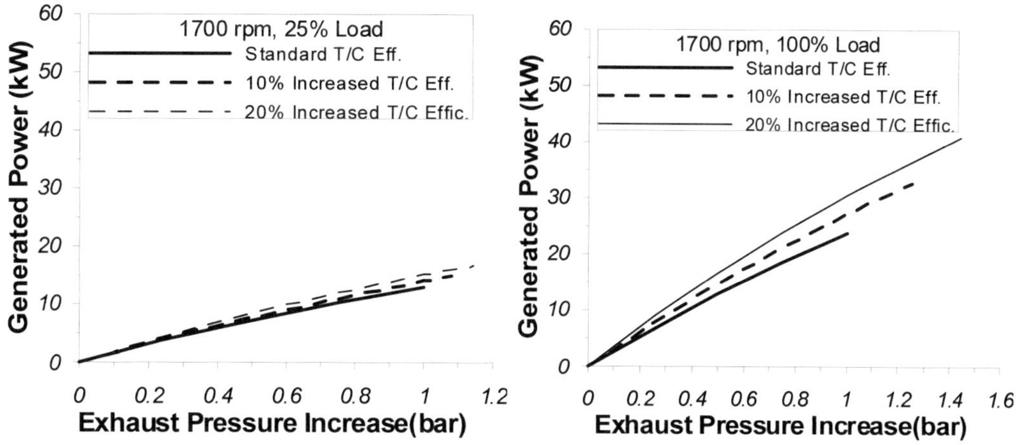

(a) (b)

Fig. 10. (a) Variation of Generated Power vs. exhaust pressure increase at 1700 rpm and 25% load for various T/C efficiencies. (b)Variation of Generated Power vs. exhaust pressure increase at 1700 rpm and 100% load for various T/C efficiencies.

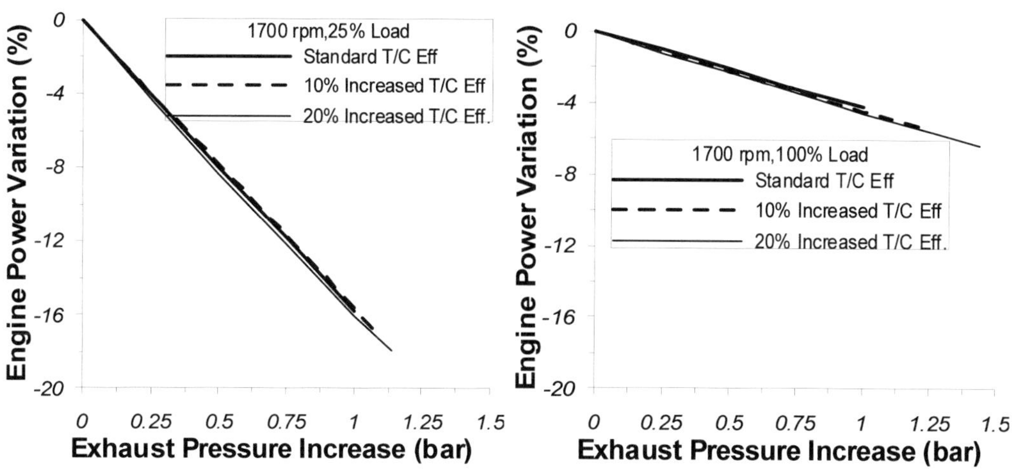

(a) (b)

Fig. 11. (a) Relative decrease of net engine power vs. exhaust pressure increase at 1700 rpm and 25% load for various T/C efficiencies. (b) Relative variation of net engine power vs. exhaust pressure increase at 1700 rpm and 100% load for various T/C efficiencies.

5. Implementation of Rankine cycle

5.1 General description

In the case of heavy duty diesel engines, suitable for truck applications, one of the most promising technical solutions for exhaust gas waste heat utilization appears to be the use of a "Bottoming Rankine Cycle" [17,18]. A systematic approach for the use of Rankine Cycle installation for truck applications is dated back in the early 70's where a research program funded by US Department of Energy (DOE) was conducted by Mack Trucks and the Thermo Electron Corporation [19-21]. During this program, an Organic Rankine Cycle System (ORCS) was installed on a Mack Truck diesel engine. Lab test results revealed a peak bsfc improvement of 10% to 12.7% which as claimed was verified from highway tests even though no detailed additional information was provided. During the next years similar research programs were performed by other research institutes and vehicle manufacturers [22,23]. However, recently the solution of Rankine Cycle systems has increased its potential competitiveness in the market [24-27]. This is the result of technical advancements in a series of critical components for the operation of such an installation (heat exchanger, condenser, expander etc.), the highly increased fuel prices and the global thermal problem. Due to the last two, the installation and use of a Rankine Cycle is not only considered nowadays as a feasible solution for efficiency improvement of heavy duty truck diesel engines [28,29] but also for smaller applications i.e. passenger cars [30].

As expected the introduction of such an advanced concept in today's HD diesel engine technological status apart from the benefits creates several problems that have to be attained and resolved during the early design stage in a competitive way so that the concept retains its feasibility. One of the most important issues [26,27] is packaging of the additional devices and components within the existing engine and within the available space on a truck. However the increment of the bsfc of the new system (Rankine cycle and diesel engine) at a typical amount of 10% triggers automatically the need to reject increased amounts of heat to the environment through the low temperature reservoir system (condenser, engine radiator in the present case). Studies performed up to date by the authors [14] and other research groups [25-27,31] have demonstrated that when a Rankine cycle is coupled to an existing truck engine installation the use of a larger radiator, compared to the existing one, is necessary. The limitations for the dimensions of the radiator installed on a truck are clear and as a consequence the use of a second radiator appears a feasible solution to cover the increased heat rejection because of the bottoming cycle [32].

A representative layout of this technology is provided in Fig 12, where an exhaust gas heat exchanger is employed after the turbocharger turbine to provide heat to the Rankine cycle working medium. The Rankine cycle system includes an expander which drives a "feeder" pump and a fan mounted on the rear of the radiator-condenser.

The thermodynamic processes in a Rankine cycle are the following: The working medium enters the circulation (feeder) pump at the saturated liquid state and exits at the high pressure p_H. Then the working medium flows into the exhaust gas heat exchanger where it is preheated from the subcooled region at the exit of the pump to the saturated liquid state. As depicted in Fig. 12 the required thermal energy for working medium evaporation and superheating is covered by the main exhaust gas stream and partially by EGR heat. Utilization of EGR heat is favorable for cycle efficiency because of its superior thermodynamic characteristics (i.e. higher temperature level). The superheated vapor expands from the high pressure p_H to the lower pressure p_L. Finally, the expanded working medium is condensed and heat is rejected to the ambience. The working fluid at its saturated liquid state enters the pump.

A variety of working media have been considered during Rankine cycle applications. They are typically separated in two main categories: i.e. water (steam) and typical organic media

(like R245ca). In an effort to reduce the dimensions of the Rankine system one of the basic ideas is the utilization of heat from other sources apart from the exhaust gas which is rejected to the environment during normal engine operation. In existing and future heavy duty diesel engine applications such sources for additional heat recuperation are CAC (Charge Air Cooler) and most likely EGR (Exhaust Gas Recirculation) Cooler Heat [31]. Significant amounts of heat are expelled to the ambient from both devices during the operation of a HD truck engine. Using the simulation models developed, it is clearly demonstrated the potential benefits arising for both system efficiency and reduction of radiator dimensions when heat from these additional energy sources is utilized.
Finally using models developed especially for the simulation of the heat exchange processes taking place in a HD diesel truck radiator estimation can be performed for the geometrical dimensions of both the diesel engine and Rankine cycle radiators that are necessary to fulfill the heat rejection demands of the installation. Different radiator arrangements have also been investigated to propose an optimized solution with minimum radiator surfaces considering for the cooling air mass flow rate.

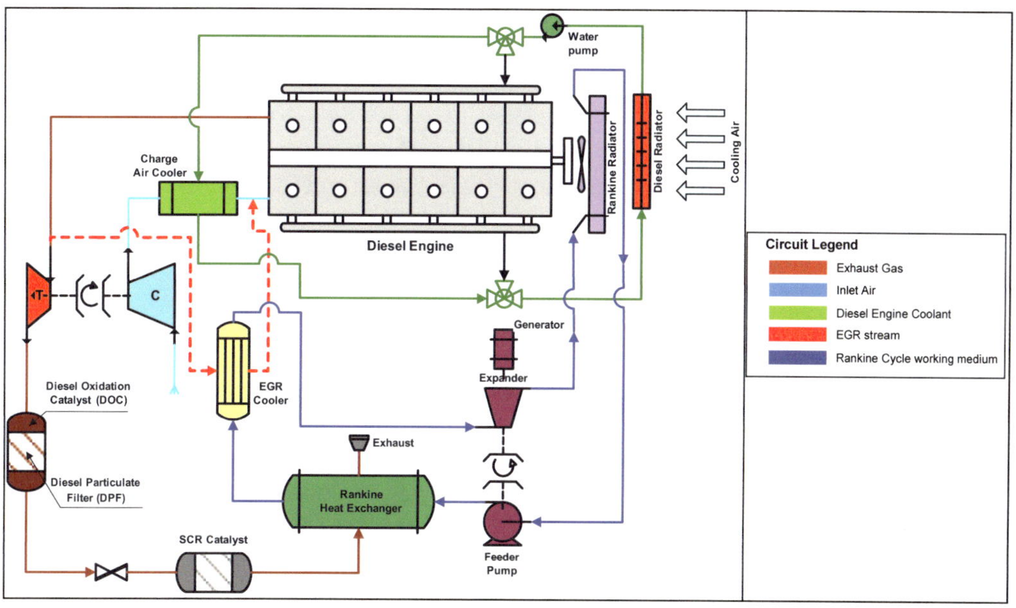

Fig. 12. General layout for the arrangement of the Rankine cycle system components on a HD DI Diesel engine installation.

5.2 Simulation model for the Rankine cycle

A detailed simulation model has been developed by the authors in NTUA for the evaluation of the Rankine cycle system which is used to recover energy from the exhaust gas and other sources of waste heat available in the diesel engine installation examined. The required thermodynamic and transport properties for the water and organic working medium have been calculated from "REFPROP", an electronic database developed by NIST (National Institute of Science and Technology) [33]. The exhaust gas properties have been calculated using polynomial expressions. On the other hand the exhaust gas composition is estimated from the diesel engine simulation (described in section 2). The various components of the

Rankine cycle have been simulated by developing appropriate sub-models based on the basic principles of thermodynamics and heat transfer [34].

Two cases for exhaust gas heat recuperation have been examined. The first is the basic configuration where heat is recovered only from the main exhaust gas. In the second it is considered the simultaneous heat recovery from the main exhaust gas stream (from the Rankine main heat exchanger), the recirculated exhaust gas stream (existing EGR cooler) and the charge air stream after the compressor (existing charge air cooler, CAC). The Rankine heat exchanger used for heat recuperation in all cases is of the shell and tube type with two tube passes.

One of the most important targets of the simulation is to estimate the optimum design of the Rankine cycle at each engine operating point to achieve the maximum efficiency gain. For this reason, a parametric analysis is performed at each operating point examined where the key parameter is Rankine cycle high pressure P_H The second parameter of the study is the superheated vapor temperature T_{win_exp} before the expander. The minimum value of temperature T_{win_exp} is carefully selected to secure that the working media after expansion has a vapor content above 90% which is accepted as the upper limit (max. liquid content 10%) for the safe operation of a reciprocating expander. The minimum value of superheated vapor temperature obviously depends on the Rankine cycle high pressure P_H.

All components of Rankine cycle are simulated based on their basic thermodynamic and heat transfer principles. Thus the model includes the following elements:

- Simulation of Exhaust Gas Heat Exchanger, EGR Cooler and Charge Air Cooler
- Expander Simulation
- Simulation of the Circulation Pump
- Calculation of Rankine Cycle Performance and Overall Energy Balance
- Simulation Model for System Radiators

5.3 Effect of the Rankine cycle on engine power and efficiency improvement

The investigation for both steam and organic has been conducted at 1700 rpm engine speed at four different loads ranging from 25%-100%. This covers the investigation for overall system performance, the effect of EGR and CAC heat utilization on total bsfc and heat exchanger size. The set of engine operating points considered in the present investigation together with the values of several characteristic operating parameters are presented in Table 2. Additionally, in Table 3 are presented the respective heat amounts rejected to the ambience during normal diesel engine operation.

Engine Speed (rpm)	Engine Load (%)	Power (kW)	$\dot{m}_{exh}$ (kg/s)	$\dot{m}_{EGR}$ (kg/s)	λ (-)	P_{EGR} (bar)	T_{exh_in} (°C)	T_{EGR_in} (°C)	T_{AIR_in} (°C)
1700	100	366.6	0.4945	0.0982	1.5404	4.14	397.8	581.0	201.46
1700	75	277.8	0.4058	0.1046	1.6410	3.87	354.3	518.2	180.42
1700	50	183.6	0.2993	0.1314	1.6930	3.79	306.6	455.3	161.17
1700	25	90.0	0.1784	0.1194	1.7781	2.47	285.3	383.4	111.13

Table 2. Engine operating conditions considered for the present investigation.

Engine Speed (rpm)	Engine Load (%)	$\dot{Q}_{COOLANT}$ (kW)	$\dot{Q}_{REJ\_EGR}$ (kW)	$\dot{Q}_{REJ\_AC}$ (kW)	$\dot{Q}_{TOT}$ (kW)
1700	100	130.0	46.0	93.8	269.8
1700	75	109.6	41.8	69.6	221.0
1700	50	88.6	43.4	47.0	179.0
1700	25	57.8	30.0	18.8	106.6

Table 3. Engine heat amounts rejected to the environment.

The main questions that arise after the introduction of the Rankine cycle in a heavy duty diesel truck refer to the expected efficiency improvement of the new power plant and the additional mechanical power that the new system is capable to provide. In the present work it is examined the effect of EGR gas and charge air heat utilization on both parameters. The results for steam are given in Figs 13 and 14 (a-b) at 1700 rpm engine speed and three engine loads namely 50, and 100% (a-b). In Fig. 13 are given the results without EGR and CAC heat utilization and in Fig. 14 the corresponding ones when both EGR cooler and CAC heat amounts are partially utilized as function of the cycle high pressure In the same figures is given the variation of the total heat amount absorbed from the Rankine cycle system vs high cycle pressure.

From Figs 13 and 14 it is observed an improvement of the overall efficiency of the Rankine-Engine power plant (reduction of specific fuel consumption) when compared to the diesel engine under the same conditions. The bsfc improvement increases slightly with engine load (from a to b) in both cases presented in Figs 13 and 14. The maximum bsfc improvement for steam at 1700 rpm is observed at 100% load case and is approximately ~ 5.5% when heat is extracted only from the main exhaust gas stream. However it should be noted that a higher bsfc reduction has been observed at 1300 rpm engine speed estimated at ~ 8-8.5% (where exhaust gas temperature characteristics are more favorable). But when heat from the EGR and CAC is utilized in the Rankine cycle, then bsfc reduction for the present operating condition (i.e. 1700 rpm, 100% load) is increased by approximately 40-50% to ~ 9% (absolute value). This trend is maintained almost the same for the entire engine operation field. The corresponding respective increase in net power benefit from the Rankine cycle is obviously similar and is ~40-50% higher with the recuperation of heat from EGR and CAC at full and part load respectively. From these results it becomes obvious that the contribution of both EGR and CAC coolers to the total amount of recuperated heat is significant.

As Rankine cycle high pressure increases, it would be expected a continuous increase of both bsfc improvement and generated power because of Rankine Cycle efficiency improvement. However, from the results of Figs 13 and 14 it is obvious that this is not the case. The reason is that the increase of cycle high pressure favors Rankine Cycle efficiency but on the other hands makes heat exchange between the working medium and the exhaust gas stream more difficult. As a result generated power output and thus bsfc reduction remain fairly constant after a certain high pressure and even start to decline at very high values. This trend is more obvious in Fig. 14 (a-b) where the EGR and CAC heat amounts are introduced into the system. The utilization of both EGR and CAC heat amounts makes it possible to extend considerably the Rankine cycle high pressure range which is favorable as for reduction of heat rejection, without negative impact on bsfc reduction potential.

The effect of cycle high pressure increase on bsfc improvement and Rankine cycle generated power is easily explained if we consider that the thermodynamic properties of the exhaust gas remain constant while the ones of the working medium (pressure, temperature) during the various phases (preheating, evaporation, superheating) increase. The increase of working medium temperature reduces the amount of heat which can be extracted from the exhaust gas as displayed in Figs 13 and 14 (a-b). Thus the increase of Rankine cycle efficiency, through the increase of cycle high pressure, has a simultaneous negative impact on the heat transfer mechanism between the exhaust gas and working medium (steam in the present case) since the temperature difference between them is decreased. The combination of these two reverse trends results to a neutral effect on overall system bsfc and generated power after a certain cycle high pressure.

Therefore from the simulation it is revealed that the use of a cycle high pressure value above the optimum one for bsfc improvement and generated power could be considered for the reduction of rejected heat especially at high load where the capacity of engine cooling system may be exceeded.

The corresponding results for 1700 rpm engine speed concerning Rankine cycle net power, efficiency (bsfc reduction) and total heat absorbed when R245ca organic working medium is used are displayed in Figs 15 and 16 (a-b). Figure 15 provides results for the case without EGR and CAC heat utilization while Fig. 16 provides the corresponding results when these heat amounts are partially utilized. In this case it is observed that the potential efficient improvement offered by the Rankine cycle (reduction of specific fuel consumption) is higher compared to steam. The difference in efficiency improvement compared to steam is ~1% in absolute units for the case without utilization of EGR and CAC heat. However when these heat amounts are partially utilized the difference is increased to 3% in absolute units at the same conditions. This is the result of the organic media thermodynamic properties which favors heat exchange from the charge air, that has a relatively low temperature, to the working fluid.

It should also be mentioned that the mass flow rate of the organic media is approximately 10 times higher compared to steam, at the same operating conditions, due to the significant differences of physical properties between the two substances. This makes possible the use of a turbine expander instead of a reciprocating one which appears to be favorable as far as packaging and cost is concerned.

As in the case of steam, efficiency improvement for R245ca increases with the increase of engine load (Figs 15 and 16 a-b) and the maximum bsfc improvement observed for 100% load is ~ 11.3% when both EGR and CAC heat amounts are partially utilized (Fig 16b). Compared to the case where EGR and CAC heat is not utilized both efficiency gain and generated power are increased by ~50% (this value is even exceeded at part load operation). Therefore, as in the case of steam, the contribution of EGR and CAC heat to the total recuperated amount is significant.

As mentioned, there exist significant differences in the thermo physical properties between steam and organic working medium. Of special interest is the opposite inclination of the saturation curve of the organic R245ca in the vapor area compared to steam, which provides higher flexibility for the thermodynamic conditions at the expander inlet since at the end of the expansion stroke the organic medium is always inside the vapor area. Thus there is no possibility for condensate formation inside the expander which allows initiation of the expansion stroke at a position favoring the remaining thermodynamic phases of the Rankine cycle.

However the limitation concerning the increasing difficulty of heat transfer with the increase of cycle high pressure (vaporization pressure), which has been already described, is still valid. Observing the corresponding results provided in Figs 15 and 16 (a-b) it is obvious that this effect exists but is not so intense due to the low critical pressure for R245ca ~ 39.25 bar which does not allow variation of high pressure up to the point where generated power will start to decline. As already mentioned supercritical cycles are not within the scope of the present analysis, even though they are interesting, because they require further considerations that are not within the scope of the present investigation. However they will be the subject of future work from the present research group.

From the results of the present work (Figs 15 and 16 it is concluded that for the organic medium considered the vaporization pressure should be kept at its highest possible level (below the critical) at any operating condition regardless of the utilization of EGR and CAC heat.

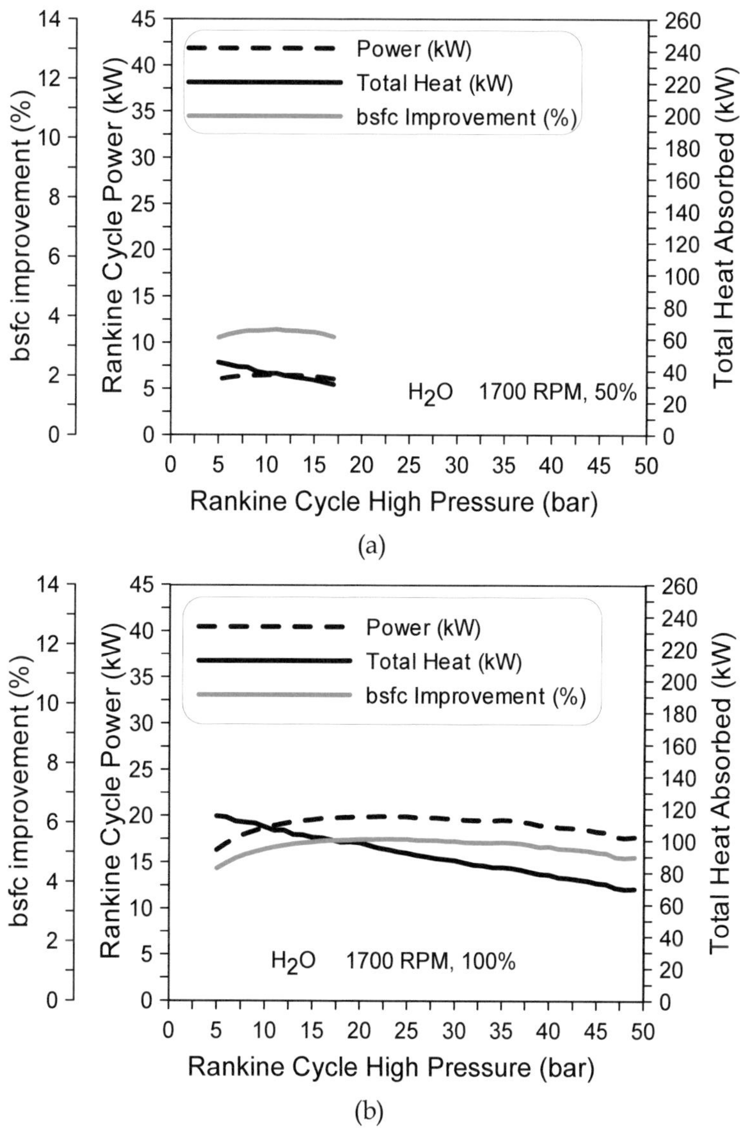

Fig. 13. bsfc improvement, generated power and total recuperated heat vs cycle high pressure, for 1700 rpm engine speed and three different loads 50% (a) and 100% (b) without EGR and CAC heat utilization.

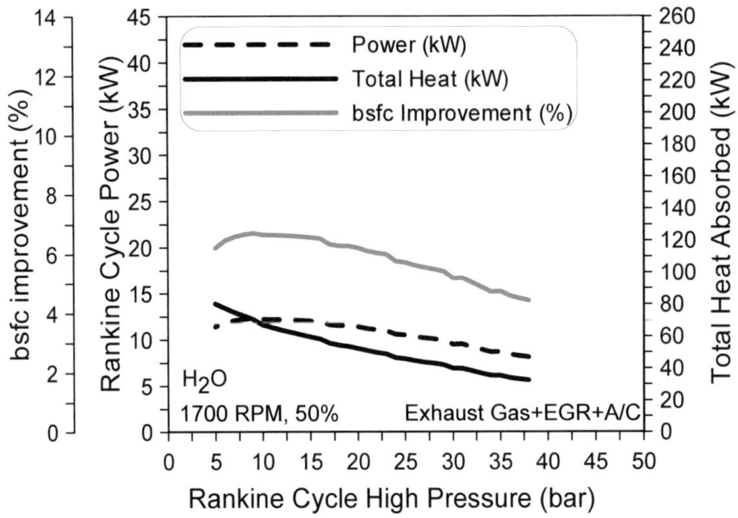

(a)

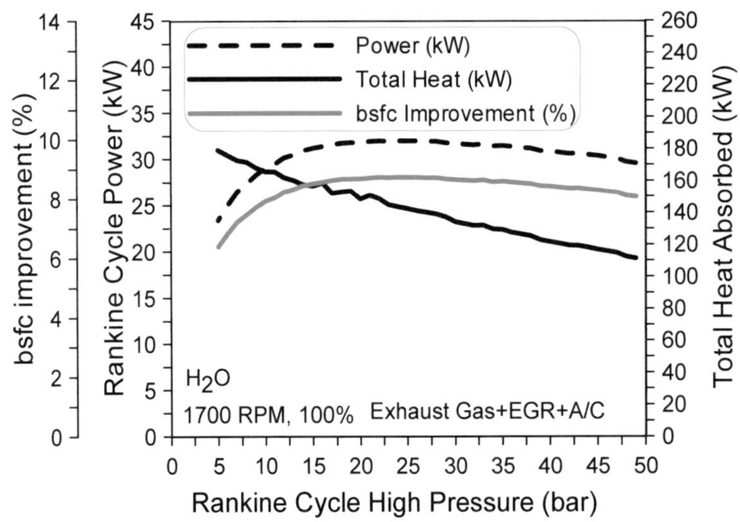

(b)

Fig. 14. bsfc improvement, generated power and total recuperated heat vs cycle high pressure, for 1700 rpm engine speed and three different loads50% (a) and 100% (b) with partially utilized EGR and CAC heat.

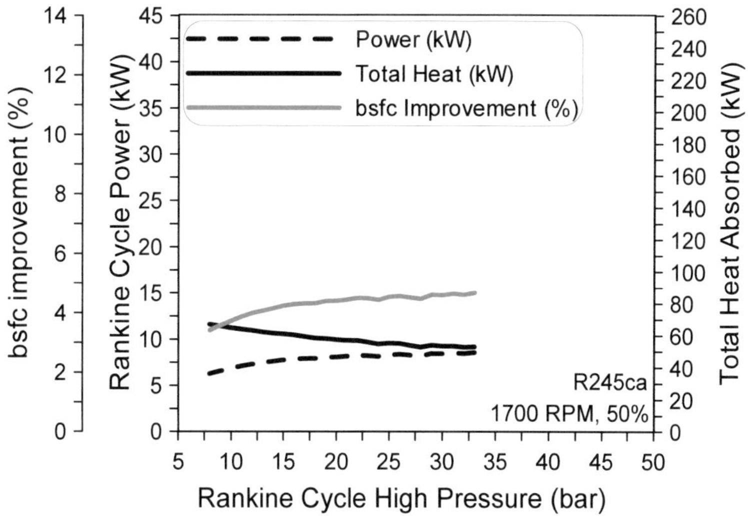

(a)

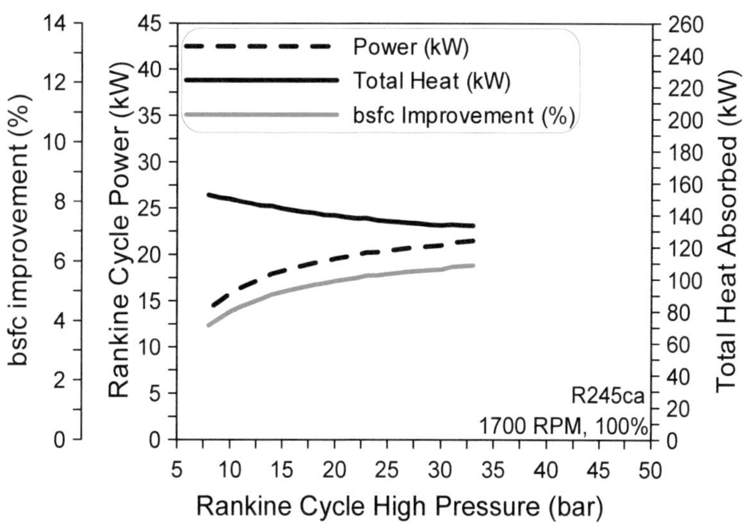

(b)

Fig. 15. bsfc improvement, generated power and total recuperated heat absorbed by the organic (R245ca) Rankine cycle vs cycle high pressure, for 1700 rpm engine speed and three different loads 50% (a) and 100% (b) without EGR and CAC heat utilization.

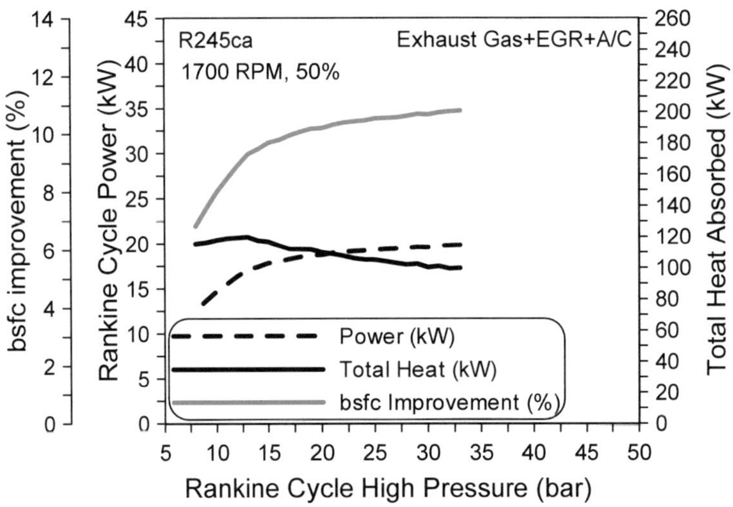

(a)

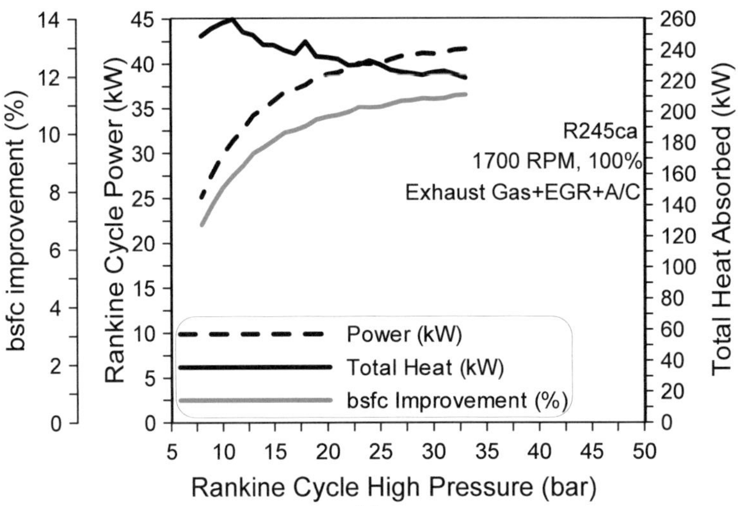

(b)

Fig. 16. bsfc improvement, generated powerand total recuperated heat absorbed by the organic (R245ca) Rankine cycle vs cycle high pressure, for 1700 rpm engine speed and three different loads of 50% (a) and 100% (b) with partially utilized EGR and CAC heat.

6. Benefits from the use of exhaust gas recuperation in HD diesel engines

Detailed investigations have been conducted to examine the effect of most promising technologies for exhaust heat recovery i.e., mechanical, electrical turbocompounding and Rankine cycle, on overall engine bsfc and power output. The results obtained from these detailed analyses have lead to the following conclusions:

- At a constant fuelling rate, both mechanical and electrical turbocompounding technologies result to reduction of primary engine power output. The effect on a percentage basis is higher at low engine load. However, both result to an improvement of overall power output due to the power generated from the waste exhaust gas.

- For mechanical turbocompounding, the optimum expansion ratio of the power turbine increases with engine load. But after a certain value in the range of 1.7-1.9 additional improvement is low considering the corresponding increase of exhaust pressure.
- As revealed from the analysis, mechanical turbocompounding can offer a maximum bsfc reduction of 0.5% at 25% load and 4.5% at full load for a power turbine efficiency of 80%. The overall bsfc benefit appears to be in the range of 2.5-3.4% at average engine load (75% engine load).
- Electrical turbocompounding can provide a maximum bsfc reduction of 6.5% using a highly efficient turbocharger. The improvement for a conventional T/C is smaller especially at low load. As revealed the maximum bsfc benefit is attributed by 2/3 to turbocompounding and 1/3 to the increase of T/C efficiency at full load. The situation is reversed at low load.
- Heat rejected from both the EGR cooler and CAC represents approximately 50% of the total heat rejected through the radiator of a heavy duty diesel. For this reason it should be considered the partial utilization of these heat amounts when attempting to reduce fuel consumption through heat recuperation. Apart from the obvious potential benefits in fuel consumption and power output the utilization of EGR cooler and CAC heat is promising for the reduction of the additional rejected heat from the Rankine cycle.
- As shown from the simulation results, a 20% increase in the case of steam and a 30% increase in the case of R245ca of radiator heat rejection capacity is adequate to successfully fulfill the additional demand for heat rejection of the diesel-Rankine power plant when both CAC and EGR heat is partially utilized. This is significantly lower compared to the case where EGR cooler and CAC heat is not utilized where the corresponding values are approximately double.
- For both steam and organic, when heat from both the EGR cooler and CAC is partially utilized the improvement in efficiency is almost 50% higher resulting to an equivalent increase of the excess generated power. This improvement is attributed more to the utilization of EGR cooler heat because it has a higher temperature allowing significant superheating of the steam.
- In the case of steam, the possibility to extend the high pressure (vaporization) in a much broader range allows better optimization of Rankine cycle operation compared to the organic.
- The maximum improvement in bsfc was observed in the case of the organic Rankine cycle and is ~ 11.3% when both EGR and CAC heat amounts are utilized. The corresponding value for steam is ~ 9%.
- The calculations reveal that recuperation of EGR and CAC heat improves system packaging significantly and that the dimensions of the additional radiator which also acts as a condenser for the Rankine cycle is almost the same with the existing diesel engine radiator.

7. Summary of results concerning exhaust gas recuperation in HD diesel engines

7.1 Expected overall efficiency gain from each technology

Considering the results presented in the previous sections as an outcome from detailed computational investigations, the potential bsfc reduction using the aforementioned exhaust heat recovery techniques is summarized in Table 4.

Technology	Current Realistic bsfc Reduction
Mechanical Turbocompounding	• **1%** at low engine load • **4.8%** at full engine load
Electrical Turbocmpounding	• **3.0-6.0%** for standard T/C efficiency • **6.0%-9.0%** with T/C efficiency of **65%**
Rankine With Steam	• bsfc reduction with steam is estimated at **6.0%-9.0%**.with an **expander efficiency of 70%** and max. high pressure value of **40** bar.
Rankine with Organic	• **9.0%-11.0%** expander efficiency of **70%** and max. high pressure value of 36 bar. • Potential improvement of bsfc with another working media.

Table 4. Potential bsfc Benefit of Various Technologies.

7.2 Benefits & drawbacks of each technological solution

Each of the aforementioned technologies presents various benefits and drawbacks. These are summarized in Table 5 below and should be considered for practical application.

Technology	Benefits	Disadvantages
Mechanical Turbocompounding	• Simplicity • Low volume & Cost.	• Interaction with engine • Relatively Low bsfc benefit. • Lower bsfc improvement at low load.
Electrical Turbocompounding	• Relative Simplicity • Low volume • Relatively good bsfc benefit	• Interaction with engine • Possible problem with electric power generation
Rankine With Steam	• Good bsfc reduction potential. • Low mass flow rate (negative impact on expander efficiency). • Small effect of engine load on bsfc benefit. • No interaction with engine.	• Complexity. • Volume and Weight. • Small expander efficiency with common technology. Microturbine technology can offer substantial improvement (Under investigation). • Possible fouling of heat exchanger without soot trap.
Rankine with Organic	• Highest bsfc reduction potential. • Small effect of engine load on bsfc benefit. • High mass flow rate (favorable for expander efficiency). • Lower heat exchanger area compared to steam.	• Complexity. • Volume and Weight. • Thermal and Chemical stability of organic fluid. • Toxicity considerations of working fluid. • Possible fouling of heat exchanger without soot trap.

Table 5. Benefits & Disadvantages of Exhaust Heat Recovery Techniques.

7.3 Most promising technologies for future investigation

Considering the results from the analysis and the advantages and disadvantages of the various technological solutions examined it is given in Table 6 their ranking with maximum of "*****".

Technology	bsfc benefit	Effect on Engine	Volume Weight	Cost	Applicability
Mechanical Turbocompounding	**	****	**	*	*****
Electrical Turbocompounding	***	***	**	**	****
Rankine With Steam	****	*****	*****	****	**
Rankine with Organic	*****	*****	****	*****	***

Table 6. Benefits & Disadvantages of Exhaust Heat Recovery Techniques.

8. Final conclusion

From the technological solutions considered, it results that each technology offers various benefits when considered for practical application. However we have to consider that the primary criterion in case of choice is in most cases, efficiency, i.e. "overall bsfc reduction".
Thus, the most promising technology appears to be Rankine Cycle with organic, followed by Rankine Cycle using Steam and finally Electric Turbocompounding.
Considering the first two solutions the problem appears to be complexity and size of required heat exchanger. The organic is favorable in this issue due to better thermodynamic characteristics. For steam, a problem exists resulting from its low mass flow rate which does not favor turbine efficiency. On the other hand this working media is favorable i.e water having no toxicity and easily replaceable (leakages). The aforementioned problem with steam could possibly be overcome using microturbine technology, which is currently under consideration. If this problem is solved and heat exchanger volume is reduced then Rankine Cycle with steam appears to be an attractive solution. On the other hand, electric turbocompounding offers relative simplicity but considerably lower bsfc reduction. The reduction in the size of exhaust gas heat exchanger when using Rankine cycle is highly favored by taking advantage of the (significant) heat amounts rejected in EGR cooler and CAC.
In conclusion, the most favorable technology appears to be Rankine Cycle with organic fluid. If the problem with expander efficiency using steam can be solved then Steam Rankine Cycle also becomes attractive. Electric turbocompounding should be considered if size and weight of previous components becomes a serious problem.

9. References

[1] Rakopoulos, C.D. and Hountalas, D.T., "Development and Validation of a 3-D Multi-Zone Combustion Model for the Prediction of a DI Diesel Engines Performance and Pollutants Emissions". SAE Transactions, Journal of Engines, Vol.107, pp.1413-1429, 1998.

[2] Hountalas, D.T., Mavropoulos, G.C., Zannis, T.C. and Schwarz, V. (2005a) Possibilities to achieve future emission limits for HD DI diesel engines using internal measures, SAE Paper No 2005-01-0377.

[3] Heywood, J.B., Internal Combustion Engine Fundamentals, McGraw-Hill, New York, 1988.

[4] Annand, W.J.D., "Heat Transfer in the Cylinders of Reciprocating Internal Combustion Engines", Proc. Inst. Mech. Engrs, 177, 973-990, 1963.

[5] Dent, J.C. and Derham, J.A., "Air Motion in a Four-Stroke Direct Injection Diesel Engine", Proc. Inst. Mech. Engrs, 188, 269-280, 1974.

[6] Ramos, J.I., Internal Combustion Engine Modeling, Hemisphere, New York, 1989.

[7] Jung, D and Assanis, D.N., "Multi-zone DI Diesel Spray Combustion Model for Cycle Simulation Studies of Engine Performance and Emissions", SAE Paper No 2001-01-1246.

[8] Borman, G.L. and Johnson, J.H., "Unsteady Vaporization Histories and Trajectories of Fuel Drops Injected into Swirling Air", SAE Paper No. 598C, National Powerplant Meeting, Philadelphia PA, 1962.

[9] Ueki, S. and Miura, A., "Effect of Difference of High Pressure Fuel Injection Systems on Exhaust Emissions from HD DI Diesel Engine", JSAE Review, 20, 555-561, 1999.

[10] Wickman, D.D., Senecal, P.K., Tanin, K.V., Reitz, R.D., Gebert, K., Barkhimer, R.L., Beck, N.J., "Methods and Results from the Development of a 2600 Bar Diesel Fuel Injection System", SAE Paper No 2000-01-0947.

[11] Tennant, D.W.H. and Walsham, B.E. (1989) "The turbocompound diesel engine", SAE Paper No. 89064.

[12] Wilson, D.E. (1986) The design of a low specific fuel consumption turbocompound engine, SAE Paper No.860072.

[13] Brands, M.C, Werner, J. and Hoehne, J.L. (1981) Vehicle Testing of Cummins Turbocompound Diesel Engine, SAE Paper No. 810073.

[14] Hountalas D.T., Katsanos C.O., Rogdakis E.D., Kouremenos D., Study of available exhaust gas heat recovery technologies for HD diesel engine applications, International Journal of Alternative Propulsion, 2006.

[15] Hopmann, U. (2004) "Diesel engine waste heat recovery utilizing electric turbocompound technology", Catterpillar, DEER Conference, San Diego, California, USA.

[16] Hopmann, U. and Algrain, M. (2003) "Diesel engine waste heat recovery utilizing electric turbocompound technology", Caterpillar Inc., Presentation in 2003 DEER Conference Newport Rhode Island.

[17] Teng H, Regner G, Cowland C. Waste heat recovery of heavy-duty diesel engines by organic Rankine cycle part I: hybrid power system of diesel and rankine engines. SAE paper no. 2007-01-0537; 2007.
[18] Teng H. Achieving high engine efficiency for heavy-duty diesel engines by waste heat recovery using supercritical organic-fluid Rankine cycle. SAE paper no. 2006-01-3522; 2006.
[19] Parimal PS, Doyle EF. Compounding the truck diesel engine with an organic Rankine cycle system. SAE paper no. 760343; 1976.
[20] Dibella FA, Di Nanno LR, Koplow MD. Laboratory and on-highway testing of diesel organic Rankine compound long-haul vehicle engine. SAE paper no. 830122; 1983.
[21] Doyle E, Di Nanno LR, Kramer S. Installation of a diesel-organic Rankine compound engine in a class 8 truck for a single-vehicle test. SAE paper no. 790646; 1979.
[22] Hideyo O, Shigeru O. Waste heat recovery of passenger car using a combination of Rankine bottoming cycle and evaporative engine cooling system. SAE paper no. 930880; 1993.
[23] Chen SK, Lin R. A review of engine advanced cycle and Rankine bottoming cycle and their loss evaluations. SAE paper no. 830124; 1983.
[24] Kadota M, Yamamoto K. Advanced transient simulation on hybrid vehicle using Rankine cycle system. SAE paper no. 2008-01-0310; 2008.
[25] Hounsham S, Stobart R, Cooke A, Childs P. Energy recovery systems for engines. SAE paper no. 2008-01-0309; 2008.
[26] Nelson C. Exhaust energy recovery. In: Proceedings of Directions in Energy-Efficiency and Emissions Research (DEER) conference, Dearborn, Michigan, August 3-6; 2009.
[27] Obieglo A, Ringler J, Seifert M, Hall W. Future efficient dynamics with heat recovery. In: Proceedings of Directions in Energy-Efficiency and Emissions Research (DEER) Conference, Dearborn, Michigan, August 3-6; 2009.
[28] Nelson C. Exhaust energy recovery. In: Proceedings of Diesel Engine-Efficiency and Emissions Research (DEER) Conference, Dearborn, Michigan, August 4-7; 2008.
[29] Kruiswyk RW. An engine system approach to exhaust waste heat recovery. In: Proceedings of Diesel Engine-Efficiency and Emissions Research (DEER) Conference, Dearborn, Michigan, August 4-7; 2008.
[30] Ringler J, Seifert M, Guyotot V, Huebner W. Rankine cycle for waste heat recovery of IC engines. SAE paper no. 2009-01-0174; 2009.
[31] Teng H. Improving fuel economy for HD diesel engines with WHR Rankine cycle driven by EGR cooler heat rejection. SAE paper no. 2009-01-2913; 2009.
[32] Charyulu DG, Singh G, Sharmac JK. Performance evaluation of a radiator in a diesel engine-a case study. Applied Thermal Engineering 1999;19:625-39.
[33] Lemmon EW, Huber ML, McLinden MO. NIST standard reference database 23: Reference Fluid Thermodynamic and Transport Properties-REFPROP. Version 8.0,

National Institute of Standards and Technology, Standard Reference Data Program, Gaithersburg, 2007.
[34] Bejan A, Kraus A. Heat transfer handbook. New Jersey: John Wiley & Sons; 2003.

18

New Trends in Efficiency Optimization of Induction Motor Drives

Branko Blanusa
University of Banja Luka, Faculty of Electrical Engineering
Bosnia and Herzegovina

1. Introduction

Scientific considerations presented in this paper are related to the methods for power loss minimization in induction motor drives.

The induction motor is without doubt the most used electrical motor and a great energy consumer. Three-phase induction motors consume 60% of industrial electricity and it takes considerable efforts to improve their efficiency (Vukosavic, 1998). The vast majority of induction motor drives are used for heating, ventilation and air conditioning (HVAC). These applications require only low dynamic performance and in most cases only voltage source inverter is inserted between grid and induction motor as cheapest solution. The classical way to control these dives is constant V/f ratio and simple methods for efficiency optimization can be applied (Abrahamsen et al.,1998). From the other side there are many applications where, like electrical vehicles, electric energy has to be consumed in the best possible way and use of induction motors in such application requires an energy optimized control strategy (Chis et al., 1997.).

The evolution of the power digital microcontrollers and development of power electronics enables applying not only methods for induction motor drives (*IMD*) control, like vector control or direct torque control, but also development of different functions which make drives more robust and more efficient. One of the more interesting algorithm which can be applied in a drive controller is algorithm for efficiency optimization.

In a conventional setting, the field excitation is kept constant at rated value throughout its entire load range. If machine is under-loaded, this would result in over-excitation and unnecessary copper losses. Thus in cases where a motor drive has to operate in wider load range, the minimization of losses has great significance. It is known that efficiency improvement of IMD can be implemented via motor flux level and this method has been proven to be particularly effective at light loads and in a steady state of drive. Also flux reduction at light loads gives less acoustic noise derived from both converter and machine. From the other side low flux makes motor more sensitive to load disturbances and degrades dynamic performances (Stergaki & Stavrakakis, 2008).

The published methods mainly solve the problem of efficiency improvement for constant output power. Results of applied algorithms highly depends from the size of drive (fig. 1) (Abrahamsen et al., 1998) and operating conditions, especially load torque and speed (Figs. 2 and 3). Efficiency of IM changes from 75% for low power 0,75kW machine to more then 95% for 100kW machine. Also efficiency of drive converter is typically 95% and more.

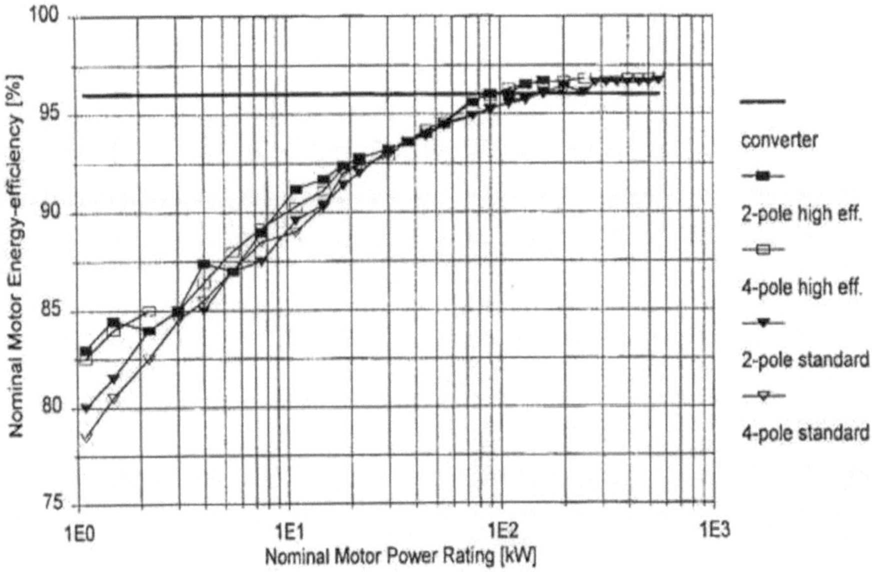

Fig. 1. Rated motor efficiances for ABB motors (catalog data) and typical converter efficiency.

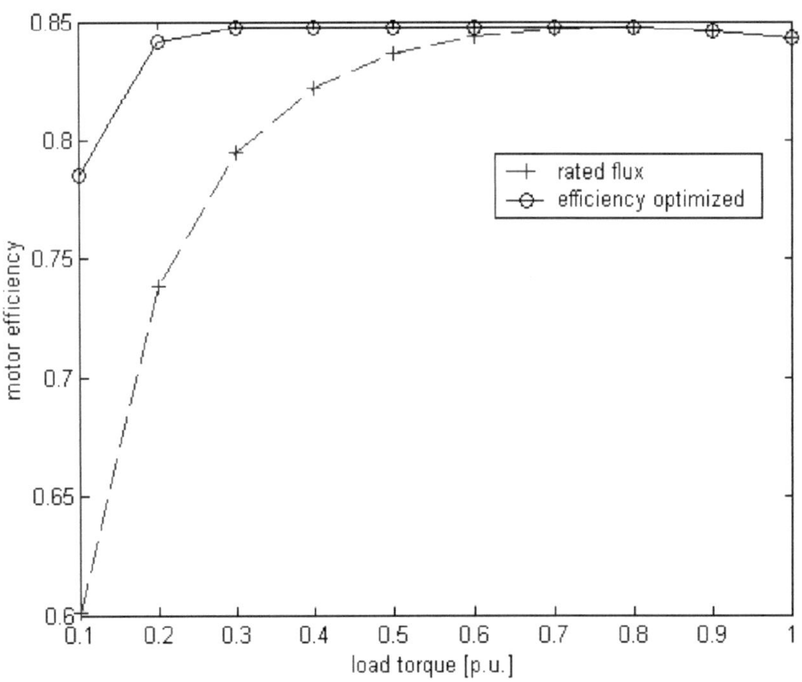

Fig. 2. Measured standard motor efficiences with both rated flux and efficiency optimized control at rated mechanical speed (2.2 kW rated power).

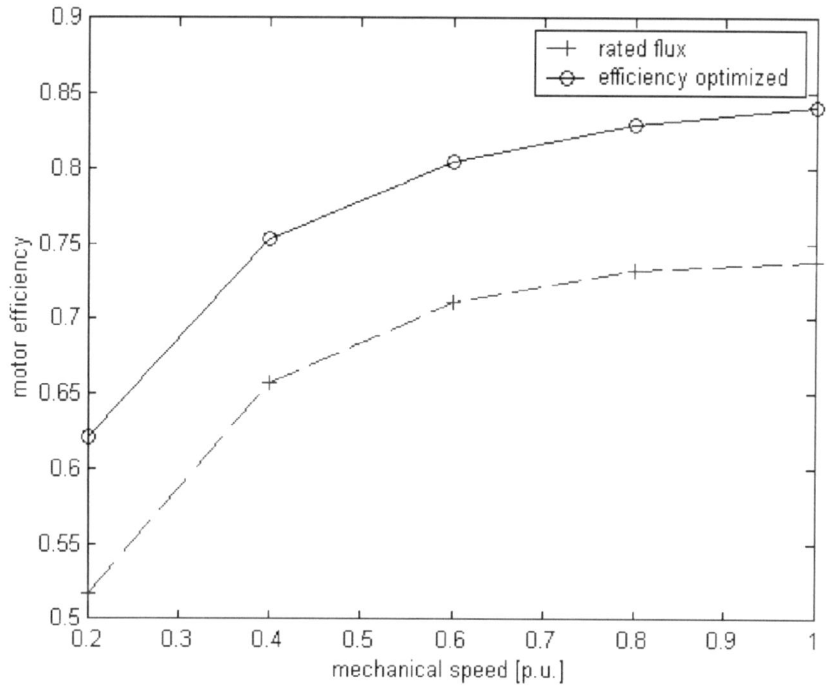

Fig. 3. Measured standard motor efficiences with both rated flux and efficiency optimized control at light load (20% of rated load).

That's obvious, converter losses is not necessary to consider in efficiency optimal control for small drives. Best results in efficiency optimization can be achieved for a light loads and steady state of drive.

Functional approximation of the power losses in the induction motor drive is given in second section.

Basic concepts strategies for efficiency optimization of induction motor drive what includes its characteristics, advantages and drawbacks are described in third section.

Implementation of modern technique for efficiency optimization of IMD based on fuzzy logic, artificial neural networks and torque reserve control are presented in fourth section.

Efficiency optimized control for closed-cycle operation of high performance IMD is presented in fifth section. The mathematical concept for computing optimal control, based on the dynamic programming approach, is described.

At the end, conclusion summarises the results achieved, implementation possibilities and directions of further research in this field.

2. Functional approximation of the power losses in the induction motor drive

The process of energy conversion within motor drive converter and motor leads to the power losses in the motor windings and magnetic circuit as well as conduction and commutation losses in the inverter.

The overall power losses (P_{tot}) in electrical drive consists of converter losses (P_{inv}) and motor losses (P_{mot}), while motor power losses can be divided in copper (P_{Cu}) and iron losses (P_{Fe}) (Uddin & Nam, 2008):

$$P_{tot} = P_{mot} + P_{inv}$$
$$P_{mot} = P_{Cu} + P_{Fe} \tag{1}$$

Converter losses: Main constituents of converter losses are the rectifier, DC link and inverter conductive and inverter commutation losses. Rectifier and DC link inverter losses are proportional to output power, so the overall flux-dependent losses are inverter losses. These are usually given by:

$$P_{inv} = R_{inv} \cdot i_s^2 = R_{inv} \cdot \left(i_d^2 + i_q^2\right), \tag{2}$$

where $i_{d,}$, i_q are components of the stator current i_s in d,q rotational system and R_{inv} is inverter loss coefficient.

Motor losses: These losses consist of hysteresis and eddy current losses in the magnetic circuit (core losses), losses in the stator and rotor conductors (copper losses) and stray losses. At nominal operating point, the core losses are typically 2-3 times smaller then the cooper losses, but they represent main loss component of a highly loaded induction motor drives (Vukosavic & Levi, 2003). The main core losses can be modeled by (Blanusa, et al., 2006):

$$P_{Fe} = c_h \Psi_m^2 \omega_e + c_e \Psi_m^2 \omega_e^2, \tag{3}$$

where ψ_d is magnetizing flux, ω_e supply frequency, c_h is hysteresis and c_e eddy current core loss coefficient.

Copper losses are due to flow of the electric current through the stator and rotor windings and these are given by:

$$p_{Cu} = R_s i_s^2 + R_r i_q^2, \tag{4}$$

The stray flux losses depend on the form of stator and rotor slots and are frequency and load dependent. The total secondary losses (stray flux, skin effect and shaft stray losses) usually don't exceed 5% of the overall losses. Considering also, that the stray losses are of importance at high load and overload conditions, while the efficiency optimizer is effective at light load, the stray losses are not considered as a separate loss component in the loss function. Formal omission of the stray loss representation in the loss function have no impact on the accuracy algorithm for on-line optimization (Vukosavic & Levi, 2003).

Based on previous consideration, total flux dependent power losses in the drive are given by the following equitation:

$$P_{tot} = \left(R_{inv} + R_s\right) i_d^2 + \left(R_{inv} + R_s + R_r\right) i_q^2 + c_e \omega_e^2 \psi_m^2 + c_h \omega_e \psi_m^2 . \tag{5}$$

Efficiency algorithm works so that flux in the machine is less or equal to its nominal value:

$$\psi_D \leq \psi_{Dn}, \tag{6}$$

where ψ_{Dn} is nominal value of rotor flux. So linear expression for rotor flux can be accepted:

$$\frac{d\psi_D}{dt} = \frac{R_r}{L_r} L_m i_d - \frac{R_r}{L_r} \psi_D, \tag{7}$$

where $\Psi_D = L_m i_d$ in a steady state.
Expression for output power can be given as:

$$P_{out} = d\omega_r \psi_D i_q, \tag{8}$$

where d is positive constant, ω_r angular speed, ψ_D rotor flux and i_q active component of the stator current. Based on previous consideration, assumption that position of the rotor flux is correctly calculated, q component of rotor flux is equal 0 (Ψ_Q =0) and relation $P_{in}=P_{tot}+P_{out}$,output power can be given by the following equation:

$$P_{in} = ai_d^2 + bi_q^2 + c_1\omega_e^2\psi_D^2 + c_2\omega_e\psi_D^2 + d\omega_r\psi_D i_q, \tag{9}$$

where $a=R_s+R_{inv}$, $b= R_s+R_{inv}+R_r$, $c_1=c_e$ and $c_2=c_h$.
Input power should be measured and exact P_{out} is needed in order to acquire correct power loss and avoid coupling between load pulsation and the efficiency optimizer.
Total power losses can be calculated as difference between input and output drive power:

$$P_{tot} = P_{in} - P_{out}, \tag{10}$$

where

$$P_{in} = V_{dc} \cdot I_{dc} \tag{11}$$

is input drive power and

$$P_{out} = \omega_r T_{em} \tag{12}$$

is output drive power.
Variables V_{dc} and I_{dc} are voltage and current in DC link. Electromagnetic torque T_{em} is known variable in a drive and speed ω_r is measured or estimated. So, we can calculate power losses without knowledge of motor parameters and power loss calculation is independent of the motor parameter changes in the working area.

3. Strategies for efficiency optimization of IMD

Numerous scientific papers on the problem of loss reduction in *IMD* have been published in the last 20 years. Although good results have been achieved, there is still no generally accepted method for loss minimization. According to the literature, there are three strategies for dealing with the problem of efficiency optimization of the induction motor drive (Abrahamsen, et al., 1996):

1. Simple State Control (SSC) ,
2. Loss Model Control (LMC) and
3. Search Control (SC)

3.1 Simple state control

The first strategy is based on the control of one of the variables in the drive (Abrahamsen, et al., 1996), (Benbouzid & Nait Said, 1998) (fig.4). This variable must be measured or estimated and its value is used in the feedback control of the drive, with the aim of running the motor by predefined reference value. Slip frequency or power factor displacement are the most often used variables in this control strategy. Which one to chose depends on which measurement signals is available (Abrahamsen, et al., 1996). This strategy is simple, but gives good results only for a narrow set of operation conditions. Also, it is sensitive to parameter changes in the drive due to temperature changes and magnetic circuit saturation.

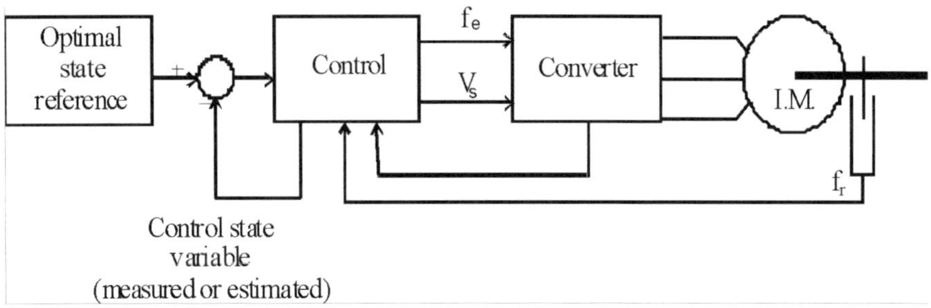

Fig. 4. Control diagram for the simple state efficiency optimization strategy.

3.2 Loss model control

In the second strategy, a drive loss model is used for optimal drive control (Fernandez-Barnal, et al., 2000), (Vukosavic & Levi, 2003) (fig. 5). These algorithms are fast because the optimal control is calculated directly from the loss model.

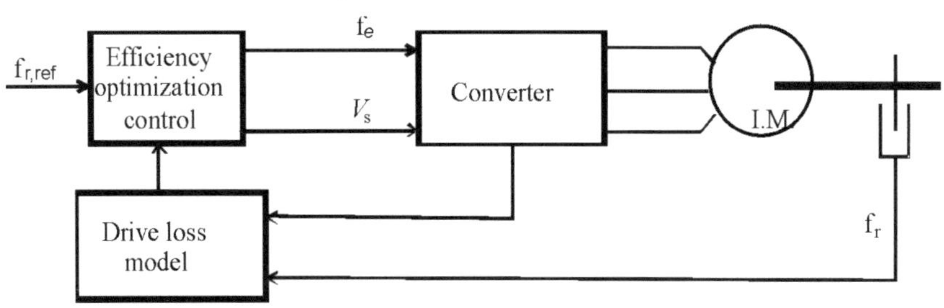

Fig. 5. Block diagram for the model based control strategy.

But, power loss modeling and calculation of the optimal operating conditions can be very complex. This strategy is also sensitive to parameter variations in the drive.

3.3 Search control

In the search strategy, the on-line procedure for efficiency optimization is carried out (Sousa at al., 1997), (Sousa at al., 2007), (Ghozzy et al., 2004) (fig. 6). The on-line efficiency optimization control on the basis of search, where the stator or rotor flux is decremented in steps until the measured input power settles down to the lowest value is very attractive.
Search strategy methods have an important advantage compared to other strategies. It is completely insensitive to parameter changes while effects of the parameter variations caused by temperature and saturation are very expressed in two other strategy.
Besides all good characteristics of search strategy methods, there is an outstanding problem in its use. When the load is low and optimal operating point is found, flux is so low that the motor is very sensitive to load perturbations. Also, flux convergence to its optimal value sometimes can be to slow, and flux never reaches the value of minimal losses then in small steps oscillates around it.

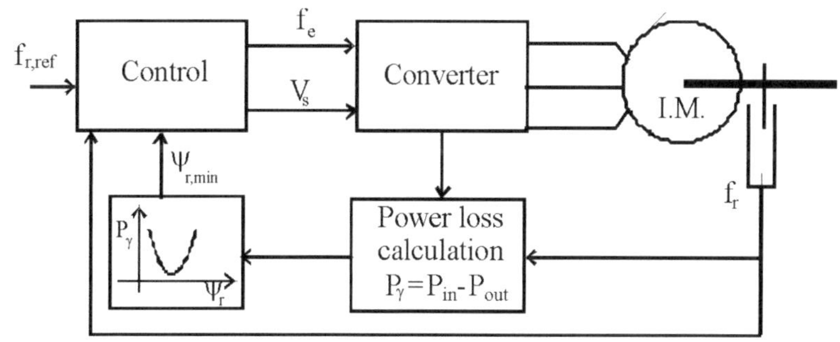

Fig. 6. Block diagram of search control strategy.

There are hybrid methods (Stergaki & Stavrakakis, 2008), (Chakaborty & Hori, 2003) which combine good characteristics of two optimization strategies SC and LMC and it was enhanced attention as interesting solution for efficiency optimization of controlled electrical drives .

4. Modern technique for efficiency optimization of IMD

Power loss model is very attractive, because it is fast and magnetizing flux which gives minimum power losses can be calculated directly from loss model. Based on expression (8), (9) and (10) power losses can be expressed in terms related to i_d, T_{em} and ω_e as follows

$$P_{tot}\left(i_d, T_{em}, \omega_e\right) = \left(a + c_1 L_m^2 \omega_e^2 + c_2 L_m^2 \omega_e\right) i_d^2 + \frac{b T_{em}^2}{\left(d L_m i_d\right)^2}. \tag{13}$$

Assuming absence of saturation and specifying slip frequency:

$$\omega_s = \omega_e - \omega_r = \frac{i_q}{T_r i_d}. \tag{14}$$

power loss function can be expressed as function of current i_d and operational conditions (ω_r, T_{em}):

$$P_{tot}\left(i_d, T_{em}, \omega_r\right) = \left(a + c_1 L_m^2 \omega_r^2 + c_2 L_m^2 \omega_r\right) i_d^2 + \frac{\left(2 c_1 \omega_r + c_2\right) L_m T_{em}}{d T_r} + \left(c_1 \frac{T_{em}^2}{\left(d T_r\right)^2} + \frac{b T_{em}^2}{\left(d L_m\right)^2}\right) \frac{1}{i_d^2}. \tag{15}$$

where $T_r = L_r / R_r$.

Based on equation (14), it is obvious, the steady-state optimum is readily found based upon the loss function parameters and operating conditions. Substituing $\alpha = \left(a + c_1 L_m^2 \omega_r^2 + c_2 L_m^2 \omega_r\right)$ and $\gamma = c_1 \frac{T_{em}^2}{d^2 T_r^2} + \frac{b T_{em}^2}{d^2 L_m^2}$ value of current i_d which gives minimal losses is:

$$i_{dLMC}^* = \left(\frac{\gamma}{\alpha}\right)^{0.25}. \tag{16}$$

If the losses in the drive were known exactly, it would be possible to calculate the optimal operating point and control of drive in accordance to that. For the following reasons it is not possible in practice (Sousa et al., 1997).

1. Even though efficiency optimization could be calculated exactly, it is probably that limitation in computation power in industrial drives would make this impossible.
2. A number of fundamental losses are difficult to predict: stray load, iron losses in case of saturation changes, copper losses because of temperature rise etc.
3. Due to limitation in costs all the measurable signals can not be acquired. It means that certain quantities must be estimated which naturally leads to an error.
4. Parameters in the loss model are very sensitive to temperature rise, magnetic circuit saturation, skin effect and so on.

For above mentioned reasons it is impractically to calculate power losses on the basis of loss model.

Search algorithms do not require the knowledge of motor parameters and these are applicable universally to any motor. So there are very intensive research of these methods, especially on academic level. Search algorithms are usually based on the following methods (Moreno-Eguilaz, et al., 1997)

4.1 Rosenbrock method

The flux is changed gradually in one direction if ($\Delta P_{tot}<0$).When algorithm detects change of power losses ($\Delta P_{tot}>0$), flux is changed in other direction, until the required accuracy is achieved:

$$\psi(n+1)=\psi(n)+k\Delta\psi(n);\; k=\begin{cases} k=1;\; \Delta P_{\gamma}(n)<0 \\ k=-1;\; \Delta P_{\gamma}(n)>0 \end{cases},$$

where $\Delta P_{tot}(n) = P_{tot}(n+1) - P_{tot}(n)$ and $\Delta\psi(n) = \psi(n+1) - \psi(n)$. This method is simple, but flux convergence can be to slow.

4.2 Proportional method

To accelerate flux convergence to its optimal value is possible to use not only the sign of the consumed power, but also the module of the input power. This can be expressed by:

$$\psi(n+1)=\psi(n)-k\,\mathrm{sgn}(\Delta\psi(n)),$$

where k is positive number. This algorithm presents convergence problems and oscillations if k is constant value. Better results are obtained if k is a nonlinear functions varying with system conditions.

4.3 Gradient method

This algorithm is based on the gradient directions search methods, using the gradient of the input power. The gradient is computed using a 1st order liner approximation.

$$\psi(n+1)=\psi(n)-k\nabla P_{\gamma}(n).$$

This problem has problems around the optimum flux due to difficulty to obtain a good numerical approximation of the gradient.

4.4 Fibonacci method

This method consists of sampling the input power of the motor working at different fluxes are function Fibonacci's series.

4.5 Search methods based on Fuzzy Logic

Search controller is used during the steady states of drive. Based on expression (13) it can be concluded that function of power loss is nonlinear. Also controller of efficiency improvement should follow known rules. These are reasons why fuzzy logic is often used in realization of efficiency optimization controller. These obtains faster and smoothly convergence of flux to the value which gives minimal power losses for a given operating conditions. Typical SC optimization block is shown in fig. 7 (Liwei et al., 2006). Input variable in optimization controller is drive input power (P_{in}), while output variable is new value of magnetization current (i^*_{dLMC}) . Fuzzy controller is very simple and it contains only one input and one output variable.

Scaling factors, input gain P_g and output gain I_g are calculated following the next expression (Liwei et al., 2006):

$$\begin{aligned} P_g &= P_{tot\_nom} - P_{totLMC} \\ I_g &= I_{dn} - i^*_{dLMC} \end{aligned} , \tag{17}$$

where P_{tot_nom} is power loss for nominal flux, and P_{tot_opt} is power loss for optimal flux value calculated from loss model, I_{dn} is nominal and i^*_{dLMC} is optimal magnetizing current defined by (16).

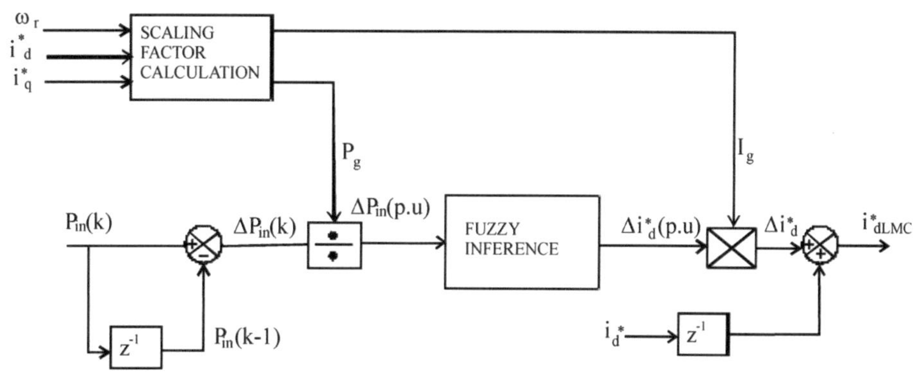

Fig. 7. SC efficiency optimization controller

4.6 Torque reserve control in search methods for efficiency optimization

One of the greatest problem of *LMC* methods is its sensitivity on load perturbation, especially for light loads when the flux level is low. This is expressed for a step increase of load torque and then two significant problems are appeared:

- Flux is far from the value which gives minimal losses during transient process, so transient losses are expressed.
- Insufficiency in the electromagnetic torque leads output speed to converge slow to its reference value with significant speed drops. Also, oscillations in the speed response are appeared.

These are common problem of methods for efficiency optimization based on flux adjusting to load torque. Speed response on the step change of load torque (from 0.5 p.u. to 1.1 p.u.), for nominal flux and when *LMC* method is applied, is presented in the fig. 8. Speed drops and slow speed convergence to its reference value are more exposed for *LMC* method.

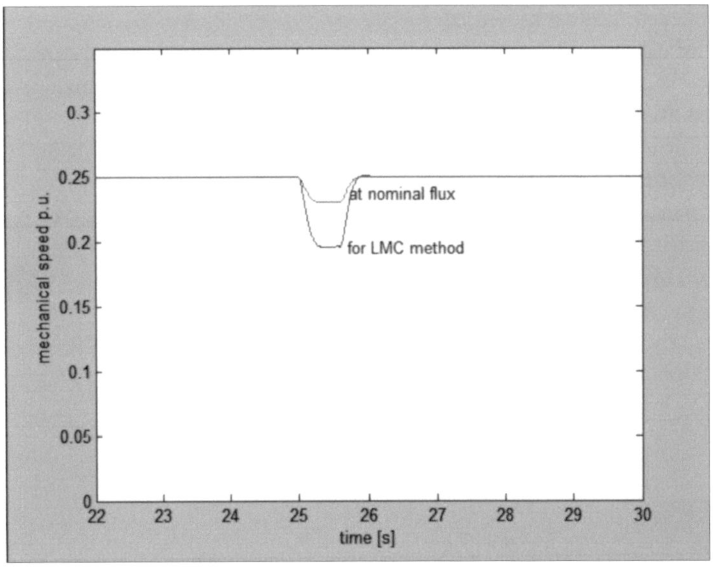

Fig. 8. Speed response on the step load increase for nominal flux and when LMC is applied.

These are reasons why torque reserve control in *LMC* method for efficiency optimization is necessary. Model of efficiency optimization controller with torque reserve control is presented in fig. 9 (Blanusa et al., 2006). Optimal value of magnetization current is calculated from the loss model and for given operational conditions (16). Fuzzy logic controller is used in determination of Δi_d, on the basis of the previously determined torque reserve (ΔT_{em}). Controller is very simple, and there is one input, one output and 3 rules. Only 3 membership functions are enough to describe influence of torque reserve in the generation of i^*_{dopt}.

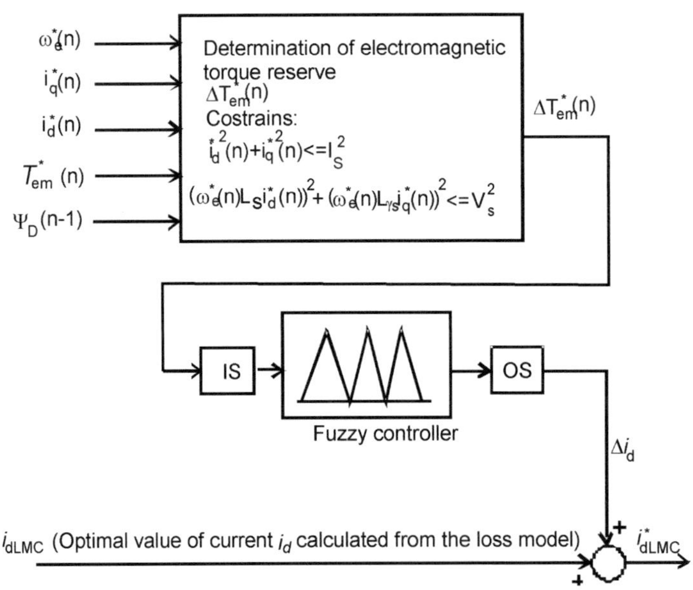

Fig. 9. Block for efficiency optimization with torque reserve control.

If torque reserve is sufficient then $\Delta i_d \approx 0$ and this block has no effect in a determination of i^*_{dLMC}. Oppositely, current i_d (magnetization flux) increases to obtain sufficient reserve of electromagnetic torque.

Two scaling blocks are used in efficiency controller. Block IS is used for normalization of input variable, so same controller can be used for a different power range of machine. Block OS is used for output scaling to adjust influence of torque reserve in determination of i^*_{dLMC} and obtain requested compromise between power loss reduction and good dynamic response.

4.7 Search methods using neural networks

To find control combination that leads to the minimum power input point an artificial neural network (ANN) based search algorithm can be employed to operate as an efficiency optimizer. One typical ANN search control block applied for direct torque controlled IMD is presented in fig. 10 (Chis et al., 1997). Also, similar method can be applied for vector controlled IMD (Prymak et al., 2002)

Input drive power is measured and difference between two successive steps is calculated. Result $\Delta P_{in}(k)$ is one input variable in artificial neural network. It is scaled to the normalized interval [0 1] in input scaling block IS. Second input variable is last step of stator flux $\Delta \Psi_s(k-1)$. The neural networks has two inputs, one output, and two hidden layers, of 4 and 2 neurons respectively. The training was done off-line, by connecting the ANN in parallel with an adaptive step minimum search system. Output variable of efficiency controller is

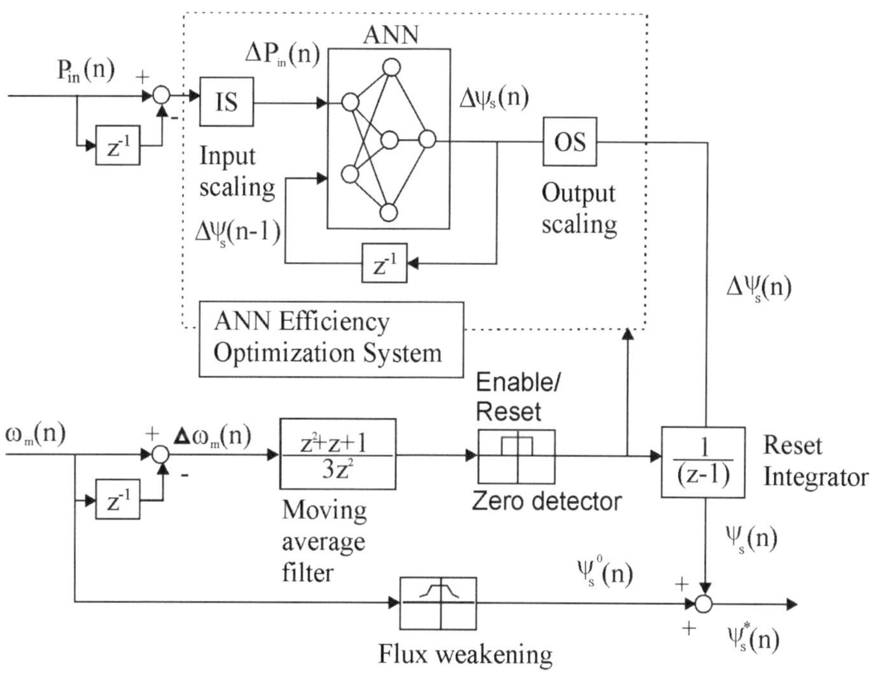

Fig. 10. ANN efficiency optimizer

new step of stator flux $\Delta\Psi_s(n)$. Also, it is normalized to interval [-1,1] and its scaling to real value is implemented in output scaling (OS) block.
Steady state of the system is detected in second part of efficiency optimization block which input is mechanical speed $\omega_m(n)$. If steady state is detected optimization block is enabled and output is $\Psi^*_s(n)=\Psi_s(n)$. Adversely, flux is set to the value given by flux weakening block and $\Psi^*_s(n)=\Psi^0_s(n)$

5. Efficiency optimization of closed cycle operation IMD

Efficiency improvement of *IMD* based on dynamic programming (optimal flux control) is an interesting solution for closed-cycle operation of drives (Lorenz & Yang, 1992). For these drives, it is possible to compute optimal control, so the energy consumption for one operational cycle is minimized. In order to do that, it is necessary to define performance index, system equations and constraints for control and state variables and present them in a form suitable for computer processing.
The performance index is as follows (Bellman, 1957), (Brayson, 1975):

$$J = \phi[x(N)] + \sum_{n=1}^{N-1} L(x(n),u(n)) \tag{18}$$

where $N=T/T_s$, T is a period of close-cycled operation and T_s is sample time. The L function is a scalar function of $\boldsymbol{x}$-state variables and $\boldsymbol{u}$-control variables, where $\boldsymbol{x}(n)$, a sequence of n-vector, is determined by $\boldsymbol{u}(n)$, a sequence of m-vector. The φ function is a function of state variables in the final stage of the cycle. It is necessary for a correct definition of performance index.
The system equations are:

$$x(n+1) = f\left[x(n),u(n)\right],\ n = 0..N-1\,, \tag{19}$$

and f can be a linear or nonlinear function. Functions L and f must have first and second derivation on its domain.
The constraints of the control and state variables in terms of equality and inequality are:

$$C\left[x(n),u(n)\right] \le 0,\ \ i = 0,1,..,N-1\,. \tag{20}$$

Following the above mentioned procedure, performance index, system equations, constraints and boundary conditions for a vector controlled *IMD* in the rotor flux oriented reference frame, can be defined as follows:

a. The performance index is (Vukosavic & Levi, 2003), (Blanusa et al. 2006):

$$J = \sum_{n=0}^{N-1}\left[ai_d^2(n) + bi_q^2(n) + c_1\omega_e(n)\psi_D^2(n) + c_2\omega_e^2(n)\psi_D^2(n)\right], \tag{21}$$

Rotor speed ω_r and electromagnetic torque T_{em} are defined by operating conditions (speed reference, load and friction).

b. The dynamics of the rotor flux can be described by the following equation:

$$\psi_D(n+1)=\psi_D(n)\left(1-\frac{T_s}{T_r}\right)+\frac{T_s}{T_r}L_m i_d(n), \tag{22}$$

where $T_r=L_r/R_r$ is a rotor time constant.

c. Constraints:

$$\begin{aligned}
&k i_d(n) i_q(n)=T_{em}(n),\ k=\frac{3}{2}\frac{p}{2}\frac{L_m^2}{L_r},\ (\textit{for torque})\\
&i_d^2(n)+i_q^2(n)-I_{s\max}^2\le 0,\ (\textit{for stator current})\\
&-\omega_{rn}\le\omega_r\le\omega_{rn},\ (\textit{for speed})\\
&\psi_D(n)-\psi_{Dn}\le 0,\ (\textit{for rotor flux})\\
&\psi_{D\min}-\psi_D(n)\le 0.
\end{aligned} \tag{23}$$

I_{smax} is maximal amplitude of stator current, ω_{rn} is nominal rotor speed, p is number of poles and Ψ_{Dmin} is minimal value of rotor flux.

Also, there are constraints on stator voltage:

$$0\le\sqrt{v_d^2+v_q^2}\le V_{s\max}, \tag{24}$$

where v_d and v_q are components of stator voltage and V_{smax} is maximal amplitude of stator voltage.

Voltage constraints are more expressed in *DTC* than in field-oriented vector control.

d. Boundary conditions:

Basically, this is a boundary-value problem between two points which are defined by starting and final value of state variables:

$$\begin{aligned}
&\omega_r(0)=\omega_r(N)=0,\\
&T_{em}(0)=T_{em}(N)=0,\\
&\psi_{Dn}(0)=\psi_{Dn}(N)=\textit{free},\\
&\textit{considering constrains in } (23)
\end{aligned} \tag{25}$$

Presence of state and control variables constrains generally complicates derivation of optimal control law. On the other side, these constrains reduce the range of values to be searched and simplify the size of computation (Lorenz & Yang, 1992).

Let us take the following assumptions into account:

1. There is no saturation effect ($\Psi_D\le\Psi_{Dn}$).
2. Supply frequency is a sum of rotor speed and slip frequency, $\omega_e=\omega_r+\omega_s$. Rotor speed is defined by speed reference whereas slip frequency is usually low and insignificantly influences on total power loss (Ionel et al., 2006)
3. Rotor leakage inductance is significantly lower than mutual inductance, $L_{\gamma r}<<L_m$.
4. Electromagnetic torque reference and speed reference are defined by operation conditions within constraints defined in equation (23).

Following the dynamic programming theory, Hamiltonian function H, including system equations and equality constrains can be written as follows (Blanusa et al., 2008):

$$\begin{gathered}H(i_d, i_d, \omega_e, \psi_D) = a i_d^2(n) + b i_q^2(n) + \\ c_1\omega_e(n)\psi_D^2(n) + c_2\omega_e^2(n)\psi_D^2(n) + \\ \lambda(n+1)\left[\psi_D(n)\frac{T_r - T_S}{T_r} + \frac{T_S}{T_r}L_m i_d(n)\right] \\ +\mu(n)\left[k i_d(n) i_q(n) - T_{em}(n)\right].\end{gathered} \tag{26}$$

In a purpose to determine stationary state of performance index, next system of differential equations are defined:

$$\begin{aligned}&\lambda(n) = \lambda(n+1)\frac{T_r - T_S}{T_r} + 2\left(c_1\omega_e(n) + c_2\omega_e^2(n)\right)\psi_D(n) \\ &2b i_q(n) + \mu(n) k i_d(n) = 0 \\ &2a i_d(n) + \mu(n) k i_q(n) + \lambda(n+1)\frac{T_S}{T_r}L_m = 0 \\ &k i_d(n) i_q(n) = T_{em}(n),\ \omega_e(n) = \omega_r(n) + \frac{L_m}{T_r}\frac{i_q(n)}{\psi_D(n)} \\ &n = 0,1,2,..,N-1,\end{aligned} \tag{27}$$

where λ and μ are Lagrange multipliers.
By solving the system of equations (27) and including boundary conditions given in (23), we come to the following system:

$$\begin{aligned}&2a i_d^4(n) + \lambda(n+1)\frac{T_S}{T_r} i_d^3(n) = \frac{2b}{k^2}T_{em}^2(n) \\ &\psi_D(n) = \frac{T_r}{T_r - T_s}\psi_D(n+1) - \frac{T_s}{T_r - T_s}L_m i_d(n) \\ &i_q(n) = \frac{T_{em}(n)}{k i_d(n)}, \omega_e(i) = \omega_r(i) + \frac{L_m}{T_r}\frac{i_q(n)}{\psi_D(n)}, \\ &\lambda(n) = 2\left(c_1\omega_e(n) + c_2\omega_e^2(n)\right)\psi_D(n) + \lambda(n+1)\frac{T_r - T_s}{T_r} \\ &n = 0,1,2,..,N-1.\end{aligned} \tag{28}$$

Every sample time values of $\omega_r(n)$ and $T_{em}(n)$ defined by operating conditions is used to compute the optimal control ($i_d(n)$, $i_q(n)$, n=0,..,N-1) through the iterative procedure and applying the backward procedure, from stage n =N-1 down to stage n =0. For the optimal control computation, the final value of ψ_D and λ have to be known. In this case, $\psi_D(N) = \psi_{Dmin}$ and

$$\lambda(N) = \frac{\partial\varphi}{\partial\psi_D(N)} = 0. \tag{29}$$

5.1 Experimental results

Simulations and experiments have been performed in order to validate the proposed procedure.

The experimental tests have been performed on the setup which consists of:

- induction motor (3 MOT, Δ380V/Y220V, 3.7/2.12A, cosϕ=0.71, 1400o/min, 50Hz)
- incremental encoder connected with the motor shaft,
- PC and dSPACE1102 controller board with TMS320C31
- floating point processor and peripherals,

The algorithm observed in this paper used the Matlab - Simulink software, dSPACE real-time interface and C language. Handling real-time applications is done in ControlDesk.

Some comparisons between algorithms for efficiency optimization are made through the experimental tests. Expressed problem in efficiency optimization methods are its sensitivity to steep increase of load or speed reference, especially for low flux level. Therefore, speed response on steep increase of load are analyzed for LMC and optimal flux control method. Torque load and speed reference for one operating cycle are shown in fig 11. Graph of power losses when nominal flux is applied and optimal flux control and one operating cycle is presented in fig. 12.

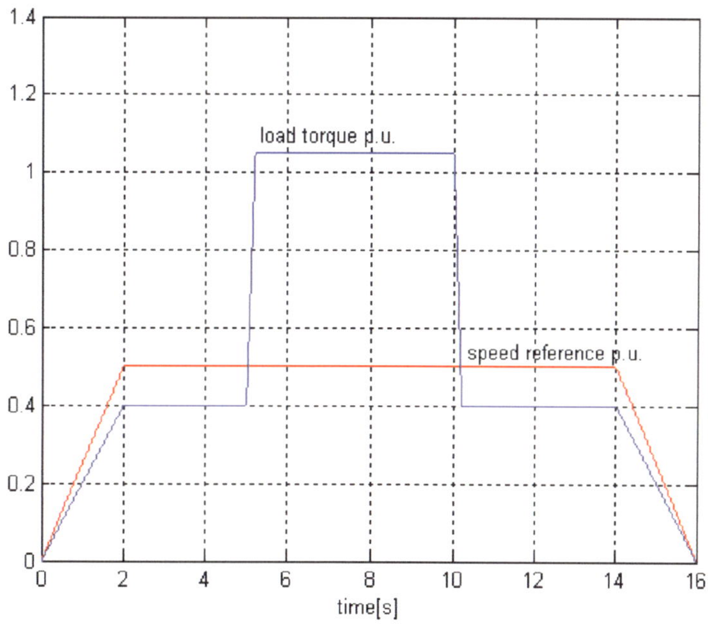

Fig. 11. Graph of speed and load torque reference in one one operating cycle.

That is obvious, for optimal flux control power loss reduction is expressed in one operating cycle

6. Conclusion

Algorithms for efficiency optimization of induction motor drives are briefly described. These algorithms can be applied as software solution in controlled electrical drives, particulary vector controlled and direct torque controlled IMD.

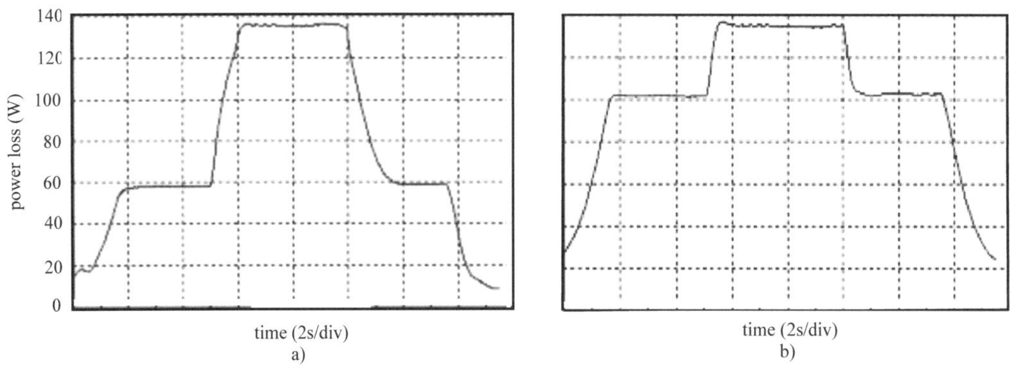

Fig. 12. Power losses in one operating cycle for a) optimal flux b) nominal flux

For a light load methods for efficiency optimization gives significiant power loss reduction (Figs. 2 and 12).

Three startegies for efficiency optimization, Simple state control, Loss model control and Search control are usually used. LMC and SC are especially interested. LMC is fastest tehnique but very sensitive to parameter variations in loss model of drive. Also, calculation of optimal control based on loss model can be to complex. SC methods can be applied for any machine and these are insensitive to parameter variations. In many applications fux change to its optimal value is too slow. Some tehniques based on fuzzy logic and artificial neural networks which obtains faster and smoothly flux convergence to the value of minimal power losses are described.

New algorithm for efficiency optimization of high performance induction motor drive and for closed-cycle operation has been proposed. Also, procedure for optimal control computation has been applied.

According to the performed simulations and experimental tests, we have arrived at the following conclusions: The obtained experimental results show that this algorithm is applicable. It offers significant loss reduction, good dynamic features and stable operation of the drive.

Some new methods for parameter identification in loss model made LMC very actual. Also, Hybrid method combines good characteristics of two optimization strategies SC and LMC were apeared. It was enhanced attention as interesting solution for efficiency optimization of controlled electrical drives. These can be very interesting for further research in this field.

7. Reference

Abrahamsen F, Pedersen JK, Blaabjerg F: *State-of-Art of optimal efficiency control of low cost induction motor drives.* In: Proceedings of PESC'96, pp. 920-924, 1996

F. Abrahamsen, F. Blaabjerg, J.K. Pedersen, P.Z. Grabowski and P. Thorgensen," On the Energy Optimized Control of Standard and High Efficiency Induction Motors in CT and HVAC Applications", *IEEE Transaction on Industry Applications,* Vol.34, No.4, pp.822-831, 1998.

Benbouzid M.E.H., Nait Said N.S., *An efficiency-optimization controller for induction motor drives,* IEEE Power Engineering Review, Vol. 18, Issue 5, pp. 63 -64, 1998.

Bellman R., *Dynamic Programming,* Prinston University Press, 1957.

Blanusa B., Matić P., Ivanovic Z, Vukosavic S.N.: *An Improved Loss Model Based Algorithm for Efficiency Optimization of the Induction Motor Drive*, In: Electronics (2006), Vol.10, No.1., pp. 49-52, 2006.

Blanusa B., Slobodan N. V.: *Efficiency Optimized Control for closed-cycle Operations of High Performance Induction Motor Drive*, Journal of Electrical Engineering, Vol.8/2008-Edition: 3, pp.81-88, 2008.

Bose B.K., Patel N.R., Rajashekara K.: A neuro fuzzy based on-line efficiency optimization control of a stator flux oriented direct vector controlled induction motor drive, In: IEEE Transaction on Industrial Electronics (1997), Vol. 44, No.2, pp. 270-273, 1997.

Brayson A. E., *Applied Optimal Control, Optimization, Estimation and Control*, John Wiley & Sons, 1975.

Chakraborty C., Ta M.C., Uchida T., Hori Y.: *Fast search controllers for efficiency maximization of induction motor drives based on DC link power measurement*. In: Proceedings of the Power Conversion Conference 2002., Vol. 2 , pp. 402 -408, 2002.

Chis M., Jayaram S., Ramshaw R., Rajashekara K.: *Neural network based efficiency optimization of EV drive*. In: IEEE-IECIN Conference Record, pp. 454-457, 1997.

Fernandez-Bernal F., Garcia-Cerrada A., Faure R.: *Model-based loss minimization for DC and AC vector-controlled motors including core saturation* .In: IEEE Transactions on Industry Applications, (2000), Vol. 36, No. 3, 2000, pp. 755 -763, 2000.

Ghozzy S., Jelassi K., Roboam X.: *Energy optimization of induction motor drive*, International Conference on Industrial Technology, pp. 1662 -1669, 2004.

Ionel D.M., Popescu M., Dellinger S.J., Miller T.J.E., Heidiman R.J., McGilp M.I.: *On the variation with flux and frequency of the core loss coefficients in electrical machines*. In: IEEE Transactions on Industry Applications (2006), Vol. 42, No. 3, pp. 658-667, 2006.

Liwei Z., Jun L., Xuhui W.: *Systematic Design of Fuzzy Logic Based Hybrid On-Line Minimum Input Power Search Control Strategy for Efficiency Optimization of IMD*, IPEMC 2006.

Lorenz, R. D. Yang. S.-M.: *Efficiency –optimized flux trajectories for closed-cycle operation of field-orientation Induction Machine Drives*. In: IEEE Transactions on Industry Applications (1992), Vol.28, No.3, pp. 574-580, 1992.

Prymak B., Moreno-Eguilaz J.A., Peracaula J.: *Neural network based efficiency optimization of an induction motor drive with vector control*, In: Proceedings of the IECON 02 Industrial Electronics Society, IEEE 2002 28th Annual Conference, Vol. 1, pp. 146-151, 2002.

Sergaki E. S., Stavrakakis G.S., *Online Search Based Fuzzy Optimum Efficiency Operation in Steady and Transient States for Dc and Ac Vector Controlled Motors*, Proceedings of the 2008 International Conference on Electrical Machine, Paper ID 1190, 2008.

Sousa D.A., Wilson C.P. de Aragao and Sousa G.C.D.: *Adaptive Fuzzy Controller for Efficiency Optimization of Induction Motors*, IEEE Transaction on Industrial Electronics, Vol. 54, No.4, pp. 2157-2164, 2007.

Sousa G.C.D., Bose B.K., Cleland J.G.: *Fuzzy logic based on-line efficiency optimization control of an indirect vector controlled induction motor drive.*, IEEE Transaction on Industrial Electronics (1995), Vol. 42, No.2, pp. 192-198, 1995.

Uddin M.N., Nam S.W.: *Development of a nonlinear and Model Based Online Loss Minimization Control of an IM Drive,* IEEE Transaction on Energy Conversion, Vol.23, No.4, pp. 1015-1024, 2008.

Vukosavic S.N., *Controlled electrical drives - Status of technology.* In: Proceedings of XLII ETRAN Conference, Vrnjacka Banja, Yugoslavia, No. 1,pp. 3-16,1998.

Vukosavic S.N., Levi E.: *Robust DSP-based efficiency optimization of variable speed induction motor drive,* IEEE Transaction of Ind. Electronics (2003), Vol. 50, No. 3,pp. 560-570, 2003

19

Modelling of Concurrent Ddevelopment of the Products, Processes and Manufacturing Systems in Product Lifecycle Context

Jan Duda
Cracow University of Technology
Institute of Production Engineering and Automation
Poland

1. Introduction

The process planning is a phase of the Product Life Cycle realized within the production preparation, which significantly influences the phases of the production and usage of the product. Therefore, the process planning should be examined in the context of the life-cycle of the product and with the consideration of new development strategies. This part presents the directions for increasing the level of integration and automation of manufacturing production preparation with the possibilities offered by PLM solutions. In the area of functional integration, the analysis of CAx and DFx systems used in the implementation of the product development strategies, processes and manufacturing systems are carried out.
In the area of information integration, the application of PDM system and the development of the workflow diagrams are discussed. The BPMN notation is used for the modelling of development processes occurring in the product lifecycle in order to implement the workflow diagrams. The conception and example of integrated process and manufacturing system planning in PLM (Product Lifecycle Management) environment are presented.

2. The trends in strategies and computer systems for concurrent product development

According to the new development strategies (Eigner, 2004), (Chlebus, 2000), the product development focuses on Fig.1:

- as much as possible parallel execution of all development related product life cycle phases, thus creating CE (Concurrent Engineering) environment. CE strategy assumes the development of resources and production facilities at the early product design phases to shorten the production start-up time.
- incorporation of the relations between business and engineering activities CEE (Cross Enterprise Engineering), securing also the access to the resources of cooperating enterprises. The parallel execution and the integration of products and processes start on the early product development stages and cover suppliers, clients, as well as internal and external IT solutions. CEE means that product components, parts, assemblies, systems and functional sub-systems are designed and developed across the local

business and engineering boundaries. Intranet and internet are here used as the means for the communications and the information exchange.

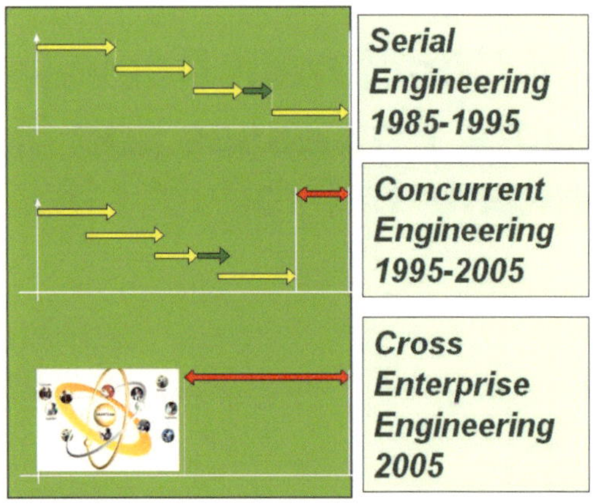

Fig. 1. Product design strategies

3. Development of CAD/CAM systems

The key condition for the efficient process planning, especially in the CE environment is computer integrated design and planning - the common platform for the CAx systems used during the product development.
The works (Dybała, Oczoś, Chlebus, & Boratyński, 2000) (Duda, 2004),presents the use of the CAx systems in the various phases of the product lifecycle fig.2.

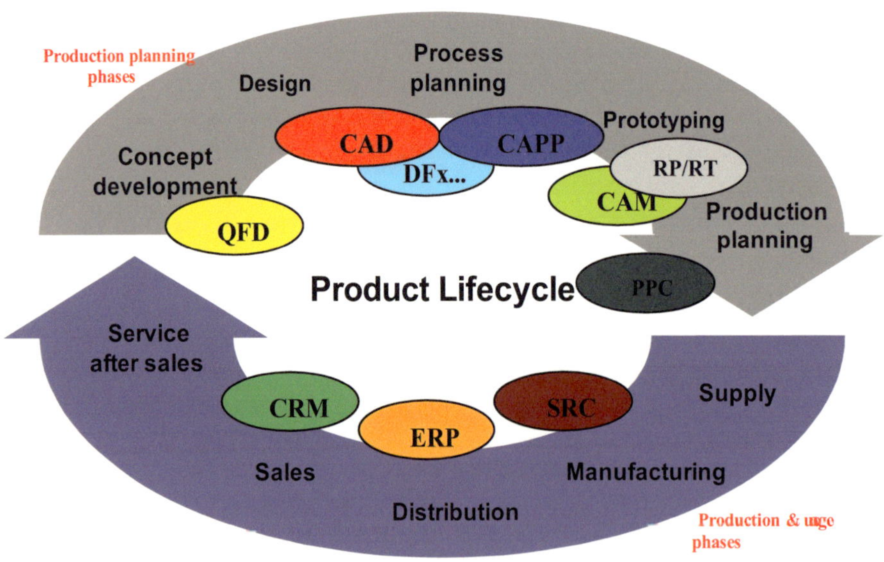

Fig. 2. The use of computer methods and systems in product life cycle

To determine the level of the automation in the area of manufacturing process planning, the selected CAD/CAM systems were analyzed in the view of:

- Assembly process planning,
- Manufacturing stock and process planning,
- Planning of the structures of the manufacturing process plans.

Based on the results of the analysis it was found that the level of automation in the area of manufacturing process planning in CAD/CAM systems is still relatively very low (Duda, 2004). The generation of assembly sequences and the development of the manufacturing process plans for the parts constituting the product is executed in the interactive manned by the manufacturing engineer.

This level is still increasing as the result of the use:

- templates allowing for process planning based on the pattern created earlier,
- modules allowing for the storage of the manufacturing knowledge and its further use during the process planning.

Because of this, CAD/CAM systems evolve toward Intelligent CAD/CAM systems fig. 3.

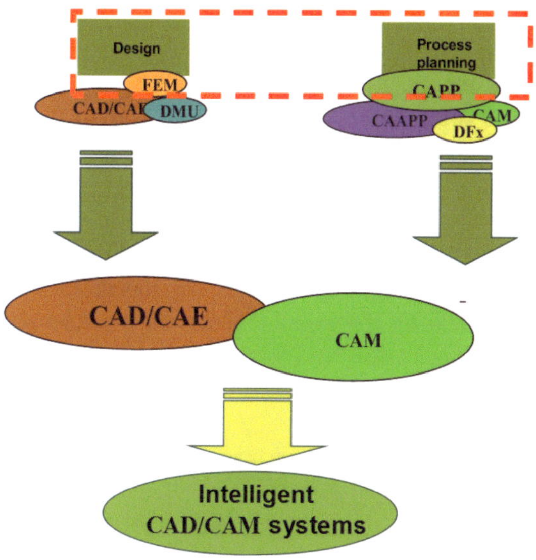

Fig. 3. Development of CAD/CAM systems

CAD/CAM systems use the processor which generates the general temporary data and then converts these data into the format fulfilling the requirements of the particular machine tools. Because of this mode, the execution of the creative decision process executed by manufacturing engineer and the planning based on the available manufacturing resources (machine tools, tools, equipments) is still very difficult.

4. Development of CAPP and CAAPP systems

This confirms the fact that manufacturing process planning is the link connecting the CAD and CAM systems, necessary to achieve the full integration of the production. The results of the executed analysis show the necessity for the intensification of works toward the development of CAPP and CAAPP systems being the key conditions for the further advancement of computer integrated product development. The interests in CAPP systems are caused by the need for the integration of design and manufacturing phases (Inyong

Ham, 1988), (Mari, Gunasekaran, & Grieve, 1995). The manufacturing process planning is the link between these systems of vital influence on the whole integration level of production preparation.

In the traditional approach, the design and manufacturing process planning were executed sequentially during the production preparations stage. The goal of the Concurrent Engineering is to convert this sequential approach in the more interactive, parallel process. The CAAPP and CAPP systems should be developed in the view of the convenient style of process planning. A lot of systems were developed based on the sequential model of product development. Such systems receive the detailed specifications of the product as the input data and the manufacturing process plan is usually created in the interactive mode – the manufacturing engineer is responsible for the final effects. Now the tendencies for the development of the systems incorporating the processing possibility of manufacturing systems at the early phase of works, for the customised planning by manufacturing engineer bases on the knowledge, without the limitations on the creative decision process can be noticed. The development of such systems is related to the aspect of the decision taking in the particular environment. The goal of the planning systems is securing the assistance for the engineer during the manufacturing process planning. The products are developed in an interactive manner in the Concurrent Engineering environment: the subsequent versions of products and its manufacturing processes are generated at the early phase of works, when the costs of the modifications are relatively very low. The computer integration of the design and manufacturing is the condition for the effective implementation of CE approach. The manufacturing process planning system should be one of the elements of such solution.

The manufacturing process planning method fulfilling the requirements of the Concurrent Engineering model should feature the possibility for the flexible generation of variants of structures of manufacturing processes and variants of its constituent elements. The semi-generative method was selected for the implementation of CAPP system fulfilling the Concurrent Engineering requirements. The process planning system working in the Concurrent Engineering environment should be one of the main elements of the computer integrated manufacturing system.

The use of modern product development strategies puts some requirements:

- for systems of assembly process planning:
  - ability to assembly processes designing for the generated acceptable assembly sequences,
  - ability to organizational forms of the assembly designing, considering the manufacture system,
- for systems of manufacture process planning:
  - ability to plan processes for the wide range of typical parts,
  - ability to plan processes taking into the account the capabilities of manufacturing systems,
  - ability to generate variants of process plans at different level of details, which is needed to evaluate the costs at various degrees of advancements, from the general concept to the finished version.

The semi-generative method was selected for the implementation of CAPP system fulfilling the Concurrent Engineering requirements. The process planning with the use of these methods requires some expert knowledge from the user. In (Duda, 2003) the process planning knowledge was divided into:

- knowledge about the structures of process plans for typical parts refereed further as 1st type knowledge,
- knowledge defining the process planning strategy imposed by the employed methodology, refereed further as 2nd type knowledge.

The manufacturing knowledge representation is based on the model of manufacturing knowledge in the form of hierarchical decision nets. The following knowledge areas were distinguished:

- the knowledge SK describing the rules for part classification in the view of manufacturing, represented in the form of frames RK and rules {set TK},
- knowledge SF describing the rules for the classification of the semi-finished product for the part types, represented in the form of frames RF and rules {Rule TF},
- knowledge describing the rules and principles for the process planning for the part types, represented in the form of frames RP and rules {Rule TP},
- knowledge SW describing the rules and principles for the semi-finished product design for the semi-finished product types, represented in the form of frames RW and rules {Rule TW}.

The generation of manufacturing process plans is executed based on the following:

- characteristics of parts stored in object oriented database of manufacturing features (Pobożniak, 2000),
- knowledgebase covering the general and detailed rules and principles of manufacturing process planning,
- database of processing capabilities of manufacturing system (Duda, Kwatera, Habel, & Samek, 1999),
- database of catalogue values.

Process planning, due to its complexity is realized in stages, from the general idea to the detailed solutions in an interactive manner. For the process planning mode three planning stages were distinguished (Duda, 2003) fig.4 :

- The selection of the general process structure (Stage I),
- Reverse design of raw material and intermediate states of the workpiece (Stage II),
- Generation of process plan (Stage III).

STAGE I. The selection of the general process structure

In first stage, the generalized structure of the manufacturing process is selected based on the manufacturing and geometrical characteristics of the product. The product is evaluated taking into consideration the values of its attributes, according to the sequence resulting from the knowledge SK describing the rules for part classification in the view of manufacturing. The classification into manufacturing class determines the manufacturing knowledge for the semi-product type selection SF and manufacturing knowledge for shaping the process plan SP.

STAGE II Reverse design of raw material and intermediate states of the workpiece

Second stage is realized in the following steps:

- determination of the type of the semi-finished product,
- planning of the intermediate shapes of the product,
- design of semi-finished product.

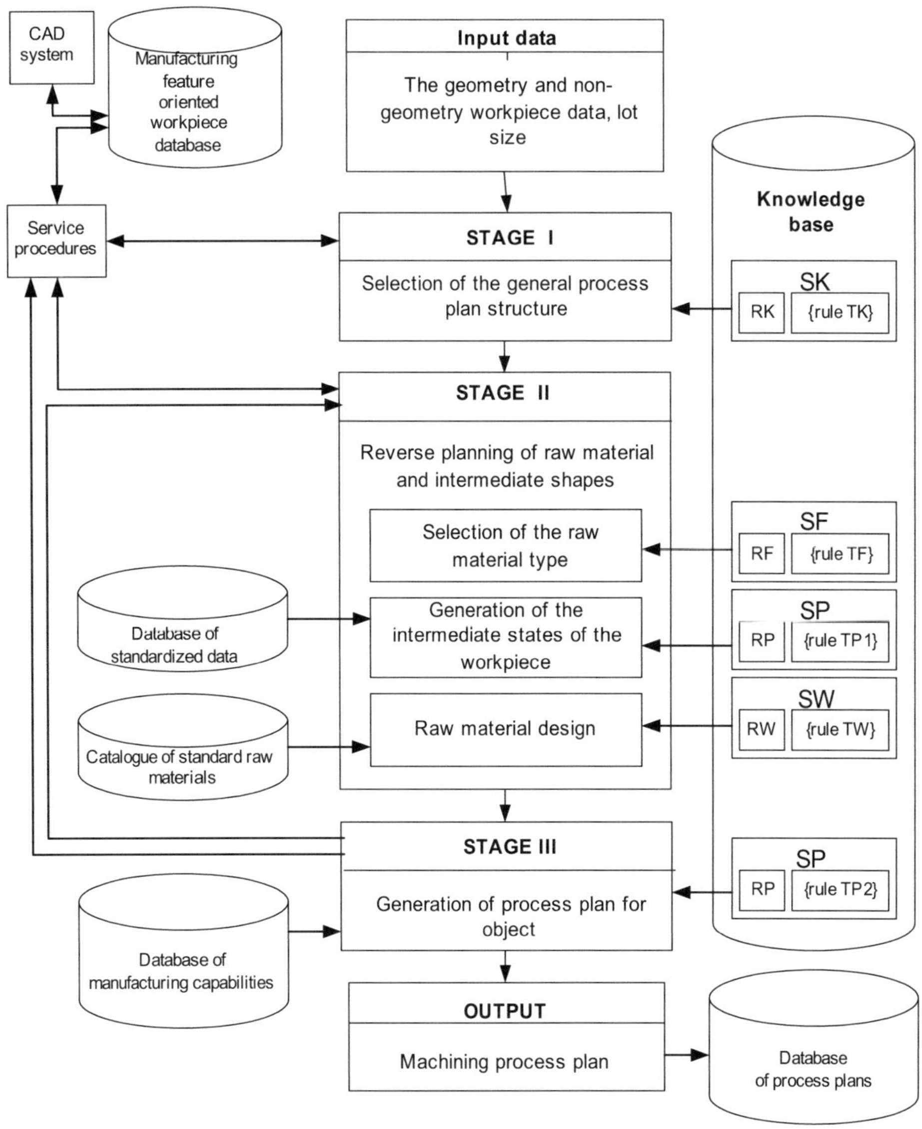

Fig. 4. The block scheme of the algorithm of machining process plan generation

The type of semi-finished product is selected on the basis of the set of rules for classification of the semi-product SF. The necessary parameters are collected from three sources:

- part type; it is the output of first stage,
- geometrical and manufacturing characteristic (material, mass, accuracy requirements, etc.) from feature/object product database,
- dimensions of standard semi- finished products from catalogue.

As the result of the reasoning the semi-finished product type is proposed (rolled, forged, etc.).

Generalized structure of manufacturing process selected during the first stage, is a base for the determination of intermediate states of the workpiece. On the basis of geometrical and

manufacturing characteristics of the product and the manufacturing knowledge the intermediate states and the semi-finished product may be determined by using the reverse (from the product to the semi-finished product) method. This method includes the identification of workpiece surfaces with characteristics matching the planning activities selected according to generalized structure, from end to start nodes on each level of hierarchical net.
The selected type of semi-product and intermediate states determined for the start node form the basis for the planning of semi-product and its process plan with knowledge RW.

STAGE III Generation of process plan

The determination of the intermediate shapes allows to decompose the overall planning goal, the development of process plan, into sub-goals. The planning goal on the III Stage can be defined as follows: Select the means (machine tools fixtures, tools) to realize the sub-goals represented by the intermediate shapes while creating the hierarchical, object oriented structure of process plan for the given workpiece.
The actions are carried out based on the following data:

- characteristics of the raw material and intermediate states of the workpiece being the outcome of the stage II,
- characteristics of the manufacturing system capabilities selected for the process plan implementation.

The algorithm for process plan generation, is realized on the subsequent levels of details covering: generation of operation, set-ups, positions and cuts.
In the first planing step, the list of manufacturing feature transformations in generated process plan PT is determined for each intermediate state.
The subsequent generation levels are realize interactively in the system „process engineer - computer" intended to identify the admissible variants of realization. The selection of the one of the presented solutions by the process engineer at the given level, limits the number of the solutions on the lower levels, allowing for rational structuring of the process plan.
In there is no appropriate solution to be selected at the given level, it is possible to change the solution selected on the higher level in the feedback manner. In this iterative way, the process plan is generated.
Also Computer Aided Assembly Process Planning (CAAPP) systems are being used. The main functions of such systems are (Duda & Karpiuk, 2008):

- definition of assembly units,
- generation of assembly sequences,
- selection of the technical means needed for the realization of the assembly,
- preparation of the technological documentation of the assembly process.

Generation of the assembly sequences has a great influence on the effectiveness of the assembly processes. The works (Łebkowski, 2000) distinguish four basic groups:

- Methods using the three-stage procedure of the sequence generation including: generation of relations between elements of the final product, generation of assembly sequences, selection of the best sequence according to accepted criterion of the optimisation,
- Methods separating the assembly into subgroups and generating the subsequence for every sub-assembly by applying simple rules,
- Methods based on expert systems created for the assembly of specific unique products,

- Methods generating different variants of assembly sequences for products by the modification of the already defined sequences.

The methods of generation of assembly sequences can be divided according to the type of applied procedures. The work (Ciszak & Żurek, 1999) distinguish:

- Algorithmic methods giving optimal sequences assembly according to given criterion,
- Heuristic methods giving solutions in the relatively short time.

5. Development of DFx systems

Market action should be directed to decrease the costs of the product manufacturing, improve the manufacturing processes and the production organization while maintaining the required quality and meeting the assumed technical requirements.
On the stage of the analysis of the design, DFx methodology is used fig 5.

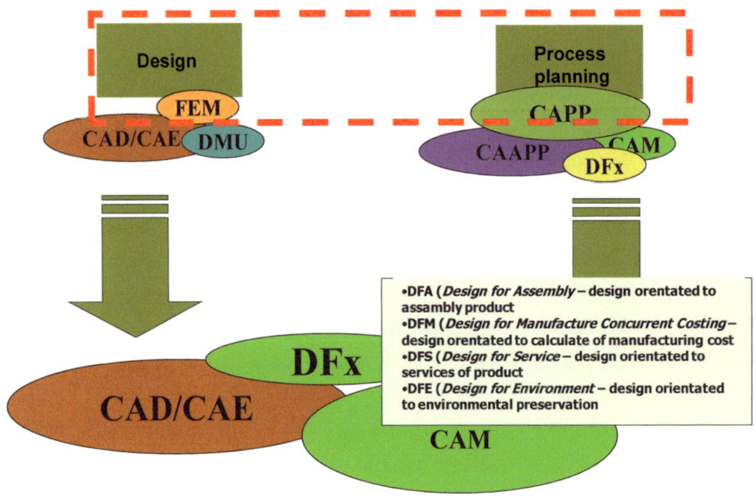

Fig. 5. Development of DFx systems

DFx techniques analyse the existing designs to propose the changes lowering the product cost. The most popular systems on the market is now Boothroyd and Dewhurst's DFMA (Design for Manufacture and Assembly) which contains the following modules:

- DFA (Design for Assembly),
- DFM (Design for Manufacture Concurrent Costing),
- DFS (Design for Service),
- DFE (Design for Environment).

DFMA methodology was developed by Prof. Boothdroyd and Dewhurst. This methodology helps to reduce the costs of manufacturing and to ensure the correctness of the assembly by analysis outputting the conclusions concerning the product design.

6. Integrated process and systems development in PLM environment

The key condition for the effective concurrent engineering and cross enterprise engineering is the computer integrated environment of design and manufacturing - the common platform for computer aided systems for the product development. Effectiveness of the CE and CEE strategies results from the better management of information in production process. The efficiency of activities depends on the provision of right information, to right places, for right peoples in the right time. PDM (Product Data Management) systems play the coordination role in the synchronization of the flow of information (Eigner, 2004). PDM systems manage the data about the product services, product structure, its documentations and processes with the ability of processing data in the electronic form.

PDM systems allow for modelling, storage and transformation of all data related to the product. Development of the integrated systems evolves toward the PLM (Product Lifecycle Management) solutions (Kahlert, 2004). PLM strategy increases the integration and automation of realized functions PLM environment includes the following applications for the product development:

- PPM (Product and Portfolio Management),
- CAx (Product Design),
- MPM (Manufacturing Process Management),
- PDM (Product Data Management).

Features of modern development strategies indicate the need for product development phase integration. Integration and parallel execution of activities were received through the separation of the conceptual design stages, allowing for the creation of the variant design solutions. Variants are then evaluated in the view of the requirements of the next development phase. The selected variant fulfilling the established criteria is next further developed in the detail design stage Fig 6.

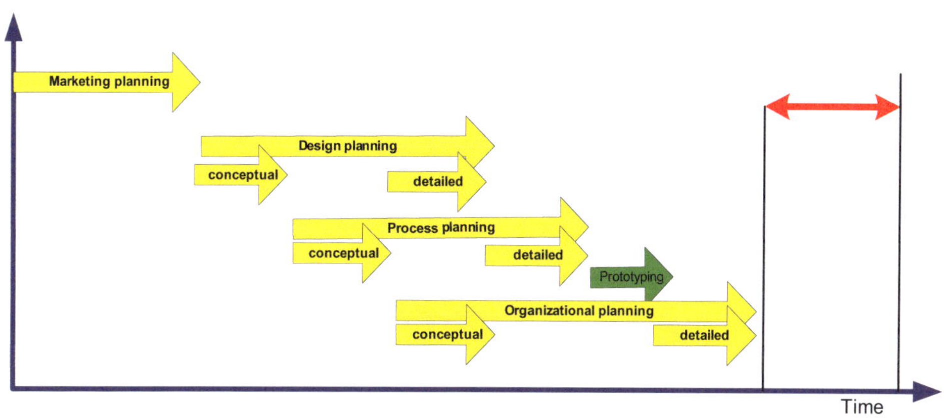

Fig. 6. The separation of the conceptual and detailed design stages

Integration in the area of design and technological product development indicates the extraction of the following phases (Fig. 7):

- conceptual process planning (PHASE III A),
- detailed process planning (PHASE III B).

The basic goal of the conceptual phase is the determination of the variants of manufacturing processes on the base of the design product features selected during the design planning stage. The multivariant nature of the process planning is due to the possibility to use different manufacturing methods and raw materials for the elements constituting the product. The multivariant nature of the assembly processes is due to the possibility to use several assembly sequences with the application of methods and manufacturing means and

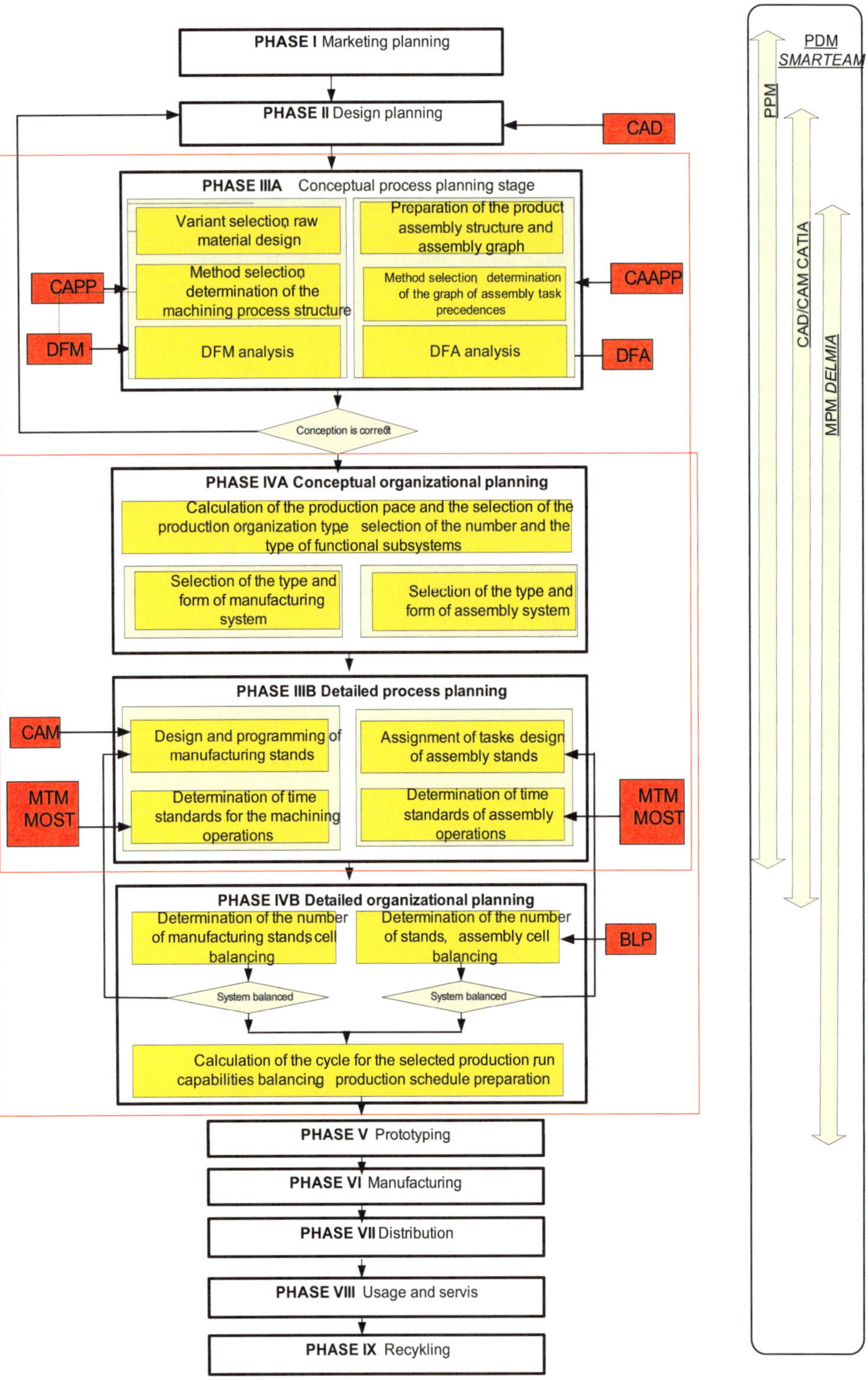

Fig. 7. Parallel execution of product development phases

different automation levels. On the base of the defined variants of manufacturing/assembly processes, the manufacturability analysis DFM and assembleability analysis DFA are carried out. The results of the subsequent iterations are used to simplify the product design (by minimization of the number of parts and the integration of parts), thus decreasing the time and costs of the assembly and to estimate the times and costs for different manufacturing methods of constituent product elements. It should be noted that the DFA and DFM analysis should be applied together to check whether the simplification of the product structure and elimination of some parts does not lead to the higher manufacturing costs. The result of the analysis is the selection of the product design variant and the variant of the manufacturing and assembly plans for constituent product elements with the level of details allowing for the necessary organizational calculations.

Integration in the area of technological and organization product development indicates the extraction of the following phases:

- conceptual organizational planning (PHASE IV A),
- detailed organizational planning (PHASE IV B).

The information from the previously executed product development stages describing the production task (production program of products and their design features, variants of manufacturing plans) are used as the base for the organizational planning. The main goal of the conceptual production organizational planning stage is the development of the production system concept. The planning activities cover:

- for the assumed planning data: required volume of production, allowable duration of production, calculated production pace,
- selection of the production organizational variant by choosing the type and organizational form of the production,
- selection of the number and types of functional subsystems for transport, storage and quality checking.

The result of the above actions is the concept of the production system offering the possibility to execute the detailed manufacturing planning (PHASE IIIB) within the scope of developing the concept of the manufacturing stands, selection and design of the manufacturing assembly stands. The design activities on the detailed process planning stage cover the activities resulting in the preparation of the instruction sheets for the manufacturing/assembly.

The planning of the process plan operation covers:

- selection and design of the manufacturing machines, equipment and tooling,
- selection of the structure of the process plan, generation of set-ups, machining cycles and technological parameters,
- preparation of the control programs for the operations executed on the programmable devices,
- selection of the time-standard elements (Duda, 2010).

The planning of the assembly plan operation covers:

- selection and design of the manufacturing machines, equipment and tooling,
- selection of the structure of assembly operation, and the assembly activities,
- preparation of the control programs for the assembly devices, robots and manipulators,
- estimation of the assembly operation times.

The results are the base for the selection of the organizational variant of the manufacturing stands and the base for the detailed organizational planning stage (PHASE IVB) covering:

- calculation of the production pace, selection of the production type (serial, parallel or serial-parallel),
- synchronization of the isolated functional subsystems of manufacturing cells by calculating the number of stands and line balancing,
- preparation of the production schedule, balancing of the throughput for the designed manufacturing systems.

The presented process of integrated product development is executed iteratively. The subsequent versions of product, process and manufacturing systems are created using digital techniques which allows for the complete analysis of the generated solutions.

7. Modelling of the product development in production preparation phases with the use of the BPMN methodology

The aims of the Concurrent Engineering are achieved through the parallel realization, integration and standardization of development phases of the whole product lifecycle. Parallel execution of the design and process planning phases leads to the reduction of production preparation time. In the area of information integration, the application of PDM system and the development of the workflow diagrams are used. The successive development phases occurring in the product lifecycle can be decomposed into the constituent elements. The BPMN notation is use for the modelling of development processes occurring in the design and process planning phases in order to implement the workflow diagrams fig 8.

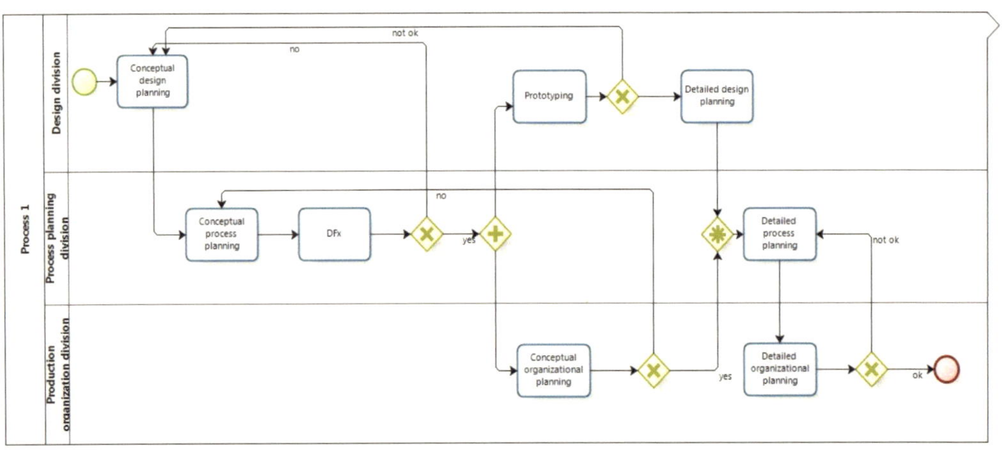

Fig. 8. Model BPMN of product development in production development phases

8. Verification concept in PLM environment

The verification was done with PLM solutions offered by Dassault Systemes. These solutions include the following systems (Fig. 9,10):

- CAD/CAM CATIA for product, manufacturing process and resource design,
- PDM SMARTEAM and ENOVIA for the development process management,
- MPM DELMIA for process and manufacturing system design.

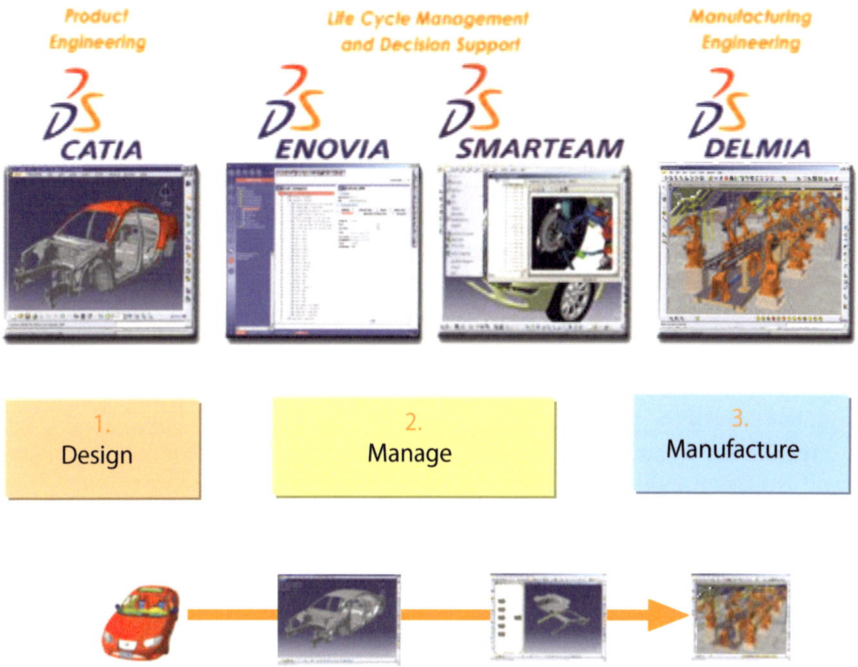

Fig. 9. Areas covered by PLM Delmia solution

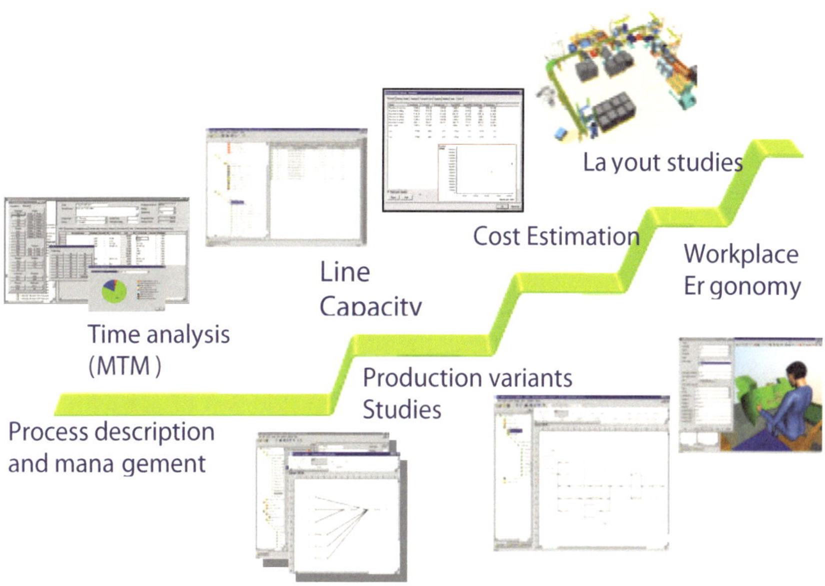

Fig. 10. Process Engineer functions

The projects of gear reducer and car water pump were used to verify the presented concept and CATIA/DELMIA software deployed at Institute of Production Engineering.
The tests included the following:

- creation of the digital product model,

- creation of the assembly documentation,
- creation of the manufacturing concept including the definition of process and assembly system library, and the preparation of the digital models of assembly system,
- creation of the process simulation and the ergonomic analysis of work stands.

8.1 Conceptual process planning phase

The assembly process is created on the basis of the digital product model prepared with CATIA system. The process planning activities includes:

- the development of the product assembly structure - separation of the assembly units (assemblies, subassemblies and parts),
- development of the assembly process plan including the basic parts for separated assembly units, methods and hierarchical order of assembly of these units to receive the design features of the product,
- mounting of subassemblies, assembles and parts prepared with Part Design module in Assembly Design module based on the developed assembly plan.

The results of the above actions are necessary to make the assembleability analysis of the product and for iterative improvement of the design form in view of the assembly requirements. The product design resulting form the subsequent iterations and its assembly plan form the base for defining the graph of the assembly activities, representing the admissible variants of the execution of product assembly.

8.2 Conceptual production organization planning phase

The conceptual manufacturing organization phase is used to select the appropriate form of the production organization, production pace, and the initial calculation of the number and type of functional subsystems. On this phase, also the type and organizational form of assembly system is selected.

8.3 Detailed process planning phase

The selected type, organizational form of the assembly and the graphical product assembly plan verified with DMU Kinematics module of CATIA are the basis for the assembly process planning. The assembly process planning includes the selection of the operations and activities, selection of the assembly equipment, conceptual design of the assembly stations as well as the decisions on the concept and the selection or design of the transport system components. Based on the above process, the digital model of the assembly system and work places is created. The data required for the making the detailed production organization calculation and for the calculation of the duration of assembly activities are gathered from the components of the digital model of the system. The duration times of the assembly activities were determined using the MTM and MOST methods.

8.4 Detailed production organization planning phase

The selected structure of the assembly process and the determined duration times of assembly activities covered by the operations are the base for the synchronization within the isolated assembly subsystem. For analysed production organization forms, for example production lines, these steps include the calculation of the number of stands and line balancing.
The activities for the selected organization forms (assembly stations) include ergonomic analysis.
The outcome of the detailed production organization phase is the digital model of the manufacturing system linking all the components of the assembly system with assembly

process plan stored in library and process schedule (Fig. 11). The further test on simulation model are used to analyze and improve the system being developed.

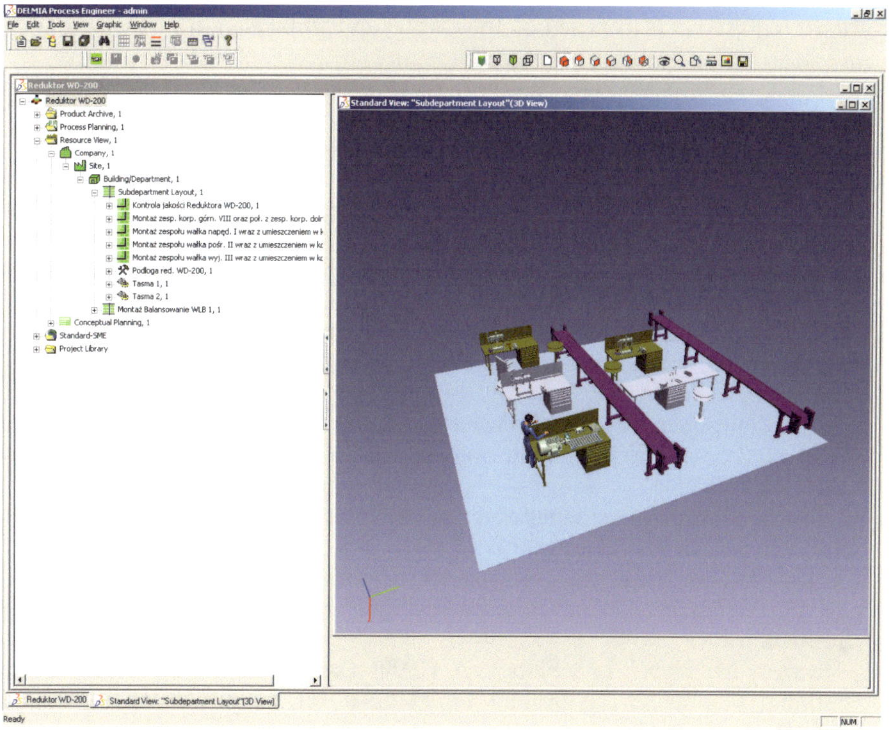

Fig. 11. Assembly station design

9. Conclusion

Design and process planning integration is achieved through:

- Creation of the interdisciplinary teams consisting of: designers, manufacturing engineers and other specialists.
- Use of DFMA (Design for Manufacturing and Assembly) systems.
- Better cooperation with suppliers and clients.
- Integration of data and functions.

The model of the development process is the key element for the creation of the workflow diagrams. These diagrams form the basis for the effective functioning of the PDM systems working according to the selected development strategy. Further works shall be directed toward the preparation of the notation for the modelling of workflow diagrams and the generation of the source code for the module managing the workflow in PDM system. This will accelerate the implementation of PDM (Product Data Management) systems which evaluate toward the PLM solutions (Product Lifecycle Management).

Integration of the design and process planning phases leads to:

- Reduction of information losses.
- Cooperation of different divisions to achieve the clearly defined common goals.
- Reduction of the competition between the company divisions.

Implementation of PLM solutions with the ERP, SCM, CRM systems and easily available means of communication like internet and intranet allow to use CE and CEE for the product development in distributed environment. New methods of product development (CE and CEE) fulfil the requirements imposed by the market and globalisation.

10. References

Chlebus, E. (2000). *Techniki komputerowe CAx w inżynierii produkcji*. Warszawa: PWN.

Chlebus, E., Oczoś, K., Dybała, B., & Boratyński, T. (2000). Nowoczesne technologie w rozwoju wyrobu. Wrocław: Prace Naukowe Instytutu Technologii Maszyn i Automatyzacji Politechniki Wrocławskiej.

Ciszak, O., & Żurek, J. (1999). *Modelowanie oraz symulacja części i zespołów maszyn za pomocą teorii grafów*. Poznań: Wydawnictwo Politechniki Poznańskiej.

Duda, J. (2004). Computer Aided Assambly and process planning in the integrated product development environment. *7th Int. Conference "New Ways in Manufacturing Technology"*. Presov.

Duda, J. (2010). Computer Aided Work Time Standardization in Integrated Development of Product, Process and manufacturing System. *New ways in manufacturing technologies 2010*. Presov.

Duda, J. (2003). *Wspomagane komputerowo generowanie procesu obróbki w technologii mechanicznej*. Kraków: Politechnika Krakowska.

Duda, J., & Karpiuk, M. Integration of systems for design and assembly process planning. *Int Conf. Flexible Automation and Intelligent Manufacturing, FAIM2008*. Skövde, Sweden.

Duda, J., Kwatera, M., Habel, J., & Samek, A. (1999). Data Base with Open Architecture for defining Manufacturing System Capabilities. *Int. Conference FAIM' 99*. Tilburg, Holland.

Eigner, M. (2004). Product Lifecycle Management - The Backbone for Engineering. *Virtual Design and Automation*. Poznań: Wydawnictwo Politechniki Poznańskiej.

Evershaim, W., Rozenfeld, H., Bohtler, W., & Graessler, R. (1995). A Methodology for Integrated Design and Process Planning on a Concurrent Engineering Reference Model. *Annals of the CIRP*.

Inyong Ham, S. C. Lu(1988, 37/2). Computer-Aided Process Planning The Present and the Future. *Annals of the CIRP*, pp. 591-601.

Kahlert, T. (2004). From PDM to PLM - from a workgroup tool to on enterprise -wide strategy. *First International Conference, "Virtual Design and Automotion"*. Poznań: Wydawnictwo Politechniki Poznańskiej.

Kiritis, D. (1995). A Review of Knowledge - Based Expert Systems for Process Planning. Methods and Problems. *Int. J. Advanced Manufacturing Technology, Springer - Verlag, London*, pp. 240-262.

Łebkowski, P. (2000). *Metody komputerowego wspomagania montażu mechanicznego w elastycznych systemach produkcyjnyc*. Kraków: AGH Uczelniane Wydawnictwa Naukowo - Dydaktyczne.

Mari, H., Gunasekaran, A., & Grieve, R. (1995, 14). Computer Aided Process Planning: State of the Art. *Int. J. Advanced Manufacturing Technology*, pp. 261-268.

Pobożniak, J. (2000). Modelowanie przedmiotów w środowisku współbieżnego projektowania procesów technologicznych. *Konf. Automatyzacja produkcji - wiedza technika i postęp*. Wrocław.

20

Printing as an Alternative Manufacturing Process for Printed Circuit Boards

Huw J Lewis and Alan Ryan
University of Limerick
Ireland

1. Introduction

The use of printing techniques as a manufacturing process for physical products is well established in the form of 3D rapid prototyping printing systems. Additive methods of manufacturing of this form are being considered as environmentally friendly as less material is wasted and the number of processes needed for the finished product is reduced.
Printed manufacturing is also being established in the electronics industries as conductive inks and electrical conductive adhesives have been developed (Gomatam & Mittal 2008). This has allowed for the creation of printed circuit boards, on various substrates using screen printing methods and developments towards the use of desktop printing for circuit design (Ryan & Lewis 2007). Coupled with this, a range of printing methods can now be used to produce components such as transistors, diodes, sensors and RFID tags, and batteries (www.plasticelectronics.org 2009).
Mass printing processes fall into approximately 6 main types (Table 1), with the properties of each analysed in Table 2. All of the processes excluding inkjet printing require some form of stencil or impression plate on which the required pattern is formed. Ink is either forced through the stencil in the case of screen printing, or is held within the impression plate for other processes. The ink is deposited onto the substrate surface when the impression plate and substrate come into contact. The means of inking the plates vary depending on the process.
Printing systems offer high resolution and speeds, and are thus suited to Mass manufacture. The resolution of the patterns formed depend on the process and the substrate, with roller based systems being able to repeatedly print lines at 50μm, screen printing to approximately 75μm, and pad printing to 10μm. The speed of printing ranges from 50 m/min to 250m/min using the roller based lithography, rotogravure and flexography methods, with screen, pad and inkjet printing up to 75K units an hour on high speed machines.
Printing processes by their very nature apply thin layers of ink/resin onto a substrate (16 microns) (Raghu et al 2008), which can be built up to form the required shape or pattern. This inherently involves the over printing of existing patterns to develop the required thicknesses. Hence two problems occur, the alignment of the substrate for subsequent printing, and the amount of time required to continually over print to get a defined thickness. For rapid prototype systems, the alignment is held by keeping the component in

one position, hence fixing the datum, and applying a single type of ink/resin which is usually cured by ultra violet light. This tends to be an accurate but slow process utilizing ink jet type technology, and is utilized for physical components.

Alignment of substrates isn't as critical if a single print pass is used. This is the norm for printed products such as newspapers, wall papers, cloth etc. The printing process is a high speed production system, thus a single print is desirable rather than a series of runs building up the ink thickness. To this end the range of product that can be developed using the printing process is limited. However, as has been mentioned this type of process is ideal for the electronics industry.

Screen printing has long been used to print solder paste onto circuit boards, thus it is an obvious step to develop a screen printing process that will also print the interconnecting track of the circuit thus removing an etching process. This Chapter will describe the methods of utilizing a screen printing processes to develop a printed circuit board using a degradable substrate and screen printing process.

Type	Schematic	Type	Schematic
Screen Printing	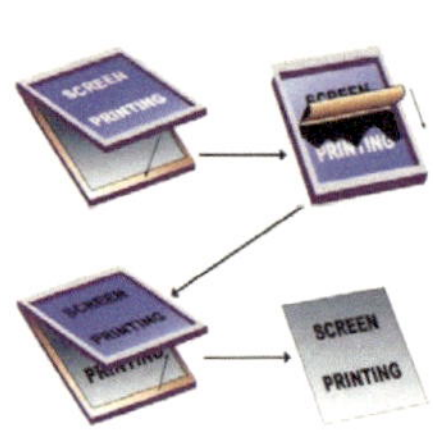	Flexography	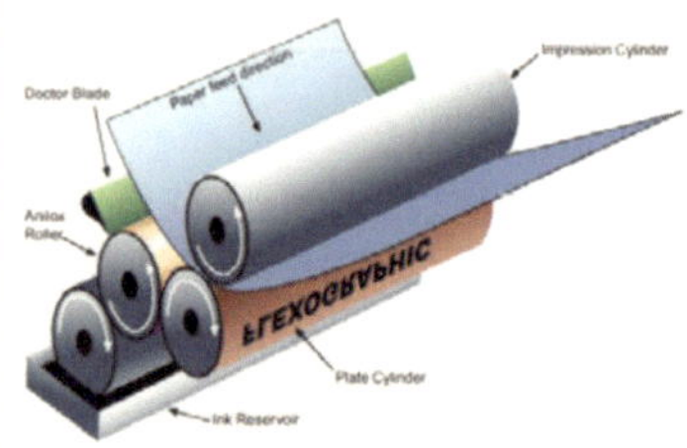
Rotogravure	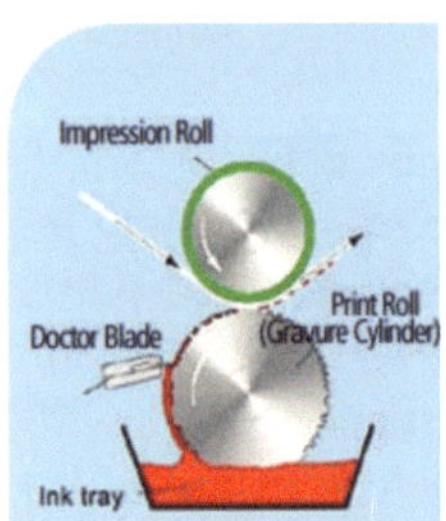	Pad Printing	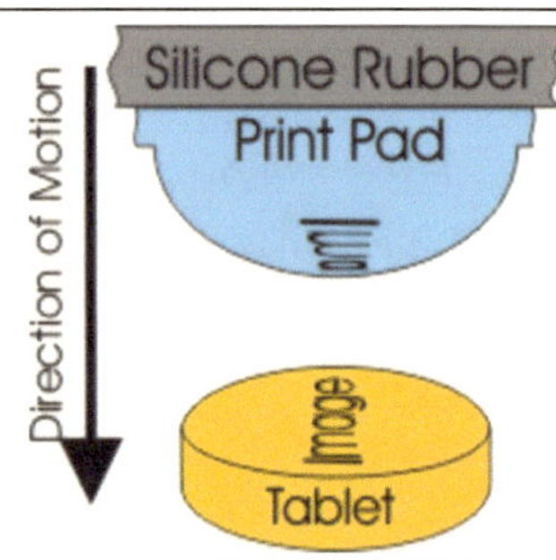
Offset Lithography	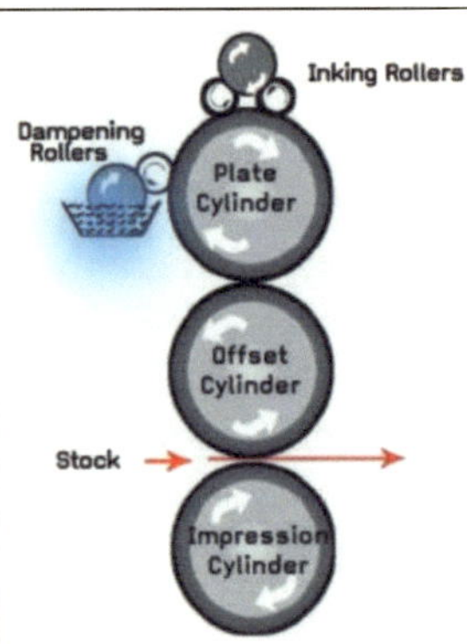	Inkjet	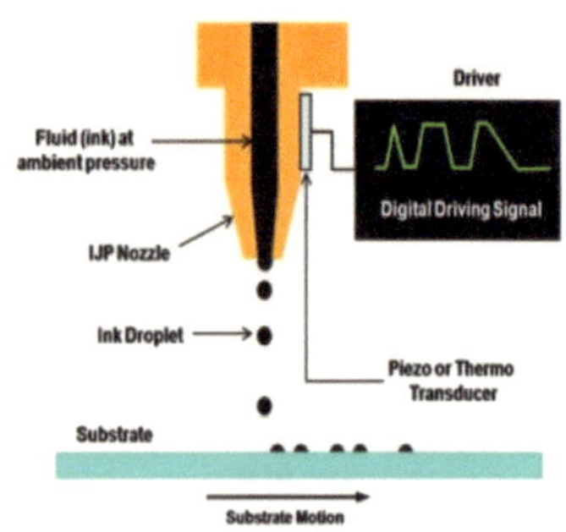

Table 1. Common Printing Methods

2. Printed circuit boards

The first printed circuit was designed and developed by an Austrian engineer Paul Eisler, as part of a radio set in 1936. His invention consisted of rectangular sections of thin copper supported on a dielectric substrate. These were used as a direct replacement for the discrete wiring or point-to-point construction, which was the popular method of developing an electrical circuit at this time.

The printed circuit board consists of 3 basic elements

- The substrate containing the electrical interconnecting tracks
- The components
- The medium (traditionally solder) for attaching the components to the track.

2.1 Substrate

A modern PCB consists of a copper coated epoxy glass laminate substrate (Herbert 1965), which provides both the physical structure for mounting electrical components and the electrical interconnection between these components (DRI-WEFA 2001), (Figure 1). Thus the laminates must have considerable strength.

0= High 5=Low	Screen Printing	Offset Lithography	Flexo-graphy	Roto-gravure	Pad Printing	Inkjet
Speed	2	4	3	5	2	1
Resolution	1	4	3	5	2	2
Film Thickness	5	2	2	2	2	1
Viscosity	2-5	5	2	2	2	1
Substrate Flexibility	5	2	3	2	4	5
Pressure	4	1	3	2	3	4
Ink volume required	4	3	3	1	4	5
Initial Waste	5	1	3	2	4	5
Image Carrier Cost	3	4	2	1	4	5
Hardware Cost	4	3	3	1	4	5

Table 2. Summary of the Parameters of Printing Processes (Adapted from Jewell & Bould 2008))

2.1.1 Production of a rigid substrate for a PCB

Currently a printed circuit Board (Figure 1) consists of a substrate (epoxy resin) populated by various numbers of components which are attached to the interconnecting pattern on the board using solder. Within the PCB family there are typically 3 types (Figure 2): the first being single sided boards where components are attached to one side of the board only. The single sided is commonly used for commercial and military circuitry and offers the lowest cost and simplest processing, (Gilleo 1992). When circuit density demands exhaust the routing limits of a single sided circuit, a double-sided circuit is the next full level of density increase where conductor layers are placed on both sides of the substrate. Circuit patterns are then formed on each conductive layer, (Gilleo 1992). The third family member is a multilayered board, which consists of alternating layers of conducting foil and insulation. In all three structures, the insulation layers are constructed of multiple laminates of epoxy-glass sheets bonded together to form a strong and rigid structure. Multilayered boards are used for complex circuit assemblies in which a large number of components must be interconnected with many track routings, thus requiring more conducting paths than can be accommodated in one or two copper layers. (Groover 2006)

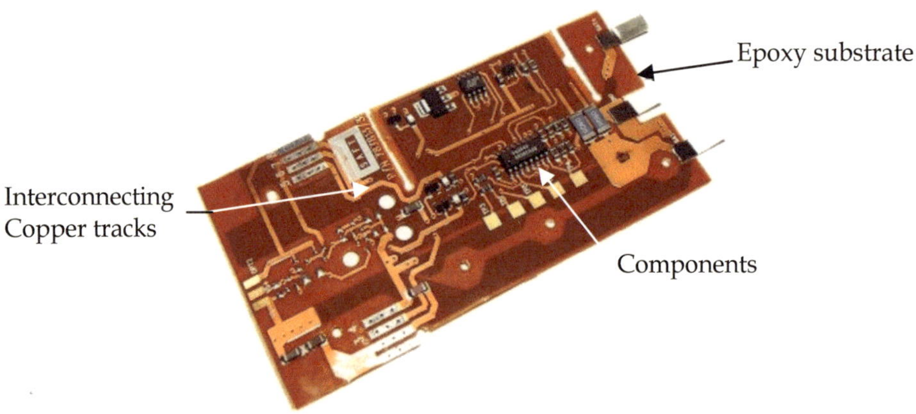

Fig. 1. Populated Printed Circuit Board

Circuit fabrication involves a number of various stages from initial preparation, to hole drilling, through to circuit imaging and etching and finally the mounting of components, this process in its entirety can be termed circuitisation. Three methods of circuitisation can be used to determine which regions of the board will be covered in copper and ultimately result in the completed interconnecting pattern, these are the subtractive method, the additive method and a semi-additive method. In the subtractive method proportions of the copper cladding of the starting board are etched away from the surface, so that the desired interconnecting pattern remains, it is called subtractive, as copper is removed from the board. The additive method uses electrolysis plating techniques to deposit the interconnecting pattern on the unclad starting board, while the semi-additive uses a combination of the two techniques where the starting board has an initial thin deposition of copper on the surface of the board (Groover 2006). The typical method of circuit fabrication is the subtractive method of manufacture where plated copper is etched from the substrate to leave the desired interconnecting pattern. In the subtractive method a photosensitive material is applied to both sides of the laminate. Once this is allowed to dry, the board is exposed to light through a transparent photographic plate, (known as artwork), which carries the desired interconnection pattern, this turns the photosensitive solution opaque where the copper is to remain and is transparent where the copper is to be etched away. The laminate is then exposed to a chemical resistant solution, which clings to the opaque areas of the laminate, thus protecting the copper from the etching solution.

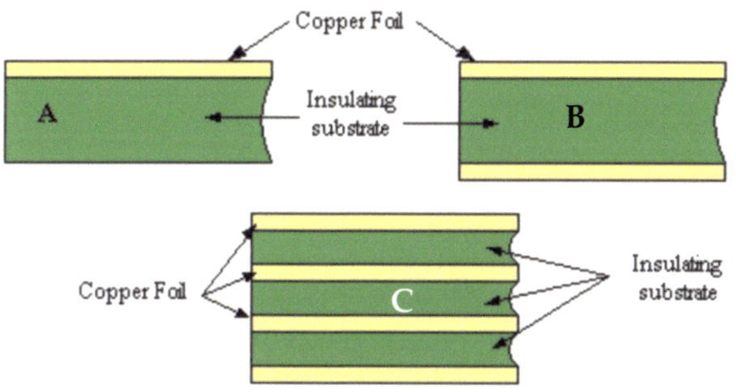

Fig. 2. Three types of PCB structure, (a) Single-sided, (b) Double sided and (c) Multilayer

The laminate is then placed into the etching solution and all exposed copper is removed leaving only the desired conductive tracks, the plating in the holes and the lands around the holes (Pardee & Pennino 1988). Once the interconnecting pattern has been created the next phase is to mount and connect the electrical components and thereby assemble the finished board.

Small light components may be attached to the board on the surface (surface mounted) (Figure 3) while larger components can be mounted using predrilled holes in the substrate, (through hole) (Figure 4).

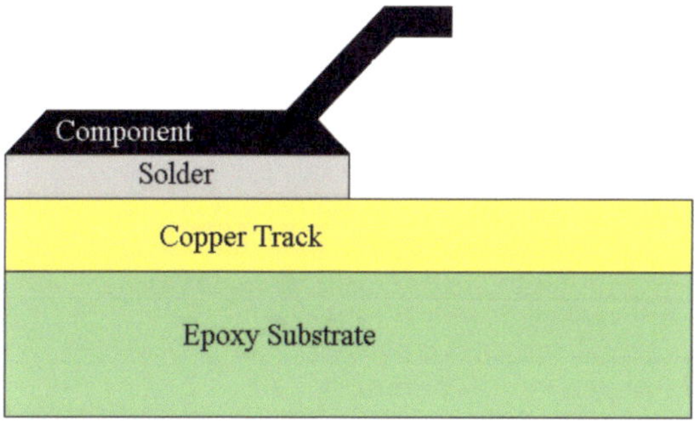

Fig. 3. Surface Mounted Component

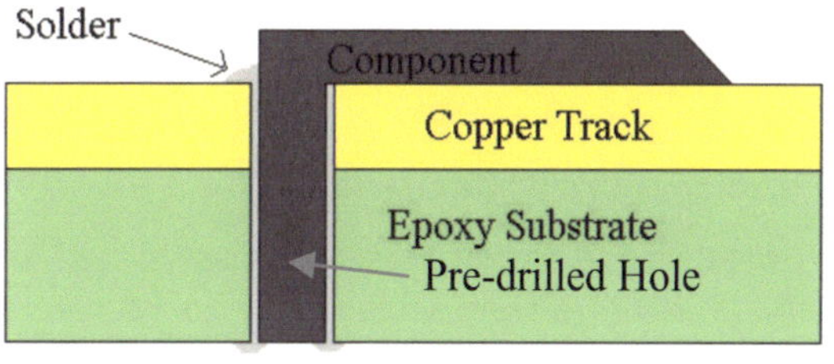

Fig. 4. Through Hole Component

Surface mount technology (SMT) (Figure 3) makes it possible to produce reliable assemblies at reduced weight, volume and cost. The surface mount components are placed on the circuit after a deposition of adhesive or solder paste on to the substrate. Accuracy requirements mandate the use of automated machines for placing surface mount components on the board as manual placement of surface mount components is operator dependant and as a result is unreliable, inaccurate and uneconomical. For prototyping purposes it is possible to manually mount components however for large-scale production of even the simplest circuits it is impractical (Prasad 1997). On attachment of the components the assembled boards are passed through a reflow oven for permanent connection.
From Figure 4 it can be seen that in through hole technology the leads of components are inserted into pre-drilled holes in the PCB. A single board may be populated with hundreds of separate through hole components all of which need to be inserted into the board. This is accomplished, for the most part, through the use of automatic insertion machines, while a small proportion are completed by hand for non-standard components that cannot be accommodated by the machines. The boards are then passed through a wave flow solder process, which connects the components to the tracks.
When examining a current circuit board and the manufacturing processes involved from an Ecodesign perspective a number of potential problems are discovered. During the manufacturing process, there are a number of effluent emissions (Table 3) generated by the subtractive process, these include the unwanted copper etched from the laminate, sulphates and acids included in the etching and protective solutions and polymer strains from the laminate. These potential effluents if exposed to the environment would result in contamination.

Organic solvents	Tin
Vinyl polymers	Lead
Stannic oxide	Palladium
Copper	Gold
Nickel	Cyanides
Iron	Sulphates
Chromium	Acids

Table 3. Effluents from PCB manufacture (World Bank 1998)

Sulphuric	Ammonia
Hydrochloric	Organic solvent vapours
Phosphoric	Isopropanol
Nitric	Acetone
Acetic	Petroleum distillates
Chlorine	Ozone-depleting substances

Table 4. Potential Air Emissions in PCB Manufacture (World Bank 1998)

A second item to be considered is the air emissions associated with the manufacturing process (Table 4). These potential air emissions have to be carefully considered by the designer, when selecting products and manufacturing processes.
As Holt explains (Holt 1994),

> "Electronics designers can have a major impact on how natural resources are used and consumed through the selection of the correct materials and through the establishment of an eco-friendly design plan."

Thus in developing environmentally acceptable PCBs it is necessary to eliminate as many of these potentially damaging areas as possible while still maintaining the integrity of the final product.

2.1.2 Proposed PCB manufacturing process

The proposed alternative method of producing a PCB with environmental considerations involves using a sustainable degradable fibre based (paper) substrate to replace the existing thermoplastic epoxy substrate. This substrate will be used in conjunction with commercially available conductive materials (inks and adhesives) to generate a comparative PCB, which will be tested to current industrial specifications in order to determine if the developed board is a viable alternative.

Table 5 outlines the differences between the proposed environmentally conscious board and the existing subtractively manufactured board. From this table it can be seen that there are essentially 3 changes in the process, the substrate, the interconnecting method and the adhesion technique.

The benefits of the proposed system are that there are no effluents from the manufacturing process and the use of a sustainable degradable substrate, will offer the potential for improving the recycling and recovery of the conductive material, substrate and componenets.

Thus the three elements that can be investigated are the substrate/track, the adhesive medium and the components. It is not in the scope of this project to investigate the individual components, only the possibility of their recovery after the end of useful life of the PCB. Therefore this investigation will examine the use of and processes required to utilise conductive materials, in the forms of conductive inks and electrically conductive adhesives on a degradable substrate, as an alternative to current manufacturing methods, utilising a screen printing process in their manufacture.

The developed board will be compared against mass manufactured PCBs and current IPC (Association of connecting electronics industries 1998) and international standards in order to identify if the proposed alternative is a credible alternative.

3. Conductive inks and adhesives

The method proposed to replace existing PCB manufacturing processes is to print the desired track pattern, using a screen printing process, and conductive ink, and to attach components with electrically conductive adhesives (ECA). Whilst the function of the track and adhesive is different, the makeup of each is similar in that a conductive medium is added to a fluid (binder) to make it viscose and hence printable, the fluid or binder then "sets" to give a track or bond.

3.1 Conductive material formation

In this context "conductive materials" refer to composites of polymer binders and conductive fillers. This makes these materials different from other interconnecting or joining materials, be they copper tracks or metallurgical solder (Gilleor 1999). They consist of three

Existing Manufacturing Process	**Proposed Alternative Process**
Step 1 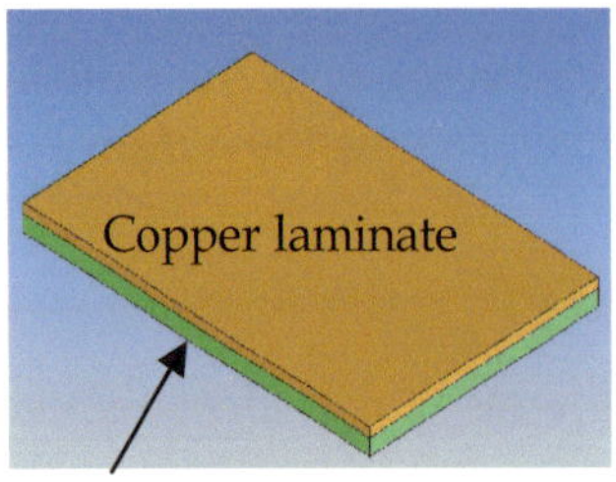 FR 4 Substrate FR 4 Epoxy, thermosetting polymer substrate laminated with copper.	**Step 1**  Fibre based paper substrate. Substrate is sustainable and degradable
Step 2 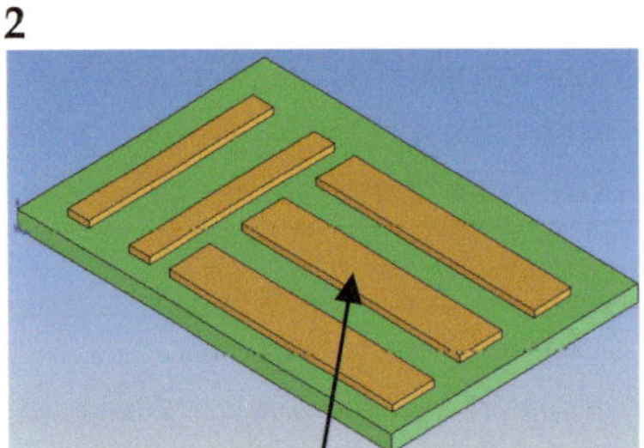 Interconnecting Pattern Unwanted copper etched away to leave interconnecting pattern, subtractive manufacture.	**Step 2**  Electrically conductive medium added to substrate where required to form the interconnecting pattern, additive manufacture.
Step 3 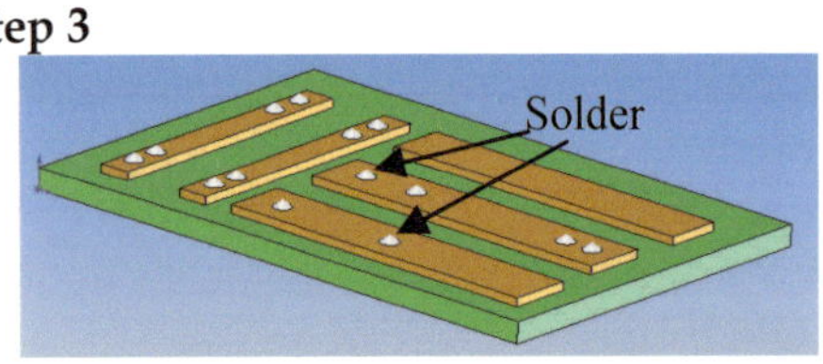 Solder used as the joint medium between components and interconnecting pattern	**Step 3** Electrically Conductive Adhesive  Electrically conductive adhesive used as the joint medium between components and interconnecting pattern
Step 4 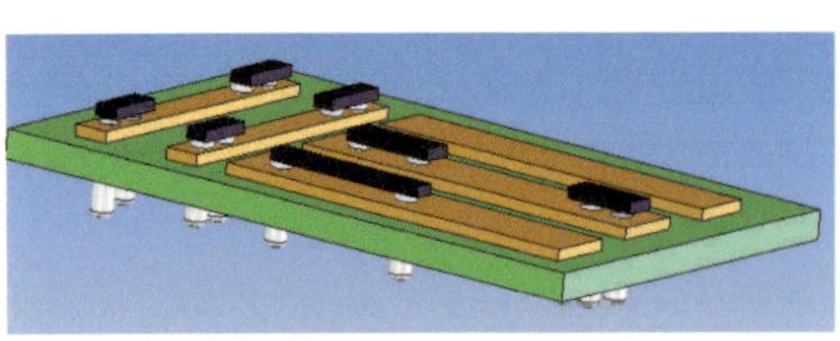 Mass manufacturable populated board	**Step 4** 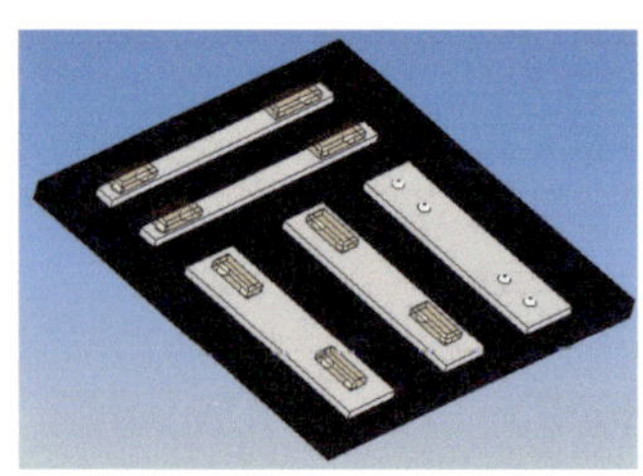 Mass manufacturable populated board

Table 5. Comparison of Manufacturing Methods

ingredients a binder material, a solvent and an inorganic material with the desired conductive, resistive or insulating properties (Kosloff 1980). The inorganic material is the metal or metal oxide that will make the material act as a conductor, resistor or dielectric (typically silver platelets). The binder material is a glass powder with thermal, mechanical and electrical properties tailored to provide adhesion to the substrate. The solvent contains plasticizers and binding solution, which control the rheological properties of the conductive material during the printing process (Lumpp 2000).

3.1.1 Binding material

In conductive ink, the binder is usually a thermoplastic material. Thermoplastics are solid materials that can be dissolved in solvent or heated until they melt and turn liquid. The solvent in an ink system turns the thermoplastic binder into a liquid so that fillers can be added and a viscosity suitable for printing achieved. Once the solvent is evaporated or the thermoplastic is allowed to cool it will turn into a solid once more, however it can then be melted or dissolved again, thus binders in solvent-based inks cannot withstand high temperatures or exposure to some solvents once they are applied and dried (Banfield 2001).
Epoxy is the most commonly used material as an ECA binder with silicone being the most unacceptable. The binder material however is only one constituent of the conductive material with the filler material providing the means of electrical conduction.

3.1.2 Filler material

The binder material is an insulator, thus adding a filler material with the desired electrical properties converts this insulating material into a conductor. All commercial conductive adhesives and inks are made with a non-conductive binder that is loaded with fillers that have the desired electrical characteristics. The most common filler materials are silver, aluminium, gold and copper. Each has different properties that determine how suitable the metal is for a particular application. These determining properties include electrical conductivity, aging mechanisms and cost (Wong 2000).
Silver is the most commonly used conductive filler for conductive materials with 80% silver loading being the limit in epoxy adhesives (Gleditch et al 2000). An alternative to silver is carbon, either graphite or carbon black, both of which are electrically conductive and are used in making electronic materials. However carbon-based materials are only used in special applications because of their poor conductivity; up to 3 orders of magnitude lower than for silver. According to Gilleo silver would seem at first a poor choice because of the cost, however, silver is unique among affordable metals due to the high conductivity of its oxide. This means that there is almost no change in conductivity as silver particles oxidise. Copper, which would appear to be the logical choice, produces oxides that become non-conductive after exposure to heat and humidity. Attempts to use copper in inks and adhesives have been under way for many decades. Some of the earliest printed circuit processes used copper powder with adhesive binder. The challenge for copper-based conductive materials is that of inhibiting oxidation under heat and humidity conditions. Copper oxidises so quickly that oxide will form unless chemical inhibitors are present (Gilleo 1999), but non-oxidising metals, such as gold, are cost prohibitive. Indeed, better conduction is achieved with silver than with gold because of the next important filler material attribute, particle formation.
Silver particles are easy to form and to fabricate therefore silver can be manipulated into a wide range of controllable sizes and shapes. The Danish EPA states that, the stability of the

electrical characteristics of conductive materials is dependent on the type and geometry of the metal filler (Danish EPA 2001). There are three different filler geometries, spheres, flakes and needles. The filler material should be of optimum geometry that provides minimum critical filler concentration for low resistance, the best contact between neighbouring metallic particles and the strongest adhesion to the polymer.

Conductive materials result in a higher resistance as compared to pure metals as even though an electron will flow easily through a metal particle, it encounters tremendous resistance when it has to jump from one metal particle to another. Even if the metal particles are pressed together tightly, the resistance to cross the gap between the particles is high compared to the resistance within the metal particles themselves. Given the fact that electrons must travel across thousands of the particle-to-particle gaps in a metal-filled ink or adhesive, it is easy to see why the resistance is greater than the resistance of a solid metal. (Banfield 2001)

Earlier versions of silver inks and adhesives used metal spheres as the conductive filler. When these spheres touch one another, there is only one small point of contact between the particles (Figure 5A). Filler flakes (Figure 5B), have more contact points and tend to be more conductive than spheres, however, Shimada & Wong claim that mixed filler shows the highest conductivity and offers the best alternative (Shimada & Wong 2000). When the flakes nest together, there is much greater surface contact between particles and more paths available for electrons to move from particle to particle.

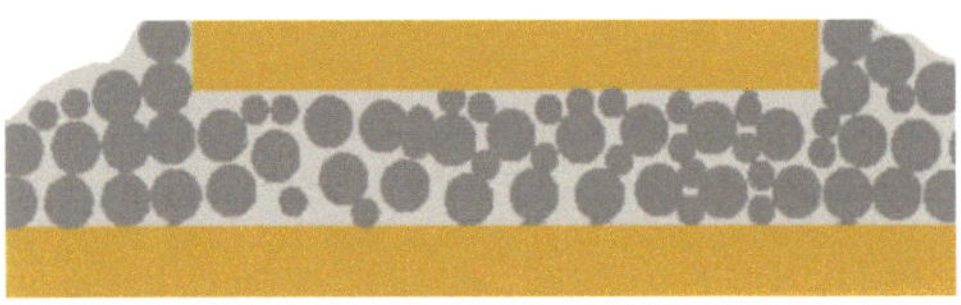

Fig. 5A. Spherical formations

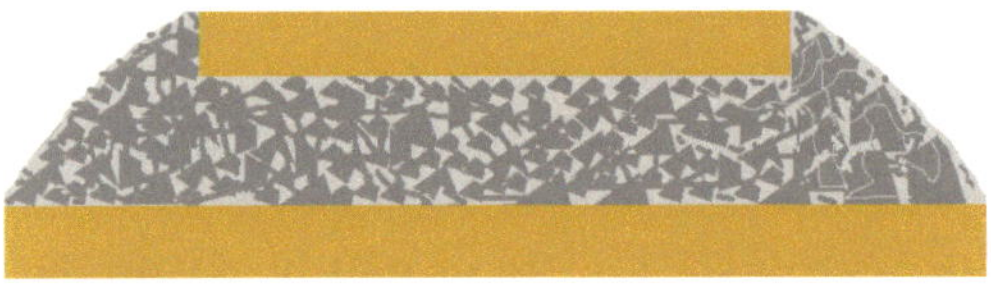

Fig. 5B. Flake formations

In general it is best to view conductive materials as a complex composite where filler, binder and additives can be selected to provide formulations having a wide range of useful properties. When examining conductive materials made from conductive particles suspended in a binding solution it is important to note that there are two distinct types, isotropic conductive materials and anisotropic conductive materials, where each distinct type has an alternative method of conducting the electrical signal through the material.

3.1 Conduction within electrical conductive materials

3.1.1 Isotropic conductive materials

These are conductive materials that conduct in all directions, isotropic conductive adhesives are currently in use in the electronic industry, primarily as die-attach adhesives (Wong 2000), whilst isotropic conductive inks are used to form the conductive layers in RFID tags, (Cuming 2006). They can be classified as non-filled or filled materials depending on how the polymer is made conductive.

3.2.1.1 Non filled conductive materials

Non-filled materials are polymers that are either inherently electrically conductive or doped and are known as intrinsically conductive polymers (ICP). Doping is the process of adding a small percentage of foreign atoms to a regular crystal lattice for the development of new properties. They are mainly polymerised compounds that derive their electrical properties from their molecular structure (Wong 2000). The most common polymers are based on polyacetylene, polyaniline and polypyrole. Gilleo (Gilleo 1999) claims that a few of these materials have been commercialised for end application such as battery electrodes and that doped polyacetylene has been found to achieve a conductivity of nearly 70% that of copper. However, he also explains that conductive polymers are extremely brittle and sensitive to oxidation and thus are not suitable to act as a replacement to solder as an adhesive technique, thus there will not be perused as part of this investigation.

3.2.1.2 Filled isotropic conductive materials

These conductive materials consist of approximately 60-80% metallic properties in a polymer/epoxy binding solution. When the material is cured the particles are distributed and a conductive polymer network is formed. By this network electrons flow across particle contact points by direct metallic contact making the mixture electrically conductive, (Figure 6). Due to the nature of the network formed, i.e. no fixed structure, the current is able to flow through the material in all directions (Wong 2000). Applications of the isotropic adhesives include the attachment of dies to lead-frames and the attachment of surface mount components to flexible circuits and ceramics (Chang 1993).

Fig. 6. SEM Micrographs illustrating contact between silver flakes

The second type of conductive material is the anisotropic conductive material.

3.2.2 Anisotropic conductive materials

Anisotropic materials provide unidirectional electrical conductivity in the vertical direction, This directional conductivity is achieved by using relatively low volume loading of the conductive filler well below the percolation threshold (Basteki 1999).

Anisotropic material, in particular adhesives, have been used for a number of years in attaching chips to package lead frames. The advantage offered by this method of connection are high density interconnects, a low temperature process and low cost. More recently, ACAs have been used to connect flip chips to other substrate materials, with varying degrees of success (Li et al 1993). The main driver towards the use of ACA materials in the area of SMT is the prospect of achieving extremely fine pitch connections at a low cost. However a disadvantage is that pressure and heat must be supplied simultaneously while the polymer matrix is hardened, otherwise the conductive pathway is lost.

With the two families of conductive material outlined it is necessary to explain how these families are formed.

Plastic polymer binders have a high resistance, if a small amount of highly conductive metal filler (silver) is added to the binder, the resistance will not change, since the metal particles will have large gaps between them. As filler is added to the binder, the resistance will remain high until a point is reached where the metal particles start to contact one another (Banfield 2001). Thus what is termed "the percolation threshold" or "point" has been reached. This is when a sufficient amount of filler has been loaded into the polymer to transform the composite from an insulator to a conductor (Morris 1999). Thus percolation theory predicts a critical filler concentration at which a three dimensional network is established and conductivity increases suddenly by several orders. Thereafter, conductivity changes slowly with increase in filer concentration.

This is illustrated in Figure 7 .

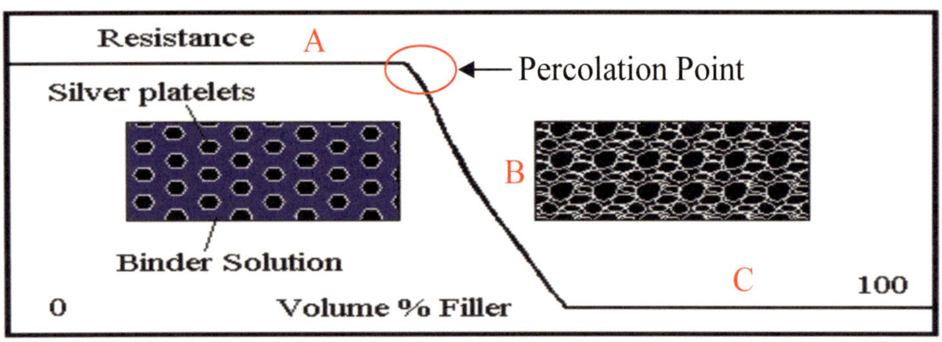

Fig. 7. Illustration of percolation theory (Gleditch etal 2000)

After the percolation point, adding a small amount of metal filler will produce a rapid drop in the resistivity of the conductive material. From Figure 7, Point A represents an area with large volume of binding polymer in relation to the quantity of conductive metal flakes. The resistance at Point A is quite high as there is a low percentage of filler material thus preventing inter-particle contact and resulting in an anisotropic conductive material being created. However as the loading of conductive material increases the resistance falls thus creating greater inter-particle contact and forming an isotropic conductive material, Point B (Banfield 2001).If metal filler continues to be added, a point is reached (Point C) where the resistivity levels off, and increased loading of metal filler will not improve the conductivity of the material. In fact, higher loadings of filler beyond this point will start to degrade the adhesion and scuff resistance properties of the material and can actually start to increase the resistance of the material as the conductive particles are not able to nest together efficiently (Banfield 2001). It is thus feasible that the conductive inks/adhesives could be used to develop the conductive track and component bonding method for a PCB, once printed on a suitable substrate.

3.3 Substrate

The substrate chosen is an important aspect of the total circuit board assembly, since it must hold the components, withstand vibration and shock and other processing in assembling (Kosloff 1980). There are a number of sub-factors of the substrate, which can adversely affect the final print quality these include:

- Substrate cleanliness
- Geometry size
- Surface roughness

Substrate Cleanliness: The cleanliness of the substrate is essential to an accurate print. Dust and dirt on the substrate can cause defects such as bridging ultimately leading to a short circuit in the PCB (Bentzen 2000). It is important therefore that the substrate is kept free from contamination and to ensure that the manufacturing environment is clean.

Geometry Size: It is important that the substrate is large enough to receive the interconnecting design but not so big as would make uneconomic sense. This factor is relatively insignificant as the printing parameters for a large substrate will be the same as for a small substrate (providing the same substrate materials are used in both cases).

Substrate Roughness: The surface roughness of the substrate is an important variable within the printing process. It is possible to establish the roughness of the substrate by moving a fine-tipped stylus across the surface in question. The deviations in the substrate surface create interference patterns that are used to calculate the roughness (Sergent 2001).

Once the roughness has been calculated it is possible to generate a profile of the surface roughness, (Figure 8). Here for a given length (L) along the surface of the substrate the variations in roughness are clearly visible. Points a and b represent peaks of surface roughness while points c and d represent valleys.

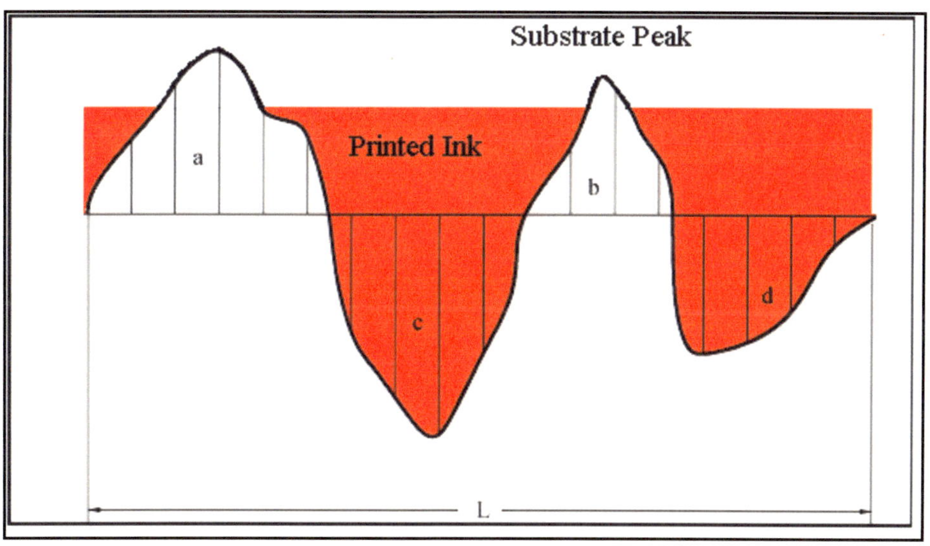

Fig. 8. Example of potential problem of substrate roughness

According to Sergent surface roughness

> "Has a significant impact on the adhesion and performance of thick film and thin film depositions. For adhesion purposes it is desirable to have a reasonably high surface roughness to increase the effective interface area between the film and the substrate. However for repeatability in the print integrity the thickness of the deposited film should be much greater than the variations in the surface" (Sergent 2001).

From the illustration of surface roughness it is apparent that if the substrate used is too rough, peaks in the substrate will interrupt the printed track. Thus selection of a substrate, which is of sufficient roughness, is critical to the final print integrity.

However as Waldvogel stated (Waldvogel 1977),

> "The substrate cannot be too smooth; a certain roughness is required for good adhesion of the conductive screened layers."

Therefore a medium is required where the substrate is sufficiently rough to achieve good ink adhesion to the substrate and not so rough as will result in large discrepancies in the ink thickness printed. In this case a standard paper card was utilised. Having chosen the substrate the method of applying the ink/adhesive needs to be developed. In this case screen printing was used, as it is a current standard in the electronic PCB manufacturing process.

4. Screen printing

Traditional screen-printing can be traced as far back as the 900's, where in China and Japan, artistic patterns were printed using screens made from human hair. Since this era the process has naturally progressed culminating in the system currently used in industry which was designed by Samuel Simon in 1907 (Bellis 2005).

Traditionally the screen printing process was used to print coloured inks onto large posters for advertising, however, in the mid 1960's with the development of conductive inks the electronics industry was quick to adopt and adapt this technology. Initially screen-printing technology was used to form resistors and dielectrics. However, since its introduction, screen-printing has been the dominant method of film deposition within the electronics industry because of the low cost associated with the process (Pan et al 1999).

The use of screen-printing is commonly associated with use in thick-film circuit production. A thick-film circuit is produced by

> "applying inks, pastes or coatings on substrates or bases and firing the printed electronic patterns to change their properties" (Kosloff 1980).

Screen-printing as a printing process is a basic, simple and efficient method of reproducing patterns on a variety of substrates. The process consists of the printing medium i.e. conductive ink being forced through open areas of a mesh-reinforced screen.(Figure 9)

These include:

- The screen chosen
- The squeegee chosen
- The substrate to be used
- The printing process parameters
- The conductive medium

Screen printing plates or printing screens are an essential part of the printing process as these carry the pattern to be printed. A printing screen consists of a frame or support on which a screen, of uniform mesh apertures, is stretched taut while the design or image to be printed is left open in the fabric with the rest of the screen filled in or blocked out (Kosloff 1980). (Figure 10).

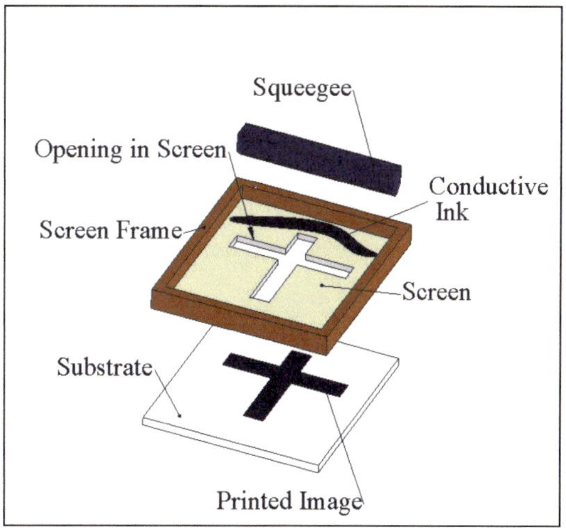

Fig. 9. Screen Printing Diagram Screen

A number of sub factors within the screen, which can influence the final print quality. These sub factors include:

- Screen Mesh Count
- Mesh and emulsion thickness
- Mesh Material.

Mesh Count: The mesh count is the number of openings or apertures in the screen per linear inch. The calculation of the mesh count allows the screen printer to determine the open area percentage of the screen, the greater the open area percentage, the greater the volume of ink printed. However care needs to be taken with this factor as too great an open area can result in print flooding and therefore destroying the final print resolution (Pan et al 1999).

Fig. 10. Screen Printing Screens

Mesh and Emulsion thickness: The mesh thickness is derived from the thickness of the wire strands within the screen. The greater the diameter of the wire, the thicker and more rigid the screen becomes. The thickness of a metal screen is typically 150 microns, but 100, 125 and 200 microns are widely available. For very fine pitch such as 0.3 mm a 100 or 125-micron screen could be used and for a pitch down to 0.5 mm 150 micron screens can be used. The screen thickness together with the aperture sizes also determines the amount of ink, which will be needed to perform the task (Pan et al 1999).
As the average width or pitch of track to be printed for a PCB is 0.3mm a 125-micron screen is suitable. However Robertson et al (Robertson etal 1999) has shown that modern screen printing techniques can achieve dimensions down to 50 µm track widths and spaces.

4.0.1 Screen vs. Stencil

Either a screen or a stencil can be used to apply the conductive material to the substrate. In general the frames of the screen and stencils are similar; the differences lie in the construction of the individual openings used for depositing the printing medium.
A screen will contain open wire mesh around which the conductive material must flow to reach the substrate surface, a stencil aperture is fully etched and therefore does not obstruct the flow of the conductive material (Prasad 1997).
In this case a steel wire mesh screen was used to apply the initial interconnecting layer of conductive ink while a steel stencil was used to apply the conductive adhesive, which is used to secure the surface, mount components onto the substrate.
When printing conductive adhesives the choice of stencil and squeegee is pertinent to a good adhesive joint, researchers recommend metal stencils with apertures 80% of the pad size (Ryszard &Andrzej 2000). The thickness of the stencil should be smaller for adhesives than for solder as a thinner joint is favourable, however it is recommended that conductive adhesives should have a thicker printed deposition than solder as they are less electrically conductive (Terstagge 2003). The viscosity of the adhesives is typically lower than that of the solder paste, (between 55,000 cps and 70,000 cps for the adhesive compared to 500,000 cps for the solder paste) thus less is required as the adhesive will slump easily and cause bridging (Lui & Lundstorm1999). A typical stencil thickness is 0.10mm to 0.15mm however it is application dependent.
Having examined the role of the screen and stencil in the screen-printing process and outlined the various screen parameters which should be considered before selecting a screen, the following section involved investigating the parameters influencing the choice of squeegee in the printing process.

4.1 Squeegee

Screen-printing squeegees come in varying different designs and are made from different materials.

4.1.1 Squeegee hardness and material

Squeegees are made from either polyurethane or metal, the selection of the squeegee is typically dependant on the material of the screen, which is employed in the printing process.

When using an emulsion screen the squeegee needs to push the conductive medium in front of it and to pump it trough the mesh in the open print areas. This is achieved by using a soft polyurethane squeegee (70-75 shore), which allows it to deform (breakaway) slightly where it contacts the screen. (McPhail,1996).
With a metal screen the squeegee rolls the conductive medium in front of it allowing the conductive medium to flow freely through the apertures of the screen without being pumped. It is neither necessary nor desirable for the squeegee to breakaway so a harder squeegee (80-90) shore or a metal squeegee is used (McPhail,1996).
In general metal squeegees should be used for metal screens as soft polyurethane squeegees used on stainless steel screens will wear out quickly and can result in scooping of material from large apertures. Metal squeegees must have a very smooth and non-sticking surface and at all times a sharp printing edge. This ensures that the ink will roll easily on top of the screen and help prevent clogging of the apertures.
A compromise of squeegee hardness and pressure must be achieved. If there is too little pressure, the conductive medium will escape under the squeegee and smear across the screen. If there is too much pressure or the squeegee is too soft then the squeegee will dip into larger apertures' in the screen and scoop the conductive medium out of them.

4.1.2 Size and shape of squeegee

When examining the squeegee used in screen-printing it important to first note that there are two distinct types of squeegees, a diamond type and a trailing edge type,.
Diamond Pattern: This pattern consists of a square section of polyurethane approximately 10mm by 10mm, which is clamped in a holder at an angle of 45° on both directions. This type of squeegee, allows the printer to print in two directions. However, as the squeegee is rigid and clamped into position it does not facilitate uneven substrates. If the substrate is warped in any way the use of this squeegee will result in incomplete prints (McPhail,1996).
Trailing edge pattern: This type of squeegee consists of a rectangular section of polyurethane supported in a holder. Two are needed, one for each direction of stroke, there is no need to hop over the printing medium as the second squeegee will drag it across once the initial cycle is reversed (McPhail,1996). The size of the squeegee is determined both by the size of the pattern to be printed and the clamp, which fastens the squeegee to the printing machine. Thus the size and shape of the squeegee are pre-set due to the equipment being employed.
Squeegee Angle: The squeegee angle must be between 45 and 60 degrees and the rolling ink should have a diameter of 15 to 20 millimetres for optimum conditions (Kosloff 1980).

4.2 Printing process

The objective when printing the conductive ink onto the PCB substrate is to establish a conductive track along which electrical signals will travel. To reach this objective the ink print must be aligned correctly, the correct amount of ink for each track must be present and the print should form an even layer of ink along the substrate.
The parameters of the equipment to print the circuit have been defined , it is now necessary to investigate the printing process The printing medium, or ink, is applied to the upper surface of the screen and a squeegee is dragged across the pattern area. The movement of the squeegee presses the screen into contact with the substrate surface and forces the ink

through the open meshes of the screen. Behind the squeegee, the screen because of its tension peels away from the substrate leaving a printed pattern of ink on the substrate, Figure 11.

To ensure the integrity of the final print is not compromised the printing process and its parameters need to be monitored and controlled. Control of the print coating is obtained by controlling the variables detailed below:

- Squeegee down stop.
- Squeegee pressure.
- Squeegee attack angle.
- Print speed.
- Print direction.
- Snap-off distance.
- Quantity of Ink before the squeegee.

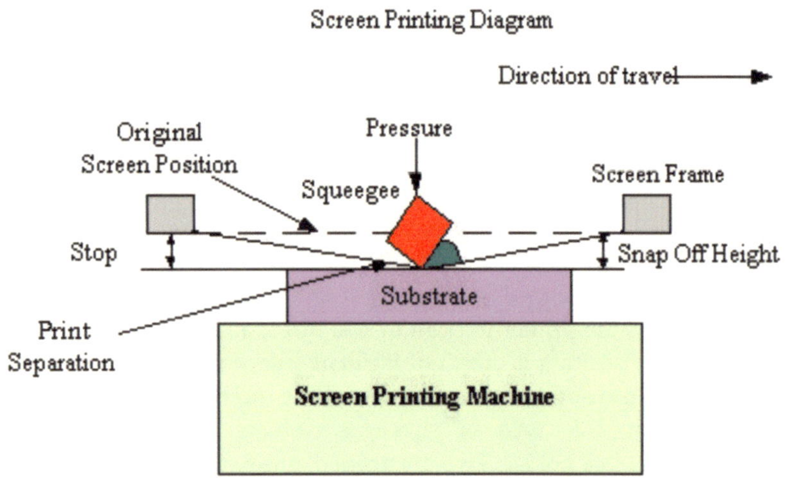

Fig. 11. Screen Printing Process

Squeegee Down Stop: This must be adjusted so as the squeegee only to just touches the screen surface. However, if the squeegee axis and the screen are not parallel it can be necessary to over-adjust the down stop to compensate. If the down stop is adjusted too far down, both screens and squeegees will wear out rapidly (Bentzen 2000).

Squeegee Pressure: The squeegee pressure should be as little as necessary to scrape the screen clean of ink when printing. If adjusted correctly, a thin layer of ink will remain on top of the screen. The printing speed and the screen type determine the amount of pressure required or indeed desired (Bentzen 2000). However if a thicker print is required an increase in the squeegee pressure will facilitate this (Parikh et al 1991).

Printing speed: The supplier of the conductive ink normally provides guidelines on printing speed. The possible printing speed depends on the ink's thixotrophic behaviour. The ink must be fluid when printed but jelly-like and stable once printed onto the PCB substrate. The more fluid the ink is when moved and rolled the higher the print speed can be used. The printing speed must be set so that the ink rolls perfectly on top of the screen. It is a major factor in the printing cycle time, the highest possible speed without compromising the print quality should be chosen (Bentzen 2000). By increasing the printing speed the likelihood of the final ink layer being thinner and patchy increases, therefore it is essential that a printing speed compatible with the chosen ink is employed (Parikh et al 1991).

Snap Off: This is the distance between the screen underside and the top of the substrate. From Figure the snap-off distance, i.e. the distance between the top of the substrate and the bottom of the screen at the start of the run, and the screen deformation can be clearly seen. The snap off distance is constrained by the tension of the screen. If the snap-off distance is too small, ink remains in the screen mesh due to insufficient tension and this results in a thinner print height. Conversely if the distance is too great, excessive squeegee pressure is required to deflect the screen resulting in a thicker pattern but a damaged screen and a worn squeegee (Parikh et al 1991).

Separation Printing: This is the term given to the separation of the screen from the substrate after printing. Figure 11 gives a graphical illustration of this term. It can be seen that during printing the force of the squeegee causes the screen to come into contact with the substrate. After printing, the screen returns to it original position giving rise to the term print separation. The speed of separation between screen and PCB after printing is important. Too rapid a separation speed when printing fine pitch will result in clogging of the screen apertures. Too fast a separation will also result in tailing and form high edges around the ink deposits. The ideal separation speed depends on the ink and the screen aperture wall smoothness. On the other hand a slow separation speed will slow down the printing process When printing electrically conductive adhesives contact printing is preferred to non-contact printing as less adhesive is printed. Contact printing involves the stencil being lowered onto the substrate before printing so that there is no snap-off or separation during printing. The above section has examined the screen-printing process and has illustrated the various process parameters and the numerous of distinct options available within each of the parameters. The following section outlines the process parameters chosen for screen-printing electrically conductive ink on to a degradable substrate. In developing a PCB using a card substrate and conductive inks/adhesives, the parameters indicated above were optimised through a number of empirical tests. This resulted in a number of PCB being developed and tested to an industrial standard.

5. Investigating industrial compatibility

The board is a single sided board and is sufficient to examine the structural, electrical and mechanical properties of the conductive ink, the conductive adhesive, and the interaction between these two and the substrate itself, when exposed to extreme environmental conditions. The test vehicle components are detailed in Table 6 and the component positions are illustrated in Figure13/14. To investigate the industrial compatibility of the proposed additive method of manufacture the initial step is to replicate the etched copper-interconnecting pattern by screen-printing the conductive ink onto the degradable substrate. In developing an additive system for PCB manufacture it must be compatible with existing processes i.e. screen-printing machines, surface placement technology, conveyor convection ovens.

This was investigated using a dedicated production line in a printed circuit board-manufacturing environment. The proposed process, Figure 12, adds an extra step to existing processes, where solder paste is printed to secure components. The test vehicle chosen was a circuit board currently in production.

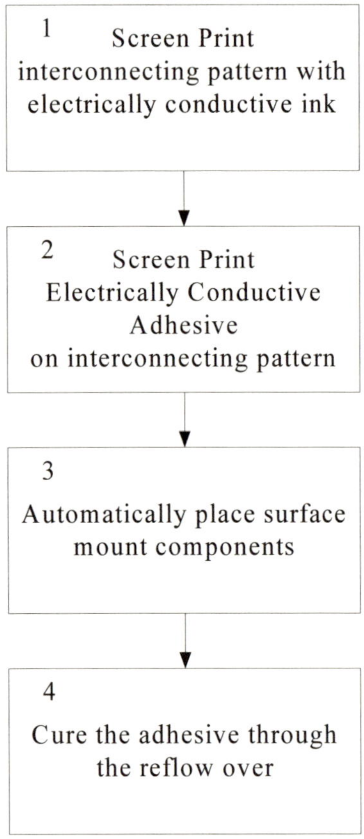

Fig. 12. The Manufacturing Process

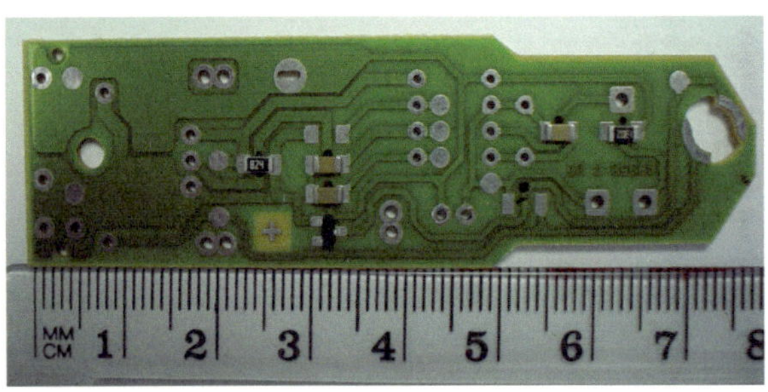

Fig. 13. Test Vehicle, single sided PCB

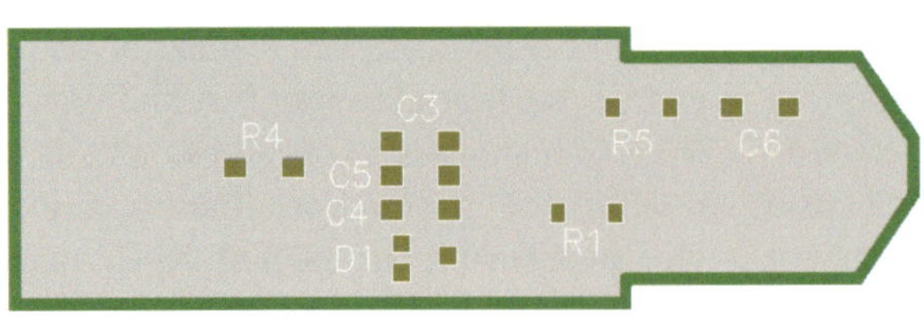

Fig. 14. Component Positioning

Part Identifier	Description
R4	820K ± 5%
C3	47 nF 50V/63V
C5	10 nF 50V/63V
C4	100 nF 50V/63V
D1	BAS 21
R1	150 Ω ± 5%
C6	10 nF 50V/63V
R5	13 kΩ ± 1%

Table 6. Test Vehicle Components

The substrate selected was a rigid paperboard material, with a surface roughness of 1.61 μm, the conductive ink was used Coates conductive ink 26-8204, (Coates data sheet 2000) and ECA: Ablebond 8175.both being off the shelf products. This resulted in 100% accuracy and completion of the printed circuit pattern, Figure 15.

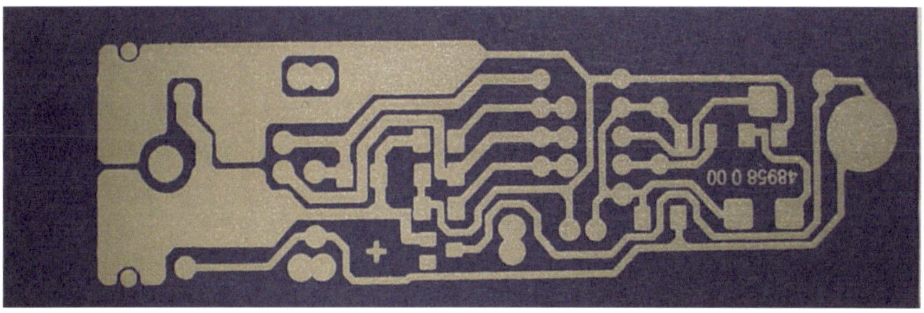

Fig. 15. Successful printing of Conductive ink

To secure the surface mount components a layer of electrically conductive adhesive, Ablebond 8175, was printed onto the interconnecting pattern using the same industrial screen printer, as used for the tracks (Figure 16).The adhesive was printed using a stainless steel stencil 150 μm thick with desired laser cut apertures as suggested by the manufacturers. It was found, through experimentation that a printing speed of 89mm per second, a squeegee pressure of 0.97 bar and a downstop of 1.9mm allowed for successful printing of the adhesive. To ensure that the conductive adhesive and components are deposited correctly, vision systems were employed. These systems use feducials (known reference points) on the board to position the board and stencil when printing the ECA and to ensure the board is correctly positioned during automatic component placement Figure 17. The shape of the feducial is not critical, two differently shaped feducials were used when printing the adhesive and mounting the Component, Figure 18. Once the feducials are located, the board is correctly aligned and printing or placement can begin. It has been shown that the degradable substrate is compatible with the screen-printing equipment, surface placement and conveyor systems encountered in manufacturing environments.

However as is common with the development of new technologies and/or manufacturing processes there are potential problem areas, which can influence the quality of the assembled boards.

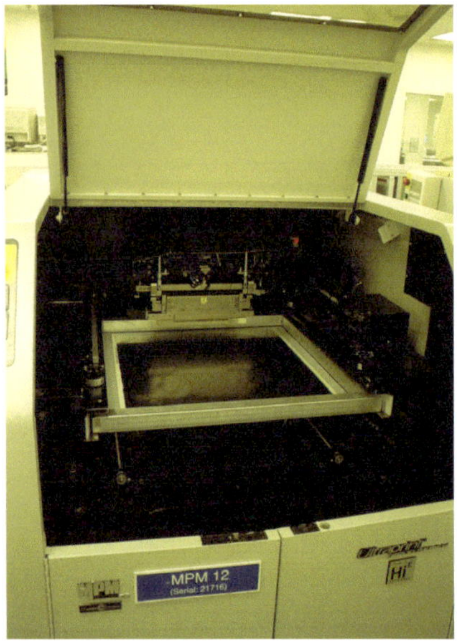

Fig. 16. MPM Screen

Fig. 17. Automatic Component Machine

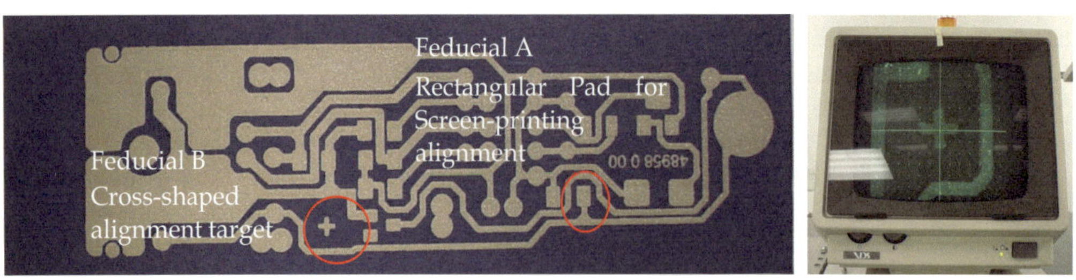

Fig. 18. Feducial used for Vision Alignment

5. Analysis of industrial trials

5.0.1 Importance of defined feducials

For large scale production of even the simplest circuits, automated processes are needed. To accommodate this it is essential that the feducial is sharp, clearly defined and provides a good contrast with the colour of the board material. Figure 20 and Figure 21 show the differences between a successfully printed and a poorly printed feducial. A number of different colour substrates were employed to determine if there was a difference between the number of accurate prints. Experimental investigations established that a black paperboard background enabled the vision system to clearly identify the feducial, as the contrast between silver and black was greater than with other colours. While the contrast of the feducial is important the definition of the feducial is also critical. When the feducial is printed incorrectly i.e. with smudged edges or bridging other tracks it is impossible for the vision alignment system to locate the fed and therefore to position the board correctly, (Figure 22). The solution to the problem of hazy feducials is uncomplicated; the substrate should be clamped into position and should not be bowed or flexed in anyway. The screen should be cleaned after 5 prints, after this time it was found that the apertures of the screen become blocked and also the conductive ink had migrated across the bottom of the surface of the screen.

Fig. 19. Successful positioning of Components

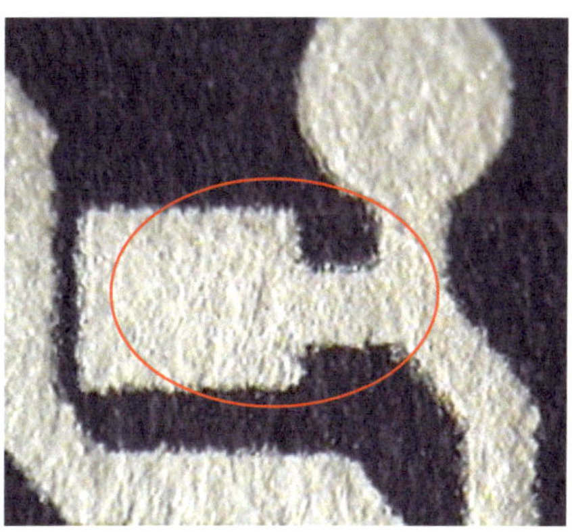

Fig. 20. Image of Defined Feducial

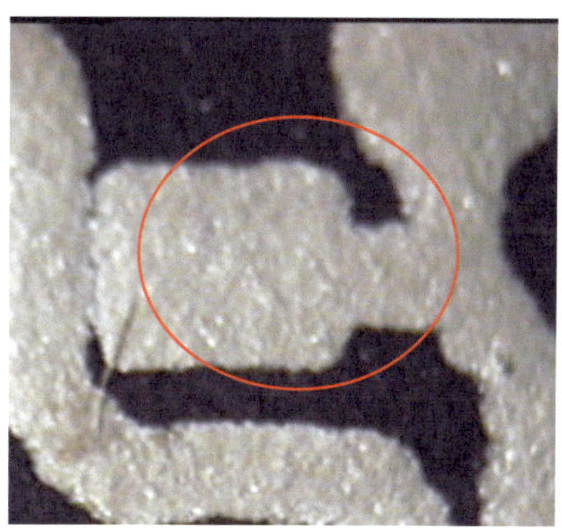

Fig. 21. Image of Poor Quality Feducial

5.0.2 The effect of poor ECA distribution

If the ECA is deposited in the incorrect position when the component is be placed there is no ECA in position to secure that component. If excess ECA is deposited it can spread out across the substrate resulting in bridging of tracks thus damaging the circuit (Figure 22). Excess adhesive can be deposited if the adhesive is at an incorrect viscosity or temperature as it will not flow smoothly across the stencil and will result in the adhesive being dragged through the apertures of the stencil and along the substrate. From Figure 22 it can be seen that the component does not form a satisfactory bond with the adhesive however when the ECA is deposited correctly, (Figure 23) the component sits correctly on the substrate forming an electrical and mechanical bond with the board.

Fig. 22. ECA Spreadage after Printing

Fig. 23. ECA Printed Correctly

Fig 24. Surface Mount Laboratory

It has been shown that it is possible to mass manufacture a degradable printed circuit board using conductive inks/adhesives and a card substrate on current manufacturing processes and industrial equipment, (Figure 24).

It is now necessary to test these alternative boards against industrial standards and boards, to establish the alternative board's viability as a replacement for existing PCBs.

5.1 Testing boards to existing standards

The previous selection has shown that the alternative manufacturing process offers the ability to mass manufacture printed circuit boards using existing industrial equipment and processes. In order to determine the acceptability of the proposed alternative manufacturing process it is now necessary to examine these alternative circuit assemblies under existing international and industrial standards, temperature cycling (MIL-STD-833F 2004), temperature storage (MIL-STD-833F 2004) and humidity ageing (Wong et al 1999) The standard tests and requirements are outlined in Table 7, together with the test results on the developed boards

6. Discussion & conclusion

It has been shown that it is possible to utilise a degradable substrate and conductive inks/ adhesives to produce a PCB that is comparable to existing PCB. In doing so the system reduces the environmental impact of the manufacturing process. This is achieved by changing the manufacturing method from being subtractive, i.e. producing waste material, to an additive method, i.e. utilising only the material that is required. Further to this the pollution threat is removed as there isn't the requirement to produce the waste by etching, hence saving cost. The printing screen printing process is well established in the electronics industry, and by using a degradable substrate, recovery of components and even tracks is enhanced (Ryan 2006), leading to further environmental benefits.

Parameter	**Test**	**Pass/Fail**	
Electrical Requirements			
Volume Resistivity	$<1X10^{-3}\mu\Omega cm$	Conductive Ink	Pass
		Conductive Adhesive	Pass
Shift of joint resistance	<±20% change after humidity aging	Conductive Ink	Pass
		Conductive Adhesive	Fail
Mechanical requirements			
Impact Strength	Required to sustain 6 drops from 1.524m	Prior to humidity aging	Pass
		Post humidity aging	Pass
Shear Strength	>3KgF	Prior to humidity aging	Fail
		Post humidity aging	Fail
Structural Requirements			
Joint structure	Section 12 of IPC-A-610C	Result of examination of Joint	Pass
Functionality			
Functionality performance	Direct comparison to normal board as per company standard	5 test points	Point 1 84% pass Point 2 88% pass rate Points3-5 100% pass rate
Flammability			
Flammability Examination	BS 61189-2	Pass subject to coating with a flame retardant	

Table 7. Requirements of Proposed Circuit Assemblies

7. References

Banfield, D.(2001), Understanding and Measuring Electrical Resistivity in Conductive Inks and Adhesives. www.*SGIA.org*

Bastecki, C. (1999) A benchmark process for the Lead-free assembly of Mixed Technology PCBs. A report to Alphametals Corp.,.

Bellis, M., (2005) *The History of Printing and Printing Processes.*.

Bentzen, B.,(2000) *Solder Paste Printing.*, SMT in Focus. p. 1-5.

Chang, D.D.,(1993) An Overview and Evaluation of Anistropically Conductive Adhesive Films for Fine Pitch electronic Assembly. *IEEE Transactions on Components, Hybrids and Manufacturing Technology*, . 16(8): p. 828-835 Coates Ltd(2000) Product Data Sheet. Massachusetts US

Components, (2000) I.f.I.a.P.E., *IPC-A-610C.*: Association of connecting electronics Industries. (www.Ipc.org)

Cuming, E. (2006)., *RFID INTERCONNECT ADHESIVES.* 2006. p. 1-4

Defense, D.o.,(2004) *MIL-STD-883 F, Method 1010.8 Temperature Cycling.* p. 3

Defense, D.o.,(2004) *MIL-STD-883 F, Method 1008.2 Stabilisation Bake..* p. 1

DRI-WEFA,(2001) *Printed Circuit Boards Industry Yearbook 2001-2002.*, DRI-WEFA

EPA, D., *Environmental and Technical Characteristics of Conductive Adhesives versus Soldering.* 2001: p. 13-15

Gilleo, K.,(1992) *Handbook of Flexible Circuits.*, New York: Van Nostrand Reinhold

Gilleo, K.,(1999) Introduction to Conductive Adhesive Joining Technology, in *Conductive Adhesives for Electronic Packaging*, L. Johan, Editor. Chapter 1 Electrochemical Pubs Ltd.

T. Gleditch, H. Hvims, R. Rorgren, and J. Vahakangas, (2002)*The Nordic Electronics Packaging Guideline – Chapter C: Conductuve Polymers, Level 2: Introduction, [Online].* Available: http://extra.ivf.se/ngl/CPolymerBonding/ChapterC.htm [May 25, 02].

Gomatam,R & Mittal,K,L (2008) *Electrically Conductive Adhesives, VSP,* Leiden,

Groover, M., (2006) Fundamentals of Modern Manufacturing. 3rd ed. 2006: *Wiley*

Herbert, R.J.,(1965) *Impact of Microminiaturisation on Printed Circuit Boards.* Available from http://ieeexplore.ieee.org

Holt, HR.,(1994), *A First Step in Electronic Ecodesign.* Proceedings of the IEEE International Symposium on Electronics and the Environment, San Fransisco CA: p.191-195

Jewell E and Bould D(2008), *Comparing the print process for micro manufacture,* Program notes, *An introduction to print for micro manufacture,* Swansea University 2008. www.ukdispay.net

Kosloff, A., (1980) *Screen Printing Electronic Circuits.* 2nd ed. 1980, Cincinnati: The Signs of the Times Publishing Company. *Cincinnatti*

Li, L., Lizzul, C.; Kim, H.; Sacolick, I.; Morris, J.E (1993)*Electrical, Structural and Processing Properties of Electrically Conductive Adhesives.* IEEE Transactions on Components, Hybrids and Manufacturing Technology,. 16(8): p. 843-850

Liu, J. and P. Lundstrom,(1999) Manufacturability, Reliability and Failure Mechanisms in conductive Adhesive Joining for Flip-Chip and Surface Mount Applications, in *Conductive Adhesives for Electronic Packaging*, L. Johan, Editor. Electronchemical Pubs Ltd. p. Chapter 9.

Lu, D., Q.K. Tong, C.P. Wong (1999), Mechanisms Underlying the Unstable Contact Resistance of Conductive Adhesives. *IEEE Transactions on Electronic Packaging Manufacturing,*. 22(3): p. 228-232

Lumpp, J.,(2000) Hybrid Assemblies, in *The Electronic Packaging Handbook,* G.R. Blackwell, Editor. CRC Press LLC.

McPhail,(1996) *Screen Printing is a Science not an Art.* Soldering and Surface Mount Technology, . 8(2): p. 25-28

Morris, J.E.,(1999) Conduction Mechanisms and Microstructure development in Isotropic, Electrically Conductive Adhesives, in *Conductive Adhesives for Electronic Packaging,* J. Johan, Editor., Electrochemical Pubs Ltd

Pan, J., G. Tonkay, and A. Quintero, (1999) Screen Printing process design of Experiments for fine line printing of Thick Film Ceramic Substrates. *Journal of Electronics Manufacturing,* 1999. 9(3): p. 203-213.

Pardee, S. and T.P. Pennino, (1988) Software tools speed circuit board design. *IEEE Spectrum,*. 25(9): p. 40-44

Parikh, M., W. Quilty, and K. Gardiner,(1991) *SPC and Setup Analysis for Screen Printed Thick Films. IEEE Transactions on Components, Hybrids and Manufacturing Technology,*. 14(3): p. 496-498.

Prasad, R.P., (1997) Surface Mount Technology Principles and Practice. *2nd ed. 1997,*: Chapman & Hall. New York

Raghu D, Harrop P, &. Holland G, (2008) *Encyclopaedia of Printed Electronics* IDTechtEx.com,

Robertson, C., Shipton R.D, &Gray. D.R,(1999) Miniature Sensors using high Density Screen Printing. *Sensor Review*. 19(1): p. 33-36.

Ryan A & Lewis H, (2007) *Manufacturing an environmentally friendly PCB using existing industrial processes and equipment,* J. Robotics & Computer Integrated Manufacture. Vol. 23, pp 720-726.

Ryan, A, (2006) An Alternative Manufacturing System for Printed Circuit Boards, *PHD dissertation University of Limerick*

Ryszard, K. & Andrzej (2000)M. Electrically Conductive Adhesives in SMT- the Influence of Adhesive Composition on Mechanical and Electrical Properties. in *IMAPS*. Poland

Sergent, J. (2001), Advanced Ceramics and Composites, in *Book of Ceramics and Composites,* H. Harper.Mcgraw hill Newyork

Shimada, Y., L. D., & C.P. Wong.(2000) Electrical Characterisations and Considerations of electrically conductive adhesives. in *International Symposium on Advanced Packaging Materials.*

SMTinfo,*Electronic Production Part 5: Conductive Adhesives, Terstagge, D, (Online).* Available: http://www.smtinfo.net/docs/Electronic%20Production/f.htm.

Waldvogel, C. (1977)Computer designed multilayer hybrid substrate using thick film technology. in *Annual ACM IEEE Design Automation Conference Proceedings of the 14th conference on Design automation.* New Jersey: IEEE Press

Wickham, M., *Electrically Conductive Adhesives Workshop*. 2005, National Physics Laboratory C.P. Wong and D. Lu, (2000) Recent Advances on Electrically Conductive Adhesives for Electronic Applications (Piscataway, NJ: IEEE,), pp. 121–129.

World Bank Group (1998) Electronics Manufacturing, in *Pollution Prevention and Abatement Handbook.*, *World Bank Group*. p. 302-306 (available at www.ifc.org)
www.plastic-electronics.org (accessed January, 2009)

21

Recent Development of Fast Numerical Solver for Elliptic Problem

Mohammad Khatim Hasan[1], Jumat Sulaiman[2], Samsul Ariffin Abdul Karim[3], and Mohamed Othman[4]
[1]*School of Information Technology, Universiti Kebangsaan Malaysia*
[2]*School of Science and Technology, Universiti Malaysia Sabah*
[3]*Department of Fundamental and Applied Science, Universiti Teknologi Petronas*
[4]*Faculty of Computer Science and Information Technology, Universiti Putra Malaysia*
Malaysia

1. Introduction

Most elliptic solvers developed by researchers need long processing time to be solved. This is due to the complexity of the methods. The objective of this paper is to present new finite difference and finite element methods to overcome the problem.

Solving scientific problems mathematically always involved partial differential equations. Two recommended common numerical methods are mesh-free solutions (Belytschko et al, 1996; Zhu 1999; Yagawa & Furukawa, 2000) and mesh-based solutions. The mesh-based solutions can be further classified as finite difference method, finite element method, boundary element method, and finite volume method. These methods have been widely used to construct approximation equations for scientific problems.

The developments of numerical algorithms have been actively done by researchers. Evans and Biggins (1982) have proposed an iterative four points Explicit Group (EG) for solving elliptic problem. This method employed blocking strategy to the coefficient matrix of the linear system of equations. By implementing this strategy, four approximate equations are evaluated simultaneously. This scenario speed up the computation time of solving the problem compared to using point based algorithms.

At the same time, Evans and Abdullah (1982) utilized the same concepts to solve parabolic problem. Four years later, the concept has been further extended to develop two, nine, sixteen and twenty five points EG (Yousif & Evans, 1986a). These EG schemes have been compared to one and two lines methods. As the results of comparison, the EG solve the problem efficiently compared to the lines methods.

Utilizing higher order finite difference approximation, a method called Higher Order Difference Group Explicit (HODGE) was developed (Yousif & Evans, 1986b). This method have higher accuracy than the EG method. Abdullah (1991) modified the EG method by using rotated approximation scheme. The rotated scheme is actually rotate the ordinary computational molecule by 45° to the left. By rearranging the new computational molecule on the solution domain, only half of the total nodes are solved iteratively. The other half can be solved directly using the ordinary computational molecule. This method was named

Explicit Decoupled Group (EDG). He use this new method to solve the two dimensional Poisson problem and was proven to be faster solver than the EG method by 50%. The performance of the EDG method was further tested by Ibrahim. He implements the method to solve boundary value problem (Ibrahim, 1993) and two dimensional diffusion problem (Ibrahim & Abdullah, 1995). The EDG method was then extended to six and nine points (Yousif & Evans, 1995).

This fast Poisson solver have been challenged by a method called four points Modified Explicit Group (MEG) method (Othman & Abdullah, 2000). The concept utilised in MEG method was created from modification of concept used in EDG method. In the MEG method, only a quarter of node points are solved iteratively, and the remaining points are solved directly using the standard and rotated algorithm (Othman & Abdullah, 2000). MEG method has successfully saving about 50% of EDG computational time and 75% of EG computational time. An additional advantage is the MEG method also has higher accuracy compared to EDG method.

In this chapter, we will demonstrate newly develop finite difference and finite element method based on the concept mentioned above for the solution of elliptic problem.

2. Finite difference method with red black ordering

We developed two practical finite difference techniques utilising the concept proposed by Abdullah (1991) and Othman & Abdullah (2000) for solving elliptic problem.

Consider the 2D Hemholtz equation as follows.

$$\frac{\partial^2 U}{\partial x^2}+\frac{\partial^2 U}{\partial y^2}-\wp U=f(x,y),(x,y)\in[a,b]\times[a,b], \tag{1}$$

Subject to Dirichlet boundary conditions

$$\begin{aligned}U(x,a)&=g_1(x),\, a\le x\le b,\\ U(x,b)&=g_2(x),\, a\le x\le b,\\ U(a,y)&=g_3(y),\, a\le x\le b,\\ U(b,y)&=g_4(y),\, a\le x\le b.\end{aligned}$$

In this article, we will only consider uniform nodes. Utilising the concept from MEG (Othman & Abdullah, 2000), we develop a method called Quarter Sweep Successive Over-Relaxation using red black ordering strategy(QSSOR-RB). Utilising the concept in EDG (Abdullah, 1991) and the red black ordering strategy, we develop a method called Half Sweep Successive Over Relaxation (HSSOR-RB). Employing the Successive Over-Relaxation (SOR) method using the same concept in MEG, QSSOR-RB method only solve quarter node point iteratively and utilising the concept in EDG, HSSOR method only solve half of the node points iteratively by SOR method. Beside that the nodes are arranged in a Red-Black ordering manner (Figure 3).

There are so many approached can be used to approximate problem (1). For instance, Rosser (1975) and Gupta et al. (1997) have proposed low and high order schemes. Both schemes can be rewritten in the forms of systems of linear equations. However, both system of equations will have distinct properties of coefficient matrix from each other.

Based on second order schemes, the full and quarter sweeps approximation equations can be generally stated as

$$U_{i-p,j}+U_{i+p,j}+U_{i,j-p}+U_{i,j+p}-(4+\wp(ph)^2)U_{i,j}=(ph)^2 f_{i,j}. \qquad (2)$$

Value of p=1 represent the full sweep scheme and is used to solve all black nodes in discrete solution domain given in Figure 1a iteratively. While p=2 represent the quarter sweep schemes and is used to solve all black nodes in discrete solution domain given in Figure 1b iteratively.

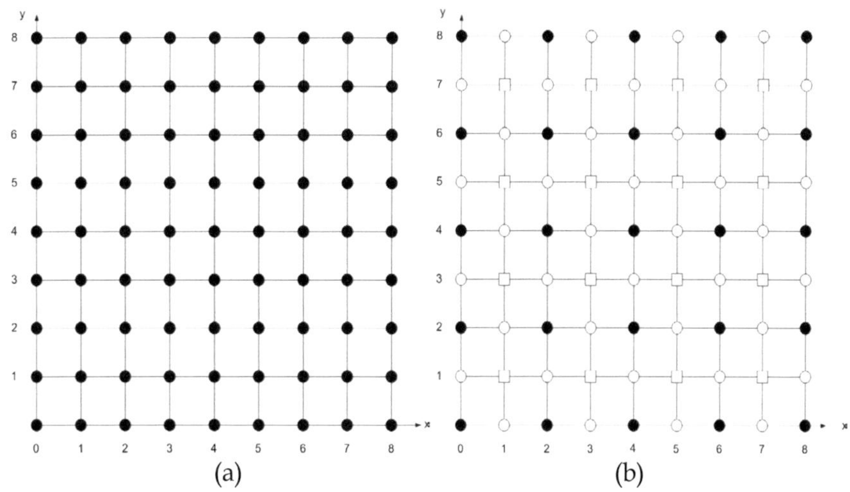

Fig. 1. Discrete solution domain for (a) full sweep and (b) quarter sweep schemes

By employing the concept used in EDG (Abdullah, 1991), the five points rotated finite difference approximation equation can be formed. The transformation processes are as follows.

$$\begin{aligned}&i,j\pm 1\rightarrow i\pm 1,j\pm 1,\\&i\pm 1,j\rightarrow i\pm 1,j\mp 1,\\&\Delta x,\Delta y\rightarrow \sqrt{(\Delta x)^2+(\Delta y)^2}=\sqrt{2}h,\Delta x=\Delta y=h\end{aligned}$$

Applying this transformation, approximation equation (2) can be rewritten as

$$U_{i-1,j-1}+U_{i+1,j-1}+U_{i-1,j+1}+U_{i+1,j+1}-(4+2\wp h^2)U_{i,j}=2h^2 f_{i,j}. \qquad (3)$$

The approximation equation (3) is applied on solution domain displayed in Figure 2 to solve all black nodes iteratively. The white box nodes in Figure 1b are solved via

$$U_{i-1,j-1}+U_{i+1,j-1}+U_{i-1,j+1}+U_{i+1,j+1}-(4+2\wp h^2)U_{i,j}=2h^2 f_{i,j}.$$

directly and the white bullet nodes in Figures 1b and 2 by

$$U_{i-1,j}+U_{i+1,j}+U_{i,j-1}+U_{i,j+1}-(4+\wp(ph)^2)U_{i,j}=(ph)^2 f_{i,j},$$

directly.

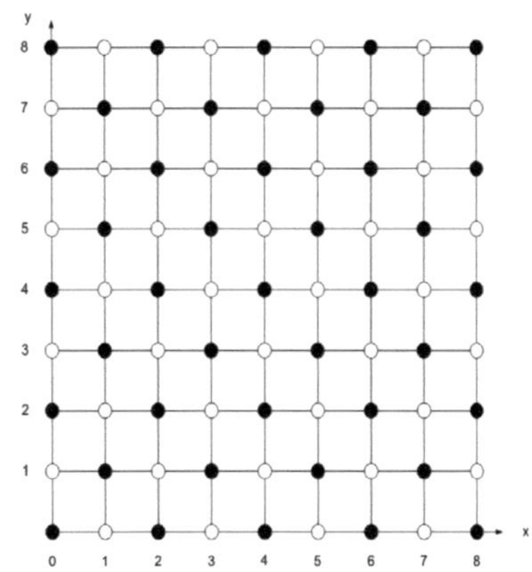

Fig. 2. Discrete solution domain for half sweep scheme

Red-Black ordering strategies have been shown to accelerate the convergence of many numerical algorithm (Parter, 1998; Evans & Yousif, 1990; Zhang, 1996). Hence, we apply this ordering strategy to further increase the speed of our computation. The Implemention of the Red-Black ordering strategy are shown in Figure 3.

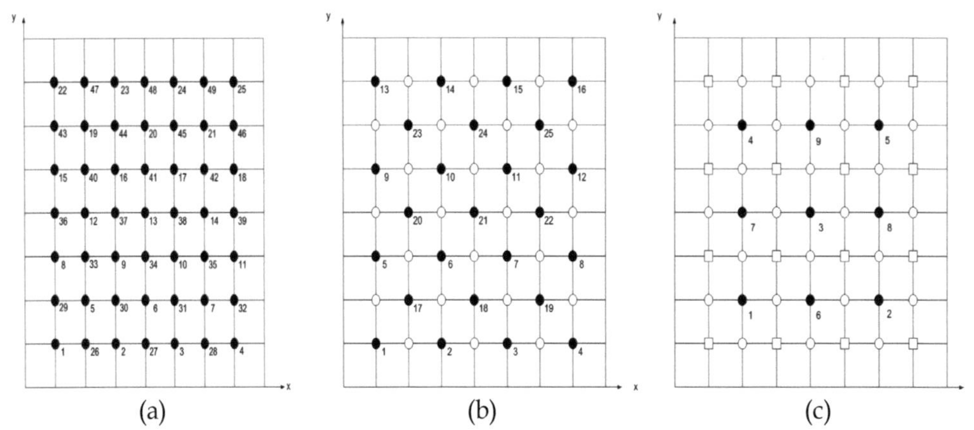

Fig. 3. Implementation of the RB ordering strategy for (a) full-sweep, (b) half-sweep, and (c)quarter-sweep cases.

The efficiency of the FSSOR-RB, HSSOR-RB and QSSOR-RB are analysed via the following two dimensional Helmholtz equation.

$$\frac{\partial^2 U}{\partial x^2}+\frac{\partial^2 U}{\partial y^2}-\rho U=6-\alpha(2x^2+y^2),(x,y)\in[0,1]\times[0,1],$$

with the boundaries and exact solution are defined by

$$U(x,y)=2x^2+y^2,0\le x,y\le 1.$$

The convergence criteria considered in these experiments is $\varepsilon=10^{-10}$. All results of numerical experiments are displayed in Figures 4 to 8.

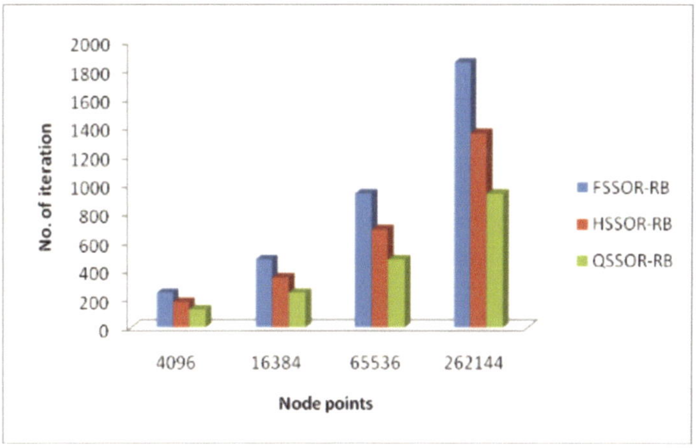

(a)

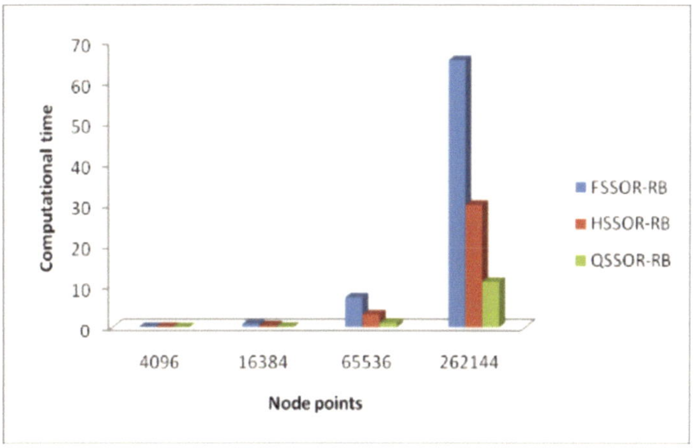

(b)

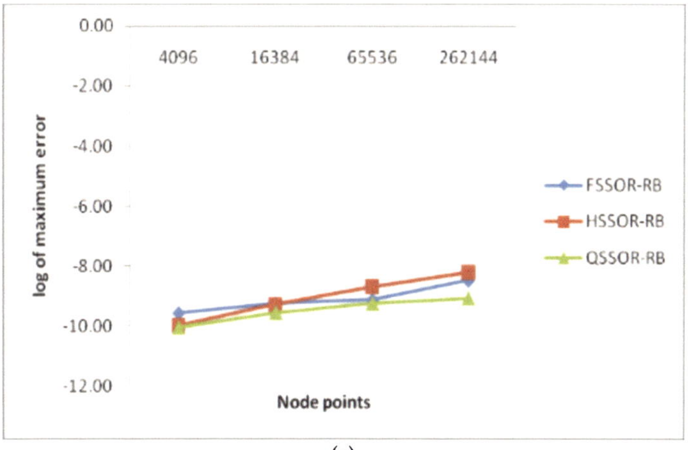

(c)

Fig. 4. Comparison of (a) iteration number, (b) computational time, and (c) accuracy for FSSOR-RB, HSSOR-RB and QSSOR-RB (α=0)

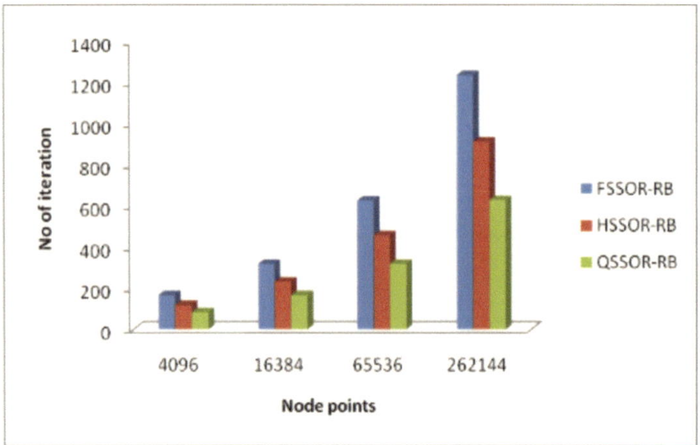

(a)

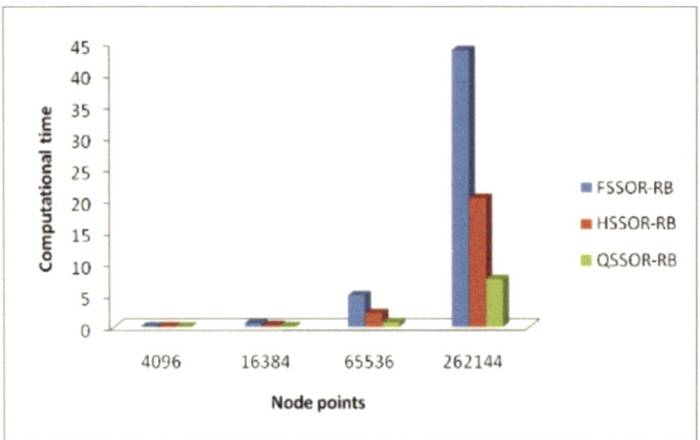

(b)

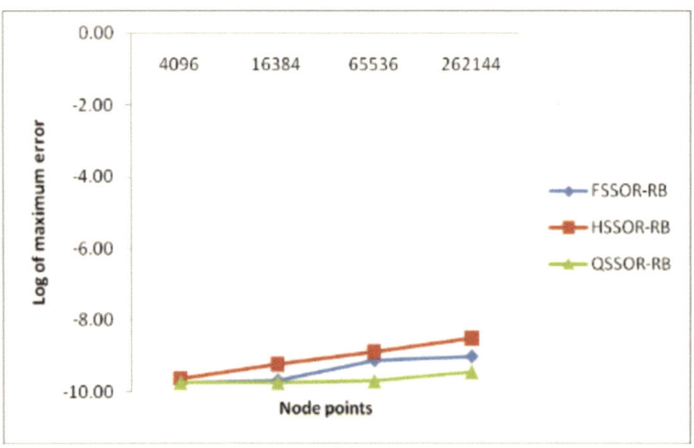

(c)

Fig. 5. Comparison of (a) iteration number, (b) computational time, and (c) accuracy for FSSOR-RB, HSSOR-RB and QSSOR-RB (α=25)

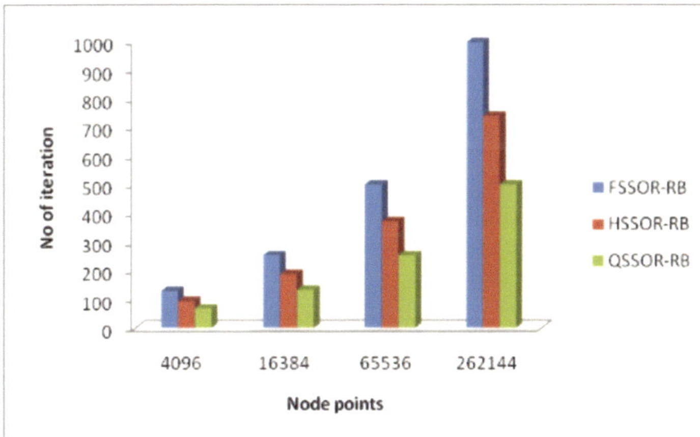

(a)

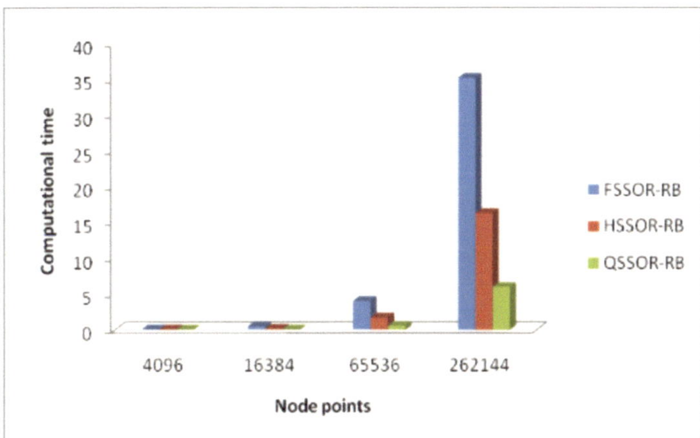

(b)

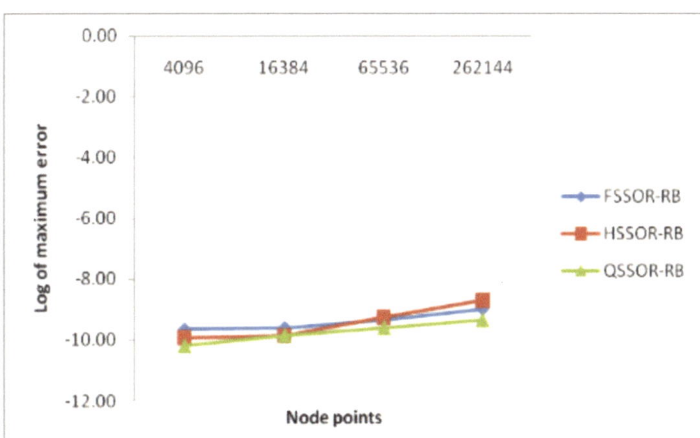

(c)

Fig. 6. Comparison of (a) iteration number, (b) computational time, and (c) accuracy for FSSOR-RB, HSSOR-RB and QSSOR-RB (α=50)

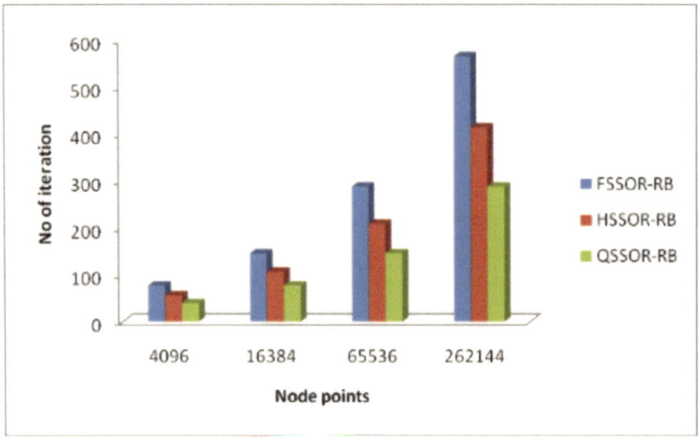

(a)

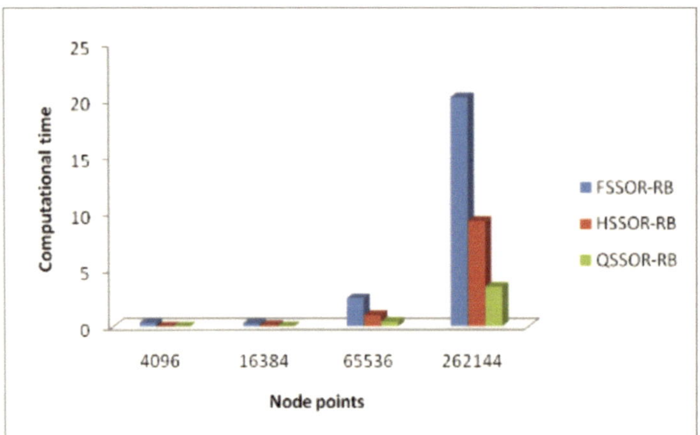

(b)

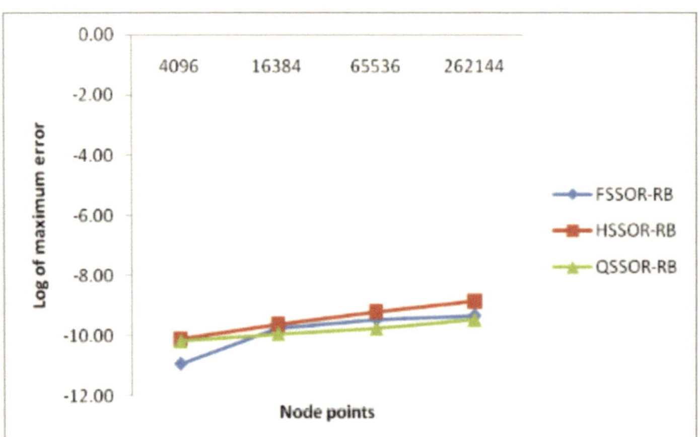

(c)

Fig. 7. Comparison of (a) iteration number, (b) computational time, and (c) accuracy for FSSOR-RB, HSSOR-RB and QSSOR-RB (α=200)

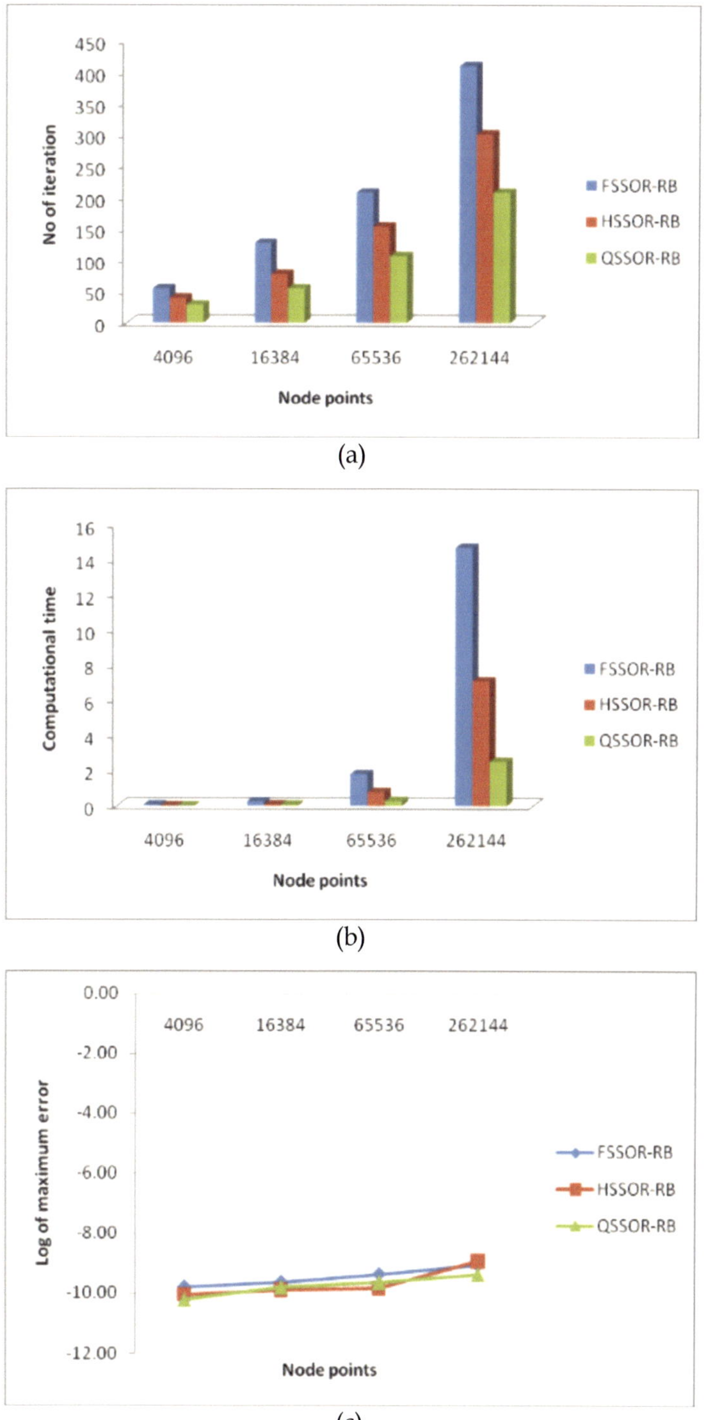

Fig. 8. Comparison of (a) iteration number, (b) computational time, and (c) accuracy for FSSOR-RB, HSSOR-RB and QSSOR-RB (α=400)

Figures 4 to 8 have shown that the results of experiments for FSSOR-RB, HSSOR-RB and QSSOR-RB methods. The results not only compares between these three methods, but also the impact of solving more node points on the iteration, computational time and accuracy of all three methods.

Results display in Figures 4(a) to 8(a) show that the numbers of iterations are impacted by the number of node points solved. The number of iterations and node points solved are the measure of complexity of methods since it refers to the number of function evaluation for each methods and problems. The number of node points solved in all experiments are 4096, 16384, 65536 and 262144. The ratio is 1: 4: 16: 64. However the ratio of number of iteration is about 1: 2: 4: 8. It means that increasing the problem sized considered by $4^i, i = 0,1,2,3$ increase the iterations by $2^i, i = 0,1,2,3$. It is the power of two relationships. By making the QSSOR-RB as the basis of comparison, the FSSOR-RB needs around 1.89 to 2.33 times and HSSOR-RB needs around 1.37 to 1.48 times number of iterations compared to QSSOR-RB. This means that FSSOR-RB is two times more complex than QSSOR-RB and HSSOR-RB is 1.5 times more complex than QSSOR-RB. This is equivalent to what we are expected since QSSOR-RB only solve a quarter of the node points in solution domain iteratively and HSSOR-RB only solve half of the node points in the solution iteratively, while FSSOR-RB have to solve every node in the solution domain iteratively.

Results displayed in Figures 4(b) to 8(b) support our description in the above paragraph. From the theoretical form of view, higher complexity method needs more computational time to solve problem. These scenarios are shown in Figures 4(b) to 8(b). However the effect of the complexity is different following the α value used. Figure 4(b) shows that FSSOR-RB needs 5 to 6.86 times more computational time compared to QSSOR-RB while HSSOR-RB only needs 2.69 to 3.5 times for α = 0. However, for α=25, FSSOR-RB needs 3 to 6.9 times more computational time to solve the problem compared to QSSOR-RB, while HSSOR-RB only need 2 to 3 times more computational time compared to QSSOR-RB (Refer Figure 5(b)). For α=50, the FSSOR-RB needs 2.5 to 7.23 times more computational time compared to QSSOR-RB and HSSOR-RB only need 1.5 to 3.09 times more computational time. As for bigger α value, the interval of computational time ratio of HSSOR-RB as compared to QSSOR-RB is narrowing and for FSSOR-RB becoming wider.

Figures 4(c) to 8(c) compares the accuracy of FSSOR-RB, HSSOR-RB and QSSOR-RB. These figures show that the accuracy of all methods are almost similar except for α=25 (refer Figure 5(c)). The figure shows that the QSSOR-RB has the highest accuracy, followed by FSSOR-RB and the last one is the HSSOR-RB.

3. Finite element method with red black ordering

In this subtopic, we will explain the development of our finite element class of method based on Galerkin scheme using triangle element discretization. Other finite element schemes are subdomain, collocation, least-square and moment.

Consider the two dimensional Poisson equation as follows.

$$\frac{\partial^2 U}{\partial x^2} + \frac{\partial^2 U}{\partial y^2} = f(x,y), (x,y) \in [a,b] \times [a,b]. \quad (4)$$

The Dirichlet boundary conditions are given by

$$U(x,a) = g_1(x),\ a \le x \le b,$$
$$U(x,b) = g_2(x),\ a \le x \le b,$$
$$U(a,y) = g_3(y),\ a \le x \le b,$$
$$U(b,y) = g_4(y),\ a \le x \le b.$$

A networks of triangle elements need to be build in order to derived the triangle element approximation equations for problem (4). The general approximation of the function, U(x,y) in the form of interpolation function is given by

$$\underline{U}^{[e]}(x,y) = N_1(x,y)U_1 + N_2(x,y)U_2 + N_3(x,y)U_3. \tag{5}$$

The shape function can be stated as

$$N_k(x,y) = \frac{1}{|A|}\left(a_k + b_k x + c_k y\right),\ k = 1,2,3 \tag{6}$$

where,

$$|A| = x_1(y_2 - y_3) + x_2(y_3 - y_1) + x_3(y_1 - y_2),$$

$$a_1 = x_2y_3 - x_3y_2,\ a_2 = x_3y_1 - x_1y_3,\ a_3 = x_1y_2 - x_2y_1,$$

$$b_1 = y_2 - y_3,\ b_2 = y_3 - y_1,\ b_3 = y_1 - y_2,$$

$$c_1 = x_3 - x_2,\ c_2 = x_1 - x_2,\ c_3 = x_2 - x_1.$$

The first order partial derivatives for the shape functions are given as follows.

$$\frac{\partial}{\partial x}\left(N_k(x,y)\right) = \frac{b_k}{|A|},\ k = 1,2,3 \tag{7}$$

in x direction and

$$\frac{\partial}{\partial y}\left(N_k(x,y)\right) = \frac{c_k}{|A|},\ k = 1,2,3 \tag{8}$$

in y direction.

Based on the definition of hat function, the approximation functions for problem (4) are given as follows.

$$\underline{U}(x,y) = \sum_{r=0}^{m}\sum_{s=0}^{m} R_{r,s}(x,y)U_{r,s} \tag{9}$$

for full sweep case,

$$\underline{U}(x,y) = \sum_{r=0(2)}^{m}\ \sum_{s=0(2)}^{m} R_{r,s}(x,y)U_{r,s} + \sum_{r=1(2)}^{m-1}\ \sum_{s=1(2)}^{m-1} R_{r,s}(x,y)U_{r,s} \tag{10}$$

for half sweep case, and

$$\underline{U}(x,y) = \sum_{r=0(2)}^{m} \sum_{s=0(2)}^{m} R_{r,s}(x,y) U_{r,s} \tag{11}$$

for quarter sweep case.
Next, let consider the Galerkin residual method as follows.

$$\iint_D R_{i,j}(x,y) E(x,y) dxdy = 0, i,j = 0,1,2,\cdots,m \tag{12}$$

With, $E(x,y) = \frac{\partial^2 U}{\partial x^2} + \frac{\partial^2 U}{\partial y^2} - f(x,y)$ is the residual function. Applying the Green theorem to equation (9) yields

$$\oint_{\lambda} \left(-R_{i,j}(x,y) \frac{\partial U}{\partial y} dx + R_{i,j}(x,y) \frac{\partial U}{\partial x} dy \right) - \int_a^b \int_a^b \left(\frac{\partial R_{i,j}}{\partial x} \frac{\partial U}{\partial x} + \frac{\partial R_{i,j}}{\partial y} \frac{\partial U}{\partial y} \right) dxdy = F_{i,j} \tag{13}$$

with $F_{i,j} = \int_a^b \int_a^b R_{i,j}(x,y) f(x,y) dxdy.$

By replacing equation (7), (8) and the boundary conditions into problem (4), can be shown that equation (13) will generate a linear system for any cases. The linear system can be stated as

$$-\sum \sum K^{*}_{i,j,r,s} U_{r,s} = \sum \sum C^{*}_{i,j,r,s} f_{r,s} \tag{14}$$

with

$$K^{*}_{i,j,r,s} = \int_a^b \int_a^b \left(\frac{\partial R_{i,j}}{\partial x} \frac{\partial R_{r,s}}{\partial x} \right) dxdy + \int_a^b \int_a^b \left(\frac{\partial R_{i,j}}{\partial y} \frac{\partial R_{r,s}}{\partial y} \right) dxdy$$

and

$$C^{*}_{i,j,r,s} = \int_a^b \int_a^b \left(R_{i,j}(x,y) R_{r,s}(x,y) \right) dxdy.$$

In stencil form, the full, half, and quarter sweep can be stated as follows.

- Full sweep stencil

$$\begin{bmatrix} & 1 & \\ 1 & -4 & 1 \\ & 1 & \end{bmatrix} U_{i,j} = \frac{h^2}{12} \begin{bmatrix} & 1 & 1 \\ 1 & 6 & 1 \\ 1 & 1 & \end{bmatrix} f_{i,j}.$$

This stencil is applied on solution domain with triangle element displayed in Figure 9.

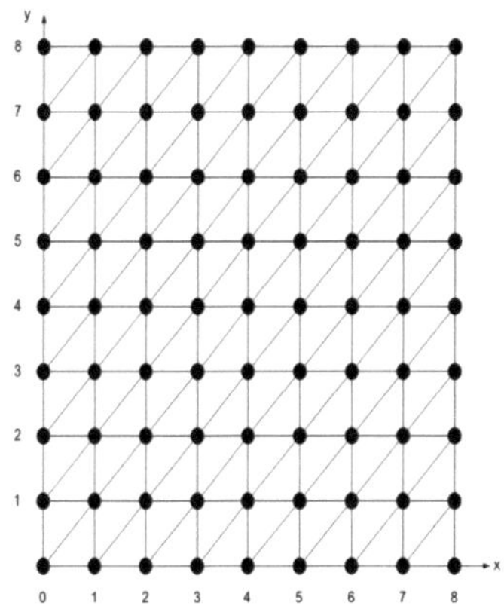

Fig. 9. Solution domain with triangle element discretization for full sweep scheme

- Half sweep stencil

$$\begin{bmatrix} 1 & & 1 & \\ & -4 & & 0 \\ 1 & & 1 & \end{bmatrix} U_{i,j} = \frac{h^2}{6} \begin{bmatrix} 1 & & 1 & \\ & 5 & & 1 \\ 1 & & 1 & \end{bmatrix} f_{i,j}, i = 1$$

$$\begin{bmatrix} & 1 & & 1 & \\ 0 & & -4 & & 0 \\ & 1 & & 1 & \end{bmatrix} U_{i,j} = \frac{h^2}{6} \begin{bmatrix} & 1 & & 1 & \\ 1 & & 6 & & 1 \\ & 1 & & 1 & \end{bmatrix} f_{i,j}, i \neq 1, n$$

$$\begin{bmatrix} & 1 & & 1 \\ 0 & & -4 & \\ & 1 & & 1 \end{bmatrix} U_{i,j} = \frac{h^2}{6} \begin{bmatrix} & 1 & & 1 \\ 1 & & 5 & \\ & 1 & & 1 \end{bmatrix} f_{i,j}, i = n$$

This stencil is applied on solution domain with triangle element displayed in Figure 10.

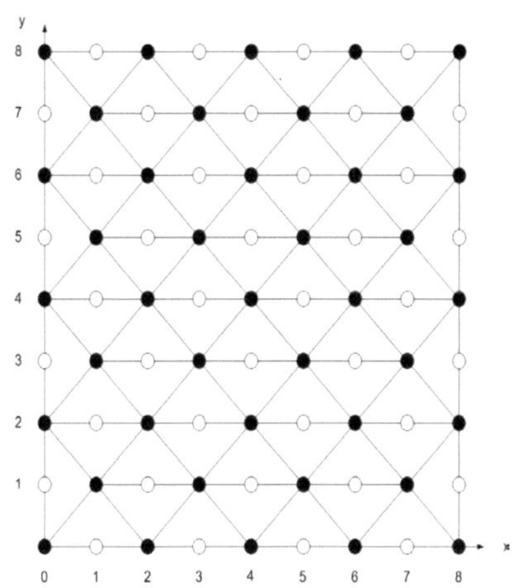

Fig. 10. Solution domain with triangle element discretization for half sweep scheme

- Quarter Sweep stencil

$$\begin{bmatrix} & & 1 & & \\ & & 0 & & \\ 1 & 0 & 4 & 0 & 1 \\ & & 0 & & \\ & & 1 & & \end{bmatrix} U_{i,j} = \frac{h^2}{3} \begin{bmatrix} 0 & & 1 & & 1 \\ & & 0 & & \\ 1 & 0 & 6 & 0 & 1 \\ & & 0 & & \\ 1 & & 1 & & 0 \end{bmatrix} f_{i,j}$$

This stencil is applied on solution domain with triangle element displayed in Figure 11.

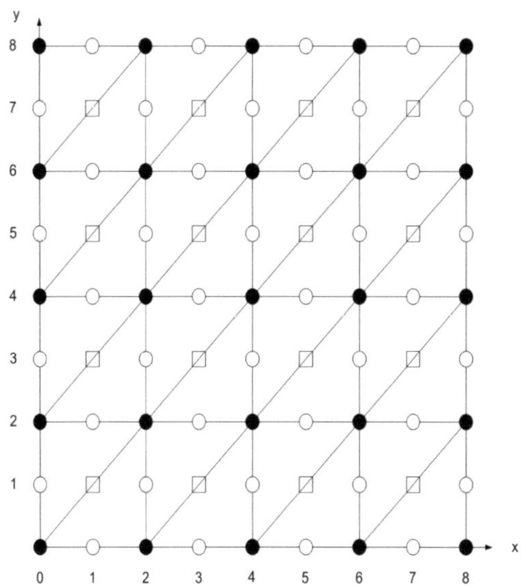

Fig. 11. Solution domain with triangle element discretization for quarter sweep scheme

All full sweep, half sweep and quarter sweep methods utilised the same red black ordering as the previous finite difference method in section 2 applied (refer Figure 3). The performance of the full sweep, half sweep and quarter sweep Gauss Seidel schemes using triangle element discritization based on Galerkin scheme are analysed for the following two dimensional Poisson equation (Abdullah, 1991).

$$\frac{\partial^2 U}{\partial x^2} + \frac{\partial^2 U}{\partial y^2} = \left(x^2 + y^2 \right) e^{xy}, \qquad (x,y) \in [0,1] \times [0,1].$$

The boundaries and the exact solution are given as follows.

$$U(x,y) = e^{xy}, \qquad 0 \le x,y \le 1.$$

The convergence criteria considered in these experiments is $\varepsilon = 10^{-10}$. All results of numerical experiments are displayed in Figure 12, 13 and 14.

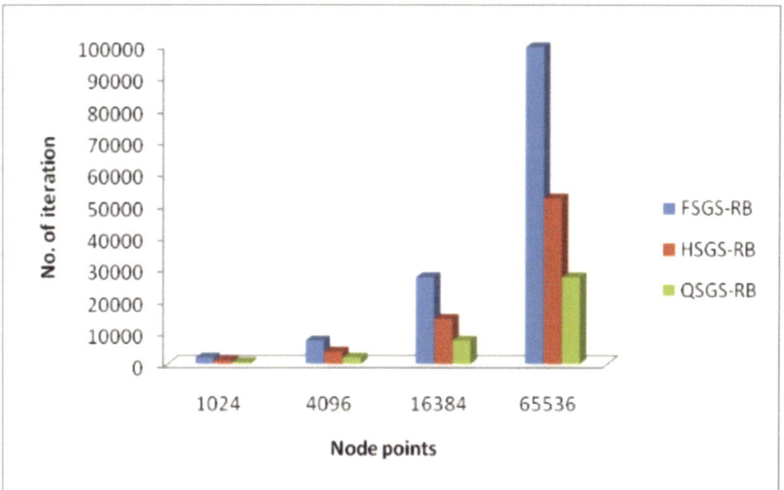

Fig. 12. Number of iterations for all compared methods.

Figures 12 to 14 have shown that the full sweep, half sweep, and quarter sweep triangle element approximation equations based on the Galerkin scheme are fast and accurate algorithms. The findings in Figure 12 shows that numbers of iteration needed by FSGS-RB are almost four times compared to QSGS-RB, while HSGS-RB is almost two times compared to QSGS-RB. The impact of increasing the number of node points to number of iterations seems too significant. In this experiment the ratio of number of points studied is 1: 4: 16: 64. However, the increment of numbers of point solves also increase the number of iterations with the ratio of 1: 4: 14: 50. This means that increasing the number of points does increase the complexity or increase the number of function evaluation. This behavior is as expected since its follows the theoretical explanation.

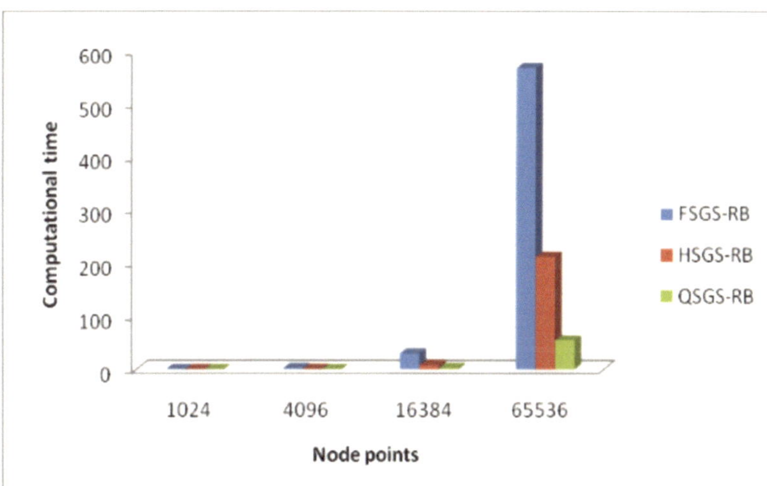

Fig. 13. Computational time in seconds for all compared methods.

Figure 13 clearly shows that QSGS-RB compute faster than the other two methods (HSGS-RB and FSGS-RB). The FSGS-RB needs 6.5 to 13.2 times more computational time compared to QSGS-RB, while HSGS-RB only need 1.5 to 3.9 times more computational time compared to QSGS-RB. This is because QSGS-RB only solved a quarter of node points iteratively, while HSGS-RB only solved half of the node points iteratively. However, the accuracy of both QSGS-RB and HSGS-RB are lower than FSGS-RB.

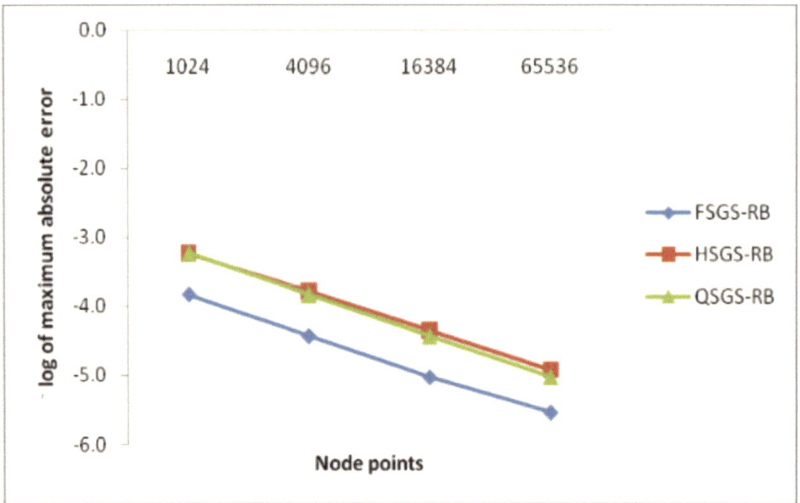

Fig. 14. Accuracy comparison for the full sweep, half sweep and quarter sweep schemes

7. Conclusion

In this chapter, we have demonstrated the development of three finite difference methods and three finite element methods. All methods utilised the red black ordering strategy and implementing the concept used in the Explicit Decoupled Group (EDG) and the Modified Explicit Group (MEG). By implementing the EDG, we developed a method called Half Sweep Successive Over-Relaxation utilising the Red Black ordering Strategy (HSSOR-RB) for finite difference scheme and Half Sweep Gauss-Seidel utilising the Red Black Strategy (HSGS-RB) for finite element scheme. Applying the concept used in MEG, we develop a method called Quarter Sweep Successive Over-Relaxation utilising the Red Black ordering strategy (QSSOR-RB) for finite difference scheme, while Quarter Sweep Gauss-Seidel utilising the Red Black ordering strategy (QSGS-RB) for finite element scheme. Both finite element schemes are developed using triangle element discretization.

The performance of both finite difference and finite element schemes are examined by comparing their number of iteration, computational time and accuracy to full sweep schemes, i.e. Full Sweep Successive Over-Relaxation utilising the Red Black Ordering Strategy (FSSOR-RB) for finite difference scheme and Full Sweep Gauss-Seidel utilising the Red Black Ordering Strategy (FSGS-RB) for finite element scheme. Helmholtz equation was used for testing the new finite difference scheme, while Poisson equation was used for testing the new finite element scheme.

From the numerical experiment, both HSSOR-RB and QSSOR-RB have shown the integrity to solve the Helmholtz equation faster than the FSSOR-RB. This is because The FSSOR-RB has the higher complexity and needs higher numbers of iteration than HSSOR-RB and QSSOR-RB methods. Having higher complexity and higher numbers of iteration makes the method required the highest number of arithmetic operation compared to HSSOR-RB and QSSOR-RB. Furthermore, solving Helmholtz equation via QSSOR-RB only needs to solve a quarter of node points in the solution domain iteratively, while solving via HSSOR-RB only required to solve half of the node points in the solution domain. This is another reason why QSSOR-RB is faster than HSSOR-RB. The best part is HSSOR-RB and QSSOR-RB not only computes the Helmholtz problem faster than FSSOR-RB but also have similar accuracy to FSSOR-RB method.

Poisson equation has been used to examine the performance of the new finite element method, i.e. QSGS-RB and HSGS-RB. The numerical experiments show that HSGS-RB needs almost two times iteration number compared to QSGS-RB, while FSGS-RB needs almost four times iteration number compared to QSGS-RB. Besides, HSGS-RB only solve half of the solution domain iteratively, while QSGS-RB solve only quarter of the solution domain iteratively. Both are the reason why QSGS-RB are faster than HSGS-RB and FSGS-RB.

As the conclusion, applying the concept used in EDG and MEG with red black strategy produces fast solvers either using finite difference or finite element approaches.

8. Acknowledgement

We acknowledge the financial support from 01-01-02-SF0677 (Malaysian Ministry of Science, Technology and Innovation Grant).

9. References

Abdullah, A. R. (1991). The Four Point Explicit Decoupled Group (EDG) Method : A Fast Poisson Solver. *International Journal Computer Mathematics* Vol. (38) : 61-70.

Belytschko, T.; Kringauz, Y. ; Organ, D. ; Fleming, M. & Krysl, P. (1996). Meshless Method : An Overview and Recent Developments. *Computer Methods in Applied Mechanics and Engineering* Vol. (139): 3-47.

Evans, D. J. & Abdullah, A. R. (1982). Group Explicit Method for Parabolic Equations. *International Journal Computer Mathematics* Vol. (14) : 73-105.

Evans, D. J. & Biggins, M. (1982). The Solution of Elliptic Partial Differential Equations by a New Block Over-Relaxation Technique. *International Journal Computer Mathematics* Vol. (10) : 269-282.

Evans, D. J. & Yousif, W. F. (1990). The Explicit Block Relaxation Method as a Grid Smoother in the Multigrid V-cycle scheme. *International Journal Computer Mathematics* Vol. (34) : 71-78.

Gupta, M. M. ; Kouatcou J. & Zhang, J. (1997). Comparison of Second and Fourth Order Discretizations for multigrid Poisson Solvers. *Journal of Computational Physics* Vol. (132 (2)) : 226-232.

Ibrahim, A. (1993). The Study of the Iterative Solution of the Boundary Value problems by the Finite Difference Methods. PhD Thesis. Universiti Kebangsaan Malaysia.

Ibrahim, A. & Abdullah, A.R. (1995). Solving the Two Dimensional Diffusion Equation by the Four Points Explicit Decoupled Group (EDG) Iterative Method. *International Journal Computer Mathematics* Vol. (58) : 253-263.

Othman, M. & Abdullah, A.R. (2000). An Efficient Four Points Modified Explicit Group Poisson Solver. *International Journal Computer Mathematics* Vol. (76) : 203-217.

Parter, S. V. (1998). Estimates for Multigrid Methods Based on Red Black Gauss-Seidel Smoothers. *Numerical Mathematics* Vol. (52) : 701-723.

Rosser, J. B. (1975). Nine Points Difference Solution for Poisson's Equation. *Computers and Mathematics with Applications* Vol. (1) : 351-360.

Yagawa, G. & Furukawa, T. (2000). Recent Developments of Free Mesh Method. *International Journal for Numerical Methods in Engineering* Vol. (47): 1419-1443.

Yousif, W. S. & Evans, D.J. (1986a). Explicit Group Over-Relaxation Methods for Solving Elliptic Partial Differential Equations. *Mathematic and Computer in Simulation* Vol. (28) : 453-466.

Yousif, W. S. & Evans, D.J. (1986b). Higher Order Difference Group Explicit (HODGE) Methods. *Communications in Applied Numerical Methods* Vol. (2) : 449-462.

Zhang, J. (1996). Acceleration of Five Points Red Black Gauss-Seidel in Multigrid for Poisson Equations. *Applied Mathematics and Computation* Vol. (80(1)) : 71-78.

Zhu, T. (1999). A New Meshless Regular Local Boundary Integral Equation (MRLBIE) approach. *International Journal for Numerical Methods in Engineering* Vol. (46): 123-125.

22

The Role of Computer Games Industry and Open Source Philosophy in the Creation of Affordable Virtual Heritage Solutions

Andrea Bottino and Andrea Martina
Dipartimento di Automatica e Informatica, Politecnico di Torino
Italy

1. Introduction

The Museum, according to the ICOM's (International Council of Museums) definition, is a non-profit, permanent institution [...] which acquires, conserves, researches, communicates and exhibits the tangible and intangible heritage of humanity and its environment for the purposes of education, study and enjoyment (ICOM, 2007). Its main function is therefore to communicate the research results to the public and the way to communicate must meet the expectations of the reference audience, using the most appropriate tools available. During the last decades of 20th century, there has been a substantial change in this role, according to the evolution of culture, literacy and society. Hence, over the last decades, the museum's focus has shifted from the aesthetic value of museum artifacts to the historical and artistic information they encompass (Hooper-Greenhill, 2000), while the museums' role has changed from a mere "container" of cultural objects to a "narrative space" able to explain, describe, and revive the historical material in order to attract and entertain visitors. These changes require creating new exhibits, able to tell stories about the objects, enabling visitors to construct semantic meanings around them (Hoptman, 1992). The objective that museums pursue is reflected by the concept of Edutainment, Education + Entertainment. Nowadays, visitors are not satisfied with "learning something", but would rather engage in an "experience of learning", or "learning for fun" (Packer, 2006). Hands-on and interactive exhibitions allow their visitors to interact with archive material, to learn while they play with it (Caulton, 2002) and to transform them from passive viewers and readers into active actors and players (Wojciechowski et al., 2004). As a result, curators are faced with several new problems, like the need to communicate with people from different age groups and different cultural backgrounds, the change in people attitude due to the massive and unexpected diffusion of technology into everyday life, the need to design the visit by a personal point of view, leading to a high level of customization that allows visitors to shape their path according to their characteristics and interests.
Recent advances in digital technologies offer archivists and exhibit designers new interesting and captivating ways to present and disseminate cultural objects, meeting the needs of personalization and interactivity requested by visitors (Addison, 2000). In particular, Virtual Reality (VR) and Mixed Reality (MR) allow creating novel exhibition paradigms, rich in those informative and emotional contents often missing in the classic

ones. The objects in the museums have lost their original context, which can be re-created through a multimedia environment, where visualization, sounds and perfumes can be used to increase the sense of immersion and presence. VR allows enjoying unavailable objects as well. Such objects can be lost in time, or cannot be shown without compromising their preservation, or are simply "buried" into dusty archives. The same concept applies to architectural objects, allowing to virtually restore lost buildings and their rooms.

The integration of cultural heritage and ICT technologies, in order to develop powerful tools to display cultural contents, is often referred to as *Virtual Heritage* (VH). VH allows the development of *virtual museums*, which, according to the Encyclopaedia Britannica, can be defined as "*a collection of digitally recorded images, sound files, text documents, and other data of historical, scientific, or cultural interest that are accessed through electronic media*". This general definition takes different realizations according to the application scenario, the available technologies and the users' involvement, ranging from the presentation of a digital collection over the Web, to the development of interactive immersive installations on the museum site (Sylaiou et al., 2009).

Despite its advantages, the diffusion of virtual museums and VH approaches has not been as wide as expected, due to several drawbacks. For instance, the first generation of virtual environments was characterized by low resolution, poor graphic displays and inadequate interaction mechanisms, which resulted in a poor user experience and, therefore, in a rapid disaffection towards the VR applications (Bowman et al., 2008). Such limitations have been overcome with new technological devices, like high definition displays, 3D monitors, holographic projectors and holotouchs, immersive environments, natural interaction through gestures, expressions and movements (Burdea & Coiffet, 2003). However, such devices are often expensive, while museums are confronted to limited budgets for designing and creating the exhibit, and therefore, especially for small realities, they often cannot afford to buy and maintain costly technological structures or apply them on a large scale. The same limitations apply to creating contents and developing real-time interactive environments, which require the use of specific, and often expensive, authoring and management tools.

Fortunately, the horizon is slowly changing, mainly due to two factors:

- the computer games industry drives technological advancement which is then used by other industry sectors. Demanding players require a constant increase of the visual realism and immersion, which in turns require a constant improvement of graphics hardware and CPU capabilities, and the availability of novel interaction devices that provide a better control, in terms of naturalness and usability, of the game reality. Furthermore, in order to allow a widespread diffusion of such products in the consumer market, their cost is kept as low as possible and often the hardware is sold at cost price while profits are made on the contents
- the increasing availability of Open Source software solutions, free to use and modify, allows to drastically reduce the costs of developing applications and creating the digital elements that will be displayed in the virtual exhibits, and facilitates a seamless integration of different interaction devices to provide a reach user experience

The rest of the paper is organized as follows. In section 2, we will show how the most recent contributions in these areas enable the development of affordable multimedia interactive applications for enhancing the fruition of archive material and for building effective edutainment exhibits. In section 3 the various types of virtual museums will be presented and analyzed. In section 4 we will discuss the steps required to design and develop them. Finally, in section 5 we will present the conclusions.

2. Affordable Virtual Heritage solutions: the enabling factors

Before tackling this point, let's first briefly introduce some basic concepts all the readers might not be familiar with. Virtual Heritage is mainly related with some key technologies like multimedia, Virtual Reality and Mixed reality, which are based on the idea of offering different solutions for the user to interact with the technology itself, and obviously with the contents.

Multimedia applications are based on the integration of multiple types of media like text, graphics, images, audio, video, etc. and usually provide a basic level of interaction with users. On the other hand, Virtual Reality (VR) recreates an (interactive) entire artificial environment that is presented through different devices, like display screens, wearable computers and haptic devices, in order to immerse the user in a believable simulation of reality. Mixed reality (MR) refers to the merging of real and virtual worlds to produce new environments and visualizations where physical and digital objects co-exist and interact in real time. Its most well-known example is Augmented Reality (AR), where the user visualizes a composite view, given by a combination of the real and a virtual scene, which is generated by the computer in order to augment the perceived reality with additional information.

VR is by far the most compelling of these technologies and in the late eighties and early nineties there was an extraordinary level of excitement and hype about anything to do with it (Rheingold 1991). However, VR soon failed to live up with such high expectations. According to Jaron Lanier (Lanier, 2003), a VR pioneer, three of the main reasons why VR did not become commonplace are the following. First, VR applications are based on real-time and interactive simulations, where 3D geometry has to properly react to the user's input and the software has to manage scene's features, lights, shadows, fluids, dynamics, etc. This requires a lot of computational power, which only high-end specialized workstations could provide at the time. Second, the lack of high quality interfaces, especially high resolution displays and suitable interaction devices, resulted in a poor user experience. Third, proper software standard platforms for designing, developing and managing the complexity of a virtual environment were not available. As a result, there was a large disaffection of the general public, while the dream of making VR a reality kept on being carried out only by a closed group of research centers.

The late nineties witnessed a rebirth of the VR. One of the most important contributing factors was the enormous improvement in PC hardware, both in terms of CPUs and PC-based graphics cards, which became faster and faster. The capabilities of computing elements are described by the well known Moore's law, which states that their performances double every 18 months. Graphic Processing Units (GPUs) were even better than Moore's prediction, to such an extent that in 2001 the performances of an off-the-shelf consumer PC matched that of a high-end graphic workstation, at a cost of ten to hundred times lower. Nowadays, GPUs are capable of handling several hundreds of millions of textured polygons per second, have a storage capability of several gigabytes and a memory bandwidth of hundreds of GB/sec, and provide in real-time complex rendering effects, like reflections, depth of field, atmospheric effects, automatic mesh complexity modification.

Not surprisingly, one of the driving forces for such a revolution was the computer games industry. Games became more and more complex, since demanding players required a constant increase of the visual realism and immersion. On the other side, the enormous incomes of global majors in terms of sales revenues involved huge investments in research and development. The technological advancement of computer games combined to bring about the need for newer and more powerful hardware to handle them, and led to an increase in processing capabilities of consumer PCs and graphics cards paralleled by a reduction of their costs. A further development comes from the fierce competition between game console vendors. Nowadays, game consoles, like Microsoft XBOX 360, Sony PlayStation3 and Nintendo WII, outperform high end PCs for their computational and graphics processing power. Moreover, since revenues come from game sales and not from hardware, these consoles are often marketed at cost price.

Significant advances have been made also in the field of interaction devices. Especially for computer games, the development of innovative products, providing a more natural and intuitive control, allows improving the usability of computer applications and the users' experience, and ensures the availability of low-cost and consumer-grade devices. Sensing hardware, coupled with specific algorithms, allows the computer to "hear" and "see" its surrounding environment and to interpret the user actions. Recent products provide state of the art motion control, have haptic feedback and ensure freedom of movement in the environment. Many of these devices can be easily integrated into different applications and, therefore, provide a very interesting mechanism for improving the interaction possibilities in Virtual Museums (McLaughlin et al., 2000, Severson et al., 2002 and Hirose, 2006). Furthermore, like game consoles, they are sold at cost price representing a valid alternative to the often expensive devices typical of the VR, like trackers, sensing gloves and 3D probes.

Computer displays, TV screens and projectors are constantly increasing their size, resolution and quality, improving the sense of immersion in the Virtual Environment. This is further enhanced by the recent introduction of 3D displays, which are radically changing the way we enjoy digital contents. Moreover, the strong competition on the consumer market ensures a constant drop of their cost.

The last years have witnessed a dramatic and capillary diffusion in the society of more and more powerful mobile phones and PDAs. Recent products provide enhanced multimedia features, like high quality displays, video cameras, touch screens, advanced audio support and 3D rendering capabilities. Their processing power and available memory are slowly reaching that of traditional computing platforms, like notebooks and desktop computers. Their communication capabilities, including high-bandwidth internet connections, enable the development of complex networked applications. Moreover, mobile phones are personal devices that users carry with them most of the time and are, therefore, specially suited for Virtual Heritage in supporting moving users with interactive multimedia information: users can receive audio explanations, complementary videos and images, text information, 3D graphics and, on the other side, users' activities can be mapped into and interact with a 3D virtual environment (Sauer & Göbel, 2003, Farella et al., 2005, Santangelo et al. 2006, Prasolova-Førland et al., 2007, Hope et al., 2009,).

Concerning content creation and handling, until the last decade, the few available authoring and management tools were professional and often expensive software, hampering their diffusion and use by everyday users. This situation has radically changed with the introduction of the *Free Software* (FS) and *Open Source* (OS) concepts by Richard Stallman and its GNU project in 1983 (Stallman, 1999). The FS/OS idea refers to practices in software production and development in which the source code is available free of charge to the general public for use and/or modification from its original design. Typically, Open Source software is created as a collaborative effort in which programmers improve the code and share their changes within the community. It's important to point out that the term "Open Source" does not absolutely mean the contrary of "professional". Several programs developed under OS/GNU/Creative Commons licenses are comparable, in terms of available features and quality of the results, with commercial software, while OS enthusiasts often claim being even better.

Nowadays, many free applications, libraries and complex authoring systems are available. As a result, the costs of creating digital contents and designing and developing applications have reduced considerably. Furthermore, recent products emphasize the simplicity of use to allow not only skilled but also non professional users to adopt them.

Virtual Heritage can obviously significantly benefit from all these technological improvements, since they allow the development of complex environments, supporting highly realistic visualization, providing a natural interaction through pervasive devices and content fruition on different platforms, from Internet to immersive environments, PDAs and smartphones. Therefore, the possibility to manage these applications in real time on low-cost consumers' hardware, the possibility to use appropriate interaction devices and the availability of free software tools enable, on one side, the development and the execution of effective and, most of all, affordable edutainment solutions and, on the other side, a wide diffusion, in the near future, of VH approaches in museums.

3. What exactly is a Virtual Museum?

Virtual Museum is a general concept appeared for the first time in 1991 (Tsichritzis et al., 1991) that can be implemented in different ways according to exhibition design, application and end user requirements. At present, the definitions of Virtual Museum are various and the projects and creation of Virtual Heritage applications have been and continue to be very different. The ICOM itself (ICOM, 2004) suggests three different types of virtual museum - the museum brochure, the content museum and the learning museum. The brochure is nothing more than the presentation of the physical museum information through Internet. The content museum is instead a website created to divulge information about the available museum collections. Finally, the learning museum is a website that provides personalized paths to navigate its contents according to the users' characteristics (age, cultural background, interests, etc…). Typically, this kind of Virtual Museum has mainly an educational purpose, and is linked with other material in order to motivate the virtual visitor to learn more about a subject of particular interest.

The ICOM definition well describes the categories of online Virtual Museums, but does not mention the full gamut of possibilities available on-site, where technologies can be integrated into the museum exhibit as an additional tool for routine visit of the physical place.

Therefore, according to several authors (Tsichritzis & Gibbs, 1991, Lepouras & Vassilakis, 2004, Hirose, 2006, Sylaiou et al., 2009, Noh et al. 2009) a more comprehensive classification of Virtual Museums can be introduced, whose brief overview is presented in the following subsections.

3.1 Content galleries

Content Galleries are collections of browsable digital objects that visitors can enjoy through different media, discovering an easy way to query and browse the content database through a graphic interface. Content galleries can be presented both online - as a web page/web application - and on-site – as an interactive installation. Creating compelling multimedia applications, providing an engaging interface to the museum repository, is quite simple, thanks to the wide availability of dedicated software libraries and languages, like Flash and Processing.

An example is *Getty Museum's* website, which let users discover the Art in many different ways: through an alphabetical list of authors, a classification of the different types of Art, a thematic classification and other brief overviews. Web 2.0 practices, like personal galleries that can be used to create personalized visits and shared with friends, are also available. A different approach is demonstrated in the *Heilbrunn Timeline of Art History* that presents a chronological, geographical, and thematic exploration of the world Art history through the Metropolitan Museum of Art's collection (Fig. 1). The timeline allows the navigation across locations and periods in order to observe not only the artworks but also their relationship with the historical and geographical context.

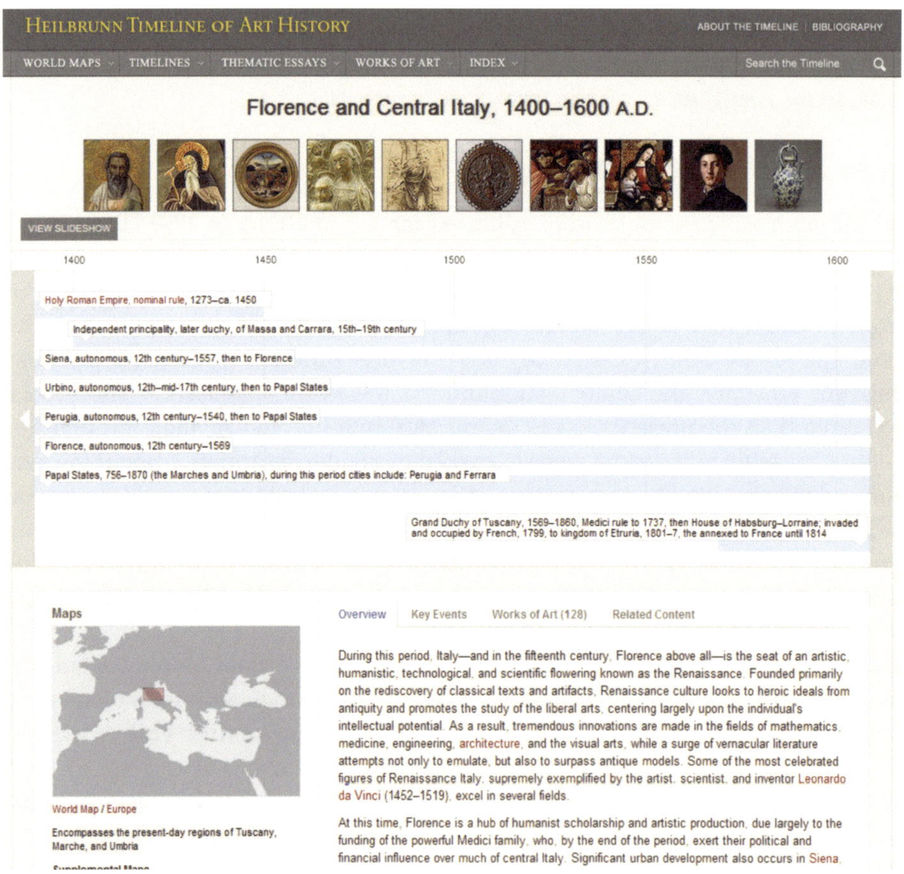

Fig. 1. The Heilbrunn Timeline of Art History in the Metropolitan Museum of Art's website.

3.2 Enhanced panoramic views

Online galleries allow visualizing cultural contents, but they are often perceived by users as a flat mean of communication. What would make visitors more involved it is to immerse them into the exhibit environment and let them have a look around. To this effect, a viewer for 360° panoramic images provides greater interactivity and engages visitors much more than single static images. The viewer allows rotation, tilt and zoom of the surrounding scene. Panoramas can be created using catadioptric cameras, consisting on a single sensor with a 360° field-of-view optic, or synthesizing them from multiple images that are projected from their viewpoints into an imaginary sphere, or cylinder, enclosing the viewer's position, and then blended with specific mosaicing software. In addition, panoramas can be made interactive embedding hotspots that allow, when selected, to invoke some actions, for example linking other types of digital media, providing an integrated multimedia experience.

The use of 360° panoramic images has become quite common, and therefore familiar to all users, after the launch on 2007 of Google Street View, a technology featured in Google Maps and Google Earth that provides panoramic views from various positions along many streets in the world. An example, applied to museums, is the virtual tour of the *Smithsonian National Postal Museum* (Fig. 2). The application is meant to show how the real museum looks like, offering a tour through its rooms, without adding any further information on the museum's collection.

Fig. 2. The Smithsonian National Postal Museum online virtual tour.

Visualized contents are not limited to real environments. As a matter of facts, it is easy to obtain photo-realistic panoramic images from high-quality rendering of any virtual environment. An interesting example is *One Day at the Sands*, developed by the Polytechnic of Turin. The project focuses on the Sands in the 1964, one of the most prestigious and oldest casino in Las Vegas. The aim of the project is to entertain visitors by giving them the chance to spend a whole day in the Sands and to discover the stories and the characters behind the Fabulous Las Vegas. Pictures and videos have been used to rebuild in 3D the original Sands atmosphere and the panoramic images of the rooms have been augmented with archive material (documents, images and videos), accessible from hotspots in the environment (Fig. 3).

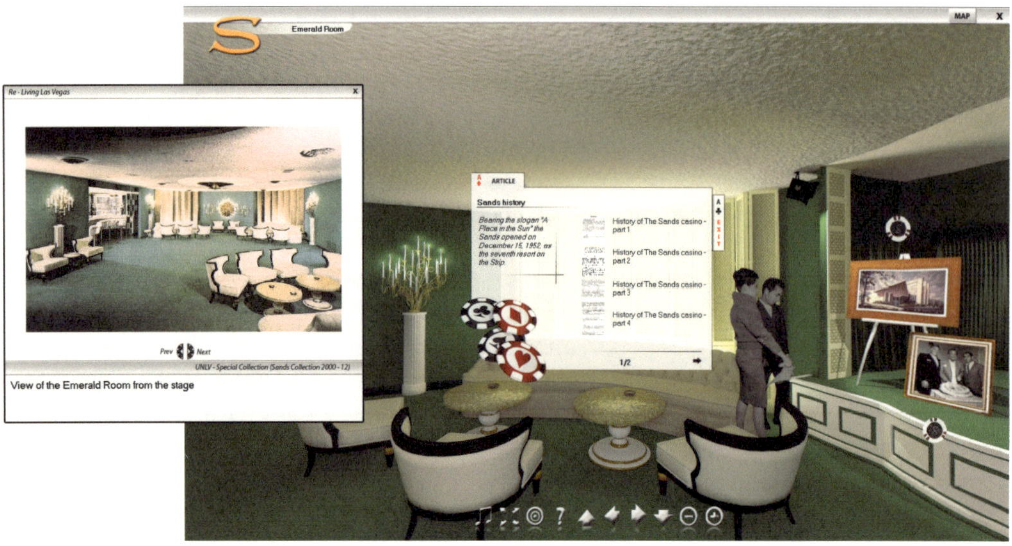

Fig. 3. The One day at the Sands online experience.

3.3 Online virtual environments

Online virtual environments (OVE) stems from the Web3D idea, which, initially, aimed at displaying and navigating the Web in 3D. So far, the term refers to all interactive 3D content that can be accessed and visualized through Internet. OVEs have a great pedagogical value, given by their ability to immerse visitors in a navigable content-rich world where they can be involved in learning (and entertaining) activities as well as collaborate in participatory activities (Scali et al., 2002).

Typically, Web3D comes in two forms. In the first, 3D contents are embedded directly into web pages. Since Web3D technology is not currently supported by browsers, they require external plug-ins to handle the 3D contents. Some of them are Java3D, the various VRML viewers, O3D, Unity3D and 3Dvia. The second solution makes use of software clients, capable of handling complex 3D interactive environments, through which the user participate in the virtual simulation. The architecture of the application can foresee a centralized server, as in SecondLife and OpenSimulator, its OS counterpart, or a peer-to-peer network of clients, as in several online 3D games.

Currently, there is still no standard language to distribute 3D contents. The most common formats are VRML, the oldest of all and first presented in 1994, X3D, an XML-based extension of VRML, and Collada, an interchange file format for interactive 3D applications that is widely supported by the most common modeling software.

OVEs are heavily influenced by the available bandwidth, since contents are downloaded while the application is running. Therefore, on slow connections, the waiting time may become long, annoying the user that will hardly continue his visit. In addition, since web contents can be consumed by potentially any user, the application cannot rely on specific

computational resources. Therefore, the developers must reach a good compromise between the complexity of the virtual environment, its visual realism and its usability.
An interesting example of OVE for VH is *Heritage Key*, which offers a SecondLife like exploration of historical sites virtual reconstructions from all over the world (Fig. 4).

Fig. 4. The Heritage Key online environment.

The Danish Centre for Urban History has presented for the centenary of the *Danish National Expo in 1909* an interactive 3D model, based on the original architectural drawings, of the international pavilions. The user can wander through the buildings, browsing, through hotspots in the environment, the additional historical documents, like photographs, text, images, videos and oral narratives (Fig. 5).

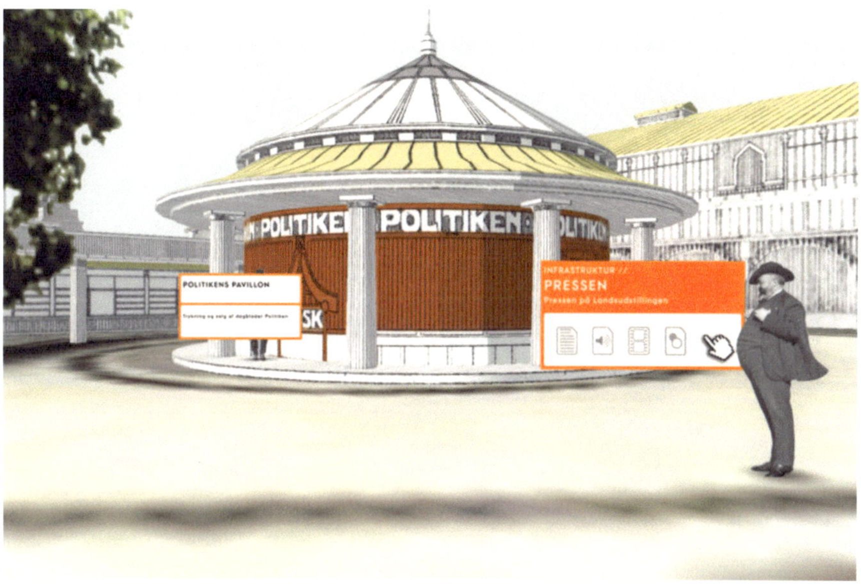

Fig. 5. The interactive 3D model of the Danish National Expo 1909.

3.4 Immersive VR environments

A typical immersive VR system consists of a VR software, managing the visualization of the virtual environment, a display system, three-dimensional audio support and the input devices that allow users to interact with the environment and the virtual objects it contains. Their essence is the illusion of "presence" inside the computer generated environment they are able to transmit to users (Schuemie et al., 2001).

There are various types of VR systems, which provide different levels of immersion and interaction, ranging from two extremes, the so called "weak VR" and "strong VR" (Styliani et al., 2009). Weak VR, also referred as Desktop VR, is typically a 3D environment displayed on a screen, either mono or stereoscopic, with a single user interacting with the virtual environment. On the other hand, strong VR, gives users a uniquely and compelling experience of "being there" in the virtual world through a converging stimulation of all human senses and the subsequent suspension of disbelief. Visualization can involve the use of virtual reality helmets (Helmet Mounted Displays, HMD) or other immersive stereoscopic displays. Users are allowed to move freely in the environment and their position and movements are captured, by means of tracking and sensing technologies, and reproduced in the 3D environment. Sometimes, users are allowed to physically touch the virtual objects using haptic interfaces, like data gloves or force-feedback devices (Burdea & Coiffet, 2003).

Unlike OVEs, here the execution hardware and the software can be tailored according to the specific needs of the application, allowing to build more realistic environments, with higher resolution models and more detailed textures, atmospheric and camera effects, 3D audio support, realistic crowd management and complex user interaction.

An example is *Live History*, an on-site installation in the Nationaal Historisch Museum in Holland. Visitors assume an active role in the history of the Netherlands, interacting with objects, documentaries and virtual representation of the main historical events through handheld multimedia devices. Users are physically immersed in a CAVE environment, a room where images of the virtual environment are displayed on walls and floor (Fig. 6).

Fig. 6. Visitors inside the CAVE of Live History installation.

The *Virtual Museum of the Ancient via Flaminia*, at the Museo Nazionale delle Terme di Diocleziano in Rome, offers a virtual tour of the via Flaminia during the Roman Empire through a combination of a virtual storytelling, in the form of audiovisual reconstructions and narrative sheets, and a free navigation, allowed only in specific points of the path. Four interactive stations allow users to explore the environment through avatars, while the rest of the audience can follow their actions on a central screen, which displays also complementary visual and informative contents (Fig. 7).

Fig. 7. A view of the Virtual Museum of the Ancient via Flaminia.

3.5 Mixed Reality environments

Mixed Reality (MR) creates environments that combine at the same time both real and virtual objects. It is possible to visualize MR/AR environments through a desktop display, see-through HMD, PDA or mobile devices. Interaction is generally managed through the display, for PDA/mobile devices and touch screens, or through tactile manipulation of visual markers/fiducials or of other sensible objects located in the real world.

An interesting example is the *Jurascope* installation at Berlin's Museum of Natural History (Fig. 8), where people, looking a dinosaur's skeleton through a telescope can see an animation that overlap the bones with organs, muscles and then skin. Finally, the virtual dinosaur starts moving inside the room. The installation is certainly impressive, but the only interaction provided to users is to activate the application pointing the telescope at the skeletons in the room.

Another interesting application is *Streetmuseum*, released by the Museum of London. Users, installing Streetmuseum on their own iPhone, transform London into a huge open-air historical museum (Fig. 9). Hundreds of images of the city from four centuries of history have been georeferenced and can be seen in the location they were taken through the iPhone, which acts as a window through time. Historical information connected to the images can be accessed tapping on the device screen.

Fig. 8. The Jurascope experience.

Fig. 9. The Streetmuseum application.

4. Developing affordable Virtual Museums

The development of a virtual museum can be roughly divided into three main steps:

- collecting, or creating, the digital versions of all the cultural objects and other elements that will populate the virtual exhibit;
- designing contents management and presentation;
- designing the interaction.

These points will be detailed in the following subsections, focusing on showing how OS software and hardware, low-cost devices and the computer games world can contribute to the development of affordable VH solutions.

4.1 Contents creation

The first step in building a virtual museum, whatever its final design and form will be, is the creation of all its digital elements. This process is often referred to as *Digital Content Creation* (DCC), and it includes also the processing and adjustments required before presenting the contents to the end-users. DCC involves several media and requires appropriate and specific software. Depending on the contents to be created, several OS tools are available. A short list of them will be presented in the following.

One important thing to underline is that the same digital contents can be exploited in different applications and displayed through different output devices. For instance, the 3D model of a building can be used to develop an immersive virtual environment that allows navigating it, to create photorealistic renderings for enhanced panoramic views or even displayed on a mobile phone for an AR application. Each environment has its own requirements, in terms of quality and resolution of the contents it can handle. However, the process of DCC, if properly handled, can be performed only once. Actually, if the contents are created at the highest level of detail and quality, they can be easily scaled down and adapted to different media, from personal computers to mobile devices, allowing to amortize in the best way the expenditure of human resources for their creation. The inverse process, that is up-scaling an original low-resolution object, is often impossible, and it usually requires creating from scratch a new version of it.

2D Image creation and editing

Digital images are at the foundation of many new two and three dimensional documentary techniques, since they provide not only the digital equivalent of traditional photography, but they can also be used to create panoramic images, videos, 3D object reconstruction and texture maps for 3D modeling. 2D images can be obtained as the result of document scanning, which converts text and graphic paper documents, photographic film, photographic paper or other files into a digital format. Operations like exposure and color correction, sharpening, noise reduction, dynamic range extension, correction of optical distortions, resizing and cropping, combining multiple images, converting between different image formats and preparation for web distribution are commonly performed on digital images. Robust image processing tools are widely available. The most known of them is Adobe Photoshop, a commercial tool that is the de-facto standard for photo retouching. However, many FS/OS valid alternatives are available, like GIMP (GNU Image Manipulation Program) and Paint.net.

Images can also be created with vector graphic drawing software, which allows representing images in computer graphics through geometrical primitives such as points, lines, curves, and shapes. Free/Open Source high-quality vector graphic editors, like InkScape and Synfig, are available. Some programs, like KToon, FlashDevelop as well as Synfig, allow also developing interactive 2D animations.

Strictly related to digital images are videos, whose educational use in museums is rapidly increasing. As we have seen, video contents can be inserted into any virtual museum, from the content gallery to VR/MR environments. At the same time, thanks to the technological development, the portability of media assets—the mobility of access to the resources and to their distribution channels—is accelerating exponentially. Digital video repositories can be built form analog (VHS) existing material, or created from scratch combining shots, images, text, 3D and 2D animations. Several OS programs are available for video capturing, processing and editing, like VirtualDub, CineFX, Cinelerra, and the Video Sequence Editor of Blender, a program for 3D modeling that will be described in details in the following.

3D contents creation

A realistic three-dimensional virtual world is the result of a composition of several 3D objects that must be properly designed and created. Some valuable OS 3D modeling software are Meshlab, Art of illusion, K-3D, Moonlight|3D and OpenFX, but the most important and famous of all is by far Blender. It is available for many different platforms (Windows, MAC OS and Linux), supporting a variety of geometric primitives. It has an internal rendering engine and it can be easily integrated with external ray tracers. Key-framed animation tools, including inverse kinematics and armature, soft and rigid body dynamics with collision detection, fluids simulation and particle systems are also available. More features can be added integrating plug-ins developed in Python. Despite being a powerful and constantly evolving tool, some of its features still need development. Complex lighting, for instance, is better handled with external rendering engines.
Digital 3D models can be also created from real artifacts with 3D scanners, which are devices that can capture their shape and appearance (i.e. color) using different techniques, like laser, modulated or structured lights, handled probes or volumetric techniques (i.e. using CT scans). While 3D scanners are often expensive devices, there is some OS software, like ARC3D, Insigth3D and Photo-To-3D, that allows the reconstruction of 3D objects from multiple images taken around them (Fig. 10). The quality and the precision of the results are not comparable with that of 3D scanners, but they are often sufficient for the majority of uses in virtual museums.

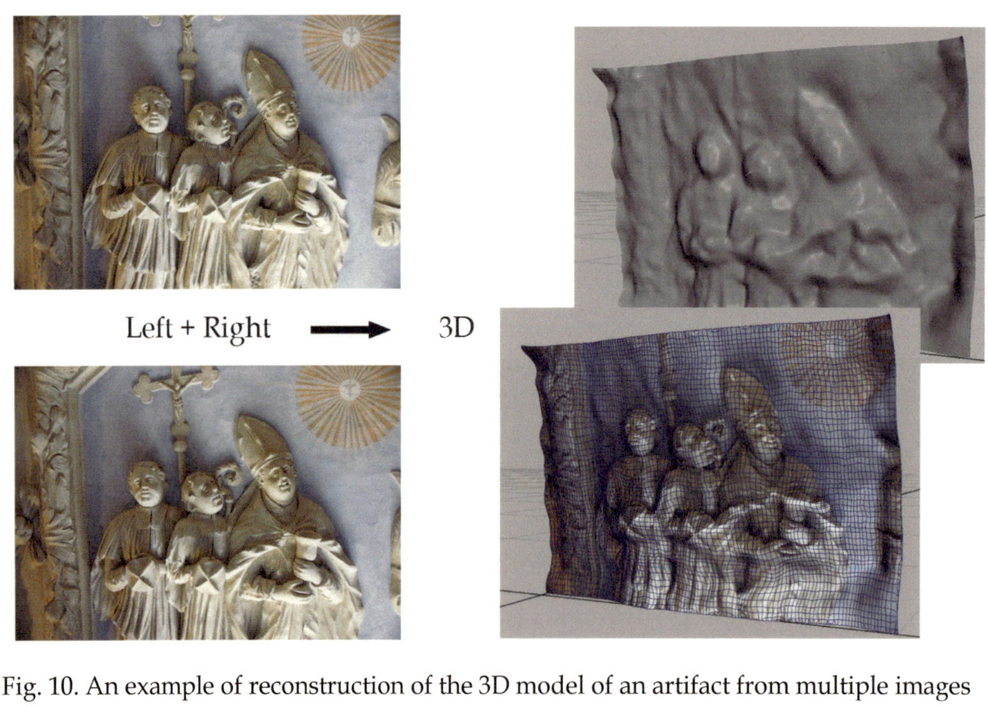

Fig. 10. An example of reconstruction of the 3D model of an artifact from multiple images

In order to produce high quality 2D images and animations from 3D models, several OS photorealistic renderers, like Yafaray, Aqsis, Indigo Renderer and POV-ray can be used. For many applications, like architectural illustration, experimental animation and educational videos for children, a non-photorealistic rendering (NPR), like the one obtainable with the Freestyle engine, is preferable. NPR is a technique that attempts to capture the elements of different expressive styles, such as painting, drawing, technical illustration, and animated cartoons, and apply them to reproducing 3D contents on screen (Masuch & Röber, 2003). In virtual heritage, this can be used, for instance, to show a cartoon-like world, probably more appealing to kids.

Audio

As our perception of an environment is not only what we see, but also what we perceive with other sensory inputs (Chalmers et al., 2009), is important to mention the role of sound. Sound and localized (3D) audio make an important aid to the museum experience and are the keys to create the atmosphere within which to immerse visitors.

To handle sound processing, a powerful tool like Audacity can be used. It is a free software, cross-platform digital audio editor and recording application. Its main features cover audio recording, sound files editing and mixing. Ardour is another OS software available for MAC and Linux distributions.

When the application requires handling interactive sounds, that is, an audio reacting to user inputs and/or changes in the environment, software platforms like MAX/MSP/Jitter, Processing and SuperCollider provide an intuitive graphic environment for combining external inputs with sound generation/processing routines.

4.2 Contents management

VR is a complex simulation of a "reality", requiring 3D graphics, physics, 3D audio, user interaction, avatar management and artificial intelligence (AI) handling. Designing such an application requires designing each of its components. This is a complex task, which involves complex programming skills. Again, OS solutions and computer games industry provide valuable contributions to this process.

Several OS packages/libraries are available for covering all the aspects of the development of a VR application. Virtual Engines, like OGRE, OpenSceneGraph, OpenSG and Ultimate 3D, can be used to render in real time the 3D environment. These engines usually provides limited features, but can be expanded by integrating other libraries. In order to manage collision detection, rigid body and fluid dynamics, physics engines that provides simulation of physical systems, like Open Dynamic Engine (ODE), SOFA (Simulation Open Framework Architecture) and Tokamak, can be used. Audio 3D can be managed by libraries that directly interface with the sound card, like Open Audio Library (OpenAL). The Artificial Intelligence component required to define avatars and characters behavior can be implemented with libraries like CHARACTERISE (Virtual Human Open Simulation Framework) and SmartBody.

The drawback is that putting all these pieces together is not always simple. Each of these libraries has been developed as a stand-alone element, using different programming languages and calling conventions, which may cause compatibility issues during the integration.

The solution to the need of all-in-one simple development environments for VR is offered by the computer game industry (Lepouras & Vassiliakis, 2005). Usually a computer game *is* a VR application. Current videogames provide reach 3D interactive environments that can be used as collaborative tools for educational and learning purposes. Therefore, they can be used not only for leisure but also as a powerful tool for supporting cultural heritage. The term "serious games" refers to applications where game technologies are used in non-entertainment scenarios, such as flight simulators, business simulation and strategic games. Seriuos games involve both entertainment and pedagogy, that is activities that educate and instruct (Zyda, 2005). Their use in cultural heritage has been emphasized by the recent works of Anderson et al., 2009, and by several international conferences and workshops.

Game development is a complex and labor intensive task. It takes years to complete a game from scratch. Gaming industries crave for reducing time to market and nowadays game development is based primarily on Game Engines (GEs). A GE is the software that forms the core of any computer game, handling all of its components: image rendering, physics and collisions management, audio 3D, and support for AI and for all other game aspects. Unreal and Cry Engine are the most famous commercial engines. The Unreal Development Kit is a version of the Unreal Engine that can be used free of charge for non-commercial and educational purposes. Many OS GEs are available as well, like Spring Engine, Nebula Device and Crystal Space. Game Engines for OVEs are also freely available over Internet.

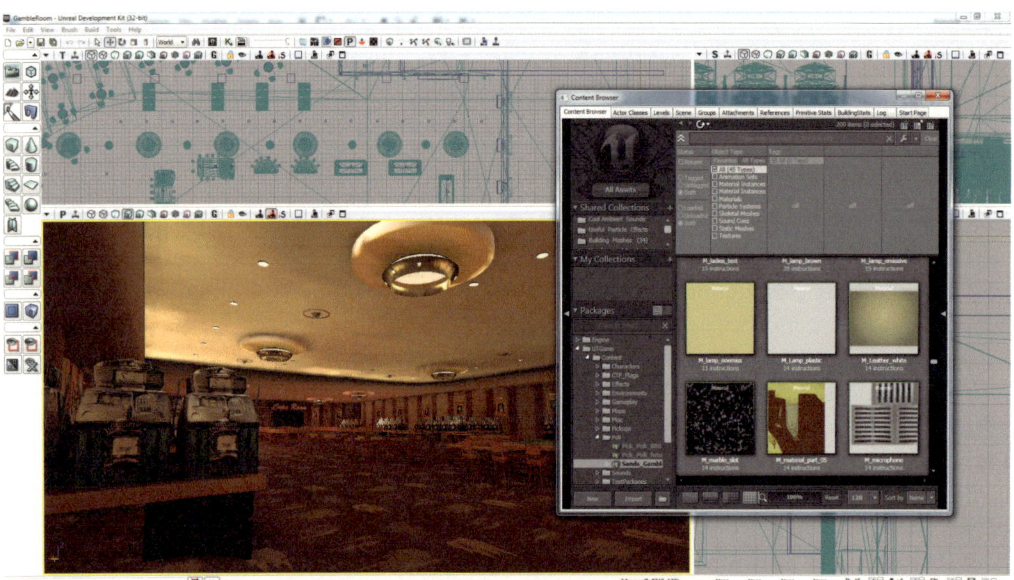

Fig. 11. The User Interface of UDK, showing the porting into the GE of the virtual reconstructions developed for the "One day at The Sands" application in Fig. 3

There are several advantages of using GEs for developing VR applications. First, while a game can be certainly developed by modifying the GE source code, many GEs can be completely customized using scripting languages, which are easy to learn and do not require an expertise in programming, allowing non-skilled users to easily adopt them. Hence, development costs are extremely reduced, since they are limited to the creation of the digital contents and of the game logic. Second, most of the GEs provide network capabilities and communication support, allowing the development of collaborative applications involving several (and possibly remote) users. Third, unlike applications designed with Virtual Engines, the developed products are often directly portable on different execution environments (such as PC, game consoles, browsers, PDA and mobile devices). Finally, UDK and many other GEs have a built in support for stereoscopic displays and they also can be easily modified (like in Jacobson, 2002, and Juarez et al., 2010) to support complex immersive environments like CAVEs.

Using GE to develop virtual environments for VH applications have proven to take almost half the effort than with traditional techniques, requiring a heavy programming effort especially for developing the necessary interaction support (Lepouras & Vassilakis, 2005).

As for MR/AR applications, developing them usually requires:

- tracking the position of the user in the real world, and
- combining the rendering of 3D and multimedia contents with images taken with a video camera.

While the second component can be easily managed by Virtual Engines like OGRE and OpenSceneGraph, the first usually involves direct user input, GPS or images of markers/fiducials taken with wearable camera and recognized by proper algorithms. Some OS libraries for marker recognition and tracking are ARToolKit, ARToolKitPlus, SSTT and BazAR. Some of them have been developed for different platforms, in order to allow the implementation of MR/AR applications over Internet (like NyARToolKit, FLARToolkit and SLARToolkit, porting in, respectively, Java, Flash and Silverlight of the ARToolKit library) or on mobile devices (like Andar and Mixare for the Android platform).

Finally we briefly mention that OS libraries and authoring tools are also available for developing enhanced panoramas viewers (PanoTools and FreePV) and content galleries (Omeka, OpenCollection and several OS Content Management Systems, like Droopal, Joomla and Plone).

4.3 Interaction design and development

The final step in creating a virtual museum it is to design and implement the user interaction. For virtual museums that are accessible through Internet the interaction is clearly limited to standard devices, like desktop displays, mice and keyboards. More challenging is the case of immersive VR or MR environments, which can benefit from the large variety of input and output devices that are currently available on the market.

Managing interaction in a virtual environment is a particularly complex task, especially in museums where technologies must convey a cultural message in the most appropriate and effective way. Recently, there has been significant development in the design of controllers, devices and software that offers new possibilities for educator and learners in 3D immersive environments. The interest in the scientific and industrial communities is in providing a *natural interaction* with machines. The basic idea is that if people naturally communicate through gestures, expressions, movements, and discover the world by looking around and manipulating physical stuff, they should be allowed to interact with technology in the same way (Valli, 2008).

An example of natural interaction is the use of spatial 3D gestures, as those provided by the WiiMote. WiiMote, the controller of Nintendo Wii, is shaped like a remote control, and therefore it is simple to use also for not regular gamers. It contains multiple accelerometers to track position and orientation of the device and, therefore, to sense user's motion. It also includes several buttons and provides haptic feedback. WiiMote has a lot of potential for educational uses due to its affordability, adaptability and popularity as game controller. Moreover, thanks to its Bluetooth support and assortment of sensors, many devices, know as WiiHacks, have been built exploiting the WiiMote capabilities. Instructions on creating head and finger trackers, sensible gloves, multitouch interactive whiteboards, and MIDI instruments can be found on the many sites dedicated to the so-called *Wii Brewing*.

Sony announced a new motion controller for the PS3, which looks like a wand with a color-changing ball on top and buttons to issue commands. It's designed to work in conjunction with the PlayStation Eye2, a low cost camera, to track its position. The various Dance Pad and Wii Balance Board are other controllers that exploit human motion for interaction. These devices have been used in VH as intuitive navigational interfaces in virtual environments (Fassbender & Richards, 2008, and Deru & Bergweiler, 2009).

Using users' actions to control the interaction can be done also exploiting computer vision techniques to analyze images of the user's movements and translate them into actions in the virtual environment. In order to allow algorithms to work also in very dark environments, camera and lights in the Near-Infrared are often used (Valli, 2008). The Sony EyeToy device, using a color digital camera, was the first attempt to use computer vision for gesture recognition in computer games. Recently, Microsoft has announced the release of Kinect, a peripheral for the Xbox 360 console acting as game controller through gestures, spoken commands, or presented objects and images. The device will include a color camera, a depth sensor and multi-array microphone and is claimed to provide full body motion capture and also face recognition. Many VH applications will benefit from this device as soon as the software libraries to interface with it will be released. A different approach is that taken by CamSpace, a computer program that can turn any object into a game controller using a webcam for tracking it. The approach is cost effective, since the software can be downloaded for free, and the available SDK allows integrating the tracker in any other program.

Research on Human Computer Interaction has demonstrated that sensory immersion is enhanced and intuitive skills are best exploited through tangible user interfaces (TUI) and intuitive tangible controls (Xin et al., 2007, Butler & Neave, 2008). As we have seen, TUI is also fundamental for providing a richer interaction in MR/AR environments.

In addition to visual markers, several low cost solutions exist for developing tangible interaction. Arduino is an OS hardware platform for physical computing that can be easily integrated with any software running on a PC connected with the board. I-CubeX is another family of sensors, actuators and interfaces to construct complex interactive systems from very simple and low-cost components. Also Radio Frequency IDentification (RFID) can be used to create interactive objects. RFID tags are small sensor that can be hidden into objects. They are sensible to (or emitting) radio frequencies, and therefore can be recognized through apposite antennas integrated in the environment in apposite locations.

A remarkable example of TUI is the Microsoft Surface, a multi touch interactive table (Fig. 11) whose surface is a display where objects can be visualized and directly manipulated by the users. Its appearance on the market gave birth to a lot of projects trying to build similar devices from off-the-shelf components, like standard PC, near-infrared cameras, IR emitters and OS computer vision libraries. The pioneering development of the Reactable (Jordà et al.,

2005), the release as OS software of its tracking technology, reactiVision, as well as the definition of TUIO, an open protocol to describe properties of finger and hand gestures and controller objects, like fiducials, on the table surface (Kaltenbrunner et al., 2005), made the development of these devices available to almost anyone. Many of these OS products have been already deployed in public spaces or embedded in art installations (Hornecker, 2008, Fikkert et al., 2009, and Hochenbaum, 2010).

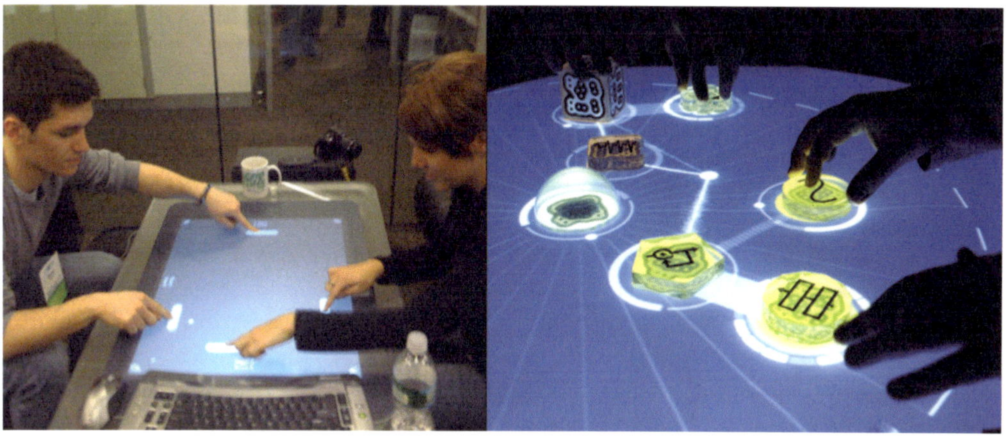

Fig. 11. Examples of interactive table (left) and use of fiducials for interaction (right)

As for the output displays, the cutthroat competition in consumer market allows the availability of large size high-definition displays at very low prices and the cost of high-resolution projectors, necessary for creating immersive VR environments, is also in constant decrease. The recent introduction of 3D in cinema and television is bringing on a revolution in consumers' market. The sense of immersion and presence produced by three-dimensional images clearly exceeds that of simple two-dimensional images, with the user becoming part of the on screen action. 3DTV and 3D projectors are becoming largely available at a price slightly higher that that of corresponding 2D products. Low cost solutions for gaming environments, like the NVIDIA 3D Vision, are available as well. The majority of these products require the use of active glasses for viewing 3D images. Such glasses are fragile, expensive and there is not yet a standard communication protocol for the synchronization with the display device, requiring using different glasses for different hardware products. This is a clear drawback for their wide diffusion, but the 3D scene is rapidly evolving and changes are foreseen in the near future. An alternative is represented by autostereoscopic displays, which can be seen in 3D without requiring any kind of glasses or external devices worn by the subject. Autostereoscopic vision is provided by means of particular optical devices laid upon the screen, like parallax barriers and lenticular lens, allowing each eye to receive a different image at an optimal distance. This technology is not yet mature, a standard format for autostereoscopic contents has not been defined yet and many of the available devices are still expensive. However, the recent announcement of the upcoming release of the Nintendo 3DS, the first autostereoscopic portable game console, shows that affordable solutions can be expected in the near future.

5. Conclusions

Virtual museums experiment the integration of cutting edge multimedia and VR/MR technologies with original archival material in order to create interactive, multi-user, mixed-reality edutainment environments. Cultural contents, enhanced through the use of technology, can be re-proposed in a captivating and engaging environment that allows visitors to learn something through an entertaining experience. This is particularly important for museums, which are facing a drastic change in their role, from mere containers of cultural objects to narrative spaces where the pedagogical level must be combined with the entertaining one.

The current potential of new technologies ensures that the digital artifacts are as communicative as the original. Therefore, ICT technologies can become a real tool for people working in museums since they are a valuable tool to help curators in the settlement of novel exhibitions, for instance allowing the creation of context and atmosphere around artifacts, or providing new dimensions and concepts for the arrangement of collections. Moreover, virtual museums allow developing versatile and scalable environments. Versatile means that they can range through contexts, events, purposes and physical environments, since the digital provides a permanent container, the technology, whose content has the possibility to be changed infinitely. Scalable means that digital contents can be adapted to different media, from personal computers to mobile devices.

However, despite their advantages and potentials, the introduction of emerging technologies in museums and cultural contexts has not yet been as wide as expected. The main reason for this small infiltration is mainly related to the affordability of both their development and execution. Virtual museums impose heavy demands in terms of computational resources, require high quality interfaces, especially high resolution displays and suitable interaction devices, and the availability of software for designing, developing and managing the complexity of a virtual environment. All this comes with a cost. On the contrary, museums are often confronted to limited budgets, and therefore the development of affordable VH solutions is sorely needed.

In this paper, we have shown that recent dramatic advances in technology, in the computer games industry and in the OS software scene can help to reach these objectives. In fact, spurred by a demanding consumer market, game industries are constantly driving technological advances in the field of CPU and graphics processors in order to support new complex games with more and more advanced visual capabilities, and are in the front line for developing new interaction devices allowing to fully exploit the potentialities of 3D environments and to improve users' experience. A Game Engine, the software that forms the core of any computer game, is a perfect and cost effective all-in-one solution for developing VR applications, since it can handle all their aspects, without requiring a specific expertise in programming. Other Free/Open Source software solutions are largely available for designing and developing every component of virtual museums, from the creation of their contents, to the implementation and deployment of the final applications.

Therefore, as we have shown in this paper, the development of effective VH solutions can be focused on using off-the-shelf components, low cost hardware and open-source software solutions. It is our firm belief that this will be the trend in VH in the next future, and that this approach, accompanied by further technological improvements, will bring unthinkable benefits, as far as curators and archivists are concerned with the development process, to the widespread diffusion of new ICT-based approaches in cultural heritage.

6. References

Addison, A.C. (2000) Emerging Trends in Virtual Heritage, *IEEE MultiMedia*, Vol. 7, No. 2, Apr-Jun, 2000, pp. 22-25.

Anderson, E. F.; McLoughlin, L.; Liarokapis, F.; Peters, C.; Petridis, P. & De Freitas, S., (2009) Serious games in cultural heritage, *VAST 2009: 10th International Symposium on Virtual Reality, Archaeology and Cultural Heritage*, pp. 29-48

Bowman, D.A.; Coquillart, S.; Froehlich, B.; Hirose, M.; Kitamura, Y.; Kiyokawa, K. & Stuerzlinger, W. (2008). 3D User Interfaces: New Directions and Perspectives, *IEEE Computer Graphics and Applications*, Vol. 28, No. 6, Nov/Dec, 2008, pp. 20-36

Burdea, G. & Coiffet, P. (2003). *Virtual Reality Technology*, Wiley, 2003, New Jersey

Butler, M. & Neave, P. (2008). Object appreciation through haptic interaction, *Proceedings of the 25th Annual Conference of the Australian Society for Computers in Learning in Tertiary Education (ascilite Melbourne 2008)*, pp. 133-141, December 2008, Deakin University, Melbourne Vic Australia.

Caulton, T. (2002). *Hands-on exhibitions: managing interactive museums and science centres.* ISBN: 978-0-415-16522-8, Routledge, London

Chalmers, A. & Debattista, K. (2009). Level of Realism of Serious Games. *Proceedings of the 2009 Conference in Games and Virtual Worlds for Serious Applications.* ISBN:978-0-7695-3588-3, Washington, DC, USA

Deru, M. & Bergweiler, S. (2009). *Surfing Google Earth on Wii Fit.* http://www.psfk.com/2009/02/surfing-google-earth-on-the-wii-fit.html, last accessed July 2010

Encyclopaedia Britannica Online, available at http://www.britannica.com/EBchecked/topic/630177/virtual-museum, last accessed July 2010

Farella E.; Brunelli D.; Benini L.; Riccò B. & Bonfigli M. E. (2005). Pervasive Computing for Interactive Virtual Heritage, *IEEE MultiMedia*, Vol. 12, No. 3, July-September, 2005, pp. 46-58, ISSN:1070-986X.

Fassbender, E. & Richards, D. (2008). Using a dance pad to navigate through the virtual heritage environment of macquarie lighthouse, Sydney. *Proceedings of the 13th international Conference on Virtual Systems and Multimedia*, pp. 1-12.

Fikkert W., Hakvoort M., Van der Vet P, Nijholt A. (2009). Experiences with Interactive Multi-touch Tables. *Proceedings 3rd International Conference on Intelligent Technologies for Interactive Entertainment (INTETAIN '09)*, pp. 193-200

Hirose, M. (2006). Virtual Reality Technology and Museum Exhibit, *International journal of virtual reality*, Vol. 5, No. 2 , pp. 31-36.

Hochenbaum, J.; Vallis, O.; Diakopoulos, D.; Murphy, J. & Kapur A. (2010). Designing expressive musical interfaces for tabletop surfaces. *Proceedings of the 2010 Conference on New Interfaces for Musical Expression (NIME 2010)*, Sydney, Australia

Hooper-Greenhill, E. (2000). Changing values in the art museum: rethinking communication and learning. *International Journal of Heritage Studies*, Vol. 6, Issue 1, (March 2000), pages 9-31.

Hope, T.; Nakamura, Y.; Takahashi, T.; Nobayashi, A.; Fukuoka; S., Hamasaki, M. & Nishimura, T. (2009). Familial collaborations in a museum. *Proceedings of the 27th international Conference on Human Factors in Computing Systems (CHI '09)*, pp. 1963-1972.

Hoptman, G.H., (1992). The virtual museum and related epistemological concerns. *Sociomedia, Multimedia, Hypermedia and the Social Construction of Knowledge*, pp. 141–159

Hornecker E. (2008). "I don't understand it either, but it is cool" - Visitor Interactions with a Multi-Touch Table in a Museum. *Proceedings of IEEE Tabletop 2008*. pp. 113-120

ICOM News, no. 3, (2004), available at: http://icom.museum/pdf/E_news2004/p3_2004-3.pdf, last accessed July 2010

Jacobson, J. (2002). Configuring Multiscreen Displays with Existing Computer Equipment, *Proceedings of Conference on Human Factors HFES'2002*

Jacobson, J. & Lewis, L. (2005). Game Engine Virtual Reality with CaveUT. *IEEE Computer*, 38(4), pp.79-82.

Jordà, S.; Kaltenbrunner, M.; Geiger, G. & Bencina, R. (2005). The reacTable, *Proceedings of the International Computer Music Conference (ICMC2005)*, Barcelona, September 2005.

Juarez, A.; Schonenberg, W. & Bartneck, C. (2010). Implementing a Low-Cost CAVE System Using the CryEngine2. *Submitted to EC journal.*

Kaltenbrunner, M.; Bovermann, T.; Bencina, R. & Costanza, E. (2005). TUIO - A Protocol for Table Based Tangible User Interfaces. *Proceedings of the 6th International Workshop on Gesture in Human-Computer Interaction and Simulation (GW 2005)*, Vannes, France, 2005

Lanier, J. (2010). *The Top Eleven Reasons VR has not yet become commonplace.* Available online at http://www.jaronlanier.com/topeleven.html, last accessed July 2010

Lepouras, G. & Vassilakis, C. (2005). Virtual museums for all: employing game technology for edutainment. *Virtual Reality*. Vol. 8(2), 2004-2005, pp. 96-106

Masuch, M. & Röber, N. (2003). Game Graphics Beyond Realism: Then, Now and Tomorrow, *Proceedings of 1st Digital Games Research Conference, Utrecht, Netherlands, 2003*

McLaughlin, M. L.; Sukhatme, G.; Shahabi, C.; Hespanha, J.; Ortega, A. & Medioni, G. (2000) The Haptic Museum. *Proceeding of the Conference on Electronic Imaging and Visual Arts (EVA 2000)*, 2000.

Milgram, P. & Kishino, A. F. (1994). Taxonomy of Mixed Reality Visual Displays, *IEICE Transactions on Information and Systems*, Vol. 12, pp. 1321-1329

Noh, Z.; Sunar, M. S.; & Pan, Z. (2009). A Review on Augmented Reality for Virtual Heritage System. *Proceedings of the 4th international Conference on E-Learning and Games*, pp. 50-61

Packer, J. (2006). Learning for fun: The unique contribution of educational leisure experiences, *Curator: The Museum Journal*, Vol. 49, No. 3, 2006, pp. 329–344, ISSN 0011-3069.

Prasolova-Førland, E.; Wyeld, T. G. & Divitini M. (2007). Battle of Stiklestad: Supporting Virtual Heritage with 3D Collaborative Virtual Environments and Mobile Devices in Educational Settings. *Proceedings of Mobile and Ubiquitous Technologies for Learning (UBICOMM 07)*, pp. 197-203

Rheingold, H. (1991). *Virtual Reality*. Summit Books

Santangelo, A.; Augello, A.; Gentile, A.; Pilato, G. & Gaglio S. (2006). A Chat-bot based Multimodal Virtual Guide for Cultural Heritage Tours. *Proceedings of the 2006 International Conference on Pervasive Systems & Computing*, pp 114-120

Sauer, S. & Göbel, S. (2003). Focus your young visitors: kids innovation - fundamental changes in digital edutainment, in: M. Bergamasco (Ed.), *Proceedings of the Conference Museums and the Web 2003*, Charlotte, USA, Toronto, pp. 131–141

Scali, G.; Segbert, M. & Morganti, B. (2002). Multimedia applications for innovation in cultural heritage: 25 European trial projects and their accompanying measure TRIS. *Proceedings 68th IFLA Council and General Conference*

Schuemie, M. J.; Van der Straaten, P.; Krijn, M. & Van der Mast, A. C. (2001). Research on presence in virtual reality: A survey. *Cyberpsychology and Behavior*, Vol. 4, No. 2, 2001, pp. 183–201

Severson, J.; Cremer, J.; Lee, K.; Allison, D.; Gelo, S.; Edwards, J.; Vanderleest, R.; Heston, S.; Kearney, J.; Thomas, G. (2002). Exploring virtual history at the National Museum of American History. *Proceedings of the 8th international conference on virtual systems and multimedia (VSMM 2002)*, Gyeong-ju, Korea, 25–27 September, pp. 61–70

Stallman, R. (1999). The GNU Operating System and the Free Software Movement. In *Open Sources : Voices from the Open Source Revolution*, O'Really publishing, 1999

Sylaiou, S.; Liarokapis, F.; Kotsakis, K. & Patias, P. (2009). Virtual museums, a survey and some issues for consideration, *Journal of Cultural Heritage*, Vol. 10, No. 4, October-December 2009, Pages 520-528

The International Council Of Museums. ICOM statutes, 2007

Tsichritzis, D. & Gibbs, S. (1991). Virtual Museum and Virtual Realities. *Proceedings of the International Conference on Hypermedia and Interactivity in Museum.*

Valli, A. (2008). The design of natural interaction. *Journal of Multimedia Tools and Applications*, Vol. 38, No. 3, July 2008, pp. 295-305

Weiser, M. & Brown J. S. (1995). Designing calm technology. *Xerox PARC*

Wojciechowski, R.; Walczak, K.; White, M. & Cellary, W. (2004). Building Virtual and Augmented Reality Museum Exhibitions. *Proceedings of the Web 3D 2004 Symposium – the 9thInternational Conference on 3D Web Technology, ACMSIGGRAPH*, Monterey, California (USA), pp.135–144.